DIFFERENTIAL ELEMENTS OF VECTOR AREA

$$d\mathbf{s} = \begin{cases} \mathbf{a}_x\,dy\,dz + \mathbf{a}_y\,dx\,dz + \mathbf{a}_z\,dx\,dy \\ \mathbf{a}_\rho \rho\,d\phi\,dz + \mathbf{a}_\phi\,d\rho\,dz + \mathbf{a}_z\,\rho\,d\rho\,d\phi \\ \mathbf{a}_r r^2 \sin\theta\,d\theta\,d\phi + \mathbf{a}_\theta r \sin\theta\,dr\,d\phi + \mathbf{a}_\phi r\,dr\,d\theta \end{cases}$$

DIFFERENTIAL ELEMENTS OF VOLUME

$$dv = \begin{cases} dx\,dy\,dz \\ \rho\,d\rho\,d\phi\,dz \\ r^2 \sin\theta\,dr\,d\theta\,d\phi \end{cases}$$

VECTOR OPERATIONS – RECTANGULAR COORDINATES

$$\nabla\alpha = \mathbf{a}_x \frac{\partial\alpha}{\partial x} + \mathbf{a}_y \frac{\partial\alpha}{\partial y} + \mathbf{a}_z \frac{\partial\alpha}{\partial z}$$

$$\nabla\cdot\mathbf{A} = \frac{\partial A_x}{\partial x} + \frac{\partial A_y}{\partial y} + \frac{\partial A_z}{\partial z}$$

$$\nabla\times\mathbf{A} = \mathbf{a}_x\left(\frac{\partial A_z}{\partial y} - \frac{\partial A_y}{\partial z}\right) + \mathbf{a}_y\left(\frac{\partial A_x}{\partial z} - \frac{\partial A_z}{\partial x}\right) + \mathbf{a}_z\left(\frac{\partial A_y}{\partial x} - \frac{\partial A_x}{\partial y}\right)$$

$$\nabla^2\alpha = \frac{\partial^2\alpha}{\partial x^2} + \frac{\partial^2\alpha}{\partial y^2} + \frac{\partial^2\alpha}{\partial z^2} \equiv \nabla\cdot\nabla\alpha$$

$$\nabla^2\mathbf{A} = \mathbf{a}_x\nabla^2 A_x + \mathbf{a}_y\nabla^2 A_y + \mathbf{a}_z\nabla^2 A_z \equiv \nabla(\nabla\cdot\mathbf{A}) - \nabla\times(\nabla\times\mathbf{A})$$

VECTOR OPERATIONS – CYLINDRICAL COORDINATES

$$\nabla\alpha = \mathbf{a}_\rho \frac{\partial\alpha}{\partial\rho} + \mathbf{a}_\phi \frac{1}{\rho}\frac{\partial\alpha}{\partial\phi} + \mathbf{a}_z \frac{\partial\alpha}{\partial z}$$

$$\nabla\cdot\mathbf{A} = \frac{1}{\rho}\frac{\partial}{\partial\rho}(\rho A_\rho) + \frac{1}{\rho}\frac{\partial A_\phi}{\partial\phi} + \frac{\partial A_z}{\partial z}$$

$$\nabla\times\mathbf{A} = \mathbf{a}_\rho\left(\frac{1}{\rho}\frac{\partial A_z}{\partial\phi} - \frac{\partial A_\phi}{\partial z}\right) + \mathbf{a}_\phi\left(\frac{\partial A_\rho}{\partial z} - \frac{\partial A_z}{\partial\rho}\right) + \mathbf{a}_z\frac{1}{\rho}\left(\frac{\partial}{\partial\rho}(\rho A_\phi) - \frac{\partial A_\rho}{\partial\phi}\right)$$

$$\nabla^2\alpha = \frac{1}{\rho}\frac{\partial}{\partial\rho}\left(\rho\frac{\partial\alpha}{\partial\rho}\right) + \frac{1}{\rho^2}\frac{\partial^2\alpha}{\partial\phi^2} + \frac{\partial^2\alpha}{\partial z^2}$$

$$\nabla^2\mathbf{A} = \mathbf{a}_\rho\left(\nabla^2 A_\rho - \frac{2}{\rho^2}\frac{\partial A_\phi}{\partial\phi} - \frac{A_\rho}{\rho^2}\right) + \mathbf{a}_\phi\left(\nabla^2 A_\phi + \frac{2}{\rho^2}\frac{\partial A_\rho}{\partial\phi} - \frac{A_\phi}{\rho^2}\right) + \mathbf{a}_z\nabla^2 A_z$$

INTRODUCTORY ELECTROMAGNETICS

INTRODUCTORY ELECTROMAGNETICS

Herbert P. Neff, Jr.

University of Tennessee

WILEY

JOHN WILEY & SONS
New York ■ Chichester ■ Brisbane ■ Toronto ■ Singapore

Copyright © 1991, by John Wiley & Sons, Inc.

All rights reserved. Published simultaneously in Canada.

Reproduction or translation of any part of
this work beyond that permitted by Sections
107 and 108 of the 1976 United States Copyright
Act without the permission of the copyright
owner is unlawful. Requests for permission
or further information should be addressed to
the Permissions Department, John Wiley & Sons.

Library of Congress Cataloging in Publication Data:

Neff, Herbert P., 1930–
    Introductory electromagnetics / Herbert P. Neff, Jr.
    Includes bibliographical references.
    ISBN 0-471-60550-6
    1. Electrical engineering.   2. Electromagnetic fields.   I. Title.
TK145.N415 1991
621.3—dc20   89-36373
                                                                     CIP

Printed in the United States of America

10 9 8 7 6 5 4 3 2 1

ABOUT THE AUTHOR

Herbert P. Neff, Jr., received his bachelor's and master's degrees in electrical engineering from the University of Tennessee at Knoxville in 1953 and 1956, respectively. He received his doctorate in electrical engineering from Auburn University in 1967. From 1956 to 1965 and from 1966 to the present he has taught courses primarily in electromagnetics at the University of Tennessee. He is also a consultant to Oak Ridge National Laboratory.

Dr. Neff has written four engineering textbooks and is the author or coauthor of numerous articles that have been published in professional journals. He is a member of Eta Kappa Nu, Sigma Xi, and Tau Beta Pi. He is a senior member of the Institute of Electrical and Electronics Engineers.

PREFACE

This text is written for a first semester junior level course in electromagnetics. The first chapter covers the basics of vector analysis, including the divergence, curl, and gradient operations. There are those who believe that the study of electromagnetics should begin with the static case (Chapter 2) allowing the student to gradually digest the material, whereas others (not including the author) believe that it is desirable to begin with the dynamic case (Chapter 3). Since all of the essentials of vector analysis are included in Chapter 1, this textbook allows the instructor to skip Chapter 2 (with the possible exception of Section 2.15) if desired. This approach allows more material to be covered toward the end of the textbook.

Chapter 2 includes electrostatics and magnetostatics. Although it certainly could be divided into two chapters, it was decided that, if combined, the topics could be covered in a shorter period of time. Something has to be reduced, in a one-semester course, and this material is an obvious candidate. As indicated in the preceding paragraph, Chapter 3 begins with the dynamic case (Maxwell's *general* equations). The treatment here is a standard one, and there are few topics that can be omitted.

The material in Chapter 4 is important for several reasons. First, uniform plane waves represent a relatively simple application of Maxwell's equations from the preceding chapter. Second, the reflection of plane waves from surfaces, such as the earth, is an important consideration on its own. The behavior of uniform plane waves is very similar to that of the fields around transmission lines (Chapter 5), and provides a natural introduction to transmission lines. Finally, the consideration of oblique incidence predicts the essential behavior of the hollow one-conductor rectangular waveguide and the dielectric slab waveguide that appear in Chapter 6. The reader who is familiar with *Basic Electromagnetic Fields*, my textbook that is designed for a two-quarter course, will recognize much of the material from that textbook in Chapters 4, 5, and 6 of this textbook.

Chapter 5 is a rather typical treatment of transmission lines. It begins with the lossless case so that the general case (beginning with Section 5.14) can be omitted if desired, or if time requires that some material be skipped. The same general things can be said for Chapter 6 with regard to waveguides, and Section 6.8, dealing with losses, can be omitted.

Radiation is ordinarily the last topic covered in material of this type. Since it is expected (based on past experience) that this material cannot be covered because of time limitations in a one-semester course, it was decided that a full chapter would not be devoted to it. Instead, Appendix D is devoted to the basics of radiation, including a brief treatment of the *Hertzian dipole* and the *half-wave dipole*. If time allows, this material can be utilized.

A first semester course in electromagnetics (for which this textbook would be appropriate) often appears *concurrently* with a course in basic *linear system analysis* where the unit-impulse function would normally first appear. In this circumstance the student would not be familiar with the use of this function, and, for this reason, it does not appear in this textbook. This is in contrast to *Basic Electromagnetic Fields*, where the unit-impulse function is used in several places to verify fundamental relations or to equate superposition and convolution integrals.

It is obvious from the preceding considerations that this text could be used in a two-quarter scheme with very little material omitted. It is believed that *Basic Electromagnetic Fields* would be more suitable for a two-semester or three-quarter scheme. A suggested timetable for a one-semester course is outlined below (assuming a 15-week semester).

Chapter 1	2 weeks
Chapter 2	4 weeks
Chapter 3	2 weeks
Chapter 4	2 weeks
Chapter 5	3 weeks
Chapter 6	2 weeks

Vector relations are conveniently summarized on the inside front and back covers for convenience. The symbols and units (MKSA) are the same as those used in *Basic Electromagnetic Fields*. They are more or less conventional, and are listed following this preface. Answers to selected problems are listed following the appendixes. Problems at the end of the chapters that are somewhat more difficult are marked with an asterisk.

I express my sincere thanks to those reviewers and others whose suggestions have contributed to this textbook. In particular, I thank Professors Leonard Taylor, Matthew Sadiku, Randy Jost, and Kondagunta Sivaprasad. The typing of the original manuscript for this textbook was done by Roberta Campbell, and she is due much praise for her fine work.

HERBERT P. NEFF, JR.

CONTENTS

CHAPTER 1

VECTOR ANALYSIS 1

1.1 Vector Addition **1**
1.2 Vector Multiplication **2**
1.3 Coordinate Systems **4**
1.4 Circulation and Flux **9**
1.5 Divergence **13**
1.6 Gradient **17**
1.7 Curl **20**
1.8 Some Fundamental Vector Identities **24**
1.9 The Divergence Theorem **26**
1.10 Stokes' Theorem **28**
1.11 Concluding Remarks **30**

CHAPTER 2

ELECTROSTATIC AND MAGNETOSTATIC FIELDS 34

2.1 Charge Configurations **35**
2.2 Coulomb's Law and Electric Field Intensity **37**
2.3 The Electrostatic Potential Field **47**
2.4 Gauss's Law **53**
2.5 Dielectrics **56**
2.6 Poisson and Laplace Equations **61**
2.7 Image Theory **67**
2.8 Electric Current and Conservation of Charge **69**
2.9 The Biot-Savart Law and Magnetic Field Intensity **72**
2.10 Ampere's Circuital Law **80**
2.11 Ampere's Law in Point Form (Maxwell's Equation) **85**

2.12 Magnetic Scalar and Vector Potential 87
2.13 Magnetic Materials 91
2.14 Force and Torque 100
2.15 Capacitance, Resistance, and Inductance 102
2.16 Concluding Remarks 111

CHAPTER 3
MAXWELL'S EQUATIONS 126

3.1 Conservation of Charge 126
3.2 Other Field Quantities 128
3.3 Faraday's Law 130
3.4 Maxwell's Second Equation 137
3.5 A Summary of Maxwell's Equations 140
3.6 Poynting's Theorem 143
3.7 Potentials 150
3.8 Voltage and Potential Difference 159
3.9 Boundary Conditions 159
3.10 Circuit Theory from Field Theory 162
3.11 Concluding Remarks 165

CHAPTER 4
UNIFORM PLANE WAVE PROPAGATION 171

4.1 The Undamped Uniform Plane Wave 172
4.2 Wavelength and Phase Velocity 178
4.3 Power Density and Velocity of Energy Flow 179
4.4 Wave Impedance 180
4.5 Uniform Damped Plane Waves 183
4.6 Reflection of Plane Waves, Normal Incidence 187
4.7 Oblique Incidence 205
4.8 Polarization 217
4.9 Dispersion 220
4.10 Concluding Remarks 220

CHAPTER 5
TRANSMISSION LINES: TEM MODES 228

5.1 The Parallel Plane Guiding System 228
5.2 The General Lossless Line 230
5.3 The Lossless Line Equivalent Circuit 234

CONTENTS

5.4 General Solutions for V and I—Lossless Line 238
5.5 Voltage Standing Wave Ratio 243
5.6 Input Impedance 244
5.7 Special Load Impedances 246
5.8 Transmission Parameters 252
5.9 Multiple Loads 254
5.10 Transmission Line Measurements 256
5.11 The Smith Chart 257
5.12 Transmission Line Matching 268
5.13 Pulses on the Lossless Line 278
5.14 Losses 281
5.15 Special Two-Wire Lines 287
5.16 Quality Factor 290
5.17 Distributed Parameters 299
5.18 The Coaxial Cavity 302
5.19 Concluding Remarks 307

CHAPTER 6

WAVEGUIDES AND CAVITIES 317

6.1 The Rectangular Waveguide, TE_{10} Mode 317
6.2 Properties of the TE_{10} Mode 320
6.3 Waveguide Current, TE_{10} Mode 329
6.4 Waveguide Vector Potentials 334
6.5 The Rectangular Waveguide, Higher Order Modes 337
6.6 Dielectric Waveguides 345
6.7 Lossless Cavity Resonators 355
6.8 Waveguide and Cavity Losses 358
6.9 Concluding Remarks 366

APPENDIX **A** PHYSICAL CONSTANTS 372
APPENDIX **B** MATERIAL PARAMETERS 373
APPENDIX **C** SPECIAL FUNCTIONS 375
APPENDIX **D** RADIATION 388

SYMBOLS AND UNITS 396

ANSWERS TO SELECTED PROBLEMS 401

INDEX 409

1
VECTOR ANALYSIS

This introductory chapter will begin with a brief review of the vector algebra to which the junior level engineering student has previously been exposed. More complicated operations, such as *divergence* of a vector, *gradient* of a scalar, *curl* of a vector, line integral, flux of a vector, and others are also presented and explained. More meaningful uses for these vector operations will occur in the following chapters when Maxwell's equations are presented and used in more practical applications such as lines, guides, and antennas.

The three most common coordinate systems, Cartesian, circular cylindrical, and spherical, that are used almost exclusively in this text, are presented along with the information necessary to move from one coordinate system to another. Vector relations and coordinate relations are conveniently summarized on the inside front and back covers.

1.1 VECTOR ADDITION

A *scalar field* may be thought of as an ensemble of point scalars, and temperature is an example that we can all appreciate. A *vector field* may, in a similar manner, be thought of as an ensemble of point vectors, and the earth's gravitational field is an example that we are all familiar with. On the other hand, a (point) *vector* is independent of any coordinate system that is being used to describe it, and is therefore unique.

Two vectors **A** and **B** are equal (**A** = **B**) if they have the same magnitude, direction, and sense. Note that this means that they are not necessarily coincident in space, but one may be translated so that they can become coincident. On the other hand, if two vector *fields* are equal, then they are coincident. When adding two vectors **A** and **B** to produce the sum **C**, we mean that one of the two vectors may be moved parallel to itself (translated) so that its initial point (tail of the arrow)

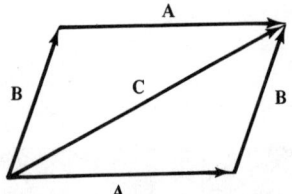

Figure 1.1. Vector addition: $C = A + B$.

coincides with the final point (head of the arrow) of the other. This is illustrated in Figure 1.1, where we see

$$C = A + B = B + A \quad (commutative\ law), \tag{1.1}$$

demonstrating that vector addition is *commutative*. If we add another vector D to $C = A + B$, obtaining a new vector E, then we get

$$E = D + C = D + (A + B)$$

or

$$E = D + (A + B) = (D + A) + B \quad (associative\ law). \tag{1.2}$$

Thus, vector addition is also *associative*.

The subtraction of vector r' from vector r to produce vector R means we have

$$R = r - r' = r + (-r'). \tag{1.3}$$

This is demonstrated in Figure 1.2.

1.2 VECTOR MULTIPLICATION

The results of Section 1.1 indicate that we have $A + A = 2A$ and $B + B + B = 3B$. More generally, if α is a scalar, then αA is a vector in the direction of A if $\alpha > 0$ or in the direction of $-A$ if $\alpha < 0$. The magnitude (length) of the vector αA is $|\alpha|$ times the magnitude of A, and this is symbolized by

$$|\alpha A| = |\alpha||A| = |\alpha|A. \tag{1.4}$$

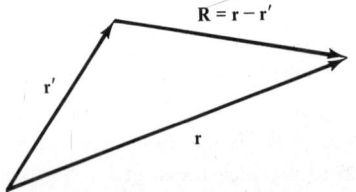

Figure 1.2. Vector subtraction: $R = r - r'$.

1.2 VECTOR MULTIPLICATION

Thus, the multiplication of a vector by a scalar is unambiguous. This is not the case when multiplying a vector by another vector, and the manner in which this multiplication is to be carried out must be stated.

The *scalar* or *dot* ($\cdot$) *product* of two vectors is defined to be a scalar such that

$$\mathbf{A} \cdot \mathbf{B} = |\mathbf{A}||\mathbf{B}| \cos [\mathbf{A}, \mathbf{B}] \quad (scalar\ product), \tag{1.5}$$

where $[\mathbf{A}, \mathbf{B}]$ is the angle between $\mathbf{A}$ and $\mathbf{B}$ measured from $\mathbf{A}$ to $\mathbf{B}$. Since $\cos [\mathbf{A}, \mathbf{B}] = \cos [\mathbf{B}, \mathbf{A}]$, it follows that

$$\mathbf{B} \cdot \mathbf{A} = |\mathbf{B}||\mathbf{A}| \cos [\mathbf{B}, \mathbf{A}] = |\mathbf{A}||\mathbf{B}| \cos [\mathbf{A}, \mathbf{B}],$$

$$\mathbf{B} \cdot \mathbf{A} = \mathbf{A} \cdot \mathbf{B} \quad (commutative\ law), \tag{1.6}$$

and the scalar product is *commutative*. It is also *distributive*:

$$\mathbf{A} \cdot (\mathbf{B} + \mathbf{C}) = \mathbf{A} \cdot \mathbf{B} + \mathbf{A} \cdot \mathbf{C} \quad (distributive\ law). \tag{1.7}$$

The scalar product is utilized very nicely in the familiar example of finding the differential work (dW) done in moving an object a distance $d\mathbf{l}$ with a force ($\mathbf{F}$) applied at an angle $[\mathbf{F}, d\mathbf{l}]$

$$dW = \mathbf{F} \cdot d\mathbf{l} = |\mathbf{F}||d\mathbf{l}| \cos [\mathbf{F}, d\mathbf{l}]. \tag{1.8}$$

Note that $|\mathbf{F}| \cos [\mathbf{F}, d\mathbf{l}]$ is the magnitude of the component of $\mathbf{F}$ in the direction of $d\mathbf{l}$.

The *vector* or *cross* ($\times$) *product* of two vectors is defined as the vector given by

$$\mathbf{A} \times \mathbf{B} = \mathbf{a}_n |\mathbf{A}||\mathbf{B}| \sin [\mathbf{A}, \mathbf{B}] \quad (vector\ product), \tag{1.9}$$

where $\mathbf{a}_n$ is a *unit vector* (unit magnitude) normal to the plane containing $\mathbf{A}$ and $\mathbf{B}$ in the *right-handed sense*. That is, if the fingers of the right hand are rotated from $\mathbf{A}$ to $\mathbf{B}$ through the smaller of the two possible angles $[\mathbf{A}, \mathbf{B}]$, then the thumb points in the direction of $\mathbf{a}_n$ and in the direction of $\mathbf{A} \times \mathbf{B}$. Since

$$\mathbf{B} \times \mathbf{A} = \mathbf{a}'_n |\mathbf{B}||\mathbf{A}| \sin [\mathbf{B}, \mathbf{A}]$$

and since $\mathbf{a}'_n = -\mathbf{a}_n$ from the preceding sentence, we have

$$\mathbf{B} \times \mathbf{A} = -\mathbf{A} \times \mathbf{B}, \tag{1.10}$$

and vector multiplication is *not commutative*. It is *distributive*:

$$\mathbf{A} \times (\mathbf{B} + \mathbf{C}) = \mathbf{A} \times \mathbf{B} + \mathbf{A} \times \mathbf{C} \quad (distributive\ law). \tag{1.11}$$

As an example of the vector product, consider a cone that is spinning about its axis with angular velocity $\boldsymbol{\omega}$ whose direction is indicated in Figure 1.3. For this

1 VECTOR ANALYSIS

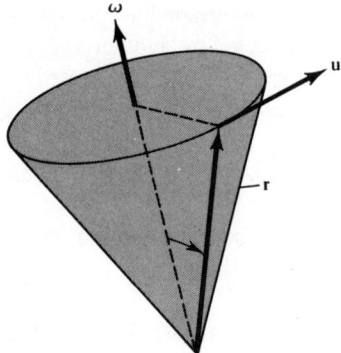

Figure 1.3. Vector product: $\mathbf{u} = \boldsymbol{\omega} \times \mathbf{r}$.

configuration, the velocity **u** of a point on the base of the cone at the periphery (tip of **r**) is given by

$$\mathbf{u} = \boldsymbol{\omega} \times \mathbf{r} \tag{1.12}$$

and **u** is normal to both $\boldsymbol{\omega}$ and **r**.

The *scalar triple product* is

$$\mathbf{A} \cdot (\mathbf{B} \times \mathbf{C}) = \mathbf{C} \cdot (\mathbf{A} \times \mathbf{B}) = \mathbf{B} \cdot (\mathbf{C} \times \mathbf{A}) \tag{1.13}$$

and is commutative. The reader can easily show how the scalar triple product can be used to give the volume of a parallelepiped. Note that the parentheses in the scalar triple product are not necessary, but in the *vector triple product*

$$\mathbf{A} \times (\mathbf{B} \times \mathbf{C}), \tag{1.14}$$

the parentheses are necessary since we have $\mathbf{A} \times (\mathbf{B} \times \mathbf{C}) \neq (\mathbf{A} \times \mathbf{B}) \times \mathbf{C}$ in general. There are many vector identities that involve vector addition and multiplication. Vector relations are summarized on the inside front and back covers.

The division of one vector by another is permissible if the vectors are parallel.

1.3 COORDINATE SYSTEMS

It is absolutely necessary that the reader be familiar with the coordinate systems we will be using in this textbook. They are the Cartesian (rectangular), circular cylindrical, and spherical coordinate systems. Figure 1.4 shows a point P described by the three coordinates associated with these systems. Also shown are coordinate transformations between the systems. These are easy to derive with Figure 1.4 and simple trigonometry.

The point P is also the locus of the intersection of the three orthogonal surfaces on which each of the coordinates is constant. This is shown in Figure 1.5. Note that

1.3 COORDINATE SYSTEMS

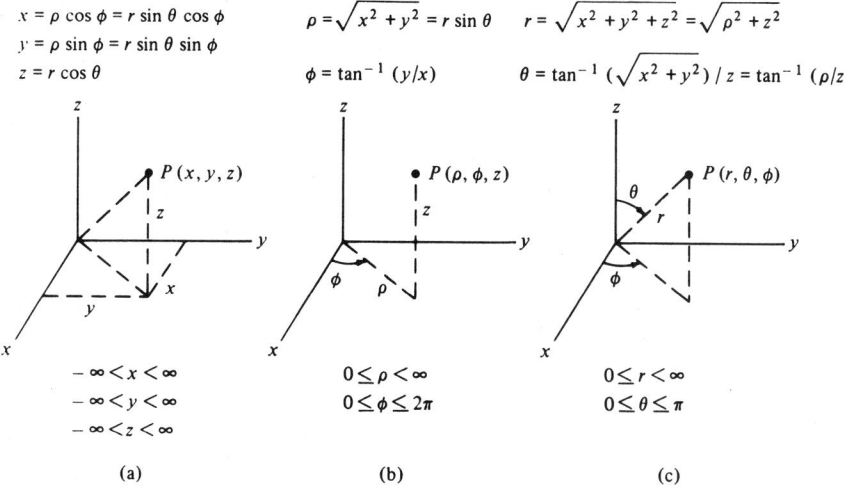

Figure 1.4. General point P described by (a) $P(x,y,z)$ in Cartesian coordinates; (b) $P(\rho,\phi,z)$ in circular cylindrical coordinates; and (c) $P(r,\theta,\phi)$ in spherical coordinates.

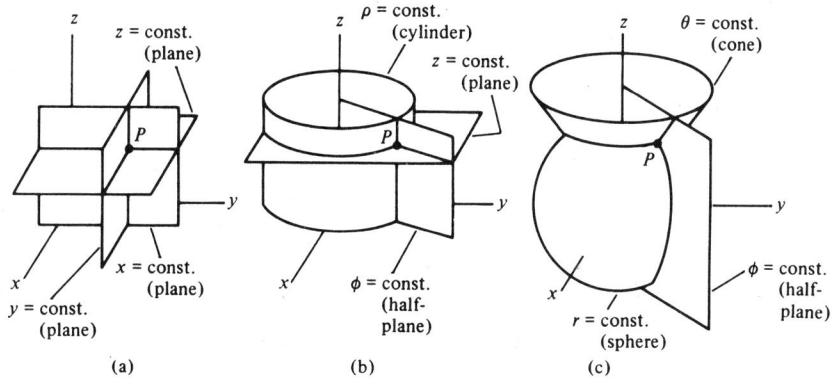

Figure 1.5. Orthogonal surfaces for (a) Cartesian, (b) cylindrical, and (c) spherical coordinates.

in Cartesian coordinates $z = $ constant is the same infinite plane as in cylindrical[1] coordinates, and in cylindrical coordinates $\phi = $ constant is the same half-plane as in spherical coordinates.

Figure 1.6 shows the three orthogonal unit vectors for each system. These systems are all right handed. That is, if the fingers of the right hand are rotated from one unit vector toward the next unit vector in the correct order ($\mathbf{a}_x$ to $\mathbf{a}_y$, $\mathbf{a}_y$ to $\mathbf{a}_z$, $\mathbf{a}_z$ to $\mathbf{a}_x$; $\mathbf{a}_\rho$ to $\mathbf{a}_\phi$, $\mathbf{a}_\phi$ to $\mathbf{a}_z$, $\mathbf{a}_z$ to $\mathbf{a}_\rho$; $\mathbf{a}_r$ to $\mathbf{a}_\theta$, $\mathbf{a}_\theta$ to $\mathbf{a}_\phi$, $\mathbf{a}_\phi$ to $\mathbf{a}_r$), then the correct direction for the third unit vector is obtained from the direction of the thumb. This is also demonstrated by the vector products listed in Figure 1.6. Note that each unit vector is normal to its respective coordinate surface, and the direction of the unit vector is that for which its coordinate is *increasing*. It

---

[1] When we refer to cylindrical coordinates in what follows, we mean circular cylindrical coordinates. There are other types of cylindrical coordinates which we will not use.

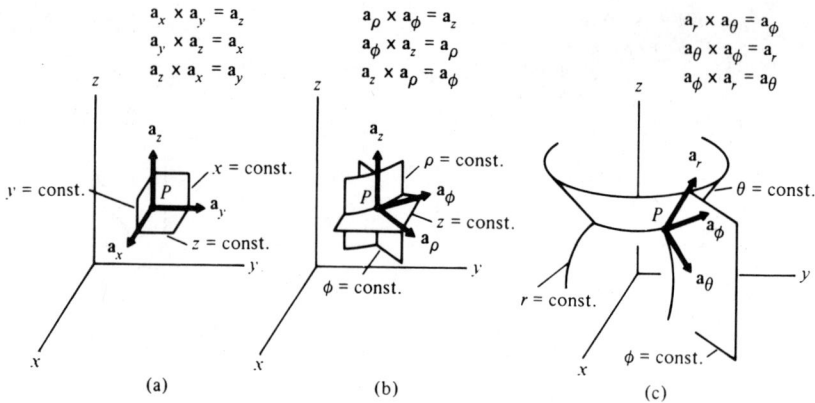

Figure 1.6. Unit vectors for (a) Cartesian, (b) cylindrical, and (c) spherical coordinates.

is important to recognize that the Cartesian coordinate system is the only one for which *all three unit vectors are constant*. All other coordinate systems possess at least one unit vector whose direction is not constant. When it becomes necessary (and it will) to integrate an expression containing a nonconstant unit vector with respect to a variable upon which the unit vector depends, the unit vector should be resolved into Cartesian components. Consider, for example, the term $\mathbf{a}_\rho \, d\phi$. The unit vector $\mathbf{a}_\rho$ depends on ϕ, and this must be accounted for. The component of $\mathbf{a}_\rho$ in the direction of $\mathbf{a}_x$ is $\mathbf{a}_\rho \cdot \mathbf{a}_x = \cos \phi$, the component of $\mathbf{a}_\rho$ in the direction of $\mathbf{a}_y$ is $\mathbf{a}_\rho \cdot \mathbf{a}_y = \sin \phi$, and the component of $\mathbf{a}_\rho$ in the direction of $\mathbf{a}_z$ is $\mathbf{a}_\rho \cdot \mathbf{a}_z = 0$. Thus, $\mathbf{a}_\rho = \mathbf{a}_x \cos \phi + \mathbf{a}_y \sin \phi$, and the resolution we are seeking is

$$\mathbf{a}_\rho \, d\phi = (\mathbf{a}_x \cos \phi + \mathbf{a}_y \sin \phi) d\phi.$$

It is often necessary to express the components of a vector that are expressed in one coordinate system into those for another coordinate system. The required coordinate component transformations are listed below and on the inside front cover:

$$\begin{aligned}
A_x &= A_\rho \cos \phi - A_\phi \sin \phi \\
&= A_r \sin \theta \cos \phi + A_\theta \cos \theta \cos \phi - A_\phi \sin \phi, \\
A_y &= A_\rho \sin \phi + A_\phi \cos \phi \\
&= A_r \sin \theta \sin \phi + A_\theta \cos \theta \sin \phi + A_\phi \cos \phi, \\
A_z &= A_r \cos \theta - A_\theta \sin \theta \\
A_\rho &= A_x \cos \phi + A_y \sin \phi = A_r \sin \theta + A_\theta \cos \theta, \quad (1.15) \\
A_\phi &= -A_x \sin \phi + A_y \cos \phi, \\
A_r &= A_x \sin \theta \cos \phi + A_y \sin \theta \sin \phi + A_z \cos \theta \\
&= A_\rho \sin \theta + A_z \cos \theta, \\
A_\theta &= A_x \cos \theta \cos \phi + A_y \cos \theta \sin \phi - A_z \sin \theta \\
&= A_\rho \cos \theta - A_z \sin \theta.
\end{aligned}$$

1.3 COORDINATE SYSTEMS

These component transformations are relatively easy to derive with the aid of Figures 1.4 and 1.6 and Equation (1.5). Consider, for example, the transformation $A_\rho = A_x \cos \phi + A_y \sin \phi$. In general $\mathbf{A} = \mathbf{a}_x A_x + \mathbf{a}_y A_y + \mathbf{a}_z A_z$. The component of the vector $\mathbf{a}_x A_x$ in the direction $\mathbf{a}_\rho A_\rho$ is $\mathbf{a}_x A_x \cdot \mathbf{a}_\rho = A_x \cos \phi$, the component of the vector $\mathbf{a}_y A_y$ in the direction of $\mathbf{a}_\rho A_\rho$ is $\mathbf{a}_y A_y \cdot \mathbf{a}_\rho = A_y \sin \phi$, and the component of the vector $\mathbf{a}_z A_z$ in the direction of $\mathbf{a}_\rho A_\rho$ is zero. Therefore, the component of $\mathbf{A}$ in the direction of $\mathbf{a}_\rho$ (that is, A_ρ) is

$$A_\rho = \mathbf{a}_x A_x \cdot \mathbf{a}_\rho + \mathbf{a}_y A_y \cdot \mathbf{a}_\rho + \mathbf{a}_z A_z \cdot \mathbf{a}_\rho,$$

$$A_\rho = A_x \cos \phi + A_y \sin \phi.$$

In the same way, we have

$$A_\rho = \mathbf{a}_r A_r \cdot \mathbf{a}_\rho + \mathbf{a}_\theta A_\theta \cdot \mathbf{a}_\rho + \mathbf{a}_\phi A_\phi \cdot \mathbf{a}_\rho,$$

$$A_\rho = A_r \sin \theta + A_\theta \cos \theta.$$

Example 1.1

Suppose that a vector field is given by

$$\mathbf{A}(x, y, z) = 2\mathbf{a}_x + x^2 \mathbf{a}_y + xy\, \mathbf{a}_z.$$

What is the field at the point (1, 1, 2)? Substituting, we get

$$\mathbf{A}(1, 1, 2) = 2\mathbf{a}_x + \mathbf{a}_y + \mathbf{a}_z.$$

At the given point, $\theta = \tan^{-1}(x^2 + y^2)^{1/2}/z = \tan^{-1} \sqrt{2}/2 = 35.3°$ and $\phi = \tan^{-1} y/x = \tan^{-1} 1 = 45°$, so using Equations (1.15), we get

$$A_\rho = A_x \cos \phi + A_y \sin \phi = 2.121,$$

$$A_\phi = -A_x \sin \phi + A_y \cos \phi = -0.707,$$

$$A_z = 1,$$

$$A_r = A_\rho \sin \theta + A_z \cos \theta = 2.041,$$

$$A_\theta = A_\rho \cos \theta - A_z \sin \theta = 1.155.$$

Thus, $\mathbf{A}$ can also be expressed as

$$\mathbf{A} = 2.121\mathbf{a}_\rho - 0.707\mathbf{a}_\phi + \mathbf{a}_z$$

or

$$\mathbf{A} = 2.041\mathbf{a}_r + 1.155\mathbf{a}_\theta - 0.707\mathbf{a}_\phi.$$

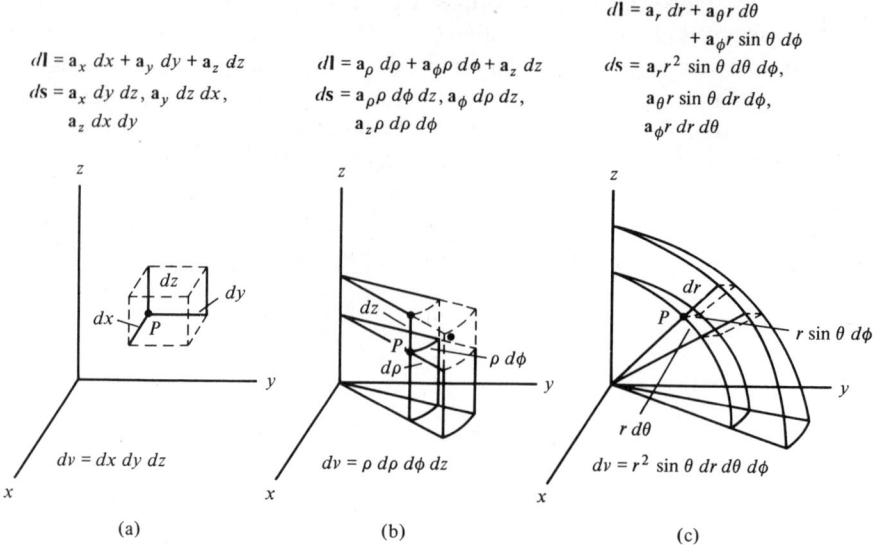

Figure 1.7. Differential elements of vector length, vector area, and scalar volume for (a) Cartesian, (b) cylindrical, and (c) spherical coordinates.

Differential elements of vector length, vector area, and scalar volume can be found using Figure 1.7. Note that each component of vector area is normal to its coordinate surface. In the case of cylindrical and spherical coordinates, the differential area and volume are obtained from the first-order approximation that the differential volumes are rectangular boxes.

■ Example 1.2

The component of differential area normal to a spherical surface is $\mathbf{a}_r \cdot d\mathbf{s} = r^2 \sin\theta \, d\theta \, d\phi$. Thus, the surface area of a sphere is

$$s = \int_{\theta=0}^{\pi} \int_{\phi=0}^{2\pi} r^2 \sin\theta \, d\theta \, d\phi = 2\pi r^2 \int_0^{\pi} \sin\theta \, d\theta = 4\pi r^2 \quad (\text{m}^2).$$

Its volume (for a radius $r = a$) is

$$v = \int_{r=0}^{a} \int_{\theta=0}^{\pi} \int_{\phi=0}^{2\pi} r^2 \sin\theta \, dr \, d\theta \, d\phi = 4\pi \int_0^a r^2 \, dr,$$

$$v = (4/3)\pi a^3 \quad (\text{m}^3).$$

Note that in this example the order of integration is unimportant. This is not always the case. ■

1.4 CIRCULATION AND FLUX

Consider the work or energy expended in moving an object along a specified path C such as that shown in Figure 1.8. The total work can be approximated as a finite sum of incremental amounts of work. The nth amount of incremental work is [see Equation (1.8)]

$$\Delta W_n = F_n \Delta l \cos \theta_n = \mathbf{F}_n \cdot \Delta \mathbf{l}_n, \tag{1.16}$$

where θ_n is the angle between $\mathbf{F}_n$ and $\Delta \mathbf{l}_n$ as shown in Figure 1.8. The total work is approximately

$$W \approx \sum_{n=1}^{N} \Delta W_n = \sum_{n=1}^{N} \mathbf{F}_n \cdot \Delta \mathbf{l}_n. \tag{1.17}$$

If we let N approach infinity in such a way that Δl_n approaches zero for any n, the limit is an exact integral expression for W, called a *line integral* in vector analysis,

$$W = \int_{P_i}^{P_f} \mathbf{F} \cdot d\mathbf{l}. \tag{1.18}$$

A force field $\mathbf{F}$ such that the line integral from P_i to P_f along the path C_1 is equal to that along the path C_2 *for any* P_i and P_f must then also be such that the line integral around *any closed path* is zero. The line integral around a closed path is called the *circulation* of $\mathbf{F}$. A field $\mathbf{F}$, whose circulation is zero is called a *conservative field* since energy is conserved in moving an object around the closed path

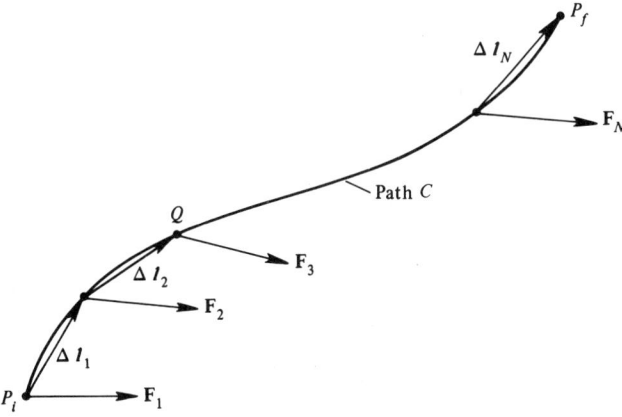

Figure 1.8. Incremental amounts of work in moving an object along a prescribed path.

$$\boxed{\oint \mathbf{F} \cdot d\mathbf{l} = 0} \quad \textit{conservative field } \mathbf{F}. \tag{1.19}$$

The differential vector length $d\mathbf{l}$ is listed in Figure 1.7 and on the inside front cover for our three standard coordinate systems.

■ Example 1.3

Suppose we desire to find the work done in moving an object from $(0, 0, 1)$ to $(2, 4, 1)$ along the parabolic path described by $y = x^2$, $z = 1$ in the nonuniform field $\mathbf{F} = 2y\mathbf{a}_x + 2x\mathbf{a}_y + z\mathbf{a}_z$ (see Figure 1.9). Since Cartesian coordinates are appropriate, we get

$$\mathbf{F} \cdot d\mathbf{l} = (2y\,\mathbf{a}_x + 2x\,\mathbf{a}_y + z\,\mathbf{a}_z) \cdot (\mathbf{a}_x\, dx + \mathbf{a}_y\, dy + \mathbf{a}_z\, dz)$$
$$= 2y\, dx + 2x\, dy + z\, dz$$

and

$$W = \left(\int_0^2 2y\, dx + \int_0^4 2x\, dy + \int_1^1 dz \right).$$

In the first integral we use the equation of the path $y = x^2$, while in the second we use $dy = 2x\, dx$ from the equation of the path. (Note that this changes the limits from those on y to those on x, i.e., 0 to 2.) The third integral is zero. Thus, we have

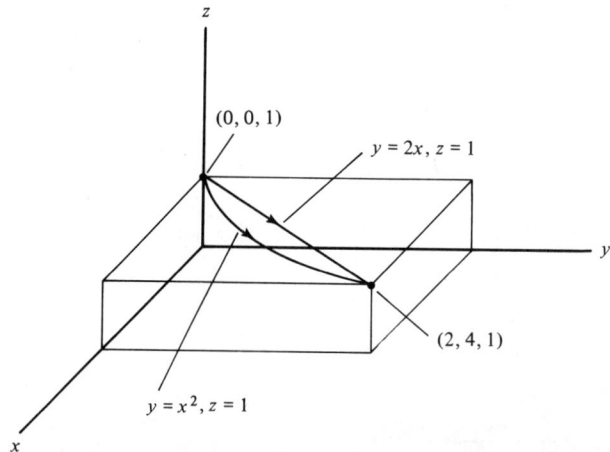

Figure 1.9. Geometry for the line integrals in Example 1.3.

1.4 CIRCULATION AND FLUX

$$W = \left(\int_0^2 2x^2 \, dx + \int_0^2 4x^2 \, dx \right),$$

$$W = 16 \quad (J).$$

If the path is changed to the straight line $y = 2x$, $dy = 2\,dx$, we have

$$W = \left(\int_0^2 4x \, dx + \int_0^2 4x \, dx \right),$$

$$W = 16 \quad (J),$$

which is the same result as before. ∎

It appears that the line integral of this particular **F** is indeed independent of the path. The preceding example offers evidence, but no proof, that the line integral is independent of the path. Proof will be given at an appropriate time.

A particular surface integral that is frequently encountered in working with vector fields is that giving the scalar *flux* or *flow* of a vector through a specified open or closed surface s. Let the flux density vector be called **D**. As shown in Figure 1.10 (for a closed surface), the normal component of the differential flux density is simply found as the component of **D** on s ($=\mathbf{D}|_s$) in the direction of $d\mathbf{s}$, so if the flux is called Ψ, then $d\Psi = \mathbf{D}|_s \cdot d\mathbf{s}$, and the total flux out of the closed surface is

$$\boxed{\Psi = \oiint_s \mathbf{D}\big|_s \cdot d\mathbf{s}} \quad \text{flux of vector } \mathbf{D}. \tag{1.20}$$

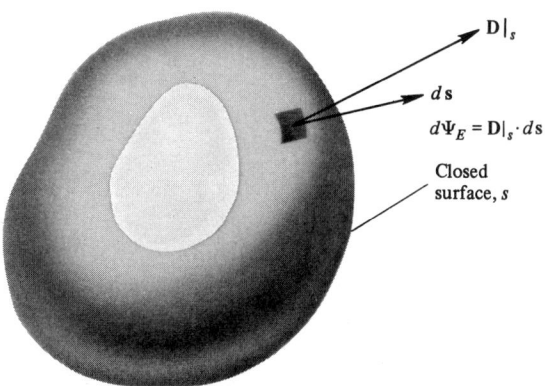

Figure 1.10. Differential flux at a point on a closed surface.

1 VECTOR ANALYSIS

Remember that the direction of $d\mathbf{s}$ is normal (outward) to the surface at the point and that the circle on the double integral symbol means that we are concerned with a closed surface, as opposed to an open surface. If we are calculating the flux through an open surface then the circle on the integral symbol in Equation (1.20) is omitted. Other integral forms that we may encounter in this text are listed on the inside back cover.

■ **Example 1.4**

Find the flux of the vector field $\mathbf{A} = \mathbf{a}_r/r^2$ out of the sphere $r = a$, $0 \leq \theta \leq \pi$, $0 \leq \phi \leq 2\pi$. We have

$$\text{Flux} = \oiint_S \mathbf{A} \cdot d\mathbf{s}\bigg|_S = \int_{\theta=0}^{\pi}\int_{\phi=0}^{2\pi} \frac{1}{r^2}\mathbf{a}_r\bigg|_{r=a} \cdot \mathbf{a}_r a^2 \sin\theta\, d\theta\, d\phi,$$

$$\text{Flux} = \int_0^{\pi}\int_0^{2\pi} \sin\theta\, d\theta\, d\phi = 4\pi.$$

Now find the flux out of a cube 1 m on a side, centered at the origin, with sides parallel to the coordinate axes. In this case the total flux is six times that out of one face because of the radial symmetry of the field. Take the face located in the $z = 0.5$ plane. The total flux is given by

$$\text{Flux} = 6\int_{x=-1/2}^{1/2}\int_{y=-1/2}^{1/2} \frac{1}{r^2}\mathbf{a}_r \cdot \mathbf{a}_z\, dx\, dy.$$

Since we have $\mathbf{a}_r \cdot \mathbf{a}_z = \cos\theta = z/r = \frac{1}{2}(x^2 + y^2 + \frac{1}{4})^{-1/2}$, then

$$\text{Flux} = 3\int_{-1/2}^{1/2}\int_{-1/2}^{1/2} \frac{dx\, dy}{(x^2 + y^2 + \frac{1}{4})^{3/2}},$$

$$\text{Flux} = 3\int_{-1/2}^{1/2}\left(\frac{x}{(y^2 + \frac{1}{4})(x^2 + y^2 + \frac{1}{4})^{1/2}}\right)\bigg|_{-1/2}^{1/2} dy,$$

$$\text{Flux} = 6\int_0^{1/2} \frac{dy}{(y^2 + \frac{1}{4})(y^2 + \frac{1}{2})^{1/2}}.$$

After using the change of variable $u^2 = y^2 + \frac{1}{4}$, we obtain

$$\text{Flux} = 12\cos^{-1}\left(\frac{1}{16u^4}\right)^{1/2}\bigg|_{1/2}^{1/\sqrt{2}} = 4\pi.$$

1.5 DIVERGENCE

It is *not* an accident that this result is identical to the first, and this will be explained when Gauss's law is considered in Chapter 2. ∎

1.5 DIVERGENCE

Consider a small box[2] with sides Δx, Δy, and Δz. Let the vector field $\mathbf{D}$ at the geometric center be given by

$$\mathbf{D} = D_1 \mathbf{a}_x + D_2 \mathbf{a}_y + D_3 \mathbf{a}_z, \quad (1.21)$$

where D_1, D_2, and D_3 are the x, y, and z components of $\mathbf{D}$. This arrangement is shown in Figure 1.11. Since the box is small, $\mathbf{D}$ is nearly constant over each of the six faces making up the box. We now calculate the flux out of the box using Equation (1.20):

$$\Psi = \oiint_S \mathbf{D} \bigg|_S \cdot d\mathbf{s}$$

realizing that we have the sum of six double integrals, one for each face. Consider the integral over the surface closest to the reader:

$$\iint_{\text{front}} \approx \mathbf{D}\bigg|_{\text{front}} \cdot \mathbf{a}_x \Delta y \, \Delta z = D_x \bigg|_{\text{front}} \Delta y \, \Delta z. \quad (1.22)$$

Equation (1.22) is approximate because $\mathbf{D}|_{\text{front}}$ is *not*, in general, constant over this face. In terms of the given flux density at the center of the box, we may write

---
[2] A rectangular parallelepiped.

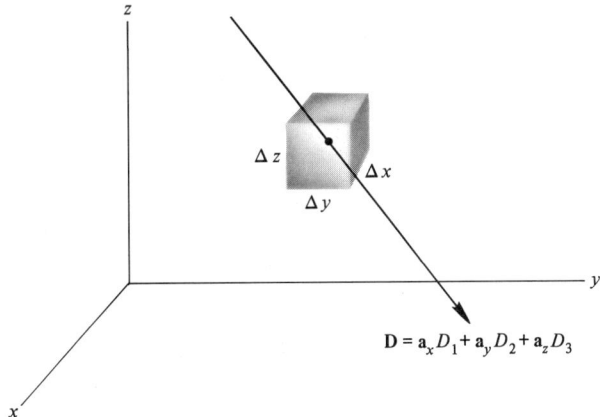

Figure 1.11. Incremental Gaussian surface for deriving Gauss's law in point form (Maxwell's equation).

$$D_x\bigg|_{\text{front}} \approx D_1 + \frac{\Delta x}{2}\frac{\partial D_x}{\partial x}. \tag{1.23}$$

In words, Equation (1.23) states that D_x on the front face is approximately equal to D_x at the center of the box (i.e., D_1) plus the rate of change of D_x with x times the distance ($\Delta x/2$) over which this change occurs. It is worth mentioning that Equation (1.23) may be obtained more rigorously as the first two terms in the Taylor's series expansion for D_x about the center of the box. We now have

$$\iint_{\text{front}} \approx \left(D_1 + \frac{\Delta x}{2}\frac{\partial D_x}{\partial x}\right)\Delta y\, \Delta z. \tag{1.24}$$

Proceeding in a similar manner for the back face, we have

$$\iint_{\text{back}} \approx \mathbf{D}\bigg|_{\text{back}} \cdot -\mathbf{a}_x \Delta y\, \Delta z = -D_x\bigg|_{\text{back}} \Delta y\, \Delta z,$$

where

$$D_x\bigg|_{\text{back}} \approx D_1 - \frac{\Delta x}{2}\frac{\partial D_x}{\partial x},$$

so that

$$\iint_{\text{back}} \approx \left(-D_1 + \frac{\Delta x}{2}\frac{\partial D_x}{\partial x}\right)\Delta y\, \Delta z$$

Thus

$$\iint_{\text{front}} + \iint_{\text{back}} \approx \frac{\partial D_x}{\partial x}\Delta x\, \Delta y\, \Delta z$$

For all six faces we have

$$\oint_{\Delta s} \mathbf{D}\bigg|_{\Delta s} \cdot d\mathbf{s} \approx \left(\frac{\partial D_x}{\partial x} + \frac{\partial D_y}{\partial y} + \frac{\partial D_z}{\partial z}\right)\Delta v.$$

Dividing by the differential volume Δv, we get

$$\frac{\partial D_x}{\partial x} + \frac{\partial D_y}{\partial y} + \frac{\partial D_z}{\partial z} \approx \frac{\oint_{\Delta s} \mathbf{D}|_s \cdot d\mathbf{s}}{\Delta v}.$$

If we now take the limit as Δv approaches zero, we obtain exactly

1.5 DIVERGENCE

$$\frac{\partial D_x}{\partial x} + \frac{\partial D_y}{\partial y} + \frac{\partial D_z}{\partial z} = \lim_{\Delta v \to 0} \frac{\oint_{\Delta s} \mathbf{D}|_s \cdot d\mathbf{s}}{\Delta v} \qquad (1.25)$$

The term on the right-hand side of Equation (1.25) is the *limit of the flux out of a small volume per unit volume as the volume becomes vanishingly small* (independent of coordinate system), and this quantity is called the *divergence of* **D**; that is,

$$\text{Divergence of } \mathbf{D} = \text{div } \mathbf{D} = \lim_{\Delta v \to 0} \frac{\oint_{\Delta s} \mathbf{D}|_s \cdot d\mathbf{s}}{\Delta v}. \qquad (1.26)$$

Consider a situation such as that shown in Figure 1.12 where flux density lines appear to *diverge* from the small volume v_a, suggesting the presence within v_a of some kind of *source*. On the other hand, flux density lines appear to *converge* (negative divergence) into the small volume v_b, suggesting the presence within v_b of some kind of *sink* (negative source). It is apparently true that the flux out of v_a is positive, and the flux out of v_b is negative. It appears that the flux out of v_c is zero. All of these conclusions could be verified by means of Equation (1.25) if we were given an explicit expression for the vector field **D** at every point in space. *A nonzero value for the divergence of a vector field at a point is intimately associated with the presence of a scalar source at that point.* Gauss's law, to be presented in Chapter 2, will shed more light on this subject.

It is apparent from Equations (1.25) and (1.26) that

$$\text{div } \mathbf{D} = \frac{\partial D_x}{\partial x} + \frac{\partial D_y}{\partial y} + \frac{\partial D_z}{\partial z} \qquad (1.27)$$

for Cartesian coordinates, but this result occurred merely because we started with a small *rectangular* box. The result given by Equation (1.26) is general, and is independent of coordinate system. In order to emphasize these points and to show explicitly that taking the divergence of a vector is a *vector operation*, we introduce the vector del operator:

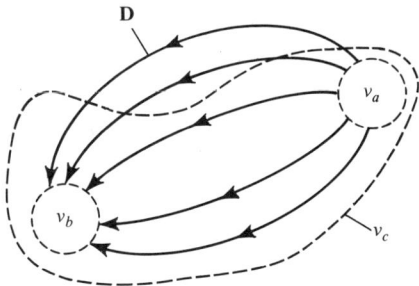

Figure 1.12. Flux density streamlines indicating flux in (or out of) volumes.

$$\boxed{\nabla \equiv \mathbf{a}_x \frac{\partial}{\partial x} + \mathbf{a}_y \frac{\partial}{\partial y} + \mathbf{a}_z \frac{\partial}{\partial z}}. \qquad (1.28)$$

The del operator must operate on some quantity, and it is easy to show (treating ∇ like any vector) that

$$\nabla \cdot \mathbf{D} = \left(\mathbf{a}_x \frac{\partial}{\partial x} + \mathbf{a}_y \frac{\partial}{\partial y} + \mathbf{a}_z \frac{\partial}{\partial z} \right) \cdot \left(\mathbf{a}_x D_x + \mathbf{a}_y D_y + \mathbf{a}_z D_z \right),$$

$$\nabla \cdot \mathbf{D} = \frac{\partial D_x}{\partial x} + \frac{\partial D_y}{\partial y} + \frac{\partial D_z}{\partial z} = \lim_{\Delta v \to 0} \frac{\oint_{\Delta s} \mathbf{D}|_s \cdot d\mathbf{s}}{\Delta v} \qquad (1.29)$$

for Cartesian coordinates.

The left and right sides of Equation (1.29) give us a completely general mathematical definition of del as an integral operator:

$$\nabla \circ \underline{} = \lim_{\Delta v \to 0} \frac{\oint_{\Delta s} d\mathbf{s} \circ \underline{}}{\Delta v} \qquad (1.30)$$

If the small circle (∘) becomes a dot, we obtain the divergence of a vector. If the small circle becomes a cross, we obtain the *curl of a vector*. Finally, if the small circle disappears, as in ordinary multiplication, we obtain the *gradient of a scalar*. The curl and gradient operations will be discussed shortly.

It is useful to list $\nabla \cdot \mathbf{D}$ for Cartesian, circular cylindrical, and spherical coordinates (also listed on the inside front and back covers):

$$\nabla \cdot \mathbf{D} = \frac{\partial D_x}{\partial x} + \frac{\partial D_y}{\partial y} + \frac{\partial D_z}{\partial z} \qquad (1.31a)$$

$$\nabla \cdot \mathbf{D} = \frac{1}{\rho} \frac{\partial}{\partial \rho} (\rho D_\rho) + \frac{1}{\rho} \frac{\partial D_\phi}{\partial \phi} + \frac{\partial D_z}{\partial z} \qquad (1.31b)$$

$$\nabla \cdot \mathbf{D} = \frac{1}{r^2} \frac{\partial}{\partial r} (r^2 D_r) + \frac{1}{r \sin \theta} \frac{\partial}{\partial \theta} (\sin \theta D_\theta) + \frac{1}{r \sin \theta} \frac{\partial D_\phi}{\partial \phi}. \qquad (1.31c)$$

As can be seen, the divergence of a vector quantity is a scalar involving partial derivatives of a particular component with respect to the variable associated with that component.

Some further remarks about the divergence are in order. An incompressible fluid, such as water, must have a velocity field **u** which has no divergence ($\nabla \cdot \mathbf{u} \equiv 0$). Fluid is neither created nor destroyed at any point, and therefore has no *sources* or *sinks*. The *magnetic* flux density field, which we will encounter later, also has no sources or sinks because no isolated *magnetic* charges have been found in nature. It too has no divergence. Such fields are said to be *solenoidal* or *sourceless*. The electrostatic flux density field, on the other hand, does have sources and sinks (the electric charges), and must have nonzero divergence at some point or points in space.

1.6 GRADIENT

The gradient of a scalar field is a vector field that lies in the direction for which the scalar field is changing most rapidly. The magnitude of the gradient is the greatest rate of change of the scalar field. A two-dimensional scalar field $\Phi(x, y)$ is shown in Figure 1.13 in terms of contours (curves in the two-dimensional case shown) of constant Φ. In the general three-dimensional case these contours are called *isothermal* surfaces when, for example, $\Phi(x, y, z)$ represents temperature, and they are called *equipotential* surfaces if $\Phi(x, y, z)$ represents *electric potential* (to be introduced in Chapter 2). The direction of the greatest rate of change of Φ will be *normal* to the contours of constant Φ (from a smaller to a larger value) since the rate of change of Φ is zero along or *tangent* to these contours. Thus, we have

$$\mathbf{grad}\ \Phi = \mathbf{a}_n \frac{d\Phi}{dn}. \qquad 1.32$$

Consider (in a completely general case) $\Phi(x, y, z)$ at a point $P(x, y, z)$ and $\Phi(x + \Delta x, y + \Delta y, z + \Delta z)$ at a nearby point $Q(x + \Delta x, y + \Delta y, z + \Delta z)$. If Δl is the distance between P and Q, then we get

$$\Delta\Phi = \Phi(x + \Delta x, y + \Delta y, z + \Delta z) - \Phi(x, y, z)$$
$$\approx \frac{\partial \Phi}{\partial x}\Delta x + \frac{\partial \Phi}{\partial y}\Delta y + \frac{\partial \Phi}{\partial z}\Delta z.$$

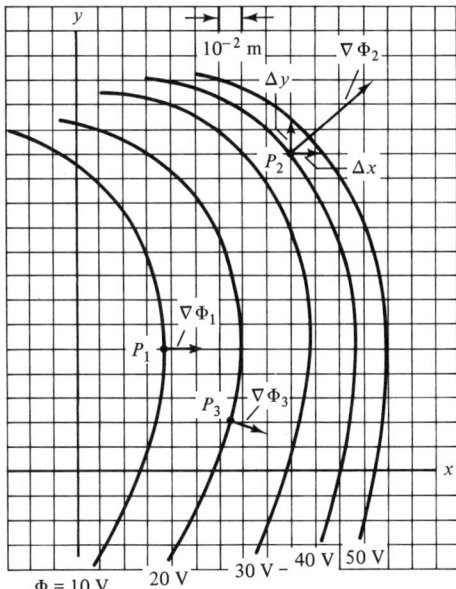

Figure 1.13. Two-dimensional potential field Φ in terms of equipotentials.

Higher-order infinitesimals that will disappear in the limiting process to follow have been dropped. Then we have

$$\frac{\Delta \Phi}{\Delta l} \approx \frac{\partial \Phi}{\partial x}\frac{\Delta x}{\Delta l} + \frac{\partial \Phi}{\partial y}\frac{\Delta y}{\Delta l} + \frac{\partial \Phi}{\partial z}\frac{\Delta z}{\Delta l}$$

and

$$\frac{d\Phi}{dl} = \lim_{\Delta l \to 0} \frac{\Delta \Phi}{\Delta l} = \frac{\partial \Phi}{\partial x}\frac{dx}{dl} + \frac{\partial \Phi}{\partial y}\frac{dy}{dl} + \frac{\partial \Phi}{\partial z}\frac{dz}{dl}.$$

This quantity is the rate of change of Φ with respect to distance at point P in a direction toward Q and is thus called the *directional derivative*. The directional derivative may also be written

$$\frac{d\Phi}{dl} = \left(\mathbf{a}_x\frac{\partial \Phi}{\partial x} + \mathbf{a}_y\frac{\partial \Phi}{\partial y} + \mathbf{a}_z\frac{\partial \Phi}{\partial z}\right) \cdot \left(\mathbf{a}_x\frac{dx}{dl} + \mathbf{a}_y\frac{dy}{dl} + \mathbf{a}_z\frac{dz}{dl}\right)$$

or

$$\frac{d\Phi}{dl} = \left(\mathbf{a}_x\frac{\partial \Phi}{\partial x} + \mathbf{a}_y\frac{\partial \Phi}{\partial y} + \mathbf{a}_z\frac{\partial \Phi}{\partial z}\right) \cdot \frac{d\mathbf{l}}{dl} \quad \text{(directional derivative)}. \tag{1.33}$$

Note that $d\mathbf{l}/dl = \mathbf{a}_l$ is a *unit* vector. From the discussion in the preceding paragraph it is apparent that the directional derivative is maximum when $\mathbf{a}_l = \mathbf{a}_n$, that is, when it is taken in a direction *normal* to the equipotential surface at P_2 (as in Figure 1.13). Thus, we have

$$\left(\frac{\partial \Phi}{dl}\right)_{\max} = \left(\mathbf{a}_x\frac{\partial \Phi}{\partial x} + \mathbf{a}_y\frac{\partial \Phi}{\partial y} + \mathbf{a}_z\frac{\partial \Phi}{\partial z}\right) \cdot \mathbf{a}_n. \tag{1.34}$$

It is now obvious that the maximum directional derivative takes place in the direction of, and has the magnitude of the bracketed term in Equation (1.34). From the definition in the preceding paragraph, the bracketed term is the gradient of the scalar field Φ, so we get

$$\mathbf{grad}\ \Phi = \nabla \Phi = \mathbf{a}_x\frac{\partial \Phi}{\partial x} + \mathbf{a}_y\frac{\partial \Phi}{\partial y} + \mathbf{a}_z\frac{\partial \Phi}{\partial z} \quad \text{(Cartesian)}. \tag{1.35a}$$

For cylindrical and spherical coordinates, we have

$$\nabla \Phi = \mathbf{a}_\rho\frac{\partial \Phi}{\partial \rho} + \mathbf{a}_\phi\frac{1}{\rho}\frac{\partial \Phi}{\partial \phi} + \mathbf{a}_z\frac{\partial \Phi}{\partial z} \quad \text{(cylindrical)} \tag{1.35b}$$

and

$$\nabla \Phi = \mathbf{a}_r\frac{\partial \Phi}{\partial r} + \mathbf{a}_\theta\frac{1}{r}\frac{\partial \Phi}{\partial \theta} + \mathbf{a}_\phi\frac{1}{r\sin\theta}\frac{\partial \Phi}{\partial \phi} \quad \text{(spherical)}. \tag{1.35c}$$

1.6 GRADIENT

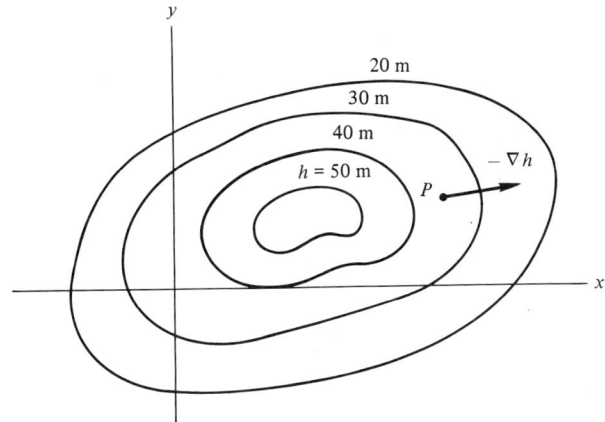

Figure 1.14. Contours of constant elevation h and $-\nabla h$.

It is important to recognize that, in general, Φ varies from point to point. A glance at a map having elevation contours (equipotentials) shows this very well. Figure 1.14 shows a hill with elevation contours around it. We now ask ourselves a question. "If we place a small ball at P, in what direction will it roll and what is the magnitude of the force acting to accelerate it in that direction?" This is equivalent to asking what is $-\text{grad } h(x, y)$?

Example 1.5

Find (approximately) the gradient of Φ at P_1 and P_2 in Figure 1.13. We have

$$\nabla \Phi = +\mathbf{a}_x \frac{\partial \Phi}{\partial x} + \mathbf{a}_y \frac{\partial \Phi}{\partial y} \approx +\mathbf{a}_x \frac{\Delta \Phi}{\Delta x} + \mathbf{a}_y \frac{\Delta \Phi}{\Delta y}$$

so, for P_1, we get

$$\nabla \Phi \bigg|_1 \approx +\mathbf{a}_x \frac{10}{0.032} = +313 \mathbf{a}_x;$$

and for P_2, we get

$$\nabla \Phi \bigg|_2 \approx +\mathbf{a}_x \frac{10}{0.013} + \mathbf{a}_y \frac{10}{0.014} = +769 \mathbf{a}_x + 714 \mathbf{a}_y.$$

Temperature is a term that is familiar to us. The temperature in a region may vary from point to point, and can, therefore, be described as a three-dimensional scalar field. Surfaces of constant temperature (the "equipotentials") are called *isothermal* surfaces. The heat flux vector $\mathbf{s}$ (W/m^2) at a point in a conducting body is given by the Fourier heat conduction law: $\mathbf{s} = -K\nabla T$, where T is the temperature and K is the thermal conductivity. Thus, heat flows in a direction opposite to the positive temperature gradient, or, put more simply, heat flows from a region of higher temperature to a region of lower temperature.

As in the case of the divergence, the gradient of a scalar field at a point P may be defined in integral form by[3]

$$\nabla \Phi \equiv \lim_{\Delta v \to 0} \frac{1}{\Delta v} \oiint_s \Phi \, d\mathbf{s}, \qquad (1.36)$$

where Δv is a small volume surrounding P, and s is the surface area of Δv.

■ Example 1.6

The vector $\mathbf{r} - \mathbf{r}'$ that appears in Figure 1.2 occurs very frequently in the superposition type integrals of field theory. This will be demonstrated in Chapter 2. If we define $\mathbf{R} = \mathbf{r} - \mathbf{r}'$, then $|\mathbf{R}| = R = |\mathbf{r} - \mathbf{r}'|$, and with $\mathbf{r} = x\mathbf{a}_x + y\mathbf{a}_y + z\mathbf{a}_z$ and $\mathbf{r}' = x'\mathbf{a}_x + y'\mathbf{a}_y + z'\mathbf{a}_z$, it follows that

$$R = \left[(x-x')^2 + (y-y')^2 + (z-z')^2\right]^{1/2}.$$

Then, we have

$$\nabla \frac{1}{R} = \mathbf{a}_x \frac{\partial(1/R)}{\partial x} + \mathbf{a}_y \frac{\partial(1/R)}{\partial y} + \mathbf{a}_z \frac{\partial(1/R)}{\partial z}$$

$$\nabla \frac{1}{R} = -\frac{\mathbf{a}_x(x-x')}{R^3} - \frac{\mathbf{a}_y(y-y')}{R^3} - \frac{\mathbf{a}_z(z-z')}{R^3}$$

$$\nabla \frac{1}{R} = -\frac{\mathbf{R}}{R^3} = -\frac{1}{R^2}\mathbf{a}_R$$

The same operation on the prime coordinates is called $\nabla'(1/R)$, and it should be obvious to the reader that we get

$$\nabla(1/R) = -\nabla'(1/R).$$

These results will be used in Chapter 2. ■

1.7 CURL

The component in the $\mathbf{a}_n$ direction (general) of the curl of $\mathbf{A}$ at a point P may be defined as the limit of the circulation of $\mathbf{A}$ per unit area as the area ($\mathbf{a}_n \Delta s$) approaches zero. Thus, we have

[3] Refer to Equation (1.30).

1.7 CURL

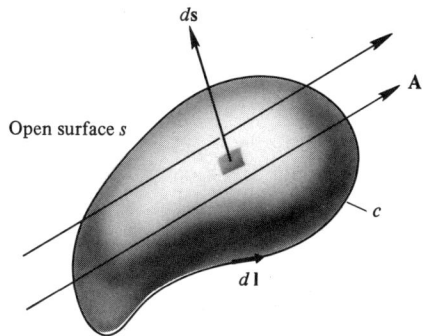

Figure 1.15. Line integral contour c and one open surface s.

$$(\text{curl } \mathbf{A})_n = \lim_{\Delta s \to 0} \frac{1}{\Delta s} \oint_c \mathbf{A} \cdot d\mathbf{l}, \tag{1.37}$$

where $\Delta \mathbf{s} = \mathbf{a}_n \Delta s$, and the closed path c defines the area Δs. The unit vector $\mathbf{a}_n$ is normal to the surface Δs, and in a direction determined by the right-hand rule as explained in Figure 1.15.

For example, following a procedure similar to that used in arriving at a formula for divergence, we may describe a rectangle parallel to the $z = 0$ plane as shown in Figure 1.16. At the center of the rectangle (whose sides are Δx, Δy), $\mathbf{A} = \mathbf{a}_x A_x + \mathbf{a}_y A_y + \mathbf{a}_z A_z$. It is easy to show by means of the Taylor series expansion that

$$A_y\bigg|_{x+\Delta x/2} = A_y + \frac{\partial A_y}{\partial x}\frac{\Delta x}{2} + \cdots, \qquad A_x\bigg|_{y+\Delta y/2} = A_x + \frac{\partial A_x}{\partial y}\frac{\Delta y}{2} + \cdots,$$

and

$$A_y\bigg|_{x-\Delta x/2} = A_y - \frac{\partial A_y}{\partial x}\frac{\Delta x}{2} + \cdots, \qquad A_x\bigg|_{y-\Delta y/2} = A_x - \frac{\partial A_x}{\partial y}\frac{\Delta y}{2} + \cdots,$$

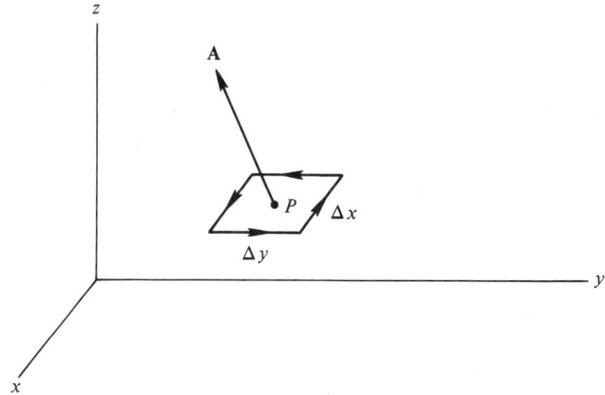

Figure 1.16. Geometry for evaluating the z component of the curl of $\mathbf{A}$.

and we get

$$\oint \mathbf{A} \cdot d\mathbf{l} = \left(\frac{\partial A_y}{\partial x} - \frac{\partial A_x}{\partial y} \right) \Delta x\, \Delta y.$$

Higher-order terms that would disappear in the limiting process have simply been dropped. In the limit then, we get

$$(\text{curl } \mathbf{A})_z = \lim_{\Delta x \Delta y \to 0} \left(\frac{\partial A_y}{\partial x} - \frac{\partial A_x}{\partial y} \right) \frac{\Delta x\, \Delta y}{\Delta x\, \Delta y}$$

or

$$(\text{curl } \mathbf{A})_z = \frac{\partial A_y}{\partial x} - \frac{\partial A_x}{\partial y}.$$

Repeating this procedure for rectangles parallel to the $y = 0$ plane and $x = 0$ plane gives

$$(\text{curl } \mathbf{A})_y = \frac{\partial A_x}{\partial z} - \frac{\partial A_z}{\partial x}$$

and

$$(\text{curl } \mathbf{A})_x = \frac{\partial A_z}{\partial y} - \frac{\partial A_y}{\partial z}.$$

Then the curl of $\mathbf{A}$, given by $\nabla \times \mathbf{A}$, may be written for Cartesian coordinates as

$$\nabla \times \mathbf{A} = \mathbf{a}_x \left(\frac{\partial A_z}{\partial y} - \frac{\partial A_y}{\partial z} \right) + \mathbf{a}_y \left(\frac{\partial A_x}{\partial z} - \frac{\partial A_z}{\partial x} \right) + \mathbf{a}_z \left(\frac{\partial A_y}{\partial x} - \frac{\partial A_x}{\partial y} \right) \quad (1.38a)$$

or, in a form more easily remembered,

$$\nabla \times \mathbf{A} = \begin{vmatrix} \mathbf{a}_x & \mathbf{a}_y & \mathbf{a}_z \\ \dfrac{\partial}{\partial x} & \dfrac{\partial}{\partial y} & \dfrac{\partial}{\partial z} \\ A_x & A_y & A_z \end{vmatrix}$$

For cylindrical and spherical coordinates, we have

$$\nabla \times \mathbf{A} = \mathbf{a}_\rho \left(\frac{1}{\rho} \frac{\partial A_z}{\partial \phi} - \frac{\partial A_\phi}{\partial z} \right) + \mathbf{a}_\phi \left(\frac{\partial A_\rho}{\partial z} - \frac{\partial A_z}{\partial \rho} \right)$$

$$+ \mathbf{a}_z \left(\frac{1}{\rho} \frac{\partial (\rho A_\phi)}{\partial \rho} - \frac{1}{\rho} \frac{\partial A_\rho}{\partial \phi} \right) \quad (\text{cylindrical}) \quad (1.38b)$$

and

1.7 CURL

$$\nabla \times \mathbf{A} = \frac{\mathbf{a}_r}{r \sin \theta} \left(\frac{\partial (A_\phi \sin \theta)}{\partial \theta} - \frac{\partial A_\theta}{\partial \phi} \right)$$

$$+ \frac{\mathbf{a}_\theta}{r} \left(\frac{1}{\sin \theta} \frac{\partial A_r}{\partial \phi} - \frac{\partial (r A_\phi)}{\partial r} \right)$$

$$+ \frac{\mathbf{a}_\phi}{r} \left(\frac{\partial (r A_\theta)}{\partial r} - \frac{\partial A_r}{\partial \theta} \right) \quad \text{(spherical)}. \quad (1.38c)$$

There are at most three components of a vector in space, and each of these three components can depend at most on three space variables. Thus, there are at most nine (first) derivatives involving a spatial vector. Six of these are involved in specifying the curl of a vector, as Equations (1.38a), (1.38b), and (1.38c) show, while the remaining three are involved in specifying the divergence of a vector, as Equations (1.31a), (1.31b), and (1.31c) show. If both the curl and the divergence of a vector are specified, then the vector can be uniquely determined.

In order to attach some physical significance to the curl of a vector, we will employ the small "paddlewheel" as suggested by Skilling.[4] Let the vector field be a fluid velocity field, regardless of what it is physically. Place the small paddlewheel in this velocity field and move it about. For every point in the field, where the curl of the field is to be found, the paddlewheel axis should be oriented in all possible directions. The maximum angular velocity of the paddlewheel at a point is proportional to the curl, while the axis of the paddlewheel points in the *direction* of the curl according to the right-hand rule. That is, if the fingers of the right hand point in the direction of the rotation of the paddlewheel blades, then the thumb of the right hand points in the direction of the axis of rotation or curl. This is demonstrated in Figure 1.17. If the paddlewheel does not rotate, the vector field is *irrotational, or has zero curl*! A simple example will help clarify this paddlewheel concept.

[4]See references at end of chapter.

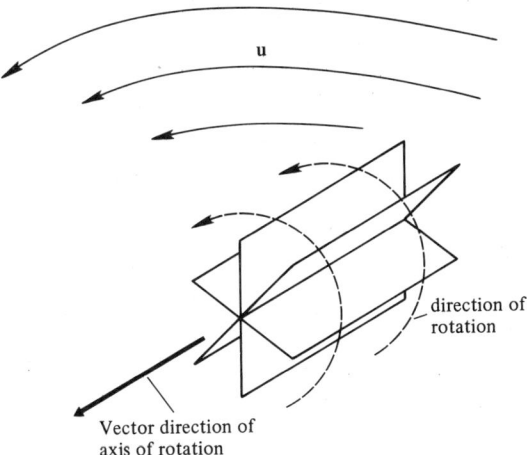

Figure 1.17. Paddlewheel rotating in a fluid velocity field and the right-hand rule.

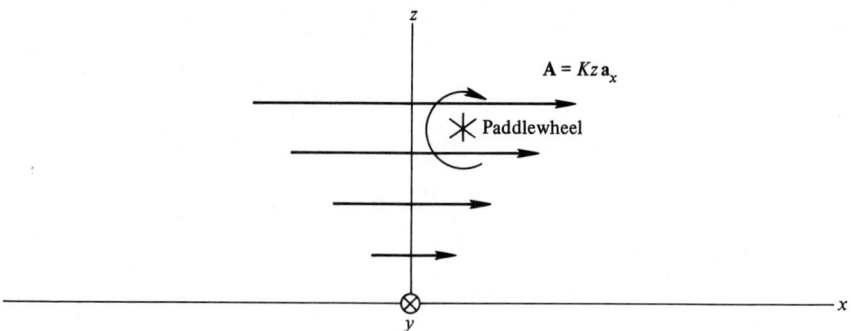

Figure 1.18. Demonstration for finding the curl of a linearly increasing vector field $\mathbf{A} = kz\mathbf{a}_x$ using a small paddlewheel.

Example 1.7

Suppose $\mathbf{A} = Kz\mathbf{a}_x$, where K is a constant. This field is sketched in Figure 1.18 for $z > 0$. It might be analogous to water flow in a river. It is obvious that the paddlewheel rotates in a clockwise direction (regardless of z) and the angular velocity of the paddlewheel is proportional to K (the slope) regardless of z. Furthermore, the maximum angular velocity occurs when the paddlewheel axis is parallel to the y axis and is zero when the paddlewheel axis is parallel to the x or z axis. According to the right-hand rule the direction of the curl is that of $+\mathbf{a}_y$.

Analytically, using Equation (1.38a), we have

$$\nabla \times \mathbf{A} = +\mathbf{a}_y \frac{\partial A_x}{\partial z} = \mathbf{a}_y K,$$

which certainly agrees with our observations with the paddlewheel! ■

1.8 SOME FUNDAMENTAL VECTOR IDENTITIES

It is easy to show in Cartesian coordinates that

$$\boxed{\nabla \times (\nabla \Phi) \equiv 0} \qquad (1.39)$$

for any scalar field Φ. Now, for example, any vector field $\mathbf{F}$ that can be written as $(\pm)$ the gradient of Φ ($\mathbf{F} = \pm\nabla\Phi$) must have zero curl ($\nabla \times \mathbf{F} \equiv 0$). This is true even if a constant C is added to Φ since $\nabla(\Phi + C) = \nabla\Phi + \nabla C = \nabla\Phi$ ($\nabla C \equiv 0$). A field that has no curl is said to be *irrotational* as mentioned at the end of Section 1.7:

$$\boxed{\nabla \times \mathbf{F} \equiv 0} \quad \textit{irrotational field } \mathbf{F} \qquad (1.40)$$

1.8 SOME FUNDAMENTAL VECTOR IDENTITIES

The *work done by an external source* in moving an object from P_i to P_f in the vector force field $\mathbf{F}$ is given by the *negative* of Equation (1.18):

$$W = -\int_{P_i}^{P_f} \mathbf{F} \cdot d\mathbf{l}.$$

Now using Cartesian coordinates, with $\mathbf{F} = -\nabla \Phi$, we get

$$\nabla \Phi \cdot d\mathbf{l} = \left(\mathbf{a}_x \frac{\partial \Phi}{\partial x} + \mathbf{a}_y \frac{\partial \Phi}{\partial y} + \mathbf{a}_z \frac{\partial \Phi}{\partial z} \right) \cdot (\mathbf{a}_x \, dx + \mathbf{a}_y \, dy + \mathbf{a}_z \, dz)$$

or

$$\nabla \Phi \cdot d\mathbf{l} = \frac{\partial \Phi}{\partial x} dx + \frac{\partial \Phi}{\partial y} dy + \frac{\partial \Phi}{\partial z} dz,$$

where the right-hand side of the equation is the differential $d\Phi$:

$$d\Phi = \nabla \Phi \cdot d\mathbf{l} = -\mathbf{F} \cdot d\mathbf{l},$$

so that we get

$$W = +\int_{P_i}^{P_f} d\Phi = \Phi_f - \Phi_i.$$

Therefore, W depends on the endpoints of the path, but not on the path itself! Furthermore, $\mathbf{F}$ must be a conservative field [see Equation (1.19) and the material preceding it]. *We conclude that any vector field that is the gradient of a scalar field is a conservative field.* We can test a field to determine whether it is conservative or not by simply determining whether or not its curl is identically zero:

$$\boxed{\nabla \times \mathbf{F} \equiv 0} \quad \text{conservative field } \mathbf{F}. \tag{1.41}$$

The electrostatic field intensity, the gravitational force field, and the heat flux vector are all examples of conservative fields.

Another useful vector identity is

$$\boxed{\nabla \cdot (\nabla \times \mathbf{M}) \equiv 0} \tag{1.42}$$

for any vector field $\mathbf{M}$. As a consequence, any vector field $\mathbf{N}$, for example, that can be written as the curl of $\mathbf{M}$ ($\mathbf{N} = \nabla \times \mathbf{M}$) must have no divergence ($\nabla \cdot \mathbf{N} \equiv 0$). Such fields are called *divergenceless* or *solenoidal*. This is true even if the gradient of any scalar field α is added to $\mathbf{M}$, since $\nabla \times (\mathbf{M} + \nabla \alpha) =$

$\nabla \times \mathbf{M} + \nabla \times (\nabla \alpha) = \nabla \times \mathbf{M}$ by Equation (1.39). The magnetic flux density field and the velocity field for an incompressible fluid are examples of solenoidal fields.

1.9 THE DIVERGENCE THEOREM

Consider a volume in space bounded by a closed surface s with a well-behaved vector field $\mathbf{F}$ present as shown in Figure 1.19. Let dv be a volume element and let $d\mathbf{s}$ be a vector surface element of the *external* surface s. Now imagine that the entire volume is partitioned into N small volumes, $\Delta v_1, \Delta v_2, \ldots, \Delta v_n, \ldots, \Delta v_N$ each having *closed* surfaces $\Delta s_1, \Delta s_2, \ldots, \Delta s_n, \ldots, \Delta s_N$. For each small volume $\nabla \cdot \mathbf{F}$ is almost uniform and equal to the value it has as Δv approaches zero at a point. According to its definition, Equation (1.26), $\nabla \cdot \mathbf{F}$ is the limit of the flux of the quantity (designated by $\mathbf{F}$) per unit volume as the volume approaches zero. That is,

$$(\nabla \cdot \mathbf{F})_n = \lim_{\Delta s_n \to 0} \frac{\oint_{\Delta s_n} \mathbf{F}_n \cdot d\mathbf{s}_n}{\Delta v_n}$$

or

$$(\nabla \cdot \mathbf{F})_n \approx \frac{\oint_{\Delta s_n} \mathbf{F}_n \cdot d\mathbf{s}_n}{\Delta v_n}$$

or

$$(\nabla \cdot \mathbf{F})_n \Delta v_n \approx \oint_{\Delta s_n} \mathbf{F}_n \cdot d\mathbf{s}_n$$

Writing similar expressions for all the small volumes and summing, we get

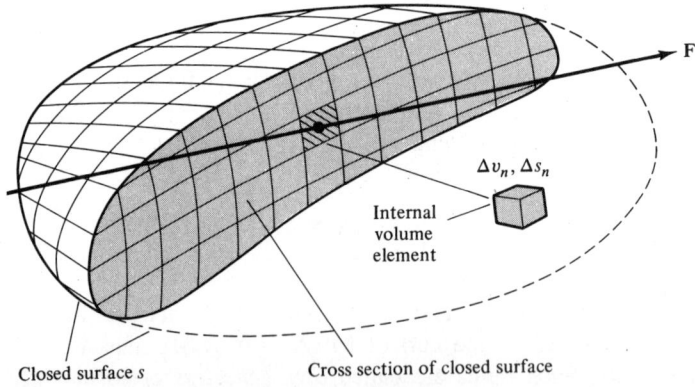

Figure 1.19. Geometry for establishing the divergence theorem (Gauss's integral theorem).

1.9 THE DIVERGENCE THEOREM

$$\sum_{n=1}^{N} (\nabla \cdot \mathbf{F})_n \Delta v_n \approx \sum_{n=1}^{N} \oiint_{\Delta s_n} \mathbf{F}_n \cdot d\mathbf{s}_n,$$

where the right-hand side of the equation is the flux out of the *external* surface s, that is,

$$\oiint_s \mathbf{F} \bigg|_s \cdot d\mathbf{s},$$

because the contributions from all the surfaces (or partial surfaces) *internal* to s cancel. They cancel, because at interface surfaces common to adjacent small volumes, $d\mathbf{s}_n$ will be the same in magnitude, but opposite in direction. Thus, we have

$$\sum_{n=1}^{N} (\nabla \cdot \mathbf{F})_n \Delta v_n \approx \oiint_s \mathbf{F} \bigg|_s \cdot d\mathbf{s}.$$

Now if we let N approach infinity so that Δv_n approaches zero, the sum on the left becomes a volume integral and the approximation becomes exact

$$\boxed{\iiint_{\text{vol} \leftarrow} \nabla \cdot \mathbf{F}\, dv = \oiint_{\rightarrow s} \mathbf{F}|_s \cdot d\mathbf{s}} \qquad \text{divergence theorem.} \qquad (1.43)$$

Equation (1.43) is known as the divergence theorem. The long arrow means that the surface on the right defines the volume on the left.

■ Example 1.8

Show that the divergence theorem holds for the vector field $\mathbf{A} = \mathbf{a}_r/r$ when the surface is that of a sphere of radius a centered at the origin. We have $\nabla \cdot \mathbf{A} = 1/r^2$ and

$$\int_{\theta=0}^{\pi} \int_{\phi=0}^{2\pi} \frac{1}{a} \mathbf{a}_r \cdot \mathbf{a}_r a^2 \sin\theta\, d\theta d\phi = \int_{r=0}^{a} \int_{\theta=0}^{\pi} \int_{\phi=0}^{2\pi} \sin\theta\, dr\, d\theta d\phi$$

or

$$4\pi a = 4\pi a.$$

Suppose the surface is that of a cone of radius a and cone angle $\theta = \theta_0$. In this case, $d\mathbf{s} = \mathbf{a}_r a^2 \sin\theta\, d\theta\, d\phi + \mathbf{a}_\theta r \sin\theta\, dr\, d\phi$, but $\mathbf{A} \cdot d\mathbf{s} = a \sin\theta\, d\theta\, d\phi$, so that

$$\int_0^{\theta_0} \int_0^{2\pi} a \sin\theta \, d\theta \, d\phi = \int_0^a \int_0^{\theta_0} \int_0^{2\pi} \sin\theta \, dr \, d\theta \, d\phi$$

or

$$2\pi a(1 - \cos\theta_0) = 2\pi a(1 - \cos\theta_0).$$

∎

1.10 STOKES' THEOREM

Consider an *open* surface s whose periphery is c with a well-behaved vector field **A** present. Figure 1.20 demonstrates the geometry and the positive sense for the vector quantities involved. The surface s (not necessarily planar) is subdivided into $\Delta\mathbf{s}_1, \Delta\mathbf{s}_2, \ldots, \Delta\mathbf{s}_n, \ldots, \Delta\mathbf{s}_N$ (incremental vector areas). If each $\Delta\mathbf{s}_n$ is sufficiently small, then $\mathbf{A}_n$ may be assumed to be constant over it. We may write, for the circulation,

$$\oint_c \mathbf{A} \cdot d\mathbf{l} \approx \sum_{n=1}^{N} \oint_{c_n} \mathbf{A}_n \cdot d\mathbf{l}.$$

This result occurs because the line integrals over segments of the c_n interior to c are zero since the line integrals along a common boundary for adjacent areas are taken in *opposite* directions. Only those parts of the c_n's that coincide with c contribute to the sum. Letting N approach infinity so that Δs_n approaches zero gives

$$\oint_c \mathbf{A} \cdot d\mathbf{l} = \lim_{N \to \infty} \sum_{n=1}^{N} \oint_c \mathbf{A}_n \cdot d\mathbf{l}. \quad (1.44)$$

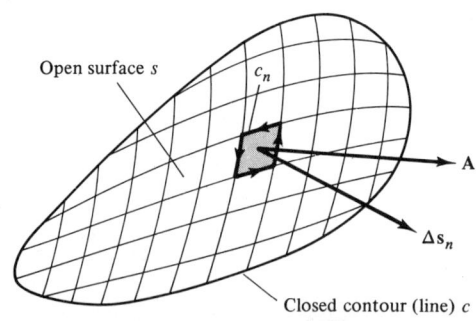

Figure 1.20. Geometry for establishing Stokes' theorem.

1.10 STOKES' THEOREM

The component in the $\mathbf{a}_n$ direction of the curl of $\mathbf{A}$, $(\nabla \times \mathbf{A})$, at a point is defined by Equation (1.37) as the limit of the circulation per unit area as the area $(\mathbf{a}_n \Delta s)$ approaches zero. Thus, we have

$$(\nabla \times \mathbf{A})_n = \lim_{\Delta s \to 0} \frac{1}{\Delta s} \oint_c \mathbf{A} \cdot d\mathbf{l}$$

or

$$\mathbf{a}_n \cdot (\nabla \times \mathbf{A}) = \lim_{\Delta s \to 0} \frac{1}{\Delta s} \oint_c \mathbf{A} \cdot d\mathbf{l}$$

or

$$\mathbf{a}_n \cdot (\nabla \times \mathbf{A}) \approx \frac{1}{\Delta s} \oint_c \mathbf{A} \cdot d\mathbf{l};$$

or, applied to Δs_n,

$$\mathbf{a}_n \cdot (\nabla \times \mathbf{A}) \approx \frac{1}{\Delta s_n} \oint_{c_n} \mathbf{A}_n \cdot d\mathbf{l}.$$

Therefore, we have

$$\mathbf{a}_n \cdot (\nabla \times \mathbf{A}) \Delta s_n \approx \oint_{c_n} \mathbf{A}_n \cdot d\mathbf{l} \tag{1.45}$$

Substituting Equation (1.45) into Equation (1.44), we get

$$\oint_c \mathbf{A} \cdot d\mathbf{l} = \lim_{\substack{N \to \infty \\ \Delta s_n \to 0}} \sum_{n=1}^{N} \mathbf{a}_n \cdot (\nabla \times \mathbf{A}) \Delta s_n$$

$$= \lim_{\substack{N \to \infty \\ \Delta s_n \to 0}} \sum_{n=1}^{N} (\nabla \times \mathbf{A}) \cdot \Delta \mathbf{s}_n;$$

but the limit is an (open surface) integral in the usual way, so

$$\boxed{\oint_c \mathbf{A} \cdot d\mathbf{l} = \iint_s (\nabla \times \mathbf{A}) \cdot d\mathbf{s}} \quad \textit{Stokes' theorem,} \tag{1.46}$$

where Equation (1.46) is known as the Stokes theorem.

Example 1.9

Show that the Stokes theorem holds for the vector field $\mathbf{F} = \rho \mathbf{a}_\phi$, $\nabla \times \mathbf{F} = 2\mathbf{a}_z$, when the closed path is a circle of radius $\rho = a$ with its center at the origin. The *open* surface is the disk $0 \leq \rho \leq a$, $z = 0$. We have $\mathbf{F} \cdot d\mathbf{l} = a\mathbf{a}_\phi \cdot \mathbf{a}_\phi a\, d\phi = a^2\, d\phi$, so that the Stokes theorem is

$$\int_0^{2\pi} a^2\, d\phi = \int_0^a \int_0^{2\pi} 2\rho\, d\rho\, d\phi,$$

$$2\pi a^2 = 2\pi a^2.$$

Suppose that the open surface is the upper hemisphere $r = a$, $0 \leq \theta \leq \pi/2$. In this case $\mathbf{F} \cdot d\mathbf{l}$ is unchanged, and the Stokes theorem gives

$$2\pi a^2 = \int_0^{\pi/2} \int_0^{2\pi} 2\mathbf{a}_z \cdot \mathbf{a}_r a^2 \sin\theta\, d\theta\, d\phi.$$

Now we have $\mathbf{a}_z \cdot \mathbf{a}_r = \cos\theta$, so that we get

$$2\pi a^2 = 2\pi a^2 \int_0^{\pi/2} 2\sin\theta \cos\theta\, d\theta = 2\pi a^2 \int_0^{\pi/2} \sin 2\theta\, d\theta,$$

and, once again, we have

$$2\pi a^2 = 2\pi a^2.$$

Any other open surface whose opening is defined by a closed path that is the circle $\rho = a$ (center at the origin) will give the same result. ∎

1.11 CONCLUDING REMARKS

Vector addition and multiplication were briefly reviewed in the first part of this chapter. This material should be familiar to the engineering student. Next, the three coordinate systems that are used exclusively in this text were introduced and related.

The concepts of circulation (line integrals) and flux were defined and given physical meaning by means of familiar examples. The vector operations of divergence, gradient, and curl that we will have much use for in the following chapters, were introduced and explained.

There are many vector identities that appear in the study of vector analysis, and the two most important of these that we will use were introduced. The divergence theorem (the Gauss integral theorem) and the Stokes theorem are fundamental to field theory. The derivation of these theorems was presented. A summary of vector relations is listed on the inside front and back covers.

REFERENCES

McQuistan, Richmond B. *Scalar and Vector Fields, a Physical Interpretation.* New York: Wiley, 1965. All of the topics of vector analysis (plus more) we shall need are included in this inexpensive paperback.

Schey, H. M. *Div, Grad, Curl, and All That.* New York: W.W. Norton Co., 1973.

Skilling, H. H. *Fundamentals of Electric Waves.* New York: Wiley, 1948. The discussion of curl and the paddlewheel idea begins on p. 23.

Spiegel, M. R. *Mathematical Handbook*, Schaum Outline Series. New York: McGraw-Hill, 1968. Many of the integrals we must evaluate are listed in this inexpensive paperback.

Spiegel, M. R. *Vector Analysis*, Schaum Outline Series. New York: McGraw-Hill, 1959. Many solved problems are included in this inexpensive paperback.

PROBLEMS

1. An object is acted on by three forces: $\mathbf{F}_1 = 10\mathbf{a}_x$, $\mathbf{F}_2 = 15\mathbf{a}_y$, and $\mathbf{F}_3 = 20\mathbf{a}_\rho$ (in the direction $\phi = 2\pi/3$ rad). Find the fourth force required to keep the object from moving.

2. If $\mathbf{A} = 4\mathbf{a}_x + 4\mathbf{a}_y - 2\mathbf{a}_z$ and $\mathbf{B} = 3\mathbf{a}_x - 1.5\mathbf{a}_y + \mathbf{a}_z$, find the angle ($< 90°$) between $\mathbf{A}$ and $\mathbf{B}$.

3. Find the angles that the vector $\mathbf{A} = 6\mathbf{a}_x - 12\mathbf{a}_y + 4\mathbf{a}_z$ makes with the x, y, and z axes.

*4. (a) Find an equation for the plane that is perpendicular to the vector $\mathbf{A} = 2\mathbf{a}_x + 3\mathbf{a}_y + 6\mathbf{a}_z$ and passes through the end point (from the origin) of the vector $\mathbf{B} = \mathbf{a}_x + 5\mathbf{a}_y + 3\mathbf{a}_z$.
   (b) What is the shortest distance from the origin to the plane?

5. If $\mathbf{A} = 10\mathbf{a}_\rho/\rho + 5\mathbf{a}_\phi + 2\mathbf{a}_z$ and $\mathbf{B} = 5\mathbf{a}_\rho + \cos\phi\,\mathbf{a}_\phi + \rho\mathbf{a}_z$,
   (a) Find $\mathbf{A} \cdot \mathbf{B}$ at the point $x = 1$, $y = 1$, $z = 1$.
   (b) Find $\mathbf{A} \times \mathbf{B}$ at the point $x = 1$, $y = 1$, $z = 1$.

6. (a) Resolve the unit vectors $\mathbf{a}_\rho$, $\mathbf{a}_\phi$, $\mathbf{a}_r$, $\mathbf{a}_\theta$ into *constant* unit vectors.
   (b) Find $\int_0^\pi \mathbf{a}_\rho\,d\phi$, $\int_0^{\pi/2} \mathbf{a}_\phi\,d\phi$, $\int_0^1 \mathbf{a}_r\,dz$, $\int_0^{\pi/2} \mathbf{a}_\theta\,d\theta$.

7. (a) Find the volume within the cylindrical wedge defined by $\rho = 1$, $0 \le \phi \le \pi/3$, $0 \le z \le 1$.
   (b) Find the total surface area of the wedge.

*8. Find the volume inside the tetrahedron defined by the intersection of the planes $x = 0$, $y = 0$, $z = 0$, and $3x + 4y + 2z = 12$.

*The more difficult problems are marked with an asterisk.

*9. (a) Find the volume of a cone whose height is h and whose base radius is a.
  (b) Find the total surface area.

10. Repeat Example 1.3 by moving the object from 0 to 2 along the x axis, then moving parallel to the y axis from $y = 0$ to $y = 4$. Prove that **F** is a conservative field.

*11. Find the flux of the vector field $\mathbf{A} = \mathbf{a}_\rho/\rho$ for:
  (a) The sphere $r = a$ centered at the origin;
  (b) The cube $2a$ m on a side, centered at the origin with sides parallel to the coordinate axes;
  (c) The cylinder $0 \le \rho \le 3a$, $0 \le \phi \le 2\pi$, $-a \le z \le a$.

*12. Show that the divergence of the vector field **A** in Example 1.4 is zero *except* at $r = 0$. Find the divergence of **A** at $r = 0$ using a sphere of radius $r = a$ and Equation (1.26) ($a \to 0$). Do the results agree? What do the results imply?

13. Find $\nabla \cdot \mathbf{C}$ if:
  (a) $\mathbf{C} = xy^2 \mathbf{a}_x + y\mathbf{a}_y + 5\mathbf{a}_z$;
  (b) $\mathbf{C} = \rho \mathbf{a}_\rho + (1/\rho)\mathbf{a}_\phi + z\mathbf{a}_z$;
  (c) $\mathbf{C} = x\mathbf{a}_\rho + y\mathbf{a}_\phi + r\mathbf{a}_z$;
  (d) $\mathbf{C} = r\mathbf{a}_r + \rho\mathbf{a}_\rho + x\mathbf{a}_x$.

14. Using a unit vector find the direction in which the scalar field $T = x^2 y + zx + 10$ is changing the most rapidly. What is this greatest rate of change at $(1,1,1)$?

15. (a) If $\Phi = 10x^2 + y$, find $\nabla\Phi$ at $(0,0,0)$ using Equation (1.35a).
  (b) Repeat using Equation (1.36) when the volume is a rectangular parallelepiped centered at the origin with sides a, b, and c parallel to the coordinate axes.

16. Determine which of the following fields are conservative:
  (a) $\mathbf{A} = \mathbf{a}_r/r^2$;
  (b) $\mathbf{A} = \mathbf{a}_\rho/\rho$;
  (c) $\mathbf{A} = \mathbf{a}_z z/|z|$;
  (d) $\mathbf{A} = x(\mathbf{a}_x + \mathbf{a}_y)$;
  (e) $\mathbf{A} = y\mathbf{a}_x + x\mathbf{a}_y$;
  (f) $\mathbf{A} = z\cos\phi\,\mathbf{a}_\phi + \rho\sin\phi\,\mathbf{a}_z$;
  (g) $\mathbf{A} = r\cos\theta\,\mathbf{a}_r - (1/2)r\sin\theta\,\mathbf{a}_\theta + \mathbf{a}_z$.

17. Use the small paddlewheel to explore the following fields to determine if the curl is zero:
  (a) $\mathbf{A} = 10\mathbf{a}_\rho/\rho$;
  (b) $\mathbf{A} = 5\mathbf{a}_r/r^2$;
  (c) $\mathbf{A} = \mathbf{a}_\phi/\rho$;

PROBLEMS

(d) $\mathbf{A} = \begin{cases} -5\mathbf{a}_z, & z < 0, \\ +5\mathbf{a}_z, & z > 0; \end{cases}$

(e) $\mathbf{A} = \begin{cases} 10\mathbf{a}_z, & -2 < z < +2, \\ 0, & \text{otherwise}; \end{cases}$

(f) $\mathbf{A} = \begin{cases} 10\mathbf{a}_x, & -2 < z < +2, \\ 0, & \text{otherwise}. \end{cases}$

*18. Obtain the circular cylindrical and spherical forms for the gradient, divergence, and curl from the Cartesian forms by using the coordinate and component transformations found on the inside front cover.

19. Using Cartesian coordinates show that

$$\alpha(\nabla \cdot \mathbf{A}) \equiv \nabla \cdot (\alpha \mathbf{A}) - \mathbf{A} \cdot (\nabla \alpha).$$

20. Using Cartesian coordinates show that
    (a) $\nabla \times (\nabla \alpha) \equiv 0$,
    (b) $\nabla \cdot (\nabla \times \mathbf{A}) \equiv 0$.

*21. Find the flux of the vector field $\mathbf{A} = \mathbf{a}_\rho/\rho$ out of:
    (a) The sphere $r = a$;
    (b) The upper hemisphere $r = a$, $0 \le \theta \le \pi/2$;
    (c) The lower hemisphere $r = a$, $\pi/2 \le \theta \le \pi$;
    (d) The cylinder $\rho = 0.1$, $-a/2 \le z \le a/2$.

22. Show that the divergence theorem holds for the field $\mathbf{A} = r\mathbf{a}_r$ when the closed surface is
    (a) A sphere $r = a$,
    (b) A cylinder $\rho = a$, $0 \le z \le h$.

23. Repeat Problem 22 when $\mathbf{A} = 5\mathbf{a}_\rho$.

24. Repeat Problem 22 when $\mathbf{A} = 10\mathbf{a}_z$.

*25. Show that the Stokes theorem holds for the field $\mathbf{A} = r\mathbf{a}_\phi$ when the closed path is the circle $\rho = a$ ($z = 0$), and the open surface is:
    (a) The disk $0 \le \rho \le a$, $z = 0$,
    (b) The lower hemisphere $r = a$, $\pi/2 \le \theta \le \pi$.

2
ELECTROSTATIC AND MAGNETOSTATIC FIELDS

The source of the electrostatic vector field is static (fixed) electric charge, and this electric charge can be spatially distributed in many ways. The source of the magnetostatic vector field is electric charge moving without acceleration (steady electric current). These fields are obtained from two famous experimental laws: *Coulomb's law* and the *Biot-Savart law*, respectively. We will introduce the fields by means of these laws. These laws lead by superposition to integrals for determining the electrostatic and magnetostatic fields from completely general charge and current distributions, respectively.

It is also possible to calculate the force on one point charge due to another point charge when these charges are in motion relative to each other and a fixed observer who detects the force.[1] This more general approach gives both the electrostatic and magnetostatic fields.

The *Gauss law*, another famous law, relates electric flux to electric charge in a very direct way. Similarly, magnetic field intensity is related to electric current by *Ampere's law*. These laws will be converted to partial differential equations employing the divergence and curl operations of Chapter 1, respectively. These differential equations make up part of a set that is normally called *Maxwell's equations* for statics. Other equations in this set are derived from the conservative nature of the electrostatic field, the absence of isolated magnetic charge, and conservation of electric charge.

Auxiliary potential fields will be derived, and these potentials give us alternate methods for determining electrostatic and magnetostatic field intensities. This is advantageous in many cases because the superposition integrals that give the potentials from the sources (charge and current) are usually simpler to treat than the superposition integrals that give electrostatic and magnetostatic fields from the sources. Superposition-type integrals are generally solutions to partial differential

---

[1]See Neff, 1981 or Shedd, 1954 in the references at the end of this chapter.

equations, and we will find these differential equations that relate the sources to the potential fields.

It will, of course, be necessary to introduce materials and how they interact with fields. This, in turn, will lead to important boundary conditions that relate field quantities at the interface between different materials.

Our primary goal in this chapter is not that of examining the details of electrostatics and magnetostatics, for much of that should have been accomplished in a prior physics course, but, rather, it is to expose the *vector* nature of these fields and their governing equations.

2.1 CHARGE CONFIGURATIONS

The most general charge configuration is the volume charge distribution $\rho_v(x,y,z)$, or simply ρ_v, in coulombs per cubic meter (C/m^3). All other charge configurations are special cases of volume charge density. Now, if Q (C) is the total charge in a region, then we have

$$\rho_v = \lim_{\Delta v \to 0} \frac{\Delta Q}{\Delta v} = \frac{dQ}{dv}, \qquad (2.1)$$

so we get $dQ = \rho_v \, dv$, and the total charge in a given volume is

$$\boxed{Q = \iiint_{\text{vol}} \rho_v(x,y,z) \, dx \, dy \, dz \quad \text{(C)},} \qquad (2.2)$$

where the limits on the triple integral must be such as to include all of the charge. That is, the integration is throughout the volume containing the charge. Some of the more useful special cases of charge distribution are discussed below.

The *point charge* is a finite charge (in coulombs) occupying a vanishingly small volume, indicating an infinite volume charge density. Thus, the concept of a point charge is primarily a mathematical convenience. It is symbolized by Q or, more explicitly, $Q(x,y,z)$, which indicates that it is located at the point (x,y,z). The total charge is obviously Q.

The line charge density ρ_l (C/m) is a filament of charge having a length (which may mathematically be infinite) but no thickness. It too is a mathematical model since its volume charge density is infinite. In many cases, a finite diameter conducting wire bearing a charge distribution, which is either known or can be approximated at every point, can be represented by a filamentary line charge density. The charged wire must look like a filament to an observer who is located several diameters away from the wire. The differential charge in a differential length is $dQ = \rho_l \, dl$; thus the total charge for a line charge density is

$$Q = \int_{\text{length}} \rho_l \, dl \quad \text{(C)}. \qquad (2.3)$$

Example 2.1

A line charge density $\rho_l = 10^{-9}$ (C/m) exists in the form of a filamentary circular loop with a radius of 0.5 (m). If the loop lies in the $z = 0$ plane with its center at the origin, then the total charge [from Equation (2.3)] is

$$Q = \int_0^{2\pi} 10^{-9}(0.5)\, d\phi = \pi \times 10^{-9} \text{ (C)}.$$

When viewed from the large distance represented by the point (100,0,0), this loop of charge would appear to the observer to be a point charge $Q = \pi \times 10^{-9}$ (C) at the origin. As we shall see later, field quantities measured at (100,0,0) will be approximately the same as if they arose from a point charge $Q = \pi \times 10^{-9}$ (C) at the origin. ■

The surface charge density ρ_s (C/m²) is a layer of charge having an area (which may mathematically be infinite) but no thickness so that its volume density is also infinite. A thin copper sheet with a known charge distribution could be approximated by an ideal surface charge density for an observer who agreed to remain several thicknesses away from the copper sheet. The differential charge for a differential area is $dQ = \rho_s\, ds$, thus the total charge for this case is

$$Q = \iint_{\text{surface}} \rho_s\, ds \text{ (C)}. \qquad (2.4)$$

Example 2.2

A nonuniform surface charge density $\rho_s = 2\rho^2 \times 10^{-5}$ (C/m²) exists in the form of a disk of radius 0.1 (m). It lies in the $z = 0$ plane with its center at the origin. The total charge [from Equation (2.4)] is

$$Q = \int_0^{0.1} \int_0^{2\pi} 2\rho^2 \times 10^{-5}(\rho\, d\rho\, d\phi),$$

$$Q = 2 \times 10^{-5} \int_0^{0.1} \int_0^{2\pi} \rho^3\, d\rho\, d\phi = 4\pi \times 10^{-5} \int_0^{0.1} \rho^3\, d\rho,$$

$$Q = \pi \times 10^{-9} \text{ (C)};$$

and when viewed from the point (0,0,100), this disk of charge would appear to the observer to be a point charge $Q = \pi \times 10^{-9}$ (C) at the origin just like the loop of Example 2.1. ■

2.2 COULOMB'S LAW AND ELECTRIC FIELD INTENSITY

Consider two point charges Q_1 and Q as shown in Figure 2.1. Here, we consider Q_1 to be the *source* of a vector force field **F**, and this force (the *response*) is exerted on the second charge Q. We will use consistent notation throughout this text in that sources will be located by primed coordinates (Q_1 is located by **r'**) *and field points will be located by unprimed coordinates* (Q is located by **r**). Coulomb's law gives the force on Q due to Q_1 as

$$\mathbf{F} = \frac{Q_1 Q (\mathbf{r} - \mathbf{r}')}{4\pi\varepsilon_0 |\mathbf{r} - \mathbf{r}'|^3} \quad (\text{N}) \quad \textit{Coulomb's law}, \tag{2.5}$$

where $\mathbf{r} - \mathbf{r}' = \mathbf{a}_x(x - x') + \mathbf{a}_y(y - y') + \mathbf{a}_z(z - z')$ is the vector directed from Q_1 toward Q whose magnitude is $|\mathbf{r} - \mathbf{r}'| = [(x - x')^2 + (y - y')^2 + (z - z')^2]^{1/2}$ (see Example 1.6) and $(4\pi\varepsilon_0)^{-1}$ is the constant of proportionality required to give a force in newtons. ε_0 is a constant called the permittivity (see Section 2.5) of a vacuum (or free space), and is given by

$$\boxed{\varepsilon_0 \approx 10^{-9}/(36\pi)} \quad (\text{F/m}) \quad \textit{permittivity of vacuum}. \tag{2.6}$$

Coulomb's law can also be written

$$\mathbf{F} = \frac{Q_1 Q}{4\pi\varepsilon_0} \frac{\mathbf{a}_R}{R^2}, \tag{2.7}$$

where $\mathbf{a}_R = (\mathbf{r} - \mathbf{r}')/|\mathbf{r} - \mathbf{r}'|$ and $\mathbf{R} = \mathbf{r} - \mathbf{r}'$. Equation (2.7) shows explicitly that Coulomb's law is an inverse square distance law. Coulomb's law further asserts

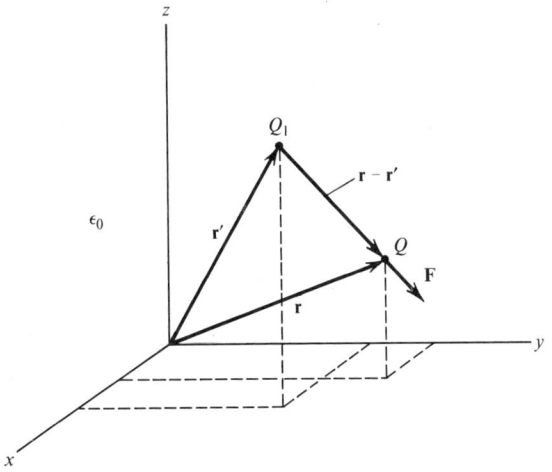

Figure 2.1. Geometry for two point charges for finding the force on Q due to Q_1 (Coulomb's law).

that the force depends *linearly* on the charge of either point charge, and thus, by the *principle of superposition* the force on Q due to several point charges Q_1, Q_2, Q_3, ... can be found as the *sum* of forces on Q due to each of the sources Q_1, Q_2, Q_3, ... acting alone.

The electrostatic field intensity **E** is defined as the force on Q when $Q = 1$ C. That is, it is the force per unit charge (newtons/coulomb = volts/meter). Thus, Equation (2.7), with $\mathbf{E} = \mathbf{F}/Q$, becomes

$$\mathbf{E} = \frac{Q_1}{4\pi\varepsilon_0} \frac{\mathbf{a}_R}{R^2} \quad \text{(V/m)} \tag{2.8}$$

Note that the electric field intensity is directed radially away from Q_1 with complete radial symmetry. The magnitude of **E** will be the same at every point on a sphere of 1-m radius ($R = 1$) centered at Q_1, but will be one-fourth as large on a 2-m radius sphere. As a reminder: $\mathbf{E} = \mathbf{E}(x, y, z) = \mathbf{E}$ at the point (x, y, z) located by **r**, and $Q_1 = Q_1(x', y', z') = Q_1$ at the point (x', y', z') located by **r**′.

As was previously mentioned Coulomb forces are additive, so that the electric field produced by N point charges is by superposition from Equation (2.8):

$$\mathbf{E}_N(x, y, z) = \frac{1}{4\pi\varepsilon_0} \sum_{n=1}^{N} \frac{Q_n(x'_n, y'_n, z'_n)}{|\mathbf{r} - \mathbf{r}'_n|^3} (\mathbf{r} - \mathbf{r}'_n) \quad \text{(V/m)}, \tag{2.9}$$

where $\mathbf{r}'_n$ is the radial vector to the nth point charge as shown in Figure 2.2. If the region containing the charge is described as being macroscopically "dense" with N point charges, *and if this region is finite in extent*,[2] we may ascribe to every point in the region a volume charge density as in Equation (2.1). We replace Q_n by ΔQ_n and then multiply and divide by the small volume Δv_n. That is,

---

[2]This is an important qualification about which more will be said.

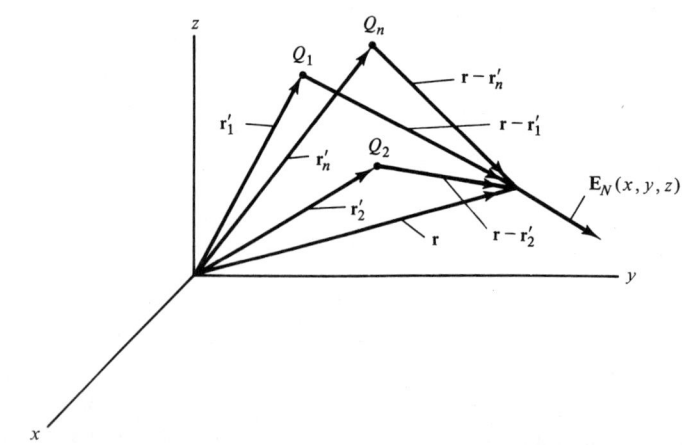

Figure 2.2. Location of several point charges and a field point (x, y, z), where **E** is to be determined.

2.2 COULOMB'S LAW AND ELECTRIC FIELD INTENSITY

$$\mathbf{E}_N(x,y,z) = \frac{1}{4\pi\varepsilon_0} \sum_{n=1}^{N} \frac{\Delta Q_n(x_n', y_n', z_n')}{\Delta v_n} \frac{\Delta v_n(\mathbf{r} - \mathbf{r}_n')}{|\mathbf{r} - \mathbf{r}_n'|^3}. \qquad (2.10)$$

We now let the volumes Δv_n approach zero as the number of charge elements N approaches infinity. In the limit, $\Delta Q_n/\Delta v_n$ becomes the volume charge density $\rho_v = dQ/dv$, as in Equation (2.1), while the sum becomes an integral in the usual way. This integral is (in general) over three primed variables, or, in other words, it is a *volume* integral. That is,

$$\mathbf{E}(x,y,z) = \lim_{\substack{\Delta v_n \to 0 \\ N \to \infty}} \frac{1}{4\pi\varepsilon_0} \sum_{n=1}^{N} \frac{\Delta Q_n}{\Delta v_n} \frac{(\mathbf{r} - \mathbf{r}_n')}{|\mathbf{r} - \mathbf{r}_n'|^3} \Delta v_n \qquad (2.11)$$

or[3]

$$\mathbf{E}(x,y,z) = \frac{1}{4\pi\varepsilon_0} \iiint_{\text{vol}'} \rho_v(x', y', z') \frac{(\mathbf{r} - \mathbf{r}')}{|\mathbf{r} - \mathbf{r}'|^3} dv', \qquad (2.12)$$

or simply

$$\boxed{\mathbf{E}(\mathbf{r}) = \frac{1}{4\pi\varepsilon_0} \iiint_{\text{vol}'} \rho_v(\mathbf{r}') \frac{\mathbf{a}_R}{R^2} dv'} \quad \text{(V/m)}, \qquad (2.13)$$

where $dv' = dx'\,dy'\,dz' = \rho'\,d\rho'\,d\phi'\,dz' = (r')^2 \sin\theta'\,dr'\,d\theta'\,d\phi'$ for rectangular, cylindrical, and spherical coordinates, respectively.[4] The integration is *throughout* the volume containing the charge density.

Figure 2.3 shows a volume containing a charge density, $\rho_v(x', y', z')$ and a point (x, y, z) where the electric field is to be evaluated. In order to find the electric field produced by ρ_v, we must integrate throughout the volume (vol′) containing this charge density. Note particularly that the integration is performed on the *primed coordinates only*! In general this integration is over three space variables. Triple integrals of this type are usually very difficult. Quite often, in solving problems of a symmetrical nature, the triple integral will simplify to a double integral or even a single integral.

The process of integration is actually the result of taking the limit of a sum as Equations (2.11) and (2.12) show. In performing calculations of the electric field with the aid of a digital computer, Equation (2.9) would actually be used. That is, the computer, because of its discrete nature, integrates by calculating a large

---
[3]This result is also a *convolution* integral as well as a *superposition* integral. This is the case for all of the superposition integrals encountered in this text. For further information the reader is referred to Neff, 1981 in the references at the end of this chapter.

[4]See Figure 1.7.

2 ELECTROSTATIC AND MAGNETOSTATIC FIELDS

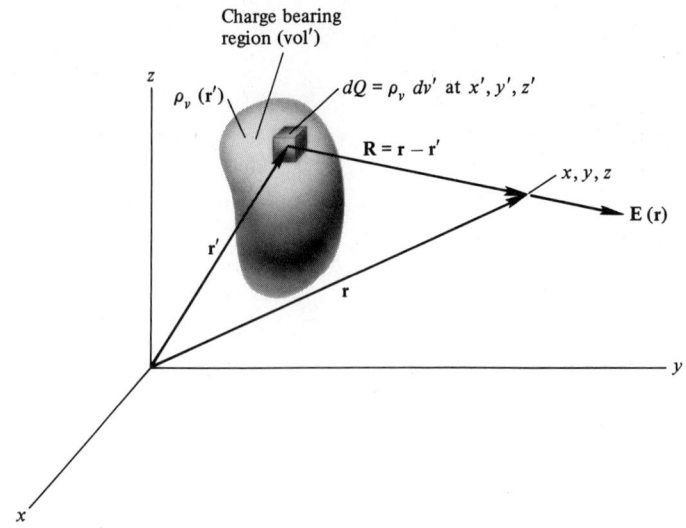

Figure 2.3. General volume charge density $\rho_v(\mathbf{r}')$ producing an electric field intensity $\mathbf{E}(\mathbf{r})$. The vector $\mathbf{r}'$ varies during the integration, but $\mathbf{r}$ is fixed.

sum. With this in mind, Equation (2.12) indicates that in order to find the electric field at some point in space, we sum the contributions to the field at the point in space from each of the many small charges which together constitute the charge distribution. This is a very fundamental concept.

If the source is surface charge density, then superposition leads us to an integral over the surface containing the charge:

$$\mathbf{E}(\mathbf{r}) = \frac{1}{4\pi\varepsilon_0} \int\int_{s'} \frac{\rho_s(\mathbf{r}')}{R^2} \mathbf{a}_R \, ds'. \tag{2.14}$$

Likewise, for line charge densities, we have

$$\mathbf{E}(\mathbf{r}) = \frac{1}{4\pi\varepsilon_0} \int_{l'} \frac{\rho_l(\mathbf{r}')}{R^2} \mathbf{a}_R \, dl'. \tag{2.15}$$

■ **Example 2.3**

We would next like to obtain an analytic solution for the electric field produced by an infinite (uniform) line charge density ρ_l. It is shown in Figure 2.4(a). Since the line charge density is placed on the z axis, we have azimuthal symmetry and z independence, so that $\mathbf{E}$ is independent of ϕ and z. This can be shown by means of a simple analogy. If we represent the uniform line charge by a very thin (but visible) uniform light source on the entire z axis, then an observer sees the same thing regardless of ϕ or z! We may then locate the field point on the x axis without loss of generality. Consider a differential charge $dQ = \rho_l \, dz'$ located at $z' = +h$ and another differential charge $dQ = \rho_l \, dz'$ located at $z' = -h$ as shown in Figure

2.2 COULOMB'S LAW AND ELECTRIC FIELD INTENSITY

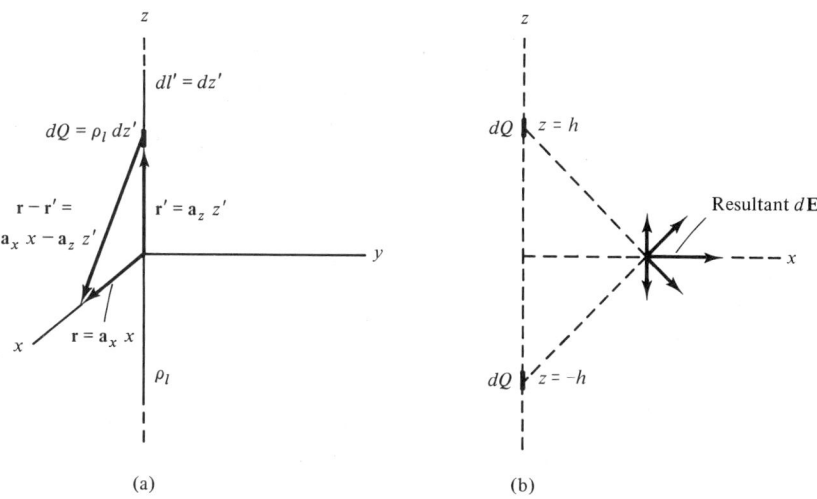

Figure 2.4. (a) Geometry for finding the electric field intensity of a uniform (infinite) line charge density on the z axis. (b) Cancellation of dE_z by symmetry.

2.4(b). Applying Coulomb's law to these charges shows that at a point on the x axis the z components of field cancel while the x components add. Since the entire charge can be divided into such pairs, we ultimately obtain $E_z = 0$ and $E_x = E_x(x)$. If the field point is located anywhere in the $z = 0$ plane ($\rho > 0$), the electric field vector will be in the cylindrical radial (ρ) direction and dependent only on ρ. Thus, in general, $\mathbf{E} = \mathbf{a}_\rho E_\rho(\rho)$. The integration is simpler when the field point is on the x axis, so using Equation (2.15) with Cartesian coordinates we have

$$\mathbf{E}(x,0,0) = \frac{1}{4\pi\varepsilon_0} \int_{-\infty}^{\infty} \frac{\rho_l(z')}{R^2} \mathbf{a}_R \, dz'.$$

From Figure 2.4(a) we see that

$$\mathbf{R} = \mathbf{r} - \mathbf{r}' = \mathbf{a}_x x - \mathbf{a}_z z', \quad R = [x^2 + (z')^2]^{1/2}.$$

Thus, we have

$$\mathbf{E}(x,0,0) = \frac{\rho_l}{4\pi\varepsilon_0} \int_{-\infty}^{\infty} \frac{\mathbf{a}_x x - \mathbf{a}_z z'}{[x^2 + (z')^2]^{3/2}} dz'$$

or

$$E_x(x,0,0) = \frac{\rho_l x}{4\pi\varepsilon_0} \int_{-\infty}^{\infty} \frac{dz'}{[x^2 + (z')^2]^{3/2}}$$

and

$$E_z(x,0,0) = \frac{-\rho_l}{4\pi\varepsilon_0} \int_{-\infty}^{\infty} \frac{z'\, dz'}{[x^2 + (z')^2]^{3/2}}.$$

We have already deduced from symmetry arguments that $E_z = 0$, so the last integral is zero. On the other hand, the mathematics will show the same thing. The integrand of the integral for $E_z(x,0,0)$ is an *odd function* of z', meaning [integrand (z')] = $-$[integrand $(-z')$]. Since the integral represents an area, and here we will have equal amounts of positive and negative area, the integral is zero! We are left with the integral for E_x where the integrand is an *even function* of z' [integrand(z') = integrand$(-z')$]. In this case the area for $-\infty$ to 0 must equal the area for 0 to ∞, and we can simply integrate from 0 to ∞ (or $-\infty$ to 0) and multiply the result by 2. In this case, we have

$$E_x(x,0,0) = \frac{\rho_l}{2\pi\varepsilon_0 x} \frac{z'}{[x^2 + (z')^2]^{1/2}}\bigg|_0^{\infty} = \lim_{z'\to\infty} \frac{\rho_l}{2\pi\varepsilon_0 x} \frac{1}{[1 + (x/z')^2]^{1/2}}$$

or

$$E_x(x,0,0) = \frac{\rho_l}{2\pi\varepsilon_0 x} \quad \text{(V/m)}.$$

Thus, $E_x(1,0,0) = \rho_l/(2\pi\varepsilon_0)$ and $E_x(2,0,0) = 0.5\rho_l/(2\pi\varepsilon_0)$.

If we move the field point off of the x axis, the radial distance becomes ρ and the radial component becomes E_ρ, so the general result is as predicted: $\mathbf{E} = \mathbf{a}_\rho E_\rho(\rho)$ or

$$\mathbf{E} = \frac{\rho_l}{2\pi\varepsilon_0 \rho} \mathbf{a}_\rho \quad \text{(V/m)}. \tag{2.16}$$

■

We could, of course, have worked this problem entirely in cylindrical coordinates with very little additional work. It is suggested that the reader carry out this calculation. Before passing on to another example let us take note of the fact that a *finite* line charge density is usually not uniformly distributed because of the Coulomb force of repulsion. Exactly how the charge would distribute itself is a topic for future concern. It was mentioned earlier that this was a useful distribution. Besides the benefits of the academic exercise, practical results for the electric field around a finite length, finite diameter, charged wire can be obtained if the field point is so close that the wire looks infinitely long, but, at the same time, far enough away so that the wire looks like a filament.

■ Example 2.4

Next we would like to obtain a numerical solution for the electric field intensity of the infinite line charge density (Example 2.3). We will represent the line charge as a linear array of point charges on the z axis. The spacing of the point charges is

2.2 COULOMB'S LAW AND ELECTRIC FIELD INTENSITY

chosen to be uniform to simplify matters, and so that the observer at the field point will see something resembling a continuous line charge, the spacing must be small compared with the distance to the field point. Figure 2.5 shows the geometry. Note that in order to obtain the proper weighting and dimensions, we have $Q_n = \rho_l \Delta z$. Using Equation (2.8), the field produced by Q_n is

$$\mathbf{E}_n = \frac{\rho_l \Delta z}{4\pi\varepsilon_0} \frac{\mathbf{a}_x x - \mathbf{a}_z n \Delta z}{[x^2 + (n\Delta z)^2]^{3/2}}, \quad (2.8a)$$

while the field produced by Q_{-n} is

$$\mathbf{E}_{-n} = \frac{\rho_l \Delta z}{4\pi\varepsilon_0} \frac{\mathbf{a}_x x + \mathbf{a}_z n \Delta z}{[x^2 + (n\Delta z)^2]^{3/2}}. \quad (2.8b)$$

It is clear from Equations (2.8a) and (2.8b) and Figure 2.5 that the z component of $\mathbf{E}$ will cancel for any pair of symmetrically located point charges, and no z component is produced by the point charge at the origin ($n = 0$). Therefore, only an x component of $\mathbf{E}$ is produced, and it is given by

$$E_x(x,0,0) = \frac{\rho_l(\Delta z)x}{4\pi\varepsilon_0} \left(\frac{1}{x^3} + 2\sum_{n=1}^{N} \frac{1}{[x^2 + (n\Delta z)^2]^{3/2}} \right)$$

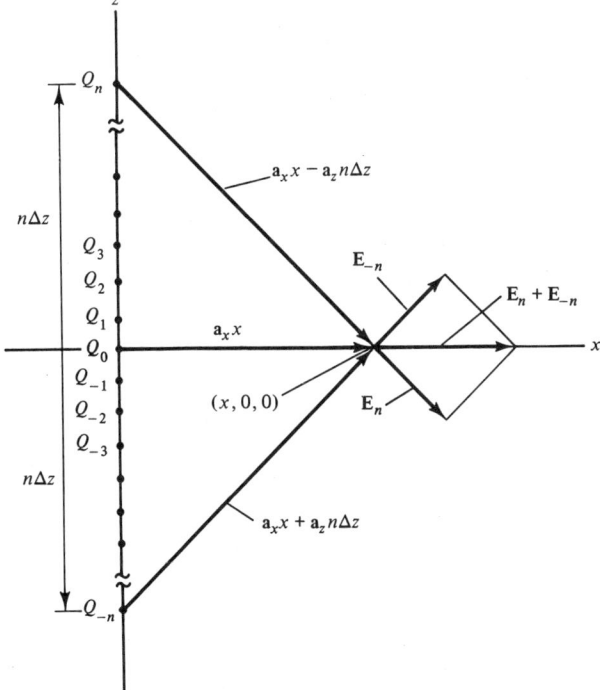

Figure 2.5. Approximation of a finite line charge density on the z axis by uniformly spaced point charges.

for a total of $2N + 1$ point charges. This result can be recognized as a prelimit form of Equation (2.15).

Suppose we choose $x = 1$ m. In light of the earlier discussion, $N \Delta z$ (half the length of the line charge) should be large compared with x, but Δz should be small compared with x. Then, let $\Delta z = 0.1$ m and $N \Delta z = 10$ m, so that $N = 100$. In this case, we have

$$E_x(1,0,0) = \frac{\rho_l}{4\pi\varepsilon_0}(0.1)\left(1 + 2\sum_{n=1}^{100}\frac{1}{[1 + 0.01n^2]^{3/2}}\right),$$

and using a calculator or computer, we get

$$E_x(1,0,0) = 0.995\rho_l/(2\pi\varepsilon_0) \quad \text{(V/m)}.$$

Moving the field point to $x = 2$ gives

$$E_x(2,0,0) = 0.491\rho_l(2\pi\varepsilon_0) \quad \text{(V/m)}.$$

These numerical results compare favorably with those from the exact solution in Example 2.3. ∎

Example 2.5

Consider an infinite sheet of uniform surface charge density ρ_s, located in the $z = 0$ plane and a field point located above it ($z > 0$) as shown in Figure 2.6.

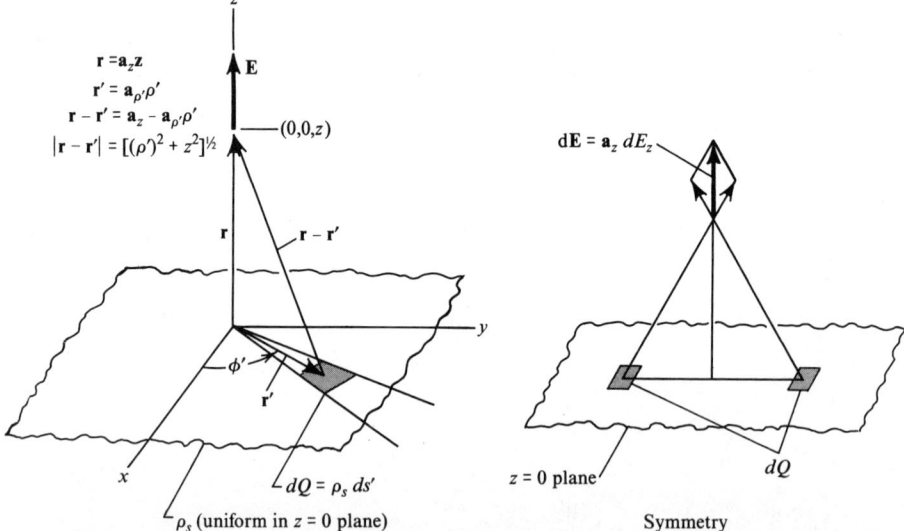

Figure 2.6. The geometry for finding the field of an infinite sheet of uniform surface charge density ρ_s in the $z = 0$ plane.

2.2 COULOMB'S LAW AND ELECTRIC FIELD INTENSITY

This is another charge distribution whose field can be determined exactly with little difficulty. Symmetry conditions are such that we immediately conclude that the electric field must be the same in amplitude in any plane parallel to the surface charge. For every differential contribution to the electric field from a differential element of surface charge, we can find a symmetrically located element of charge of the same amount. These two elements of charge produce equal components of electric field perpendicular to the surface, but produce opposite amounts of electric field parallel to the surface. Using the light source analogy of Example 2.3, we can imagine the entire $z = 0$ plane covered with a uniform light source. An observer located above this source will certainly see the same thing as far as his x or y coordinate is concerned. He will, in fact, also see the same thing as far as his z coordinate is concerned! We can now predict that the electric field intensity has only a *uniform z* component. More light is shed on this subject when the integration is carried out.

The geometry of the problem is shown in Figure 2.6. Without loss of generality the field point is located on the z axis $[\mathbf{E}(0, 0, z) = \mathbf{E}(x, y, z)]$. Using cylindrical coordinates, we get

$$\mathbf{r} = \mathbf{a}_z z, \qquad\qquad x' = \rho' \cos \phi';$$
$$\mathbf{r}' = \mathbf{a}_{\rho'} \rho', \qquad\qquad y' = \rho' \sin \phi';$$
$$ds' = \rho' \, d\rho' \, d\phi' \qquad (z' = 0)$$

and

$$\mathbf{E}(\rho, \phi, z) = \frac{1}{4\pi\varepsilon_0} \int_0^\infty \int_0^{2\pi} \frac{\rho_s (\mathbf{a}_z z - \mathbf{a}_{\rho'} \rho') \rho' \, d\rho' \, d\phi'}{[(\rho')^2 \cos^2 \phi' + (\rho')^2 \sin^2 \phi' + z^2]^{3/2}}$$

or

$$\mathbf{E}(\rho, \phi, z) = \frac{\rho_s}{4\pi\varepsilon_0} \int_0^\infty \int_0^{2\pi} \frac{(\mathbf{a}_z z - \mathbf{a}_{\rho'} \rho') \rho' \, d\rho' \, d\phi'}{[(\rho')^2 + z^2]^{3/2}}.$$

The subtle difficulty is that $\mathbf{a}_{\rho'}$ is *not* a constant unit vector! Its magnitude is certainly constant (unity) but its direction depends on ϕ', the azimuth angle (a variable). The obvious cure is to represent $\mathbf{a}_{\rho'}$ in terms of constant unit vectors: namely, the rectangular unit vectors $\mathbf{a}_x$ and $\mathbf{a}_y$. This representation was mentioned in Section 1.3. It is

$$\mathbf{a}_{\rho'} = \mathbf{a}_x \cos \phi' + \mathbf{a}_y \sin \phi'$$

Then, we get

$$\mathbf{E}(\rho, \phi, z) = \frac{\rho_s}{4\pi\varepsilon_0} \int_0^\infty \int_0^{2\pi} \frac{(\mathbf{a}_z z - \mathbf{a}_x \rho' \cos \phi' - \mathbf{a}_y \rho' \sin \phi') \rho' \, d\rho' \, d\phi'}{[(\rho')^2 + z^2]^{3/2}}$$

or

$$E_x(\rho, \phi, z) = \frac{-\rho_s}{4\pi\varepsilon_0} \int_0^\infty \int_0^{2\pi} \frac{(\rho')^2 \cos\phi' \, d\rho' \, d\phi'}{[(\rho')^2 + z^2]^{3/2}},$$

$$E_y(\rho, \phi, z) = \frac{-\rho_s}{4\pi\varepsilon_0} \int_0^\infty \int_0^{2\pi} \frac{(\rho')^2 \sin\phi' \, d\rho' \, d\phi'}{[(\rho')^2 + z^2]^{3/2}}$$

and

$$E_z(\rho, \phi, z) = \frac{\rho_s z}{4\pi\varepsilon_0} \int_0^\infty \int_0^{2\pi} \frac{\rho' \, d\rho' \, d\phi'}{[(\rho')^2 + z^2]^{3/2}}.$$

Now, since the integral of $\cos\phi'$ or $\sin\phi'$ over a range of 2π rad is zero, E_x and E_y are both zero. The third integral above gives

$$E_z(\rho, \phi, z) = \frac{\rho_s z}{2\varepsilon_0} \int_0^\infty \frac{\rho' \, d\rho'}{[(\rho')^2 + z^2]^{3/2}}$$

or, when the integral is found in the tables, we get

$$E_z = \frac{\rho_s}{2\varepsilon_0}, \quad z > 0 \quad (\text{V/m}).$$

The reader should recognize that $E_z = -\rho_s/(2\varepsilon_0)$ for $z < 0$. Thus, for all z,

$$E_z = \begin{cases} +\rho_s/(2\varepsilon_0), & z > 0 \\ -\rho_s/(2\varepsilon_0), & z < 0 \end{cases} \quad (\text{V/m}). \qquad (2.17)$$

∎

We have examined three one-component fields from three sources: the point charge, the infinite line charge density, and the infinite surface charge density. They all have infinite volume charge densities, but produce very different fields. The fields do, however, share one common trait: the vector fields are normal to the charge.

We now return to Equation (2.12) where one special case can be disposed of almost immediately. Suppose the field point is far removed from the origin, while the charge density is finite in extent and relatively close to the origin. These conditions are satisfied if $r \gg r'$. Formally, if $r \gg r'$, then

$$R = |\mathbf{r} - \mathbf{r}'| \approx r, \qquad \mathbf{R} = \mathbf{r} - \mathbf{r}' \approx \mathbf{r}$$

and

$$\mathbf{E}(\mathbf{r}) = \frac{1}{4\pi\varepsilon_0} \iiint_{\text{vol}'} \frac{\rho_v(\mathbf{r}')}{r^3} \mathbf{r} \, dv'$$

or

$$\mathbf{E}(\mathbf{r}) = \frac{\mathbf{r}}{4\pi\varepsilon_0 r^3} \iiint_{vol'} \rho_v(\mathbf{r}')\, dv'.$$

It is helpful to verify the approximations being made by re-examining Figure 2.3. Now, using Equation (2.2), we get

$$\mathbf{E}(\mathbf{r}) = \frac{Q\mathbf{r}}{4\pi\varepsilon_0 r^3}, \qquad r \gg r'.$$

The interpretation of this result is easy. From a large distance any finite charge distribution close to the origin looks like a point charge at the origin with total charge Q C! See Equation (2.8) for verification of this statement.

2.3 THE ELECTROSTATIC POTENTIAL FIELD

The electric field intensity $\mathbf{E}$ can be determined directly from the volume charge density ρ_v, as Equation (2.13) shows. It can also be determined from an auxiliary *scalar* potential function as follows.

The electric field was defined as the force on a *unit* positive charge, so the force on *any* point charge Q in an arbitrary electric field is

$$\boxed{\mathbf{F} = Q\mathbf{E}} \quad (\text{N}). \tag{2.18}$$

It is this force that must be overcome if an external force is to move the charge Q. More precisely, if we desire to move the charge an incremental distance, $\Delta\mathbf{l} = \mathbf{a}_l \Delta l$, then the component of force which must be overcome is $\mathbf{F} \cdot \mathbf{a}_l = Q\mathbf{E} \cdot \mathbf{a}_l$, so that the external scalar force required is

$$F_{\text{ext}} = -Q\mathbf{E} \cdot \mathbf{a}_l.$$

Since work is force times distance, the incremental *work done by the external source* is

$$\Delta W = -Q\mathbf{E} \cdot \mathbf{a}_l \Delta l = -Q\mathbf{E} \cdot \Delta\mathbf{l} \quad (\text{J}).$$

The unit of force is the newton and the unit of work is the joule. Suppose that the external force moves the point charge from an initial point P_i to a final point P_f along a prescribed path C (see Figure 1.8). The total work required can be approximated as a finite sum of incremental amounts of work, each being given by the preceding equation:

$$W \approx -Q \sum_{n=1}^{N} \mathbf{E}_n \cdot \Delta\mathbf{l}_n,$$

where $\mathbf{E}_n$ is the electric field intensity at the nth point on the path. If we let N approach infinity in such a way that $\Delta\mathbf{l}_n$ approaches zero for any n, the total work required in moving Q from P_i and P_f in the field $\mathbf{E}$ is given by the line integral (see Sections 1.4 and 1.8)

$$W = -Q \int_{P_i}^{P_f} \mathbf{E} \cdot d\mathbf{l} \quad (\text{J}). \tag{2.19}$$

It is a simple matter now to define the *potential difference between P_f and P_i* as the work required of an external source in moving a *unit* positive charge from P_i to P_f in the field $\mathbf{E}$. From Equation (2.19), with $Q = 1$, we get

$$\boxed{\Phi_{fi} \equiv \Phi_f - \Phi_i = -\int_{P_i}^{P_f} \mathbf{E} \cdot d\mathbf{l}} \quad (\text{J/C or V}). \tag{2.20}$$

As indicated by Equation (2.20), the symbol for potential is Φ, and it is the work per unit charge (J/C). As a matter of fact, in the *electrostatic case* (which concerns us here), the *potential difference* is the same as the *voltage difference* of dc circuit theory. Although it is usually well disguised in terms of field quantities, the circuit theory definition of potential difference is really the same as that given in Equation (2.20).

In the dynamic case, to be considered later, the relation between voltage and potential will be reexamined. Note that the potential difference $\Phi_f - \Phi_i = \Phi_{fi}$ can be interpreted as the *absolute potential* at P_f (i.e., Φ_f) minus the *absolute potential* at P_i (i.e., Φ_i). We must now explain what is meant by absolute potential.

The absolute potential at a point, $\Phi(x, y, z)$, is simply the potential difference between that point and some *reference* point where the potential is zero, commonly called the "ground" or "earth." A voltmeter (with its two leads) inherently measures the voltage or potential difference between two points. If, however, one lead is connected to the reference point or ground, the voltmeter reads the absolute potential at the other point.

Potential difference can be determined from absolute potential, regardless of the reference for zero potential, if the absolute potential is given by

$$\boxed{\Phi(\mathbf{r}) = -\int \mathbf{E} \cdot d\mathbf{l} + C} \quad (\text{V}) \quad \textit{absolute potential,} \tag{2.21}$$

where C is a constant that can be chosen to make the potential vanish wherever we like. For a point charge at the origin, Equation (2.8) with Equation (2.21) gives

$$\Phi(\mathbf{r}) = -\int \frac{Q}{4\pi\varepsilon_0} \frac{\mathbf{a}_r}{r^2} \cdot (\mathbf{a}_r \, dr + \mathbf{a}_\theta r \, d\theta + \mathbf{a}_\phi r \sin\theta \, d\phi) + C,$$

$$\Phi(\mathbf{r}) = -\int \frac{Q}{4\pi\varepsilon_0 r^2} \, dr + C,$$

$$\Phi(\mathbf{r}) = \frac{Q}{4\pi\varepsilon_0 r} + C \quad (\text{V})$$

2.3 THE ELECTROSTATIC POTENTIAL FIELD

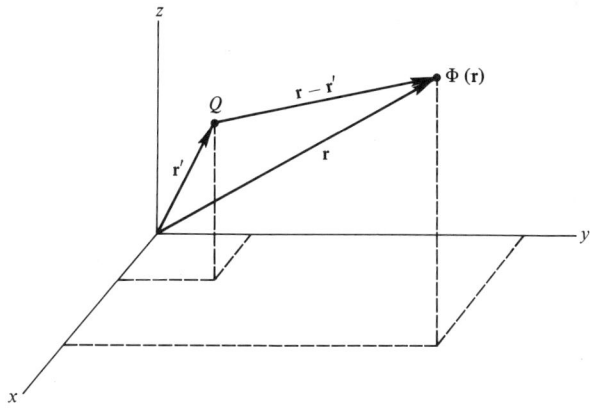

Figure 2.7. Geometry for finding the potential at **r** due to a point charge Q at **r**′.

and equipotential surfaces (surfaces of constant potential) are concentric spheres centered at the origin where Q is located. Ordinarily, we would want the potential to vanish on a sphere of infinite radius, and in this case $C = 0$, so that

$$\Phi(\mathbf{r}) = \frac{Q}{4\pi\varepsilon_0 r} \quad (V). \tag{2.22}$$

If the point charge is located at **r**′ as in Figure 2.7, rather than at the origin, we then obtain the more general result

$$\Phi(\mathbf{r}) = \frac{Q}{4\pi\varepsilon_0} \frac{1}{|\mathbf{r} - \mathbf{r}'|} \quad (V) \tag{2.23}$$

and equipotential surfaces are spheres centered at Q.

We can obtain a superposition integral for the potential due to the more general volume charge density ρ_v in exactly the same way that we did for the electric field intensity **E** in Section 2.2. In fact, Figure 2.2 still applies, except that we are seeking $\Phi(\mathbf{r})$. It is easy to see that the limiting process for $N \to \infty$, $\Delta v_n \to 0$, and $\Delta Q_n / \Delta v_n \to \rho_v$ will give[5]

$$\boxed{\Phi(\mathbf{r}) = \frac{1}{4\pi\varepsilon_0} \iiint_{\text{vol}'} \frac{\rho_v(\mathbf{r}')}{|\mathbf{r} - \mathbf{r}'|} dv'} \quad \begin{array}{l}\text{superposition integral for}\\ \text{electrostatic potential.}\end{array} \tag{2.24}$$

Note that this "derivation" depends, as did that for Equation (2.12), on the assumption that all of the charge is within a *finite* volume located a *finite distance* from the origin. If this is not true then the potential, as calculated by Equation (2.24)

---

[5]This is also a *convolution* integral.

may not vanish at infinity, as it certainly will otherwise, for if $r \gg r'$, then $|\mathbf{r} - \mathbf{r}'| \approx r$, and Equation (2.24) gives

$$\Phi(\mathbf{r}) \approx \frac{1}{4\pi\varepsilon_0 r} \iiint_{\text{vol}'} \rho_v(\mathbf{r}')\, dv', \qquad r \gg r',$$

where the integral is that for calculating the *total* charge Q, so the result is

$$\Phi(\mathbf{r}) \approx \frac{Q}{4\pi\varepsilon_0 r}, \qquad r \gg r'.$$

Comparing this result to Equation (2.22) leads once more to the conclusion that "from a large distance any finite charge distribution close to the origin looks like a point charge Q at the origin." Equation (2.24) reduces to

$$\Phi(\mathbf{r}) = \frac{1}{4\pi\varepsilon_0} \iint_{s'} \frac{\rho_s(\mathbf{r}')}{|\mathbf{r} - \mathbf{r}'|} ds' \qquad (2.25)$$

for surface charge densities ρ_s, and to

$$\Phi(\mathbf{r}) = \frac{1}{4\pi\varepsilon_0} \int_{l'} \frac{\rho_l(\mathbf{r}')}{|\mathbf{r} - \mathbf{r}'|} dl' \qquad (2.26)$$

for line charge densities.

Any constant C can be added to Equations (2.24), (2.25), or (2.26) to change the reference for zero potential, and this constant has no effect whatever on *potential difference*. We shall shortly show that it also has no effect on the electric field intensity which is calculated from the potential.

Since the potential can be found from the electric field intensity by way of a line integral, Equation (2.21), one would naturally expect that the electric field intensity could be found from the potential by some special kind of differentiation. Consider the gradient operation

$$\nabla\Phi = \nabla\left(\frac{1}{4\pi\varepsilon_0} \iiint_{\text{vol}'} \frac{\rho_v(\mathbf{r}')}{|\mathbf{r} - \mathbf{r}'|} dv'\right).$$

Since the del operation is on the unprimed coordinates and the integration is with respect to the primed coordinates, the del operator may be carried inside the triple integral symbol

$$\nabla\Phi = \frac{1}{4\pi\varepsilon_0} \iiint_{\text{vol}'} \rho_v(\mathbf{r}') \nabla\left(\frac{1}{|\mathbf{r} - \mathbf{r}'|}\right) dv'.$$

It was shown in Example 1.6 that $\nabla\{1/|\mathbf{r} - \mathbf{r}'|\} = \nabla(1/R) = -\mathbf{a}_R/R^2$, so that

2.3 THE ELECTROSTATIC POTENTIAL FIELD

$$\nabla \Phi = -\frac{1}{4\pi\varepsilon_0} \iiint_{\text{vol}'} \frac{\rho_v(\mathbf{r}')\mathbf{a}_R}{R^2}\, dv', \tag{2.27}$$

and with Equation (2.13), the right-hand side of Equation (2.27) is identified as $-\mathbf{E}$:

$$\boxed{\mathbf{E} = -\nabla\Phi} \tag{2.28}$$

Note that, as predicted earlier, $\mathbf{E}$ is obtained from Φ by differentiation, and any constant C that might be added to Φ to change the potential reference has no effect on $\mathbf{E}$ because $\nabla C \equiv 0$. Finally, note that *the potential must be a continuous function of position*, for if it is not, the gradient operation of Equation (2.28) will create impulses of electric field intensity at the discontinuities of Φ, and this is contrary to experience.

Other fundamental results can now be obtained. If we take the curl of both sides of Equation (2.28) using equation (1.39), we obtain *Maxwell's curl* $\mathbf{E}$ *equation for electrostatics*:

$$\boxed{\nabla \times \mathbf{E} = 0} \quad \textit{Maxwell's curl } \mathbf{E} \textit{ equation.} \tag{2.29}$$

Furthermore, as a consequence of the fact that $\mathbf{E} = -\nabla\Phi$, it follows that the *circulation* of $\mathbf{E}$ is zero (see Section 1.8)

$$\boxed{\oint_c \mathbf{E} \cdot d\mathbf{l} \equiv 0} \quad \textit{circulation of } \mathbf{E}, \tag{2.30}$$

and, thus, electrostatic $\mathbf{E}$ is both *conservative* and *irrotational*. A conservative field must be irrotational (*its curl must be identically zero*). Note that Equations (2.29) and (2.30) are in agreement with the Stokes theorem (see Section 1.10):

$$\oint_c \mathbf{E} \cdot d\mathbf{l} = \iint_s \nabla \times \mathbf{E} \cdot d\mathbf{s}$$

or

$$0 = \iint_s \nabla \times \mathbf{E} \cdot d\mathbf{s}.$$

This last result must hold for *any* open surface s that is bounded by *any* closed path C, and therefore the right-hand side can only be zero if $\nabla \times \mathbf{E} \equiv 0$.

Finally, note that, in accordance with Section 1.6, $\mathbf{E}$ is normal to equipotential surfaces, pointing in the direction of *decreasing* Φ. The reader can easily verify (and should attempt to do so) that the results of this section apply to the very symmetrical infinite uniform line charge density on the z axis and the infinite uniform surface charge density on the $z = 0$ plane of Section 2.2. Begin by determining the equipotential surfaces and then using Equation (2.2) to show that $\Phi = -\rho_l \ln \rho/(2\pi\varepsilon_0)$ in the former case and $\Phi = -\rho_s |z|/(2\varepsilon_0)$ in the latter case. Why does the potential not vanish at infinity? Where does it vanish? In calculating potential differences, what paths would you choose for these two cases (or any other case)? Remember that any constant can be added to Φ.

■ Example 2.6

An electrostatic dipole is shown in Figure 2.8(a). It can be described as having a moment $\mathbf{p} = Q\mathbf{d}$, or, in this example, $\mathbf{p} = Qd\mathbf{a}_z$. Superposition gives the potential as

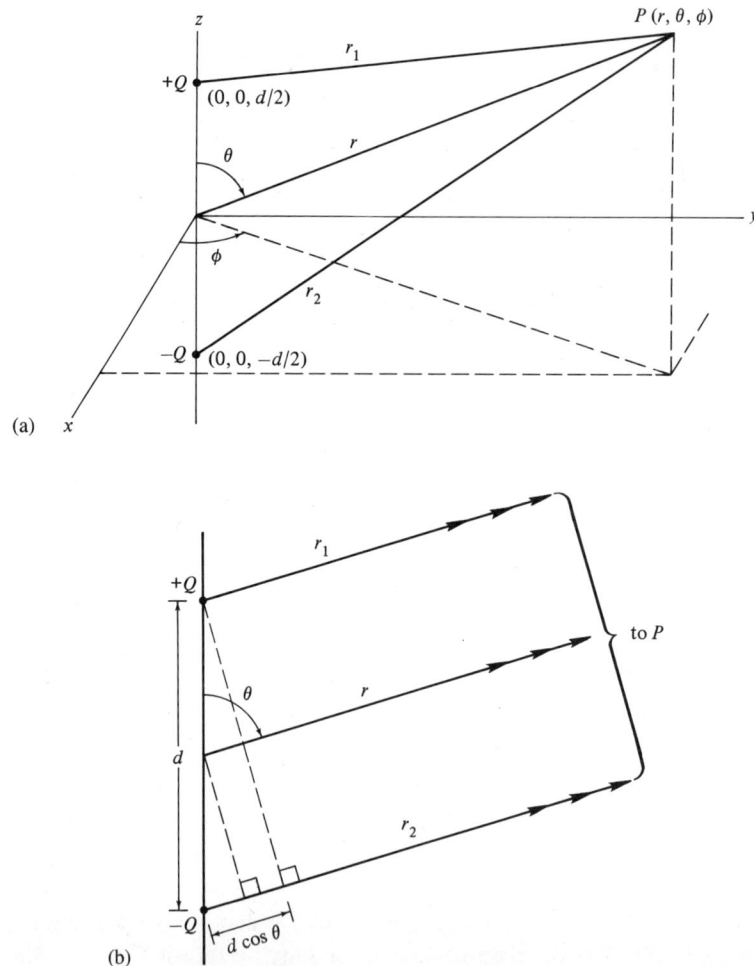

Figure 2.8. (a) Electric dipole centered on the z axis (b) Simplifications for calculating Φ at a remote point P.

$$\Phi = \frac{Q}{4\pi\varepsilon_0}\left(\frac{1}{r_1} - \frac{1}{r_2}\right) = \frac{Q}{4\pi\varepsilon_0 r_1 r_2}(r_2 - r_1).$$

For sufficiently large r ($r \gg d$), the dipole will look like a point charge at the origin with *zero* net charge. This concept is not useful to us. Instead, we allow r to be large enough compared to d so that r, r_1, and r_2 are essentially parallel. In this case, as seen in Figure 2.8(b), $r_2 - r_1 \approx d\cos\theta$, and $r_1 r_2 \approx r^2$. Thus, a good approximation for the potential when $r \gg d$ is

$$\Phi = \frac{Qd\cos\theta}{4\pi\varepsilon_0 r^2} = \frac{\mathbf{p}\cdot\mathbf{a}_r}{4\pi\varepsilon_0 r^2} \tag{2.31}$$

For a dipole whose center is located by $\mathbf{r}'$, the potential at $\mathbf{r}$ will be given by the more general result

$$\boxed{\Phi = \frac{\mathbf{p}\cdot\mathbf{a}_R}{4\pi\varepsilon_0 R^2}} \quad (\text{V}), \tag{2.32}$$

where, as usual, $\mathbf{a}_R/R^2 = (\mathbf{r} - \mathbf{r}')/|\mathbf{r} - \mathbf{r}'|^3$. This last result will be very useful when dielectrics are considered in Section 2.5. ∎

2.4 GAUSS'S LAW

Electric flux density, symbolized by $\mathbf{D}$, is related to electric field intensity in a very simple manner for a vacuum

$$\boxed{\mathbf{D} = \varepsilon_0 \mathbf{E}} \quad (\text{C/m}^2), \tag{2.33}$$

where ε_0, the permittivity of a vacuum, is given by Equation (2.6). Gauss's law states that the net *flux* of $\mathbf{D}$, or electric flux, symbolized by Ψ_E, out of any closed surface is equal to the net positive charge enclosed by that surface, or

$$\boxed{\Psi_E = Q} \quad (\text{C}) \quad \textit{Gauss's law}. \tag{2.34}$$

Using Equations (1.20) and (2.2), Gauss's law can be written

$$\boxed{\oint_S \mathbf{D}\cdot d\mathbf{s} = \iiint_{\text{vol}} \rho_v\, dv} \quad \textit{Gauss's law}, \tag{2.35}$$

where the volume for the integral on the right is that within the surface of the left integral even though ρ_v may be zero in some smaller volumes within the original. In other words, the right-hand side of Equation (2.35) calculates the *net* charge within the volume enclosed by the surface on the left. Equation (2.35) holds regardless of the size or shape of the surface and its resulting volume. That is, it holds regardless of the limits on the integrals. The divergence theorem (or Gauss's integral theorem), Equation (1.43), enables us to write

$$\oint_s \mathbf{D} \cdot d\mathbf{s} = \iiint_{vol} \nabla \cdot \mathbf{D} \, dv \quad \text{divergence theorem}$$

Thus, Equation (2.35) can be written

$$\iiint_{vol} \nabla \cdot \mathbf{D} \, dv = \iiint_{vol} \rho_v \, dv$$

and the volumes are identical, *but still arbitrary*. Under these conditions the integrands must be identical. Therefore, we obtain another of Maxwell's equations

$$\boxed{\nabla \cdot \mathbf{D} = \rho_v} \quad \text{Maxwell's divergence } \mathbf{D} \text{ equation} \quad (2.36)$$

This is the *point* form of Gauss's law. The formal solution to this partial differential equation is obtained by combining Equations (2.33) and (2.13):

$$\boxed{\mathbf{D}(r) = \frac{1}{4\pi} \iiint_{vol'} \rho_v(r') \frac{\mathbf{a}_R}{R^2} dv'} \quad (2.37)$$

It is also true that

$$\rho_v = \varepsilon_0 \nabla \cdot \mathbf{E}. \quad (2.38)$$

We take note of the fact that Equation (2.36) is completely general. It holds when material bodies are present and also in the dynamic case to be considered later.

■ Example 2.7

Consider the spherical surface $r = a$ with the uniform surface charge density $\rho_s = Q_t/(4\pi a^2)$ (C/m²) residing on it as shown in Figure 2.9. It is easy to see that two differential amounts of charge $\rho_s \, ds$ symmetrically located at equal distances from the origin produce only a radial component of field (**E** or **D**), and this field can only depend on r (not θ or ϕ). That is, $\mathbf{D} = \mathbf{a}_r D_r(r)$. Equipotential surfaces are, therefore, spheres concentric with the origin. Imagine such a sphere (or *Gaussian surface*) with $r < a$. Since this sphere encloses no charge, the

2.4 GAUSS'S LAW

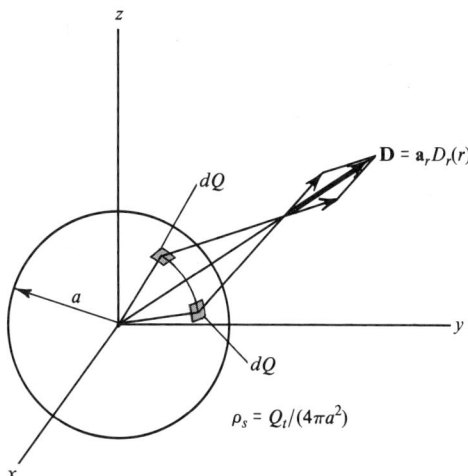

Figure 2.9. Radial symmetry for finding the field of a uniform surface charge density on the sphere $r = a$.

flux out of it is zero, and $D_r(r) = 0$. Next, imagine the same sphere, except that $r > a$, and again apply Gauss's law

$$\Psi_E = \int_0^\pi \int_0^{2\pi} \mathbf{a}_r D_r \cdot \mathbf{a}_r r^2 \sin\theta \, d\theta \, d\phi = Q_{\text{enc}} = Q_t = 4\pi a^2 \rho_s,$$

$$D_r r^2 \int_0^\pi \int_0^{2\pi} \sin\theta \, d\theta \, d\phi = 4\pi a^2 \rho_s$$

$$D_r(4\pi r^2) = 4\pi a^2 \rho_s$$

$$D_r(r) = \rho_s a^2/r^2 \quad (\text{C/m}^2)$$

or

$$E_r(r) = Q_t/(4\pi\varepsilon_0 r^2), \qquad r > a \quad (\text{V/m}).$$

Note that this field is the same (for $r > a$) as if it arose from a point charge Q_t at the origin. The result was easily obtained because we chose a Gaussian surface, $r > a$, where **D** *was normal and uniform*, and, therefore, was factored from within the integral on the left, leaving an integral that merely calculated the *area* of the Gaussian surface. ∎

Results from Example 2.7 can be generalized to some extent. For example, if $\rho_v = f(r)$, then it follows that $\mathbf{E} = \mathbf{a}_r E_r(r)$, and Gauss's law [Equation (2.35)] leads to

$$E_r(r) = \frac{1}{\varepsilon_0 r^2} \int_0^r f(\tau)\tau^2 \, d\tau, \qquad \rho_v = f(r). \tag{2.39}$$

If $\rho_v = f(\rho)$, the reader is encouraged to show that

$$E_\rho(\rho) = \frac{1}{\varepsilon_0 \rho} \int_0^\rho f(\tau)\tau\, d\tau, \qquad \rho_v = f(\rho). \tag{2.40}$$

Example 2.8 demonstrates the use of the last result.

■ Example 2.8

Suppose that a volume charge density is given by $\rho_v = \rho_{v0} e^{-\rho}$. Then, with Equation (2.40), we get

$$E_\rho(\rho) = \frac{1}{\varepsilon_0 \rho} \int_0^\rho \rho_{v0} e^{-\tau} \tau\, d\tau = \frac{\rho_{v0}}{\varepsilon_0 \rho} \int_0^\rho \tau e^{-\tau}\, d\tau.$$

The integral can be found in the tables, and results in

$$E_\rho(\rho) = \frac{\rho_{v0}}{\varepsilon_0 \rho} \big[e^{-\tau}(-\tau - 1) \big]_0^\rho = \frac{\rho_{v0}}{\varepsilon_0 \rho} \big[1 - e^{-\rho}(\rho + 1) \big].$$

Maxwell's divergence **D** equation can be used to verify this result. We have $\nabla \cdot \mathbf{D} = \rho_v$, or $\rho_v = \varepsilon_0 \nabla \cdot \mathbf{E}$, or

$$\rho_v(\rho) = \frac{\varepsilon_0}{\rho} \frac{\partial}{\partial \rho}(\rho E_\rho) = \frac{1}{\rho} \frac{\partial}{\partial \rho} \{ \rho_{v0}[1 - e^{-\rho}(\rho + 1)] \},$$

$$\rho_v = \rho_{v0} e^{-\rho} \quad (\text{C/m}^3),$$

which is the volume charge density with which we began.[6] ■

The reader can easily verify Equation (2.16) for the infinite line charge density by using the Gauss law. The proper Gaussian surface is a circular cylinder concentric with the z axis. Why?

2.5 DIELECTRICS

We have not considered what happens to field quantities when materials are present, and at this point in our development of electrostatic relationships we must do so before we proceed. Materials can be broadly classified into four types.

---

[6] It is very difficult to explicitly verify that Maxwell's divergence **D** equation holds for charge distributions that are not continuous at all points in space without using the Dirac delta function. The interested reader is referred to Neff, 1981 in the references at the end of this chapter.

2.5 DIELECTRICS

1. If the characteristics of the material do not depend on *position* (x, y, z), the material is said to be *homogeneous*. The atmosphere above us is obviously inhomogeneous because of its varying density, humidity, temperature, or ionization.
2. If the characteristics of the material are independent of the *direction* of the vector fields, the material is *isotropic*. Some very important *crystals* are *anisotropic*, and have some properties that depend on direction.
3. If the parameters of the material do not depend on the *magnitude* of the field quantities, then the material is linear. A *ferroelectric* or *ferromagnetic* material is nonlinear because its electrical or magnetic properties depend on the strength of the applied field.
4. In general the parameters of a material may be frequency-dependent. For example, the permittivity (to be introduced shortly) of polystyrene depends slightly on frequency. This dependence is discussed in more detail later.

We can consider pairs of positive and negative charges (protons and electrons) in a dielectric material to act like *dipoles*. The action of an applied field is to tend to align the dipole moments, where

$$\mathbf{p} = Q\mathbf{d} \quad \textit{electric dipole moment} \quad (2.41)$$

is the dipole moment, as in Example 2.6. Using the mechanical equation for torque $\mathbf{T} = \mathbf{R}_0 \times \mathbf{F}$, where $\mathbf{R}_0$ is the vector lever arm, it is not difficult to show that, for a single dipole centered at the origin with $\mathbf{p} = Qd\mathbf{a}_z$, $\mathbf{T} = \mathbf{a}_z(d/2) \times Q\mathbf{E} + (-\mathbf{a}_z)(d/2) \times (-Q)\mathbf{E} = -QdE_0\mathbf{a}_x$, when $\mathbf{E} = E_0\mathbf{a}_y$. This result can be put in a form which is *general*:

$$\mathbf{T} = \mathbf{p} \times \mathbf{E} \quad (\text{N} \cdot \text{m}) \quad (2.42)$$

is the torque on the dipole.

We are not in a position to examine the behavior of every atom in a dielectric that is subjected to the forces of an applied field. That is, we are not interested in the *microscopic* behavior (small d) but, rather, in the *macroscopic* behavior (large d) and in this sense, we intend to use Equation (2.32) for the dipole potential. In this case the dipole is considered to be a *point* source, and the two charges making up the individual dipoles are now *bound* charges. Dielectrics store energy under the influence of an applied field like a spring that stores energy when a force is applied to it.

If there are n dipoles per cubic meter, then the resultant dipole moment for an incremental volume Δv will be

$$\mathbf{p}_r = \sum_{m=1}^{n\Delta v} \mathbf{p}_m. \quad (2.43)$$

The *polarization* vector **P** is defined as the *dipole moment per unit volume*:

$$\mathbf{P} = \lim_{\Delta v \to 0} (\mathbf{p}_r/\Delta v) \quad (\text{C/m}^2). \quad (2.44)$$

The polarization is the general source function for bound charge in the present situation in the same way that volume charge density [Equation (2.1)]

$$\rho_v = \lim_{\Delta v \to 0} \frac{\Delta Q}{\Delta v} \quad (\text{C/m}^3)$$

is the general source function for isolated charge. Dimensionally the polarization is like a surface charge density ρ_s.

Since the potential of an isolated point charge leads by superposition to the potential for the general source [Equation (2.23) leads to Equation (2.24)]

$$\frac{Q}{4\pi\varepsilon_0 R} \to \frac{1}{4\pi\varepsilon_0} \iiint_{\text{vol}'} \frac{\rho_v}{R} \, dv',$$

the potential of a point dipole moment, Equation (2.32), must lead by superposition to the potential for the general source function $\mathbf{P}$ for bound charge

$$\frac{\mathbf{p} \cdot \mathbf{a}_R}{4\pi\varepsilon_0 R^2} \to \frac{1}{4\pi\varepsilon_0} \iiint_{\text{vol}'} \frac{\mathbf{P} \cdot \mathbf{a}_R}{R^2} \, dv'.$$

That is, for the bound charge,

$$\Phi(\mathbf{r}) = \frac{1}{4\pi\varepsilon_0} \iiint_{\text{vol}'} \frac{\mathbf{P}(\mathbf{r}') \cdot \mathbf{a}_R}{R^2} \, dv'. \tag{2.45}$$

The integration is throughout the volume containing the dielectric. Here, we are treating $\mathbf{P}$ like any other field quantity, but it should be visualized as an average value taken over many molecules. It has the effect of giving the dipoles a net alignment or polarization from their otherwise random arrangement. We continue to use ε_0 (the permittivity of vacua), since the dielectric effects have been included in the dipoles.

Using the relation $\nabla'(1/R) = \mathbf{a}_R/R^2$ (Example 1.6), Equation (2.45) can be written

$$\Phi(\mathbf{r}) = \frac{1}{4\pi\varepsilon_0} \iiint_{\text{vol}'} \mathbf{P}(\mathbf{r}') \cdot \nabla' \frac{1}{R} \, dv'. \tag{2.46}$$

By vector identity (inside back cover), we get

$$\mathbf{P} \cdot \nabla'(1/R) \equiv \nabla' \cdot (1/R)\mathbf{P} - (1/R)\nabla' \cdot \mathbf{P},$$

so we have

$$\Phi(\mathbf{r}) = \frac{1}{4\pi\varepsilon_0} \iiint_{\text{vol}'} \frac{-\nabla' \cdot \mathbf{P}(\mathbf{r}')}{R} \, dv' + \frac{1}{4\pi\varepsilon_0} \iiint_{\text{vol}'} \nabla' \cdot \frac{\mathbf{P}(\mathbf{r}')}{R} \, dv'$$

2.5 DIELECTRICS

where we can apply the divergence theorem [Equation (1.43)] to the last term in the equation giving

$$\Phi(\mathbf{r}) = \frac{1}{4\pi\varepsilon_0} \iiint_{vol'} \frac{-\nabla' \cdot \mathbf{P}(\mathbf{r}')}{R} dv' + \frac{1}{4\pi\varepsilon_0} \oiint_{s'} \frac{\mathbf{P}(\mathbf{r}')}{R} \cdot \mathbf{a}_n \, ds', \qquad (2.47)$$

where $d\mathbf{s}' = \mathbf{a}_n \, ds'$, and $\mathbf{a}_n$ is a unit vector normal to and out of s'. If we now compare Equation (2.47) to Equations (2.24) and (2.25), we see that $-\nabla \cdot \mathbf{P}$ has the effect of an *isolated volume charge density*, while $\mathbf{P} \cdot \mathbf{a}_n$ has the effect of a *closed surface charge density*. Thus, we are justified in using $\rho_v^b = -\nabla \cdot \mathbf{P}$ and $\rho_s^b = \mathbf{P} \cdot \mathbf{a}_n$ for bound charges.

Our goal is to obtain a new simple relation between $\mathbf{D}$ and $\mathbf{E}$, like $\mathbf{D} = \varepsilon_0 \mathbf{E}$, but one that includes the dielectric effects. Inside the dielectric we have both bound charges and isolated charges. This can be shown explicitly be writing the point form of Gauss's law as

$$\nabla \cdot \varepsilon_0 \mathbf{E} = \rho_v + \rho_v^b = \text{total volume charge density}, \qquad (2.48a)$$

which is valid, even though a dielectric is present, since all of the charge has been accounted for and may be treated as though it existed in a vacuum or free space. Substituting $-\nabla \cdot \mathbf{P}$ for ρ_v^b in Equation (2.48a) gives $\nabla \cdot \varepsilon_0 \mathbf{E} = \rho_v - \nabla \cdot \mathbf{P}$, or

$$\nabla \cdot (\varepsilon_0 \mathbf{E} + \mathbf{P}) = \rho_v \quad (\text{C/m}^3). \qquad (2.48b)$$

Maxwell's divergence $\mathbf{D}$ equation, the point form of Gauss's law, $\nabla \cdot \mathbf{D} = \rho_v$, is completely general, so $\mathbf{D}$ can be identified in Equation (2.48b) as

$$\mathbf{D} = \varepsilon_0 \mathbf{E} + \mathbf{P} \quad (\text{C/m}^2). \qquad (2.49)$$

Thus, the polarization vector $\mathbf{P}$, or dipole moment per unit volume is added when a dielectric is present.

The relation between $\mathbf{P}$ and $\mathbf{E}$ depends on the dielectric material. $\mathbf{P}$ and $\mathbf{E}$ are not parallel in anisotropic materials, and are not simply related. In ferroelectric materials the relation between $\mathbf{P}$ and $\mathbf{E}$ is nonlinear, and may also depend on the past history of the sample, indicating a *hysteresis* effect. The relation between $\mathbf{P}$ and $\mathbf{E}$ is usually written

$$\boxed{\mathbf{P} = \chi_E \varepsilon_0 \mathbf{E}} \quad (\text{C/m}^2) \qquad (2.50)$$

where χ_E is the *electric susceptibility* of the material. If we define $\chi_E \equiv \varepsilon_R - 1$, where ε_R (dimensionless) is the *relative permittivity* or *dielectric constant* of the material, then $\mathbf{P} = (\varepsilon_R - 1)\varepsilon_0 \mathbf{E}$, and Equation (2.49) becomes

$$\boxed{\mathbf{D} = \varepsilon \mathbf{E}} \quad (\text{C/m}^2), \qquad (2.51)$$

where

$$\varepsilon = \varepsilon_0 \varepsilon_R \quad \text{(F/m)}. \tag{2.52}$$

This is the desired result: a simple (looking) relation between **D** and **E** like that for free space. For dielectric materials that are linear, isotropic, and homogeneous, the only change required in order to use previous equations is that of replacing ε_0 with ε, a scalar constant. In anisotropic materials ε in Equation (2.51) is not a scalar, but is a *tensor*, and each component of **D**, in general, depends on every component of **E**. That is, $D_x = \varepsilon_{xx}E_x + \varepsilon_{xy}E_y + \varepsilon_{xz}E_z$, for example. We will not study anisotropic media in this text.

Dielectrics will break down if the electric field intensity is too large. This is a very important consideration in the design of electrical apparatus. For a nonuniform field, a *corona* discharge may occur before sparking occurs, but if the field is uniform, sparking or arcing may occur as soon as the critical value of **E**, called the *dielectric strength*, is reached. A list of ε_R and dielectric strengths for several dielectrics is found in Appendix B. Also, see Example 2.10.

It is both necessary and sufficient for completeness to specify what happens to the *normal* and *tangential* components of vector field quantities at the interface between different media. The behavior of tangential components is determined from the *circulation* of the vector field, and the behavior of normal components is determined from the *flux* of the vector field. Consider Figure 2.10 where two dielectrics are shown separated by an interface. Since $\Delta w \to 0$ in Figure 2.10, the circulation, $\oint \mathbf{E} \cdot d\mathbf{l} = 0$ leads to $lE_{t1} - lE_{t2} = 0$, or

$$\boxed{E_{t1} = E_{t2}, \qquad D_{t1}/\varepsilon_1 = D_{t2}/\varepsilon_2}. \tag{2.53}$$

Since $\Delta h \to 0$, all of the electrostatic flux must flow out of the top and bottom of the can in Figure 2.10, and the Gauss law gives $D_{n1}\Delta s - D_{n2}\Delta s = \rho_s \Delta s$ or

$$\boxed{D_{n1} - D_{n2} = \rho_s, \qquad \varepsilon_1 E_{n1} - \varepsilon_2 E_{n2} = \rho_s}. \tag{2.54}$$

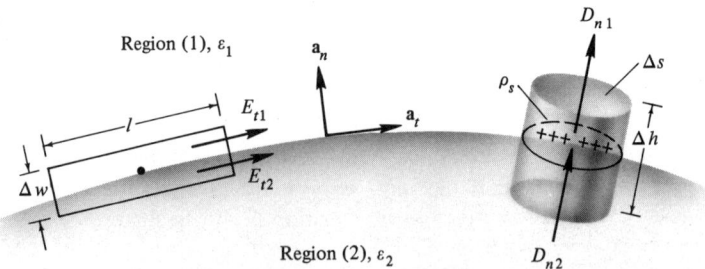

Figure 2.10. Geometry for determining boundary conditions, dielectric–dielectric interface.

2.6 POISSON AND LAPLACE EQUATIONS

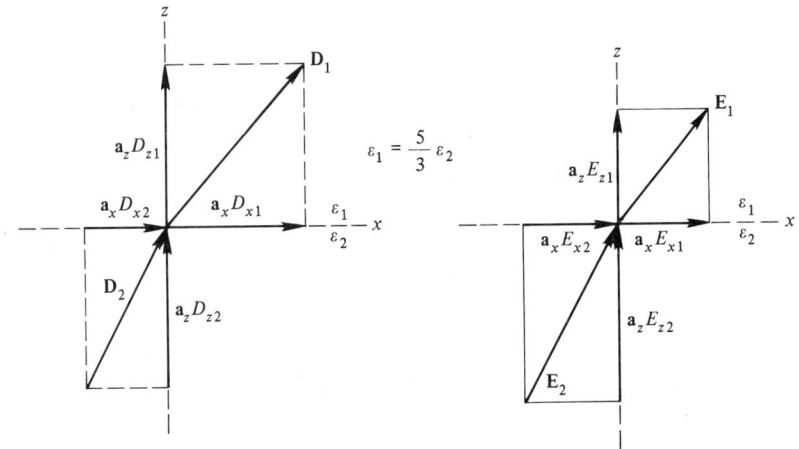

Figure 2.11. Resultant electric flux density and field intensity at a plane interface (Example 2.9).

In the usual case $\rho_s = 0$ and

$$D_{n1} = D_{n2}, \qquad \varepsilon_1 E_{n1} = \varepsilon_2 E_{n2} \qquad (\rho_s = 0). \qquad (2.55)$$

Example 2.9

Figure 2.11 shows the $z = 0$ plane as the interface between two dielectrics: $\varepsilon_1 = 5\varepsilon_0$ for $z > 0$ and $\varepsilon_2 = 3\varepsilon_0$ for $z < 0$. It is given that $\mathbf{E}_2 = 10\mathbf{a}_x + 20\mathbf{a}_z$ for $z < 0$, and our task is to determine $\mathbf{D}_2$, $\mathbf{D}_1$, and $\mathbf{E}_1$. There is no surface charge density present. We immediately obtain

$$\mathbf{D}_2 = \varepsilon_2 \mathbf{E}_2 = \varepsilon_0(30\mathbf{a}_x + 60\mathbf{a}_z).$$

Tangential $\mathbf{E}$ and normal $\mathbf{D}$ are continuous, so $E_{x1} = E_{t1} = E_{t2} = E_{x2} = 10$, and $D_{z1} = D_{n1} = D_{n2} = D_{z2} = 60\varepsilon_0$ from Equations (2.53) and (2.55). Also, $D_{x1} = (\varepsilon_1/\varepsilon_2)D_{x2} = 50\varepsilon_0$, and $E_{z1} = (\varepsilon_2/\varepsilon_1)E_{z2} = 12$. Thus, we have

$$\mathbf{D}_1 = \varepsilon_0(50\mathbf{a}_x + 60\mathbf{a}_z),$$

$$\mathbf{E}_1 = 10\mathbf{a}_x + 12\mathbf{a}_z.$$

The results are shown in Figure 2.11.

2.6 POISSON AND LAPLACE EQUATIONS

Substituting Equation (2.51) into (2.36), we obtain

$$\nabla \cdot \varepsilon \mathbf{E} = \rho_v.$$

Since $\mathbf{E} = -\nabla\Phi$, the last result can be written

$$\nabla \cdot (\varepsilon\nabla\Phi) = -\rho_v. \tag{2.56}$$

We now need a vector identity for the divergence of the product of a scalar ε and vector $\nabla\Phi$, since they both will, in general, depend on position. This identity can be found on the inside back cover, and gives

$$\varepsilon\nabla \cdot (\nabla\Phi) + \nabla\Phi \cdot \nabla\varepsilon = -\rho_v. \tag{2.57}$$

Using Equation (1.28) twice for the del operator, we find that

$$\nabla \cdot \nabla\Phi \equiv \nabla^2\Phi = \frac{\partial^2\Phi}{\partial x^2} + \frac{\partial^2\Phi}{\partial y^2} + \frac{\partial^2\Phi}{\partial z^2} \tag{2.58}$$

for Cartesian coordinates. ∇^2 is the second-order differential operator called the *Laplacian* operator, and its form for cylindrical and spherical coordinates can be found on the inside front and back covers. Equation (2.57) can now be written

$$\boxed{\varepsilon\nabla^2\Phi + \nabla\Phi \cdot \nabla\varepsilon = -\rho_v}. \tag{2.59}$$

This is the partial differential equation to be satisfied by Φ in an inhomogeneous region containing charge density. It and its equivalent, Equation (2.56), are *point* forms. They can be converted to forms pertaining to regions by integrating both sides of Equation (2.56) throughout the region (volume) of interest, applying the divergence theorem to the left-hand side, and then recognizing the right-hand side as the negative of the charge enclosed. The result is the Gauss law:

$$\boxed{\oint_s \varepsilon\nabla\Phi \cdot d\mathbf{s} = -Q} \quad \text{Gauss's law.} \tag{2.60}$$

In the special case where the dielectric is homogeneous, ε is not a function of position, $\nabla\varepsilon \equiv 0$, and Equations (2.59) and (2.60) reduce to

$$\boxed{\nabla^2\Phi = -\rho_v/\varepsilon} \quad \text{Poisson's equation} \tag{2.61}$$

and

$$\oint_s \nabla\Phi \cdot d\mathbf{s} = -Q/\varepsilon. \tag{2.62}$$

Equation (2.61) is very famous and appears in many areas of engineering, physics, and mathematics. It is called *Poisson's equation*. If, additionally, the region is charge-free, then we obtain

2.6 POISSON AND LAPLACE EQUATIONS

$$\boxed{\nabla^2 \Phi = 0} \quad \text{Laplace's equation} \tag{2.63}$$

and

$$\oiint_S \nabla \Phi \cdot d\mathbf{s} = 0. \tag{2.64}$$

Equation (2.63) is called *Laplace's equation*.

The formal solution to the Poisson *and* Laplace equations is the superposition integral for potential, Equation (2.24), with ε_0 replaced by ε:

$$\Phi(\mathbf{r}) = \frac{1}{4\pi\varepsilon} \iiint \frac{\rho_v(\mathbf{r}')}{R} dv'. \tag{2.65}$$

A solution for Φ can be obtained from Equation (2.65) *if we know* ρ_v at every point. Many times ρ_v is not known, but a solution for Φ can be found if certain *boundary conditions* are known. A *unique* solution for Φ exists[7] inside a region (closed surface s) if:

1. Φ is specified everywhere on s, or
2. $\partial \Phi / \partial n$ (normal outward derivative) is specified everywhere on s, or
3. Φ is specified on s_1 and $\partial \Phi / \partial n$ is specified on s_2, where $s = s_1 + s_2$.

Conducting surfaces are equipotential surfaces in the static case. That is, the potential *on* and *inside* the surface of a solid conductor is a constant in order to ensure that *electrostatic* **E** is zero ($\mathbf{E} = -\nabla \Phi = -\nabla C \equiv 0$) in the conductor. If the electrostatic field were not zero in a conductor, it would lead to charge motion (a conduction current density, to be discussed later in this chapter), and charge motion is not a static situation. Thus, electric fields leave (or enter) a charged conductor in a direction normal to the (equipotential) conductor surface in the static case.

■ Example 2.10

Consider the parallel plate capacitor shown in Figure 2.12. The battery and ground connections to the conducting plates assure us that $\Phi = 0$ when $z = 0$ and $\Phi = V_0$ when $z = d$. It is normally assumed that if the plate dimensions are large compared to d, then fringing of the electric field at the sides and front and back can be ignored as a minor effect. Here we need only say that no fringing is exactly equivalent to requiring that $\partial \Phi / \partial n = 0$ for the two sides and the front and the back of the capacitor. We can accomplish this by requiring that $\partial / \partial x = \partial / \partial y \equiv 0$ for *all* x, y, and z, and the resulting one-dimensional boundary-value problem is of the third type above. Laplace's equation is

---

[7] See Neff, 1981 in the references at the end of this chapter for a proof of the uniqueness theorem.

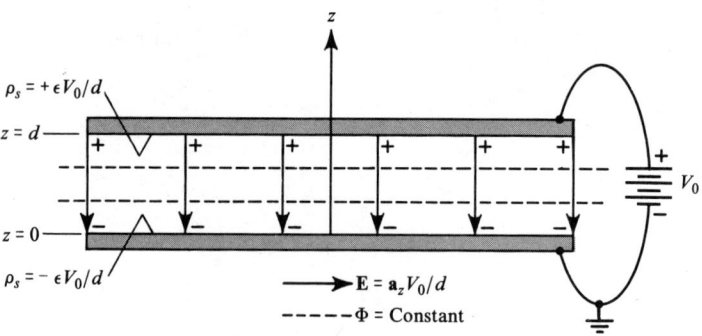

Figure 2.12. Parallel-plate capacitor (ignoring fringing of **E** at the sides).

$$\nabla^2 \Phi = \frac{\partial^2 \Phi}{\partial x^2} + \frac{\partial^2 \Phi}{\partial y^2} + \frac{\partial^2 \Phi}{\partial z^2} = \frac{\partial^2 \Phi}{\partial z^2} = \frac{d^2 \Phi}{dz^2} = 0.$$

The solution is easily found by successive integrations to be $\Phi(z) = Az + B$, and applying the two boundary conditions imposed by the battery gives

$$\Phi(z) = V_0 z/d, \qquad 0 \le z \le d.$$

Equipotential surfaces are parallel planes, and this is a unique solution for the stated boundary conditions (including all six faces).

The electric field is given by $\mathbf{E} = -\nabla \Phi$, or $E_z = -d\Phi/dz$:

$$E_z = -V_0/d, \qquad 0 \le z \le d.$$

This is a uniform electric field. The electric flux out of a closed surface that encloses an area s of the upper conductor is $\Psi_E = D_z s$ and this must equal the charge enclosed, $\rho_s s$, by the Gauss law. Thus, $\rho_s = D_z = \varepsilon E_z = \varepsilon V_0/d$. Thus, the uniform surface charge density on the upper conductor is $\varepsilon V_0/d$, while that on the lower conductor is $-\varepsilon V_0/d$.

If the dielectric between the capacitor plates is air whose dielectric strength is 3×10^6 V/m (Appendix B), and the plate spacing is 1 mm, then the maximum voltage that can be applied is $V_0 = 3 \times 10^3$ V. If the dielectric is polyethylene, then the maximum voltage is $V_0 = 20 \times 10^3$ V. ∎

A more challenging two-dimensional boundary-value problem follows.

■ Example 2.11

Consider the two-dimensional problem ($\partial/\partial z = 0$) as shown in Figure 2.13. We will follow the method of *separation of variables* to obtain a solution. We assume a product solution

$$\Phi(\rho, \phi) = R(\rho) P(\phi),$$

2.6 POISSON AND LAPLACE EQUATIONS

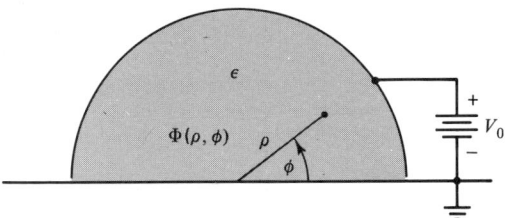

Figure 2.13. Two-dimensional boundary-value problem (cylindrical coordinates).

where, as indicated, R is a function of ρ alone, and P is a function of ϕ alone. Laplace's equation (inside front cover) with $\partial \Phi / \partial z = 0$ is

$$\nabla^2 \Phi = 0 = \frac{1}{\rho} \frac{\partial}{\partial \rho}\left(\rho \frac{\partial \Phi}{\partial \rho}\right) + \frac{1}{\rho^2} \frac{\partial^2 \Phi}{\partial \phi^2}$$

or

$$\rho^2 \frac{\partial^2 \Phi}{\partial \rho^2} + \rho \frac{\partial \Phi}{\partial \rho} + \frac{\partial^2 \Phi}{\partial \phi^2} = 0.$$

Upon substituting the assumed solution, and dividing by $\Phi = RP$, we obtain

$$\left(\frac{\rho^2}{R} \frac{\partial^2 R}{\partial \rho^2} + \frac{\rho}{R} \frac{\partial R}{\partial \rho}\right) + \frac{1}{P} \frac{\partial^2 P}{\partial \phi^2} = 0.$$

The sum of the first two terms is a function of ρ only, while the third is a function of ϕ only. That is, we have $f(\rho) + g(\phi) = 0$, and the only possible nontrivial solution is that for which $f(\rho)$ is a constant (possibly complex) and $g(\phi)$ is the negative of this constant. We choose $f(\rho) = n^2$ and $g(\phi) = -n^2$. Thus, we have separated the variables and

$$\frac{\rho^2}{R} \frac{\partial^2 R}{\partial \rho^2} + \frac{\rho}{R} \frac{\partial R}{\partial \rho} = n^2, \qquad \frac{1}{P} \frac{\partial^2 P}{\partial \phi^2} = -n^2$$

or

$$\rho^2 \frac{d^2 R}{d\rho^2} + \rho \frac{dR}{d\rho} - n^2 R = 0, \qquad \frac{d^2 P}{d\phi^2} + n^2 P = 0.$$

The second *ordinary* differential equation is the equation of harmonic motion, and its solution consists of the sum of harmonic functions.[8] The form of the first equation suggests that we try a solution of the form $R = \rho^\alpha$. This trial leads quickly to the two solutions ρ^n, ρ^{-n} $(n > 0)$. Thus, we obtain the general solutions

---

[8] $\cos n\phi$, $\sin n\phi$, $e^{jn\phi}$, or $e^{-jn\phi}$.

$$R(\rho) = A\rho^n + B\rho^{-n}, \qquad P(\phi) = C\cos n\phi + D\sin n\phi$$

and

$$\Phi(\rho,\phi) = (A\rho^n + B\rho^{-n})(C\cos n\phi + D\sin n\phi).$$

Since the z axis ($\rho = 0$) must be included in the solution, and since $\rho^{-n} \to \infty$ for $\rho \to 0$, we require that $B = 0$. The potential must vanish for $\phi = 0, \pi$, and since $\cos n\phi$ cannot satisfy this boundary condition, we require that $C = 0$. Thus, with $AD \equiv F_n$, we get

$$\Phi(\rho,\phi) = F_n \rho^n \sin n\phi$$

and the potential will vanish for $\phi = 0$ or π if $n = 1, 2, 3, \ldots$. The potential function we have found will *not* satisfy the last boundary condition: $\Phi(a,\phi) = V_0$, but we are dealing with a linear system, and a more general solution than that above is the superposition

$$\Phi(\rho,\phi) = \sum_{n=1}^{\infty} F_n \rho^n \sin n\phi.$$

The last boundary condition gives

$$\Phi(a,\phi) = V_0 = \sum_{n=1}^{\infty} (F_n a^n) \sin n\phi,$$

or, with $b_n \equiv F_n a^n$; we get

$$V_0 = \sum_{n=1}^{\infty} b_n \sin n\phi, \qquad 0 < \phi < \pi.$$

The reader may recognize that the series has the form of a Fourier trigonometric series, and this series is capable of representing the periodic *odd* square wave with zero average value and period 2π (that has the value V_0 in $0 < \phi < \pi$), as shown in Figure 2.14. The Euler formula for calculating b_n for such a series is well known

$$b_n = \frac{1}{\pi} \int_{-\pi}^{0} (-V_0) \sin n\phi \, d\phi + \frac{1}{\pi} \int_{0}^{\pi} V_0 \sin n\phi \, d\phi,$$

$$b_n = 4V_0/(n\pi), \quad n \text{ odd},$$

or, $F_n = 4V_0/(n\pi a^n)$. The final solution is

$$\Phi(\rho,\phi) = \frac{4V_0}{\pi} \sum_{\substack{n=1 \\ n\text{ odd}}}^{\infty} \left(\frac{\rho}{a}\right)^n \frac{\sin n\phi}{n}$$

2.7 IMAGE THEORY

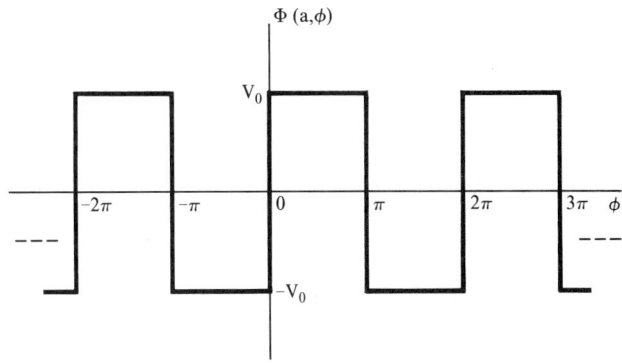

Figure 2.14. Odd periodic square wave with $\Phi(a, \phi) = V_0$ for $0 < \phi < \pi$.

Suppose that a graduate student taking a course in electromagnetics has found what he or she claims to be a *closed-form* solution to this problem

$$\Phi(\rho, \phi) = \frac{2V_0}{\pi} \tan^{-1}\left(\frac{2(\rho/a) \sin \phi}{1 - (\rho/a)^2}\right).$$

This would be highly advantageous since the closed form is much easier to work with numerically speaking. Suppose, for example, that we were required to find the positive surface charge density on the upper (semicircular) conductor of Figure 2.13. This requires that we evaluate the normal (inward) component of the flux density at $\rho = a$; that is, $\rho_s = -D_\rho|_{\rho=a} = -\varepsilon E_\rho|_{\rho=a} = \varepsilon(\nabla\Phi)_\rho|_{\rho=a} = \varepsilon[\partial\Phi/\partial\rho]_{\rho=a}$, and the latter form for Φ is easier to treat than the former. How do we know that the closed form for Φ is truly a solution? It is easy to see that it satisfies the two boundary conditions: $\Phi(a, \phi) = V_0$ and $\Phi(\rho, \phi) = 0$, $\phi = 0$ or π. It also satisfies Laplace's equation, $\nabla^2\Phi = 0$ (above), when substituted. According to the *uniqueness theorem* it too is a solution, and we have, as a byproduct, established a mathematical identity when the two solutions are equated. It is worth mentioning that the closed-form solution can be obtained by the method of *conformal transformations*, but we shall not treat the method in this text. ∎

2.7 IMAGE THEORY

Sometimes a problem can be solved by a technique called the *method of images* or *image theory*, which amounts to finding an equivalent problem whose solution we already know. Example 2.12 serves best to show how the method works.

■ Example 2.12

A point charge located at $(0, 0, d/2)$ above an infinite grounded conducting plane (located at $z = 0$) is shown in Figure 2.15(a). What is the potential at any point P $(z > 0)$? That is, what is $\Phi(x, y, z)$? Note that we require that the potential

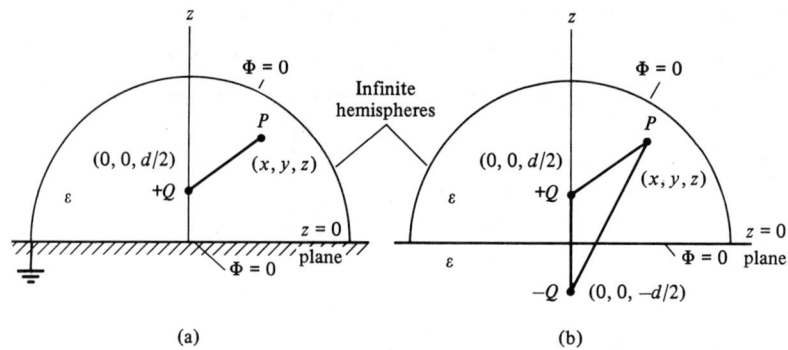

Figure 2.15. (a) Original problem (point charge above an infinite grounded plane). (b) Equivalent problem ($z > 0$) with the image of the original charge and no ground plane.

vanish at infinity on the infinite hemisphere. Now consider the electrostatic dipole located in the same medium (ε) with no ground plane as shown in Figure 2.15(b). It is easy to show by means of superposition that the potential at P for this dipole is

$$\Phi(x,y,z) = \frac{Q}{4\pi\varepsilon}\left(\frac{1}{[x^2+y^2+(z-d/2)^2]^{1/2}}\right.$$

$$\left. - \frac{1}{[x^2+y^2+(z+d/2)^2]^{1/2}}\right) \qquad (2.66)$$

For any point in the $z = 0$ plane, Equation (2.66) gives $\Phi(x,y,0) \equiv 0$. Equation (2.66) also shows that the potential on the infinite hemisphere vanishes. Comparing Figures 2.15(a) and 2.15(b), we see that for $z > 0$ the boundary conditions are the same, the media are the same, and the sources are the same. Therefore, according to the uniqueness theorem, the solutions must be the same for $z > 0$. That is, the original problem with only one charge and a ground plane can be replaced by another problem with the same charge and its (negative) *image* and no ground plane whose solution is given by Equation (2.66). Thus, for $z > 0$ (only), Equation (2.66) represents the solution for the potential in Figure 2.15(a). ∎

What is the potential for $z < 0$ in Figure 2.15(a)? What is the surface charge density at $(x,y,0+)$ in Figure 2.15(a)?

The *general image problem* for a charge configuration above an infinite grounded plane is shown in Figure 2.16. This result follows directly from Example 2.12 and the fact that any charge distribution can be resolved into point charges.

If the point charge in Figure 2.15(a) is replaced with a filamentary line charge density lying parallel to the $z = 0$ plane, then the equivalent problem for $z > 0$ is that of the same line charge density, no ground plane, and the image line charge density (negative) lying symmetrically below (parallel to) the original line charge density so that the combination produces zero potential for $z = 0$. Thus, the problem

2.8 ELECTRIC CURRENT AND CONSERVATION OF CHARGE

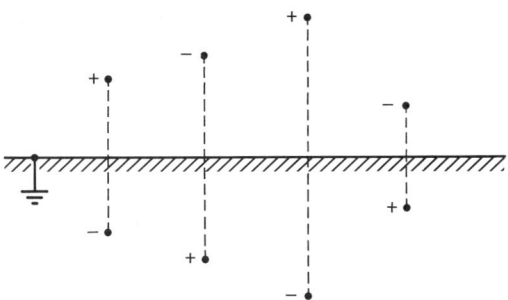

Figure 2.16. Several point charges above an infinite ground plane and their images.

of a single wire parallel to a ground plane is equivalent to that of the balanced two-wire line. Other examples of the use of image theory are found in the Problems at the end of this chapter.

2.8 ELECTRIC CURRENT AND CONSERVATION OF CHARGE

Electric current consists of charge motion due to an applied electric field whenever the charges are not constrained so that they cannot move. In circuit theory we frequently speak of the current in a branch or through an element of the circuit. The current is measured in amperes or coulombs per second, indicating that it is the rate of movement of charge passing some reference point. In field theory we are usually interested in point forms, and thus the concept of current density (which may vary over the cross section of a conductor) is more useful. Charge, in the form of a volume density ρ_v in uniform motion with velocity **u** constitutes a vector current density **J** according to

$$\boxed{\mathbf{J} = \rho_v \mathbf{u}} \quad (\text{A/m}^2). \tag{2.67}$$

Note that so far as Equation (2.67) is concerned, we have said nothing about conductors. In other words, a current density, as described by Equation (2.67) can exist in a region free of conductors, although we usually think of conductors when we speak of electric current. A current density, such as that between the cathode and plate of a vacuum diode, or that between the cathode and screen of a cathode ray tube in an oscilloscope, may be called a *convection current density*.

Let us now consider the various mathematical and physical distributions that an electric current density may assume in special cases. We found that line charge densities ρ_l and surface charge densities ρ_s were special cases of the general volume charge density ρ_v. If the differential elements of charge making up these various distributions are moving with velocity **u**, and the distributions are otherwise unchanged, then we have line (filamentary) currents, I (in amperes), surface current densities $\mathbf{J}_s$ (in amperes per meter), and the general current density **J** (in amperes per square meter) of Equation (2.67). It will be helpful here, and particularly so in the following sections, if we utilize the idea of a *differential current element*. We will find that these current elements act as sources for a new field (force) quantity. Consider a filamentary current I (a moving thread of charge with negligible cross section) which exists over a differential length dl. Current I in amperes, is not a vector (although we commonly speak of the direction of the current in an electric circuit), but the differential current element is a vector, being given by $I\,d\mathbf{l}$. The vector $d\mathbf{l}$ gives the direction of the current. Note that $I\,d\mathbf{l}$ corresponds to the differential line charge element, $\rho_l\,dl$, where the charges are in motion. If the current is in the form of a surface current density, then the differential element is $\mathbf{J}_s\,ds$. If the current is in the form of a general current density $\mathbf{J}$, then the differential element is $\mathbf{J}\,dv$. We may replace ρ_v with dQ/dv in Equation (2.67) so that $\mathbf{J} = \mathbf{u}\,dQ/dv$, or $\mathbf{J}\,dv = \mathbf{u}\,dQ$. The *current elements* then are

$$I\,d\mathbf{l}, \quad \mathbf{J}_s\,ds, \quad \mathbf{J}\,dv, \quad \mathbf{u}\,dQ \quad (\text{A}\cdot\text{m}). \tag{2.68}$$

These current elements, elementary sources, or basic "building blocks," will be useful to us in the material in future sections.

The total current flowing in a filamentary path is obviously I, and if we desire the *total* current in a nonfilamentary path, we merely need to integrate the normal component of the current density $\mathbf{J}$ over the given cross section. It is the *flux* of the current density. As shown in Figure 2.17,

$$\boxed{I = \iint_s \mathbf{J}\cdot d\mathbf{s}} \quad (\text{A}) \tag{2.69}$$

is the definition of the total current. If the current is in the form of a surface current density $\mathbf{J}_s$ (A/m), it is a current per unit width, and the total current can be found by integrating over the width. If, for example, we have a z-directed surface current density on the cylinder $\rho = a$, then $\mathbf{J}_s = \mathbf{a}_z J_{sz}$:

$$I = \int_0^{2\pi} J_{sz} a\,d\phi = a\int_0^{2\pi} J_{sz}\,d\phi;$$

and if, in addition, J_{sz} is independent of ϕ, then

$$I = 2\pi a J_{sz} \quad (\text{A}).$$

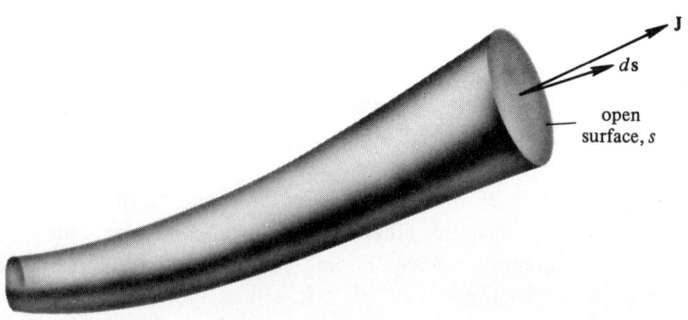

Figure 2.17. General current density $\mathbf{J}$ over a cross section (open surface) s.

2.8 ELECTRIC CURRENT AND CONSERVATION OF CHARGE

The principle of conservation of charge states that charge can be neither created nor destroyed. *Equal* amounts of *positive* and *negative* charge may, of course, be simultaneously created and destroyed. Consider a closed surface s. Conservation of charge requires that the net, steady (non-time-varying) current passing through this surface be zero. Thus,

$$\oint_s \mathbf{J} \cdot d\mathbf{s} = 0 = \iiint_{\text{vol}} \nabla \cdot \mathbf{J}\, dv \qquad (2.70)$$

by means of the divergence theorem. Since the volume in question is completely arbitrary, we have the conservation of charge equation for *steady* currents

$$\boxed{\nabla \cdot \mathbf{J} = 0} \quad (\textit{conservation of charge}), \qquad (2.71)$$

which may be recognized as the point form of Kirchhoff's current law of circuit theory. This result may also be called the *continuity equation* for *steady* currents. We will find that the right-hand side of Equation (2.71) is not zero for currents that vary with time! Equation (2.71) simply states that the steady current diverging from a small volume per unit volume is zero. This statement follows from the definition of the divergence as a limit.

A conductor is a material that allows charge (in the form of *valence* or *conduction* electrons) motion under the influence of an applied electric field. The force involved is easy to calculate (from the definition of the electric field intensity): $\mathbf{F} = Q\mathbf{E} = -e\mathbf{E}$ [$e = 1.602 \times 10^{-19}$ (C) for an electron]. An electron in a vacuum would accelerate unimpeded, but an electron in a conductor cannot do this. Instead, an electron attains a constant *average* velocity called the *drift velocity* $\mathbf{u}_d$. The ability of an electron to traverse the obstacle course created by the thermally excited crystalline lattice structure of the conductor is called the *mobility*, μ_e (m²/V · s) of the electron in the material. Thus, due to collisions and the resulting loss of energy, an electron reaches this average velocity. The equation that relates drift velocity to mobility and electric field intensity is $\mathbf{u}_d = -\mu_e \mathbf{E}$. The drift velocity for most conductors is surprisingly small for normal temperatures.[9]

Current density $\mathbf{J}$ in a conductor is thus equal to the density of valence electrons $\rho_{ve} < 0$, times their average velocity $\mathbf{u}_d = -\mu_e \mathbf{E}$, so that

$$\mathbf{J} = -\rho_{ve}\mu_e \mathbf{E}.$$

It is more common to relate $\mathbf{J}$ to $\mathbf{E}$ through the parameter *conductivity* σ (℧/m), where $\sigma = -\rho_{ve}\mu_e > 0$. Thus, we have Ohm's law in point form

$$\boxed{\mathbf{J} = \sigma \mathbf{E}} \quad \textit{Ohm's law.} \qquad (2.72)$$

Conductivity for common conductors is listed in Appendix B for room temperature.

---

[9] A few centimeters per second or less.

Temperature must affect conductivity, for a higher temperature gives rise to a more excited crystalline lattice structure and thus a lower drift velocity. On the other hand, for some conductors (aluminum, for example) the conductivity approaches infinity as 0 K is approached. Other materials have recently been discovered that exhibit (essentially) infinite conductivity at higher temperatures. Such materials are called *superconductors*, and offer much promise in improving electrical efficiency.

In order to determine the boundary conditions on **J** that must exist at the interface between different conducting media, we merely return to Figure 2.10 and imagine that the dielectrics are conductors. In the dielectric case, the equation $\nabla \cdot \mathbf{D} = 0$ ($\rho_s = 0$) led to $D_{n1} = D_{n2}$ [Equation (2.55)], therefore conservation of charge, $\nabla \cdot \mathbf{J} = 0$, leads to

$$\boxed{J_{n1} = J_{n2}}. \tag{2.73}$$

Since $\mathbf{J} = \sigma \mathbf{E}$, and since both **J** and **E** remain the same macroscopically at any instant of time, Equation (2.53) leads to

$$\boxed{J_{t1}/\sigma_1 = J_{t2}/\sigma_2}. \tag{2.74}$$

2.9 BIOT–SAVART LAW AND MAGNETIC FIELD INTENSITY

A calculation can be performed for the force on a point charge Q that is detected by a fixed observer when point charges Q_1 and Q are in uniform motion with velocities $\mathbf{u}_1$ and $\mathbf{u}$, respectively.[10] The force given by Coulomb's law appears as one term, and a new force term due entirely to the charge motion (current) also appears. At the present time we are interested in this second term, and we intend to deal with it in a manner similar to that used in Section 2.2 for Coulomb's law. For this reason we would like to present the force in differential form, being due to differential current elements. The geometry is shown in Figure 2.18 and the differential force on dQ is

$$d(d\mathbf{F}) = \frac{\mu_0}{4\pi} \frac{\mathbf{u}\, dQ \times (\mathbf{u}_1\, dQ_1 \times \mathbf{a}_R)}{R^2} \quad \text{(N)} \tag{2.75}$$

where the constant μ_0 is

$$\mu_0 = 4\pi \times 10^{-7} \quad \text{(H/m)} \tag{2.76}$$

and is called the *permeability* of free space, and where we have $\mathbf{a}_R/R^2 = (\mathbf{r} - \mathbf{r}')/|\mathbf{r} - \mathbf{r}'|^3$ as before.

Note that, except for the fact that the right-hand side of Equation (2.75) is a vector triple product, it is very similar in form to Coulomb's law [Equation (2.7)]. The magnitude of the force varies inversely with the square of the separation.

---

[10] See Neff, 1981 or Shedd, 1954 in the references at the end of the chapter.

2.9 BIOT–SAVART LAW AND MAGNETIC FIELD INTENSITY

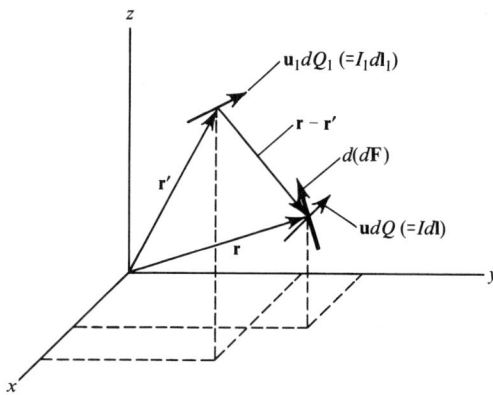

Figure 2.18. Geometry for two current elements for finding $d(d\mathbf{F})$ on $\mathbf{u}\,dQ$ (or $I\,d\mathbf{l}$) due to $\mathbf{u}_1\,dQ_1$ (or $I_1\,d\mathbf{l}_1$).

In order to relate Equation (2.75) to the Biot–Savart law as it is normally stated, we would like to change the current elements from $\mathbf{u}\,dQ$ to $I\,d\mathbf{l}$ [see (2.68)]:

$$d(d\mathbf{F}) = \frac{\mu_0}{4\pi} \frac{I\,d\mathbf{l} \times (I_1\,d\mathbf{l}_1 \times \mathbf{a}_R)}{R^2} \tag{2.77}$$

or

$$d(d\mathbf{F}) = I\,d\mathbf{l} \times \left(\frac{\mu_0}{4\pi} \frac{I_1 d\mathbf{l}_1 \times \mathbf{a}_R}{R^2} \right). \tag{2.78}$$

Note that this result cannot be verified experimentally [in the form of Equation (2.78)] because the differential current elements cannot be duplicated in practice. That is, they violate Equation (2.71) (continuity of current). On the other hand, these current elements can be (separately) part of closed loops, and (twofold) integration will give a force that can be verified experimentally.

Following the procedure that was used in Section 2.2 to define the electric field intensity (as the force per unit charge), we identify the term in parentheses in Equation (2.78) as the *force per unit current element*, and call it the differential *magnetic flux density* $d\mathbf{B}$. Thus,

$$d(d\mathbf{F}) = I\,d\mathbf{l} \times d\mathbf{B} \tag{2.79}$$

and

$$\boxed{d\mathbf{B} = \frac{\mu_0}{4\pi} \frac{I_1 d\mathbf{l}_1 \times \mathbf{a}_R}{R^2}} \quad \text{(Wb/m}^2 \text{ or tesla)}. \tag{2.80}$$

Equation (2.79) can also be written

$$d\mathbf{F} = I\,d\mathbf{l} \times \mathbf{B}, \tag{2.81}$$

which gives the differential force on a current element due to the magnetic flux density that is being produced by some other source.

The magnetic flux density **B** is related to the *magnetic field intensity* **H** (in a vacuum) by

$$\boxed{\mathbf{B} = \mu_0 \mathbf{H}} \quad (\text{Wb/m}^2) \tag{2.82}$$

$$\mathbf{H} = \mathbf{B}/\mu_0 \quad (\text{A/m}) \tag{2.83}$$

The Biot–Savart law is usually stated in terms of **H**, and using Equation (2.80), we get

$$d\mathbf{H} = \frac{1}{4\pi} \frac{I_1 d\mathbf{l}_1 \times \mathbf{a}_R}{R^2} \quad (\text{A/m}) \tag{2.84}$$

or, in our standard form, appropriate to Figure 2.19, we get

$$\boxed{d\mathbf{H}(\mathbf{r}) = \frac{1}{4\pi} \frac{I \, d\mathbf{l}(\mathbf{r}') \times \mathbf{a}_R}{R^2}} \quad (\textit{Biot–Savart law}). \tag{2.85}$$

The Biot–Savart law states that the differential magnetic field intensity is proportional to the vector product of the vector differential current element and the unit vector directed from the current element toward the field point. It is inversely proportional to the square of the distance from the source point to the field point.

Any of the current elements listed in the set (2.68) can be used in Equation (2.85). Therefore, we get

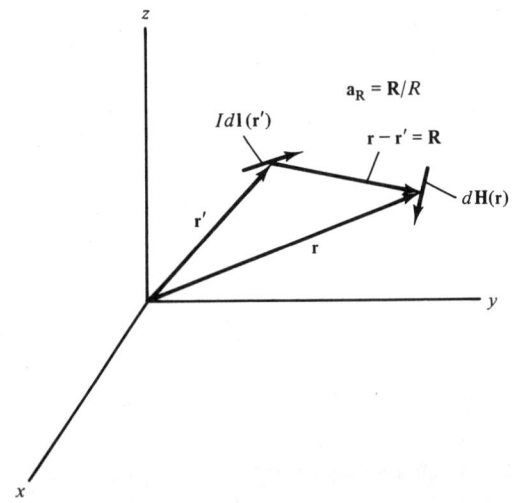

Figure 2.19. Geometry for the Biot–Savart law.

2.9 BIOT–SAVART LAW AND MAGNETIC FIELD INTENSITY

$$d\mathbf{H}(\mathbf{r}) = \frac{1}{4\pi} \frac{u \, dQ(\mathbf{r}') \times \mathbf{a}_R}{R^2} \qquad (2.86)$$

or

$$d\mathbf{H}(\mathbf{r}) = \frac{1}{4\pi} \frac{\mathbf{J}(\mathbf{r}')dv \times \mathbf{a}_r}{R^2} = \frac{1}{4\pi} \frac{\mathbf{J}(\mathbf{r}') \times \mathbf{a}_R}{R^2} dv. \qquad (2.87)$$

As we have seen before, superposition leads from Equation (2.87) to the general result

$$\boxed{\mathbf{H}(\mathbf{r}) = \frac{1}{4\pi} \iiint_{vol'} \frac{\mathbf{J}(\mathbf{r}') \times \mathbf{a}_R}{R^2} dv'} \quad \text{(A/m)}. \qquad (2.88)$$

The integration is throughout the volume that bears the current density **J** as in Figure 2.20. If the current source is a surface current density $\mathbf{J}_s$, then we get

$$\mathbf{H}(\mathbf{r}) = \frac{1}{4\pi} \iint_{s'} \mathbf{J}_s(\mathbf{r}') \times \frac{\mathbf{a}_R}{R^2} ds'; \qquad (2.89)$$

and if the current source is a filamentary current I, then we get

$$\mathbf{H}(\mathbf{r}) = \frac{1}{4\pi} \int_{l'} I \, d\mathbf{l}(\mathbf{r}') \times \frac{\mathbf{a}_R}{R^2}. \qquad (2.90)$$

Equations (2.88), (2.89), and (2.90) are superposition integrals, and are extensions of the Biot–Savart law, Equation (2.85). In order to demonstrate that this is the case, we consider first an example that is very similar to Example 2.4.

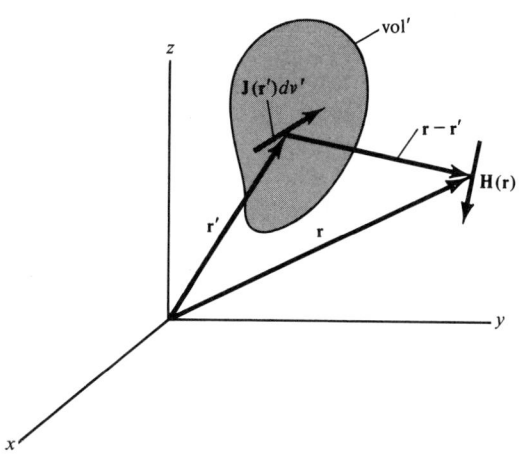

Figure 2.20. Geometry for the superposition integral for calculating the magnetic field intensity due to a general current density.

Example 2.13

Consider a long filamentary current I lying symmetrically on the z axis and a field point lying in the $z=0$ plane. The magnetic field intensity, under these symmetrical conditions, is certainly independent of azimuth angle. Thus, the field point can be placed on the x axis without loss of generality. Since we want to consider a numerical example first, we next resolve the current into a chain of filamentary current elements $I\Delta z \mathbf{a}_z$. The spacing is chosen to be uniform to simplify matters. In order that an observer at the field point $(x,0,0)$ will see something resembling a continuous filamentary current, the spacing is small compared with x. We will also choose the total length of the current filament to be large compared with x. It should be pointed out that these conditions are identical to those for Example 2.4 (the long line charge density on the z axis).

Using Equation (2.84) and Figure 2.21 the field produced by the nth current element is

$$\Delta \mathbf{H}_n = \frac{I\Delta z \mathbf{a}_z}{4\pi} \times \frac{\mathbf{a}_x x - \mathbf{a}_z n\Delta z}{[x^2 + (n\Delta z)^2]^{3/2}} = \frac{I\Delta z}{4\pi} \frac{x\mathbf{a}_y}{[x^2 + (n\Delta z)^2]^{3/2}},$$

while the field of the $-n$th current element is

$$\Delta \mathbf{H}_{-n} = \frac{I\Delta z \mathbf{a}_z}{4\pi} \times \frac{\mathbf{a}_x x + \mathbf{a}_z n\Delta z}{[x^2 + (n\Delta z)^2]^{3/2}} = \frac{I\Delta z}{4\pi} \frac{x\mathbf{a}_y}{[x^2 + (n\Delta z)^2]^{3/2}}.$$

The field of the element at the origin is

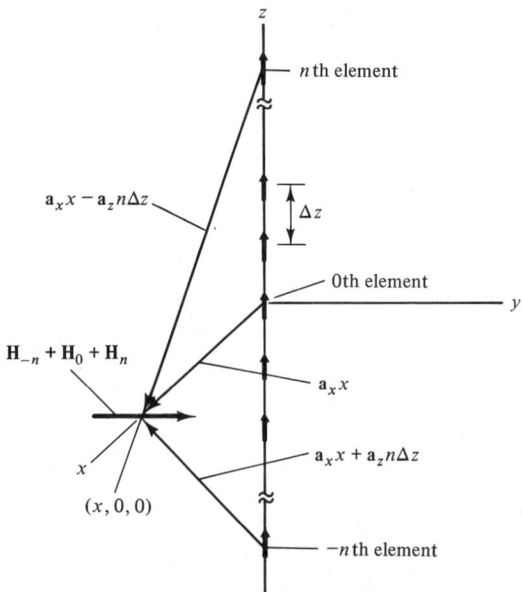

Figure 2.21. Geometry for calculating the magnetic field intensity due to a collinear array of $2N+1$ current elements located symmetrically on the z axis.

2.9 BIOT–SAVART LAW AND MAGNETIC FIELD INTENSITY

$$\Delta \mathbf{H}_0 = \mathbf{a}_y \frac{I \Delta z}{4\pi x^2}.$$

By superposition the total field for $2N + 1$ current elements is given by a y component only:

$$H_y(x,0,0) = \frac{I \Delta z x}{4\pi} \left(\frac{1}{x^3} + 2 \sum_{n=1}^{N} \frac{1}{[x^2 + (n\Delta z)^2]^{3/2}} \right).$$

Note that this result is essentially identical to $E_x(x,0,0)$ of Example 2.4.

Based on the conditions enumerated earlier, let $\Delta z = 0.1$ m, $N\Delta z = 10$ m, $N = 100$, and $x = 1$ m. With these, we get

$$H_y(1,0,0) = \frac{I}{4\pi}(0.1)\left[1 + 2 \sum_{n=1}^{100} \frac{1}{(1 + 0.01n^2)^{3/2}} \right]$$

or (from Example 2.4)

$$H_y(1,0,0) = 0.995 I/(2\pi) \quad (A/m).$$

In the same way, for $x = 2$ m,

$$H_y(2,0,0) = 0.491 I/(2\pi) \quad (A/m).$$

Note the similarity between these results and those for the line charge in Example 2.4, but also note particularly that this magnetic field intensity is in the y direction rather than in the radial direction. If the field point is moved off the x axis, but kept in the $z = 0$ plane, the y component of **H** becomes the ϕ component and x becomes ρ. The magnetic field lines apparently close on themselves, indicating that the magnetic field intensity (at least, in this example) has no divergence ($\nabla \cdot \mathbf{H} = 0$)!

One last point should be emphasized. An electric current cannot suddenly start on the negative z axis and then suddenly stop on the positive z axis as demanded in this example. ∎

■ Example 2.14

Consider the infinitely long filamentary wire lying on the z axis as shown in Figure 2.22. We must keep in mind that there must be a return current somewhere. In the present example we consider the return current to be so far removed that it will not affect our answer. Since the filament is of infinite extent, our answer will not be affected if the point at which we choose to evaluate H is placed in the $z = 0$ plane. Symmetry considerations tell us that the field must also be independent of the azimuth angle ϕ, as in Example 2.13. Using Equation (2.90), we have (in cylindrical coordinates)

2 ELECTROSTATIC AND MAGNETOSTATIC FIELDS

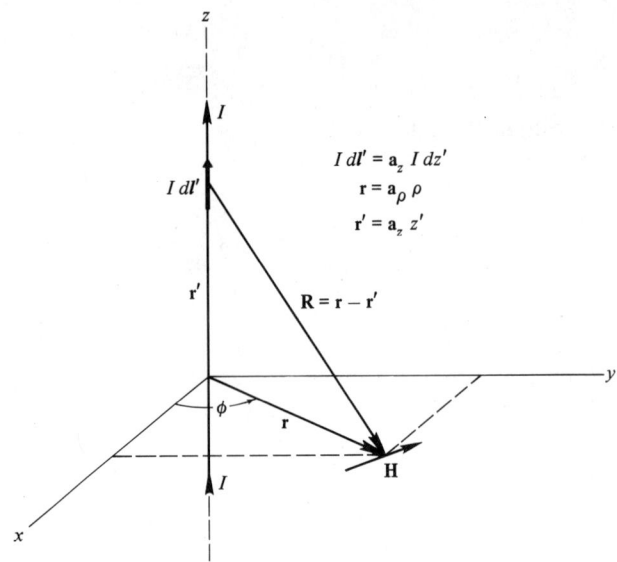

Figure 2.22. Geometry for finding the magnetic field intensity due to an infinite filamentary current on the z axis.

$$\mathbf{H} = \int_{-\infty}^{\infty} \frac{I\,dz'\mathbf{a}_z}{4\pi} \times \frac{\mathbf{a}_\rho\rho - \mathbf{a}_z z'}{[\rho^2 + (z')^2]^{3/2}}$$

or

$$\mathbf{H} = \frac{I}{4\pi}\int_{-\infty}^{\infty} \frac{\rho\mathbf{a}_\phi}{[\rho^2 + (z')^2]^{3/2}}\,dz',$$

where the only variable is z', in which case ρ and $\mathbf{a}_\phi$ are both constant, and so we have

$$\mathbf{H} = \frac{I\rho\mathbf{a}_\phi}{4\pi}\int_{-\infty}^{\infty} \frac{dz'}{[\rho^2 + (z')^2]^{3/2}}\,dz',$$

or simply

$$H_\phi(\rho) = \frac{I\rho}{4\pi}\int_{-\infty}^{\infty} \frac{dz'}{[\rho^2 + (z')^2]^{3/2}}.$$

The required integral can be found in the tables, and so we have

$$H_\phi(\rho) = \frac{I}{2\pi\rho} \quad \text{(A/m)}. \tag{2.91}$$

This result should not be at all surprising.

2.9 BIOT–SAVART LAW AND MAGNETIC FIELD INTENSITY

If the field point is on the x axis ($\phi = 0$), then ρ becomes x and H_ϕ becomes H_y:

$$H_y = \frac{I}{2\pi x}$$

and if, in addition, $x = 1$ m, we get

$$H_y(1,0,0) = I/(2\pi) \quad \text{(A/m)};$$

or, for $x = 2$ m, we get

$$H_y(2,0,0) = 0.5I/(2\pi) \quad \text{(A/m)}.$$

Compare these results with those in Example 2.13.

Equation (2.91) bears a striking resemblance to that for the electrostatic field of an infinite line charge which is repeated here for convenience.

$$E_\rho(\rho) = \rho_l/2\pi\varepsilon\rho \quad \text{(V/m)}.$$

In amplitude, these equations behave the same way (inverse distance). The electric field begins with the line charge and is radially directed. The magnetic field, however, is azimuthally directed. ∎

The reader is encouraged to show that the uniform surface current density $\mathbf{J}_s = J_{s0}\mathbf{a}_x$ (A/m) in the $z = 0$ plane produces the field

$$\mathbf{H} = \begin{cases} -\mathbf{a}_y J_{s0}/2, & z > 0; \\ +\mathbf{a}_y J_{s0}/2, & z < 0. \end{cases} \tag{2.92}$$

Compare this result to Equation (2.17).

The magnetic flux Ψ_m [in webers (Wb)] is found in the same way that we found electric flux: by integrating the normal component of the flux density over the surface of interest. The flux through the surface is

$$\Psi_m = \iint_s \mathbf{B} \cdot d\mathbf{s} \quad \text{(Wb)}. \tag{2.93}$$

In Example 2.14 the magnetic field lines close on themselves [Equation (2.91)]; that is, they do not diverge. Compared to electric lines, this behavior is easy to explain. At the present time, no isolated magnetic charges have been found in nature, and so there are no sources or sinks for magnetic field lines to begin or terminate on. Thus, for a *closed* surface,

$$\Psi_m = \oiint_s \mathbf{B} \cdot d\mathbf{s} = 0,$$

and with the divergence theorem, we get

$$\oint_s \mathbf{B} \cdot d\mathbf{s} = \iiint_{\text{vol}} \nabla \cdot \mathbf{B}\, dv = 0.$$

Since the volume and its surface are arbitrary, we obtain

$$\boxed{\nabla \cdot \mathbf{B} = 0}. \qquad (2.94)$$

2.10 AMPERE'S CIRCUITAL LAW

Ampere's law (or perhaps more correctly, Ampere's circuital law) simply states that

> *The circulation of the magnetic field intensity equals the net electric current enclosed by the path.*

The path is completely arbitrary. The direction of the current is that of an advancing right-hand threaded screw turned in the direction in which the closed path is traversed (right-hand rule). Thus, as seen in Figure 2.23, Ampere's law is simply

$$\boxed{\oint \mathbf{H} \cdot d\mathbf{l} = I_{\text{enc}} = \iint_s \mathbf{J} \cdot d\mathbf{s}} \quad \text{(Ampere's law)}. \qquad (2.95)$$

Ampere's law can be derived from the Biot–Savart law, but this derivation is not essential at this point in the development. It is worth mentioning that the "clamp-on" ammeter, used to measure ac current, is based on the principle behind Ampere's law. It measures the current in a wire or a bundle of wires when the clamp is inserted *around* the wires regardless of the orientation of the clamp thereafter.

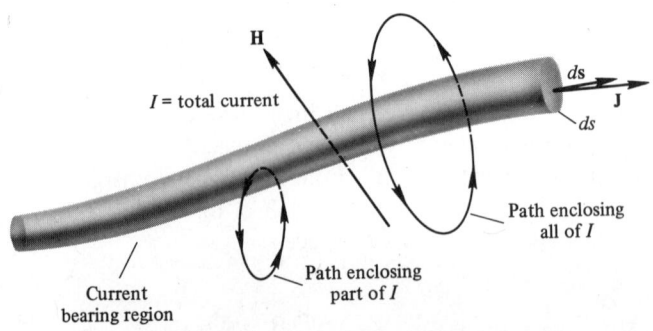

Figure 2.23. Direction of integration and current for $\oint \mathbf{H} \cdot d\mathbf{l} = I_{\text{enc}}$ (right-hand rule).

2.10 AMPERE'S CIRCUITAL LAW

■ Example 2.15

Just as the Gauss law enabled us to find electrostatic fields from charges (when the proper symmetry was present), Ampere's law enables us to determine magnetic fields when certain symmetry conditions are met. As a first example let us consider the infinite filamentary current on the z axis. This problem was solved earlier by means of the Biot–Savart law. It is obvious from the symmetry in Figure 2.24 that *whatever* field is present must be independent of ϕ and z as far as its magnitude is concerned. The elementary form of the Biot–Savart law tells us that the magnetic field is ϕ directed. Hence, we know that $\mathbf{H} = \mathbf{a}_\phi H_\phi(\rho)$. In order to obtain any advantage from this method, we must choose as our path of integration one for which the magnitude of $\mathbf{H}$ is constant and to which $\mathbf{H}$ is either perpendicular or tangent. In Figure 2.24 we have chosen a path to which $\mathbf{H}$ is tangent and for which $|\mathbf{H}|$ is constant, namely, a circle of radius ρ. Hence, we get

$$I = \oint \mathbf{H} \cdot d\mathbf{l} = \int_0^{2\pi} \mathbf{a}_\phi H_\phi \cdot \mathbf{a}_\phi \rho \, d\phi;$$

or, since H_ϕ is constant, we get

$$I = \rho H_\phi \int_0^{2\pi} d\phi = 2\pi \rho H_\phi$$

or

$$H_\phi(\rho) = \frac{I}{2\pi\rho},$$

which is Equation (2.91) and was obtained with much less effort. ■

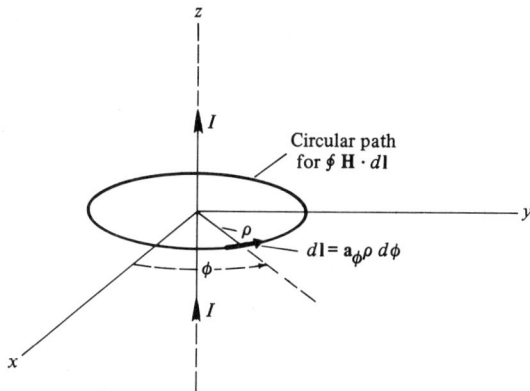

Figure 2.24. Path of integration around an infinite filamentary current for evaluating the circulation $\oint \mathbf{H} \cdot d\mathbf{l} = I$.

Example 2.16

Another example, of considerably more practical importance, is the coaxial cable. A total current of $+I$ A with uniform density exists in the center conductor (radius $=a$), and a total (return) current of $-I$ A with uniform density exists in the outer conductor (shield) whose inner radius is b and outer radius is c. This geometry is shown in Figure 2.25. The cable has infinite length so that end effects are eliminated.

Using exactly the same arguments as in Example 2.15, we can show that we only have $H_\phi = H_\phi(\rho)$. We also know from Example 2.15 that

$$H_\phi = \frac{I}{2\pi\rho}, \quad a < \rho < b.$$

For $\rho < a$ we choose a path of integration which is a circle of radius ρ. The current *enclosed* by such a path is, from Equation (2.69),

$$I_{enc} = \iint_S \mathbf{J} \cdot d\mathbf{s} = \int_0^\rho \int_0^{2\pi} J_z \mathbf{a}_z \cdot \mathbf{a}_z \rho \, d\rho \, d\phi$$

Now, for the inner conductor the uniform current density is

$$J_z = \frac{I}{\pi a^2} \quad (\text{A/m}^2)$$

Therefore, we have

$$I_{enc} = \frac{I}{\pi a^2} \int_0^\rho \int_0^{2\pi} \rho \, d\rho \, d\phi = I \frac{\rho^2}{a^2}$$

Applying Ampere's law, we get

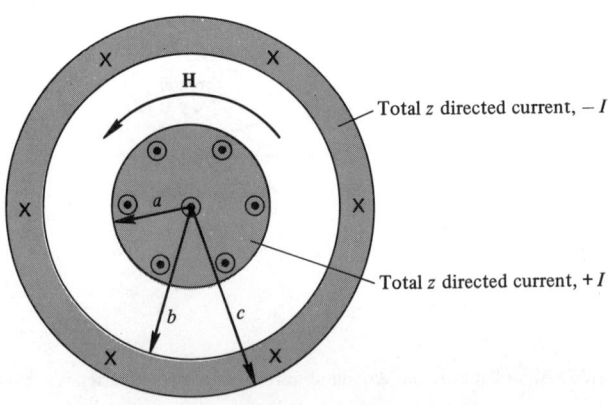

Figure 2.25. Coaxial cable with uniform, oppositely directed current densities.

2.10 AMPERE'S CIRCUITAL LAW

$$\int_0^{2\pi} H_\phi \rho\, d\phi = I_{enc} = I\frac{\rho^2}{a^2}, \qquad \rho < a;$$

or, since H_ϕ is constant, we get

$$H_\phi \rho \int_0^{2\pi} d\phi = I\frac{\rho^2}{a^2}$$

so

$$H_\phi = \frac{I\rho}{2\pi a^2}, \qquad \rho < a.$$

For $b < \rho < c$, we have

$$I_{enc} = I + \int_b^\rho \int_0^{2\pi} J_z \mathbf{a}_z \cdot \mathbf{a}_z \rho\, d\rho\, d\phi;$$

but for the outer conductor the uniform current density is

$$J_z = \frac{-I}{\pi(c^2 - b^2)}.$$

Then we have

$$I_{enc} = I - \frac{I}{\pi(c^2 - b^2)} \int_b^\rho \int_0^{2\pi} \rho\, d\rho\, d\phi$$

or

$$I_{enc} = I - I\frac{\rho^2 - b^2}{c^2 - b^2} = I\frac{c^2 - \rho^2}{c^2 - b^2}.$$

Ampere's law gives

$$H_\phi = \frac{I}{2\pi\rho}\frac{c^2 - \rho^2}{c^2 - b^2}, \qquad b < \rho < c.$$

For $\rho > c$, $H_\phi = 0$, so that the system is "shielded." A plot of H_ϕ against ρ shows that H_ϕ is a continuous function of ρ. ∎

Example 2.17

The ideal solenoid consists of the surface current density $\mathbf{J}_s = \mathbf{a}_\phi J_{s\phi}$ covering the infinite cylinder $\rho = a$. The field must be independent of z and ϕ, and can depend

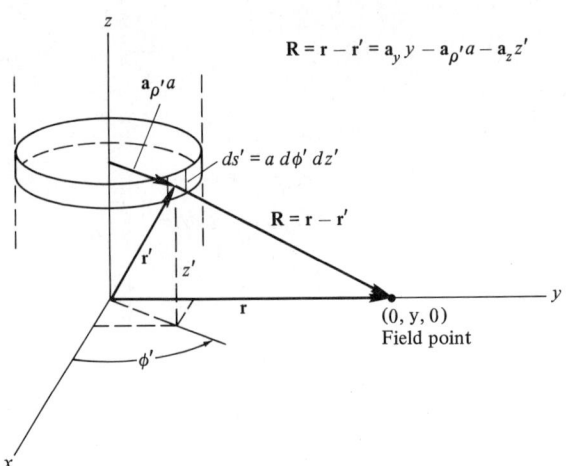

Figure 2.26. Biot-Savart law applied to the idealized solenoid.

at most on ρ. In order to show that only a z component of field is produced we use the Biot–Savart law with Figure 2.26.

The Biot–Savart law gives

$$\mathbf{H} = \frac{aJ_{s\phi}}{4\pi} \int_0^{2\pi} \int_{-\infty}^{\infty} \mathbf{a}_\phi \times \frac{\mathbf{r} - \mathbf{r}'}{|\mathbf{r} - \mathbf{r}'|^3} \, d\phi' \, dz',$$

where

$$\mathbf{r} - \mathbf{r}' = \mathbf{a}_y y - \mathbf{a}_\rho' a - \mathbf{a}_z z' = -\mathbf{a}_x a \cos\phi' - \mathbf{a}_y (a \sin\phi' - y) - \mathbf{a}_z z'$$

and

$$\mathbf{a}_{\phi'} \times (\mathbf{r} - \mathbf{r}') = -\mathbf{a}_x z' \cos\phi' - \mathbf{a}_y z' \sin\phi' - \mathbf{a}_z (y \sin\phi' - a)$$

since

$$\mathbf{a}_{\phi'} = -\mathbf{a}_x \sin\phi' + \mathbf{a}_y \cos\phi'.$$

Also, we have

$$|\mathbf{r} - \mathbf{r}'|^3 = [y^2 + a^2 - 2ya \sin\phi' + (z')^2]^{3/2}.$$

Thus, the integrands for H_x and H_y in the Biot–Savart integral above are odd functions of z', and are to be integrated over symmetric limits $(-\infty, +\infty)$. H_x and H_y are both identically zero, and $\mathbf{H} = \mathbf{a}_z H_z(\rho)$ at most.

It is easy, however, to show that H_z is constant. Consider Figure 2.27. For $\rho > a$ we see that $H_{z1} = H_{z2}$ in order that $\oint \mathbf{H} \cdot d\mathbf{l} = 0$. Therefore H_z is a constant for $\rho > a$. This constant must be zero, since H_z must vanish for large ρ. For the other path $\oint \mathbf{H} \cdot d\mathbf{l} = I_{\text{enc}}$, or $H_z l = J_{s\phi} l$ where l is the length of the path in the z direction. Remember that $H_z = 0$ for $\rho > a$. Thus, $H_z = J_{s\phi}$ for $\rho < a$, or

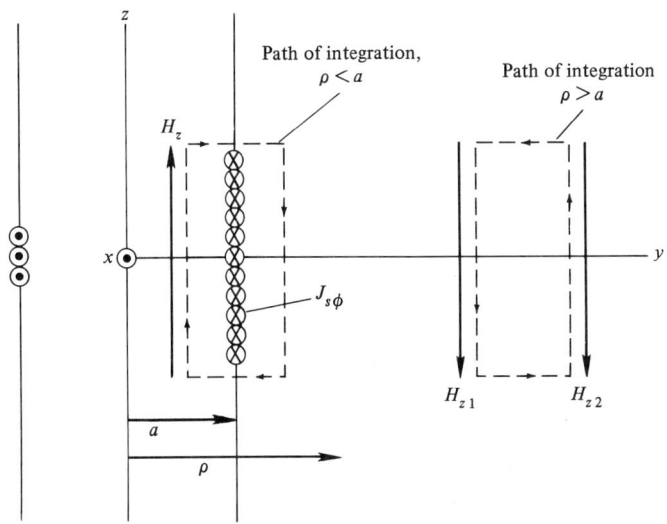

Figure 2.27. Ampere's law applied to the idealized solenoid.

$$\mathbf{H} = \begin{cases} \mathbf{a}_z J_{s\phi}, & \rho < a, \\ 0, & \rho > a, \end{cases} \quad \text{ideal solenoid} \qquad (2.96)$$

A real solenoid, which is simply a coil or helix of finite length, will have a z component of current (not just a ϕ component), since it has a nonzero pitch. Ampere's law applied to a circular path of radius $\rho > a$ gives $\oint \mathbf{H} \cdot d\mathbf{l} = I =$ coil current, and thus $H_\phi \neq 0$ for $\rho > a$. Thus a real solenoid will always produce leakage flux. ∎

2.11 AMPERE'S LAW IN POINT FORM (MAXWELL'S EQUATION)

Stokes' theorem [Equation (1.46)] may be applied to Ampere's law (2.95), giving

$$\oint_c \mathbf{H} \cdot d\mathbf{l} \equiv \iint_{s_1} \mathbf{\nabla} \times \mathbf{H} \cdot d\mathbf{s} = I, \qquad (2.97)$$

where s_1 is any one of the infinite number of possible *open* surfaces defined by the path of integration used for the line integral (see Figure 2.28). We also have Equation (2.75):

$$I = \iint_{s_2} \mathbf{J} \cdot d\mathbf{s} \qquad (2.98)$$

Combining Equations (2.97) and (2.98), we get

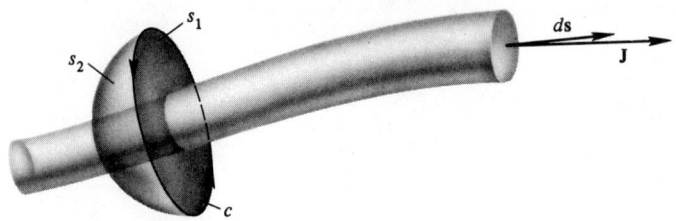

Figure 2.28. Geometry for deriving the point form of Ampere's law (Maxwell's equation).

$$\iint_{s_1} \nabla \times \mathbf{H} \cdot d\mathbf{s} = \iint_{s_2} \mathbf{J} \cdot d\mathbf{s}.$$

Since the surfaces s_1 and s_2 are arbitrary, they may be made identical, but still arbitrary. In this case, we have

$$\iint_s \nabla \times \mathbf{H} \cdot d\mathbf{s} = \iint_s \mathbf{J} \cdot d\mathbf{s},$$

where s is arbitrary. Since s is still arbitrary, we arrive at the result

$$\boxed{\nabla \times \mathbf{H} = \mathbf{J}} \quad \text{(Maxwell's curl } \mathbf{H} \text{ equation)}. \tag{2.99}$$

This is the point form of Ampere's law, and is also one of Maxwell's equations for *steady* currents. The modification to this equation for time dependent currents was probably Maxwell's most important contribution. Recall that the electrostatic field had no curl ($\nabla \times \mathbf{E} = 0$) and was conservative. The magnetic field is thus nonconservative.

Let us next take the divergence of both sides of Equation (2.99). We have

$$\nabla \cdot \nabla \times \mathbf{H} \equiv 0 = \nabla \cdot \mathbf{J}, \tag{2.100}$$

since the divergence of the curl of *any* vector is identically zero [see Equation (1.42)]. Equation (2.100) verifies Equation (2.71) which we "derived" on the basis of conservation of charge.

■ **Example 2.18**

Consider the coaxial cable of Example 2.16. The field was

$$H_\phi = \begin{cases} \dfrac{I\rho}{2\pi a^2}, & 0 \le \rho \le a, \\[4pt] \dfrac{I}{2\pi\rho}, & a \le \rho \le b, \\[4pt] \dfrac{I}{2\pi\rho}\dfrac{c^2 - \rho^2}{c^2 - b^2}, & b \le \rho \le c, \\[4pt] 0, & c \le \rho < \infty. \end{cases}$$

Maxwell's equation gives $\mathbf{J} = \nabla \times \mathbf{H}$ or

$$J_z = \frac{I}{\rho}\frac{\partial}{\partial \rho}(\rho H_\phi) = \begin{cases} \dfrac{I}{\pi}\dfrac{1}{a^2}, & 0 \le \rho \le a, \\ 0, & a \le \rho \le b, \\ -\dfrac{I}{\pi}\dfrac{1}{c^2 - b^2}, & b \le \rho \le c, \\ 0, & c \le \rho < \infty; \end{cases}$$

this is exactly the current density with which Example 2.16 began. ■

2.12 MAGNETIC SCALAR AND VECTOR POTENTIAL

In free space, where the current density is zero, Equation (2.99) gives

$$\nabla \times \mathbf{H} = 0;$$

and, as in the electrostatic case where $\nabla \times \mathbf{E} \equiv 0$ and $\mathbf{E} = -\nabla \Phi$, we can define a *magnetic scalar potential* Φ_m [in amperes (A)] such that

$$\mathbf{H} = -\nabla \Phi_m. \qquad (2.101)$$

The differential equation that Φ_m must satisfy is the Laplace equation

$$\nabla^2 \Phi_m = 0, \qquad (2.102)$$

so long as $\nabla \cdot \mathbf{H} = 0$. This magnetic scalar potential can be used like the electric scalar potential. Since $\oint \mathbf{H} \cdot d\mathbf{l} = I$, each time a complete lap around the current is made, the circulation increases by I. If the current that is enclosed is zero, then a *single-valued* Φ_m may be defined, but, in general, the path must be specified if Φ_m is to be single valued; that is,

$$\Phi_{m,fi} = -\int_{p_i}^{p_f} \mathbf{H} \cdot d\mathbf{l} \quad \text{(A)} \quad \text{(path to be specified)}. \qquad (2.103)$$

Keep in mind the fact that $\mathbf{E}$ is a conservative field, but $\mathbf{H}$ is not a conservative field.

Since the divergence of the curl of any vector is identically zero (see Section 1.8), and since it is *always* true that $\nabla \cdot \mathbf{B} = 0$, we can define a *magnetic vector potential* $\mathbf{A}$ such that

$$\boxed{\mathbf{B} = \nabla \times \mathbf{A}}. \qquad (2.104)$$

This result holds true even if $\mathbf{A}$ is replaced by $\mathbf{A} + \nabla \alpha$ for any α, since $\nabla \times \nabla \alpha \equiv 0$. The curl of $\mathbf{A}$ specifies six partial derivatives (see Section 1.7), and the divergence of **A** specifies the remaining three (nine total for three spatial variables and three components). Thus, we must specify the divergence of **A** to completely specify **A**, and we can do this in a very advantageous way. Taking the curl of both sides of Equation (2.104) with $\mathbf{B} = \mu_0 \mathbf{H}$, and also using Equation (2.99), we get

$$\nabla \times \mathbf{H} = \frac{1}{\mu_0} \nabla \times (\nabla \times \mathbf{A}) = \mathbf{J}. \tag{2.105}$$

There is an identity available for the curl of the curl of a vector (see inside back cover): $\nabla \times (\nabla \times \mathbf{A}) = \nabla(\nabla \cdot \mathbf{A}) - \nabla^2 \mathbf{A}$, where $\nabla^2 \mathbf{A}$ is the *Laplacian of a vector*. Thus, Equation (2.105) can be written

$$\nabla(\nabla \cdot \mathbf{A}) - \nabla^2 \mathbf{A} = +\mu_0 \mathbf{J}. \tag{2.106}$$

If we choose

$$\boxed{\nabla \cdot \mathbf{A} = 0}, \tag{2.107}$$

then[11] Equation (2.106) becomes

$$\boxed{\nabla^2 \mathbf{A} = -\mu_0 \mathbf{J}}, \tag{2.108}$$

whose *x* component (see inside front cover), for example, is *Poisson's equation*

$$\nabla^2 A_x = -\mu_0 J_x \quad \text{Poisson's equation.} \tag{2.109}$$

Note that the inside front and back covers show that $\nabla^2 \mathbf{A}$ is not a simple operation, except in Cartesian coordinates. The real point here, however, is that our choice of $\nabla \cdot \mathbf{A} = 0$ leads to Poisson's equation for each Cartesian component, and *we already know how to solve this equation*. In fact, since Poisson's equation in the form of Equation (2.61) is satisfied by Equation (2.65), we know by the *principle of duality* that Equation (2.109) is satisfied by

$$A_x(\mathbf{r}) = \frac{\mu_0}{4\pi} \iiint_{\text{vol}'} \frac{J_x(\mathbf{r}')}{R} dv';$$

or, more generally, by superposition of Cartesian components, the solution to Equation (2.108) is

$$\boxed{\mathbf{A}(\mathbf{r}) = \frac{\mu_0}{4\pi} \iiint_{\text{vol}'} \frac{\mathbf{J}(\mathbf{r}')}{R} dv'}. \tag{2.110}$$

---

[11] In this case, **A** is said to be in the *Coulomb gauge*.

2.12 MAGNETIC SCALAR AND VECTOR POTENTIAL

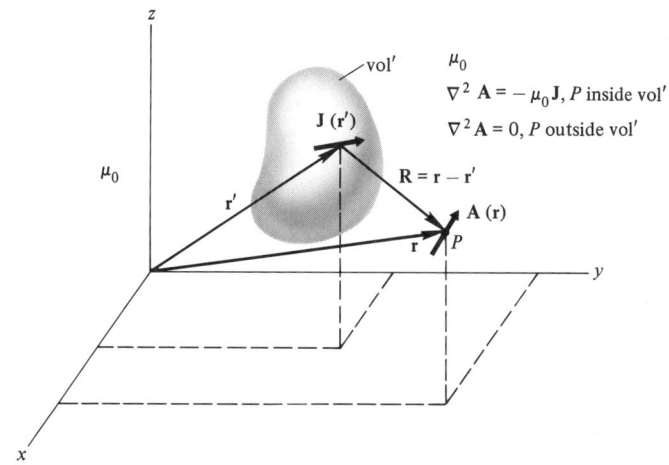

Figure 2.29. Geometry for finding the magnetic vector potential **A** from its source **J**.

The discussion with regard to boundary value problems in Section 2.5 will of course, apply here, and is not repeated. Figure 2.29 shows the geometry for Equation (2.110).

Example 2.19

Suppose that a very ambitious student has found the magnetic vector potential for the ideal solenoid (Example 2.17):

$$\mathbf{A} = \begin{cases} \mathbf{a}_\phi \mu_0 J_{s\phi} \rho/2, & 0 \leq \rho \leq a; \\ \mathbf{a}_\phi \mu_0 J_{s\phi} a^2/(2\rho), & a \leq \rho < \infty. \end{cases}$$

Then, $\mathbf{B} = \nabla \times \mathbf{A}$, or $\mathbf{H} = (\nabla \times \mathbf{A})/\mu_0$ or

$$H_z = \frac{1}{\mu_0 \rho} \frac{\partial}{\partial \rho}(\rho A_\phi) = \begin{cases} J_{s\phi}, & 0 \leq \rho \leq a; \\ 0, & a \leq \rho < \infty. \end{cases}$$

Since this is the correct magnetic field intensity, the magnetic vector potential is correct. Note that a constant added to **A** will not alter this conclusion.

Next we would like to use Equation (2.110), appropriately reduced for filamentary currents, to find the field of a distribution that will be very useful to us. The *magnetic dipole* is an electric current loop. It is called a magnetic dipole because its magnetic field is dual to the electric field of the electrostatic dipole at large distances. That is, they have the same mathematical form. We will examine only the simplest case.

Example 2.20

Consider the filamentary current loop lying in the $z = 0$ plane as shown in Figure 2.30. The reduced form of Equation (2.110) for this case is

$$\mathbf{A} = \frac{\mu_0}{4\pi} \int_{l'} \frac{I}{R} d\mathbf{l}' = \frac{\mu_0 I}{4\pi} \int_{l'} \frac{1}{R} d\mathbf{l}'.$$

Using the last identity on the inside back cover ($\mathbf{a}_n = \mathbf{a}_z$), we get

$$\mathbf{A} = \frac{\mu_0 I}{4\pi} \int\int_{s'} \mathbf{a}_z \times \nabla'\left(\frac{1}{R}\right) ds' = \frac{\mu_0 I}{4\pi} \int\int_{s'} \mathbf{a}_z \times \left(\frac{\mathbf{a}_R}{R^2}\right) ds',$$

since $\nabla'(1/R) = \mathbf{a}_R/R^2$. It can be seen in Figure 2.30 that for large distances $R \approx r$ and $\mathbf{a}_R \approx \mathbf{a}_r$, so that

$$\mathbf{A} \approx \frac{\mu_0 I}{4\pi} \int\int \mathbf{a}_z \times (\mathbf{a}_r/r^2) ds' = \frac{\mu_0 I}{4\pi r^2} \mathbf{a}_z \times \mathbf{a}_r \int\int ds',$$

$$\mathbf{A} = \frac{\mu_0 I s}{4\pi r^2} (\mathbf{a}_z \times \mathbf{a}_r).$$

Since $\mathbf{a}_z = \mathbf{a}_r \cos\theta - \mathbf{a}_\theta \sin\theta$, $\mathbf{a}_z \times \mathbf{a}_r = \mathbf{a}_\phi \sin\theta$,

$$\mathbf{A} = \mathbf{a}_\phi \frac{\mu_0 I s}{4\pi r^2} \sin\theta, \qquad (2.111)$$

and this result is independent of the *shape* of the loop, but depends on its area s.

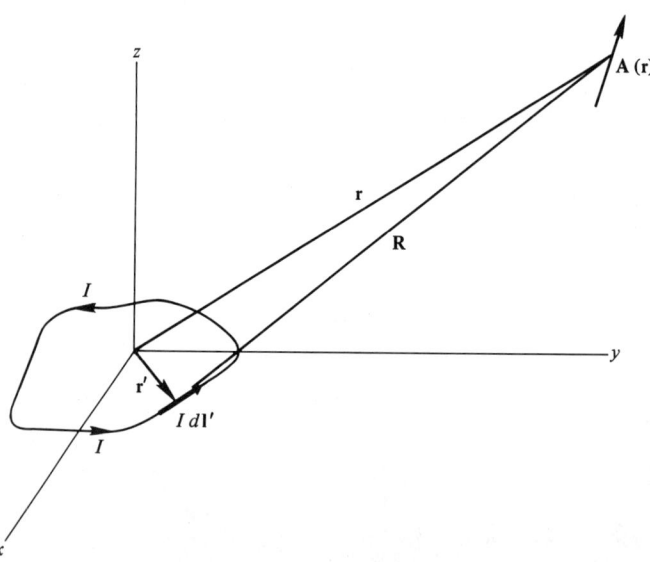

Figure 2.30. Geometry for finding the magnetic field of a planar magnetic dipole (electric current loop).

The *magnetic dipole moment* is defined as

$$\boxed{\mathbf{m} = I\mathbf{s}} \quad \textit{magnetic dipole moment,} \quad (2.112)$$

where $\mathbf{s}$ is the vector area whose direction is determined by the right-hand rule with the thumb of the right hand pointing in the direction of the current I. That is, $\mathbf{s} = \mathbf{a}_z s$ in this example. Based on Equations (2.111) and (2.112), an equation can now be written for $\mathbf{A}$ that is independent of the dipole orientation:

$$\boxed{\mathbf{A} = \frac{\mu_0}{4\pi} \frac{\mathbf{m} \times \mathbf{a}_R}{R^2}}. \quad (2.113)$$

∎

2.13 MAGNETIC MATERIALS

Magnetic dipole moment was defined as $\mathbf{m} = I\mathbf{s}$ in Equation (2.112). A magnetic dipole in the form of a filamentary circular loop with current I in the direction of $\mathbf{a}_\phi$, with radius a, and with its center at the origin has a moment $\mathbf{m} = \pi a^2 I \, \mathbf{a}_z$. When a magnetic flux density $\mathbf{B} = \mathbf{a}_y B_0$ is applied, there is a vector torque on the dipole. The differential torque is $d\mathbf{T} = \mathbf{R}_0 \times d\mathbf{F}$, where $\mathbf{R}_0 = \mathbf{a}_\rho a = a(\mathbf{a}_x \cos \phi + \mathbf{a}_y \sin \phi)$ and $d\mathbf{F} = I \, d\mathbf{l} \times \mathbf{B} = I \, a \, d\phi \, \mathbf{a}_\phi \times \mathbf{a}_y B_0$, or

$$d\mathbf{F} = I \, a \, d\phi(-\mathbf{a}_x \sin \phi + \mathbf{a}_y \cos \phi) \times \mathbf{a}_y B_0,$$

$$d\mathbf{F} = -\mathbf{a}_z I \, a \sin \phi \, B_0 \, d\phi,$$

$$d\mathbf{T} = I \, a^2 B_0 \, d\phi(\mathbf{a}_y \sin \phi \cos \phi - \mathbf{a}_x \sin^2 \phi),$$

$$\mathbf{T} = \int_0^{2\pi} d\mathbf{T} = I \, \pi a^2 B_0 (-\mathbf{a}_x).$$

This can be put in the *general* form

$$\boxed{\mathbf{T} = \mathbf{m} \times \mathbf{B}} \quad (\mathrm{N} \cdot \mathrm{m}), \quad (2.114)$$

that can be compared to the torque on an electric dipole $\mathbf{T} = \mathbf{p} \times \mathbf{E}$.

In the simplest picture of an atom, an electron or electrons orbit around the nucleus. As such, these electrons are effectively current loops, or magnetic dipoles. We should, then, be able to develop a fairly accurate theory of material behavior, at least for present purposes.

There are other vector moments present in an atom besides those attributed to the magnetic dipole (orbiting electron). For example, *electron spin* and *nuclear spin* also contribute moments. It is the combination of all the moments which determines the magnetic properties of materials. For an *isolated* current loop or magnetic dipole, the field of the loop itself always acts to add to the external field giving an *increased* field in the neighborhood of the loop. In a material, the internal magnetic field may increase or decrease, depending on *all* of the moments. Let us examine briefly some classes of magnetic materials.

1. In *diamagnetic* materials (bismuth) spin moments tend to be dominant and produce fields that *oppose* the external field. Thus, the internal magnetic field is reduced slightly compared to the external field.
2. If the dipole moments dominate slightly, then the internal field is increased slightly over the external field and the material is *paramagnetic* (tungsten).
3. Large dipole moments are produced in certain regions or *domains* for the *ferromagnetic* materials (iron). A random domain alignment exists for virgin ferromagnetic material. When an external field is applied and then removed, a net alignment occurs giving *permanent magnetization* and a *hysteresis* effect. Alloys of some of the ferromagnetic materials are also ferromagnetic (alnico).
4. In *ferrimagnetic* materials, adjacent atoms develop unequal, but oppositely directed moments, allowing a rather larger response to external fields. From the point of view of engineering applications, the *ferrites* are very important ferrimagnetic materials. Ferrites possess a very high resistance, and hence give very little eddy current loss at higher frequencies when used as transformer cores.
5. The magnetic tape used for audio and video recording is a superparamagnetic material and is composed of an array of small ferromagnetic particles.

Magnetization, perhaps more aptly called magnetic polarization, is a vector quantity that is quite analogous to the polarization vector used to explain dielectric behavior. The development here is very similar to that used for dielectrics in Section 2.5.

The magnetic dipole has been described in terms of its moment by Equation (2.112), $\mathbf{m} = I\mathbf{s}$. For present purposes, the magnetic dipole is (macroscopically) considered to be a *point source* of strength $I\mathbf{s}$. If there are n dipoles per cubic meter, then the resultant dipole moment for an incremental volume Δv will be

$$\mathbf{m}_r = \sum_{n=1}^{n\Delta v} \mathbf{m}_n. \qquad (2.115)$$

The magnetization or magnetic polarization $\mathbf{M}$ is defined as the *dipole moment per unit volume*,

$$\mathbf{M} = \lim_{\Delta v \to 0} \frac{\mathbf{m}_r}{\Delta v} \quad (\text{A/m}).$$

The magnetization is the *general source function* for *bound current* in the same way that current density $\mathbf{J}$ is the general source function for isolated current.

2.13 MAGNETIC MATERIALS

Note that dimensionally the magnetization is like ordinary surface current density (A/m).

The magnetic vector potential due to ordinary current density **J** is given by

$$\mathbf{A} = \frac{\mu_0}{4\pi} \iiint\limits_{vol'} \frac{\mathbf{J}}{R} \, dv'. \quad (2.110)$$

Therefore, we expect that the magnetic vector potential of a point dipole moment must lead by way of superposition to the magnetic vector potential for the general source function **M**. That is, from Equation (2.113),

$$\frac{\mu_0}{4\pi} \frac{\mathbf{m} \times \mathbf{a}_R}{R^2} \rightarrow \frac{\mu_0}{4\pi} \iiint\limits_{vol'} \frac{\mathbf{M} \times \mathbf{a}_R}{R^2} \, dv' \quad (2.116a)$$

or

$$\mathbf{A}(\mathbf{r}) = \frac{\mu_0}{4\pi} \iiint\limits_{vol'} \frac{\mathbf{M}(\mathbf{r}') \times \mathbf{a}_R}{R^2} \, dv' \quad (2.116b)$$

We are treating **M** like any other field quantity in Equations (2.116b), but it should be visualized as an average value taken over many molecules, and it has the effect of giving the magnetic dipoles a net alignment or polarization from their otherwise random arrangement. Note that we are still using μ_0 because we are including the magnetic effects in the molecular magnetic dipoles. We next replace $\mathbf{a}_R/R^2$ with $\nabla'(1/R)$, obtaining

$$\mathbf{A}(\mathbf{r}) = \frac{\mu_0}{4\pi} \iiint\limits_{vol'} \mathbf{M}(\mathbf{r}') \times \nabla'\left(\frac{1}{R}\right) dv'. \quad (2.117)$$

We have previously used the vector identity

$$\mathbf{C} \times \nabla'\alpha \equiv \alpha \nabla' \times \mathbf{C} - \nabla' \times (\alpha \mathbf{C}).$$

In Equation (2.117) this identity gives

$$\mathbf{A}(\mathbf{r}) = \frac{\mu_0}{4\pi} \iiint\limits_{vol'} \frac{1}{R} \nabla' \times \mathbf{M}(\mathbf{r}') \, dv' - \frac{\mu_0}{4\pi} \iiint\limits_{vol'} \nabla' \times \left[\frac{\mathbf{M}(\mathbf{r}')}{R}\right] dv'. \quad (2.118)$$

We now need a vector identity not previously encountered, but it can be found on the inside back cover:

$$\iiint\limits_{vol} (\nabla \times \mathbf{C}) \, dv \equiv -\oiint\limits_{s} \mathbf{C} \times d\mathbf{s} = -\oiint\limits_{s} \mathbf{C} \times \mathbf{a}_n \, ds \quad (\mathbf{a}_n \, ds = d\mathbf{s}).$$

With this identity Equation (2.118) becomes

$$\mathbf{A}(\mathbf{r}) = \frac{\mu_0}{4\pi} \iiint_{vol'} \frac{1}{R} \nabla' \times \mathbf{M}(\mathbf{r}') \, dv' + \frac{\mu_0}{4\pi} \iint_{s'} \frac{\mathbf{M}(\mathbf{r}')}{R} \times \mathbf{a}_n \, ds'. \qquad (2.119)$$

If we now compare Equation (2.119) to both Equation (2.110) and its reduced version for surface current densities

$$\mathbf{A}(\mathbf{r}) = \frac{\mu_0}{4\pi} \iint_s \frac{\mathbf{J}_s(\mathbf{r}')}{R} \, ds',$$

we see that $\nabla \times \mathbf{M}$ has the effect of a general current density, while $\mathbf{M}$ has the effect of a surface current density. We are then justified in writing

$$\nabla \times \mathbf{M} = \mathbf{J}^b \qquad (2.120)$$

and

$$\mathbf{M} \times \mathbf{a}_n = \mathbf{J}_s^b. \qquad (2.121)$$

These are the *bound* currents, or currents due to bound charges, which were mentioned earlier.

We now have a situation where the permeability is that of free space μ_0 and where *both* bound current and ordinary current may exist. Let us write Maxwell's equation in terms of $\mathbf{B}$ instead of $\mathbf{H}$. The reason for this is that we intend to obtain a new, more general, relation between $\mathbf{B}$ and $\mathbf{H}$ (other than the original $\mathbf{B} = \mu_0 \mathbf{H}$). We have

$$\nabla \times \frac{\mathbf{B}}{\mu_0} = \text{total vector current density.} \qquad (2.122)$$

This equation is correct, because the permeability is μ_0, and *all* of the current is considered. Equation (2.122) may be written

$$\nabla \times \frac{\mathbf{B}}{\mu_0} = \mathbf{J} + \mathbf{J}^b.$$

With Equation (2.120), we have

$$\nabla \times \frac{\mathbf{B}}{\mu_0} = \mathbf{J} + \nabla \times \mathbf{M}$$

or

$$\nabla \times \left(\frac{\mathbf{B}}{\mu_0} - \mathbf{M} \right) = \mathbf{J}. \qquad (2.123)$$

2.13 MAGNETIC MATERIALS

The right-hand side of Equation (2.123) is the current density due to ordinary processes and is not bound current. We now have a way of avoiding the use of bound current if we simply accept Maxwell's equation (or Ampere's law)

$$\nabla \times \mathbf{H} = \mathbf{J} \tag{2.99}$$

as being correct. A comparison of Equations (2.99) and (2.123) gives us the new relation between **B** and **H**:

$$\mathbf{H} = \frac{\mathbf{B}}{\mu_0} - \mathbf{M} \tag{2.124}$$

or

$$\boxed{\mathbf{B} = \mu_0(\mathbf{H} + \mathbf{M})} \tag{2.125}$$

Note that the right-hand side of Equation (2.99) is a current density due to conduction electrons, while the effect of the bound charges is included in the left-hand side in **H**.

We next need a relation between **M** and **H** to simplify Equation (2.125). For linear isotropic materials, a simple relationship exists:

$$\mathbf{M} = \chi_m \mathbf{H}, \tag{2.126}$$

where χ_m is the *magnetic susceptibility*. In this case, we have

$$\mathbf{B} = \mu_0(\mathbf{H} + \chi_m \mathbf{H}) = \mu_0(1 + \chi_m)\mathbf{H},$$

or simply

$$\boxed{\mathbf{B} = \mu_0 \mu_R \mathbf{H} = \mu \mathbf{H}} ; \tag{2.127}$$

that is,

$$\boxed{\mu = \mu_0 \mu_R} \tag{2.128}$$

is the *permeability* of the material, and μ_R is its *relative permeability*. If the material is anisotropic, μ is *not* a scalar, but is a tensor (as was ε under the same conditions). The relation between **B** and **H** is still Equation (2.127) for anisotropic material, but **B** and **H** may not be parallel.

In most of the problems we will be concerned with here, we simply replace μ_0 by μ if magnetic material is present, and then use the equations in their original form (for free space). This is a great step forward. It is rewarding to return to Section 2.5 to see that the development for dielectric effects there exactly parallels the development here. As a matter of fact, we could have invoked the principle of duality and written the results of this section after examining the results of Section 2.5.

It is a simple matter to determine the boundary conditions at the interface between two linear, isotropic, and homogeneous magnetic materials. When Ampere's law $\oint \mathbf{H} \cdot d\mathbf{l} = I$ is applied to the rectangular path of Figure 2.31, we obtain $lH_{t1} - lH_{t2} = lJ_s$ for $\Delta w \to 0$, so

$$\boxed{H_{t1} - H_{t2} = J_s, \qquad B_{t1}/\mu_1 - B_{t2}/\mu_2 = J_s} \qquad (2.129)$$

In many cases $J_s = 0$ and

$$\boxed{H_{t1} = H_{t2}, \qquad B_{t1}/\mu_1 = B_{t2}/\mu_2 \qquad (J_s = 0)} \qquad (2.130)$$

Applying the flux equation $\oiint \mathbf{B} \cdot d\mathbf{s} = 0$ to the can for $\Delta h \to 0$ gives

$$\boxed{B_{n1} = B_{n2}, \qquad \mu_1 H_{n1} = \mu_2 H_{n2}} \qquad (2.131)$$

It is also true in many cases of practical interest that $\mu_2 \gg \mu_1$ (medium 2 is silicon steel and medium 1 has $\mu = \mu_0$ for a ferromagnetic material next to air, for example). In the case of Equation (2.130), $B_{t1} = B_{t2}\mu_1/\mu_2$ or $B_{t1} \ll B_{t2}$, and so magnetic flux lines leave medium 2 at right angles to the surface for all practical purposes. The surface of medium 2 must then be an equipotential surface ($\Phi_m = C$).

Next we would like to consider *magnetic circuits*. The uniform flux density inside the ideal solenoid was found in Example 2.17, and from Equation (2.96) we get $B_z = \mu J_{s\phi}$ when the magnetic material within the solenoid (the core) has permeability μ. The flux is easily found with Equation (2.93): $\Psi_m = Bs = \mu J_{s\phi}s$, where s is the cross-sectional area of the solenoid. If we now suppose that the surface current density $J_{s\phi}$ is being produced (approximately) by an N-turn helical coil (a *real* solenoid) that is l (m) in length (l is the length of the solenoid, and is

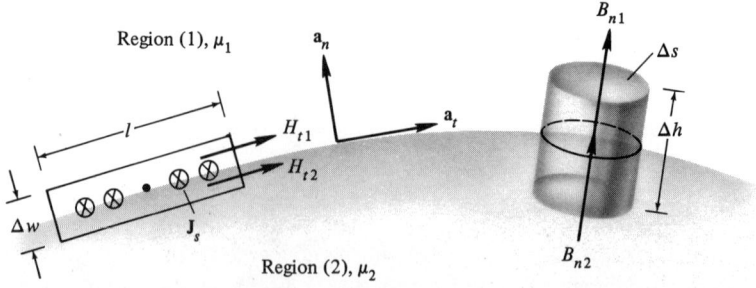

Figure 2.31. Geometry for determining the boundary conditions at the interface between simple magnetic media.

2.13 MAGNETIC MATERIALS

not the total length of wire), then $J_{s\phi} \approx NI/l$ (A/m) and the magnetic flux can be written

$$\Psi_m = \frac{NI}{l/(\mu s)} \quad \text{(Wb)}. \tag{2.132}$$

At this point we need to refer back to some information gained from a sophomore circuits course. A conductor of cross-sectional area s, length l, and conductivity σ has a resistance $R = l/(\sigma s)$ [also below Equation (2.145)] and a current given by Ohm's law:

$$I = \frac{V}{R} = \frac{V}{l/(\sigma s)}.$$

Thus, comparing Ψ_m with I, we see that in a *magnetic circuit*, magnetic flux Ψ_m is mathematically analogous to electric current I, NI (ampere-turns) is analogous to an ideal voltage source and is called *magnetomotive-force* (mmf), and

$$\mathcal{R} = \frac{l}{\mu s} \quad (\text{H}^{-1}) \tag{2.133}$$

is analogous to resistance and is called *reluctance*. These analogies are invaluable when analyzing magnetic circuits.

Next let us consider the behavior of *ferromagnetic materials* graphically since the relationship between B and H is nonlinear and μ_R for a given sample is not unique. This graph is called the B-H curve or hysteresis curve. If an mmf is applied to a virgin (unmagnetized) sample of ferromagnetic material with B and H both zero (point *a*), a magnetization curve such as that shown in Figure 2.32 is established (from point *a* to *b*). As H is increased, for example, by increasing the current in a solenoid around the sample, B increases, slowly at first, then more rapidly, then more slowly to point *b*. Further increase in H gives little increase in B, and finally B will not increase at all, indicating magnetic saturation (point *c*). The curve *a-b-c* can be called the magnetization curve.

If, in the process of magnetization, point *b* is reached and then H is reduced, the original curve is not retraced, but instead, the B-H relationship follows a curve such as *b-d*, so that at point *d* a *remnant flux density*, B_r exists even though H is zero. This is *permanent* magnetization. Now if H is decreased (negative), the B-H relation follows the curve *d-e*, and at point *e*, B is zero, but H has the value $-H_c$, where H_c is *the coercive mmf*. Further decrease of H can lead to saturation again; but if, instead, H is now increased at point *f* (which is symmetrical to *b*), then the B-H curve will follow the path *f-g-h-b*. The path *b-d-e-f-g-h-b* is called the hysteresis loop.

If, when the sample was first being magnetized, point *b'* had been reached and then H had been decreased, then a smaller hysteresis loop could be traced out. In this way a sample can be demagnetized. That is, if an ac current is applied to the solenoid around the sample, and if the amplitude of this ac signal is gradually reduced, the hysteresis loops will become smaller and eventually degenerate to point *a* where B is zero.

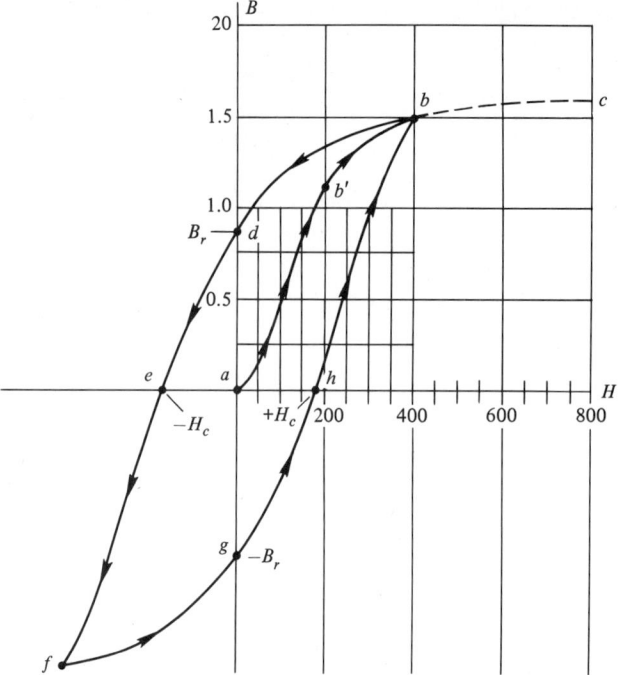

Figure 2.32. Magnetization curve and hysteresis loop for a ferromagnetic sample.

Example 2.21

Let us next solve, approximately, a simple problem involving a magnetic circuit composed of a ferromagnetic core with an air gap. Let the mmf be applied by means of a toroidal coil as shown in Figure 2.33. Let the air gap be 1 mm, N be 1000 turns, and the *mean* radius of the toroid be 22.5 cm. We would like to find the coil current required to produce $B = 0.5$ Wb/m² in the core. Some approximations must be made if we are to accomplish this. First of all we assume that B is uniform in the ferromagnetic core. Second, the air gap is very small compared to the diameter of the core, so the flux density is essentially *normal* to the interface between the air and the ferromagnetic core and must therefore be continuous [Equation (2.131)].

The small leakage flux around the periphery of the air gap is ignored. There will also be some leakage flux between the turns of the coil, and flux lines may even extend across the interior of the toroid through the air. These effects are small and are neglected, but it should be recognized that magnetic flux lines "prefer" the ferromagnetic material over air by a factor of only about 10^3, whereas in the equivalent electric problem the electric flux lines "prefer" good conductors over air by a factor of about 10^{15}. That is, the permeability of the ferromagnetic material will be roughly 10^3 times that of air, but the conductivity of a good conductor is about 10^{15} times that of air. Thus, the magnetic problem leads to a much more crude (but still useful) model than the electric problem.

2.13 MAGNETIC MATERIALS

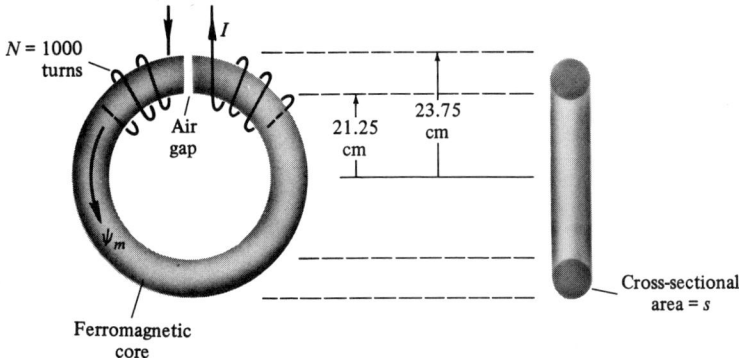

Figure 2.33. N-turn toroid with an air gap in its ferromagnetic core.

The magnetic material in this example has the B-H magnetization curve shown in Figure 2.34 so that if $B = 0.5$, then $H = 100$. The assumptions we have made lead to the equivalent electric circuit shown in Figure 2.35. For the air gap, we have

$$\mathcal{R}_{ag} = \frac{l}{\mu_0 s} = \frac{10^{-3}}{4\pi \times 10^{-7}[\pi(2.5 \times 10^{-2})^2]/4} = 1.621 \times 10^6$$

and (everywhere)

$$\psi_m = Bs = 0.5 \frac{\pi(2.5 \times 10^{-2})^2}{4} = 245.4 \times 10^{-6}.$$

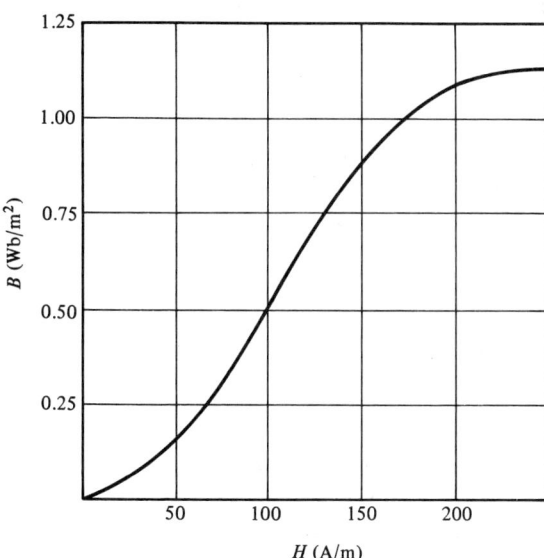

Figure 2.34. Magnetization curve for the ferromagnetic core in Example 2.21.

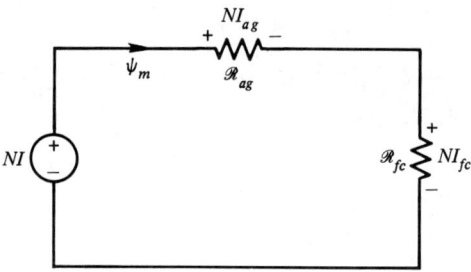

Figure 2.35. Equivalent electric circuit (Example 2.21) showing the (linear) air-gap reluctance, (nonlinear) reluctance of the core, and the applied mmf.

Therefore the mmf "drop" across the air gap is $\psi_m \mathcal{R}_{ag}$, neglecting flux fringing at the gap; so

$$(NI)_{ag} = \psi_m \mathcal{R}_{ag} = (1.621 \times 10^6)(245.5 \times 10^{-6}) = 398 \quad (A \cdot t).$$

Producing 0.5 Wb/m² in the ferromagnetic core requires $H_{fc} = 100 = (NI)_{fc}/l$, so the "core" mmf in Figure 2.33 can be calculated from $(NI)_{fc}/l = 100$ (A · t/m), or

$$(NI)_{fc} = 100l = 100 \times 2\pi \times 0.225 = 141.4 \quad (A \cdot t),$$

using the mean radius for finding l. Therefore, the "source" mmf is

$$NI = (NI)_{ag} + (NI)_{fc} = 539$$

or

$$I = \frac{539}{1000} = 0.539 \quad (A).$$

Instead of requiring the $B = 0.5$, suppose we specify that $I = 0.3$ A and find the resulting flux density. A trial and error technique will be followed to find B. The answer to the preceding problem gives us a first trial. The reader is encouraged to show that $B \approx 0.25$. ∎

2.14 FORCE AND TORQUE

If a point charge in motion is subjected to both an external electric and magnetic field, we have the *Lorentz force* equation

$$\mathbf{F} = Q(\mathbf{E} + \mathbf{u} \times \mathbf{B}) \quad \textit{Lorentz force} \tag{2.134}$$

from Sections 2.2 and 2.9. The force on a differential filamentary current (conductor) was given by

$$d\mathbf{F} = I \, d\mathbf{l} \times \mathbf{B} \tag{2.81}$$

2.14 FORCE AND TORQUE

In a *conductor* carrying a current, the force, just described, is actually a force on the electrons that are moving (rather slowly) to constitute the current. We might expect that this magnetic force would merely shift the electrons with respect to the total conductor. There are, however, very strong Coulomb forces between the electrons and positive ions in a conductor, so any electron displacement due to the magnetic force is very small. Thus, the magnetic force is transferred to the *total conductor* according to Equation (2.81).

Any filamentary current path must ultimately be closed, so Equation (2.81) results in an equation that can be verified experimentally:

$$\boxed{\mathbf{F} = \oint_l I\, d\mathbf{l} \times \mathbf{B}} \quad (\text{N}) \qquad (2.135)$$

We may extend this result, as we have previously done in similar situations, to include surface currents and general current densities

$$\mathbf{F} = \iint_s \mathbf{J}_s \times \mathbf{B}\, ds \qquad (2.136)$$

and

$$\mathbf{F} = \iiint_{\text{vol}} \mathbf{J} \times \mathbf{B}\, dv. \qquad (2.137)$$

One fact should be mentioned at this point. The magnetic field *produced* by a current in a closed circuit will *not* produce a net force on the circuit (itself) which would cause the circuit (itself) to move. (Why not?) It *will* create a force of tension in the circuit. For example, a large current in a very flexible square loop would create a force of tension which would tend to make the loop circular in shape.

■ Example 2.22

Consider the highly idealized radial gap dc motor shown in Figure 2.36. The cylinder (armature) at $\rho = a$ carries the uniform surface current density $\mathbf{J}_s = J_0 \mathbf{a}_z$ (A/m) (produced by many windings). A magnetic flux density $\mathbf{B} = B_0 \mathbf{a}_\rho$ (perhaps produced by a permanent magnet) exists at the cylinder surface. With $d\mathbf{T} = \mathbf{R}_0 \times d\mathbf{F}$ and $d\mathbf{F} = \mathbf{J}_s \times \mathbf{B}\, ds$, we obtain

$$d\mathbf{T} = a\mathbf{a}_\rho \times (J_0 \mathbf{a}_z \times \mathbf{a}_\rho B_0 a\, d\phi\, dz),$$

$$d\mathbf{T} = a\mathbf{a}_\rho \times \mathbf{a}_\phi a J_0 B_0\, d\phi\, dz = a^2 J_0 B_0\, d\phi\, dz\, \mathbf{a}_z,$$

$$\mathbf{T} = a^2 J_0 B_0 \mathbf{a}_z \int_0^{2\pi} \int_0^L d\phi\, dz = 2\pi L a^2 J_0 B_0 \mathbf{a}_z,$$

or

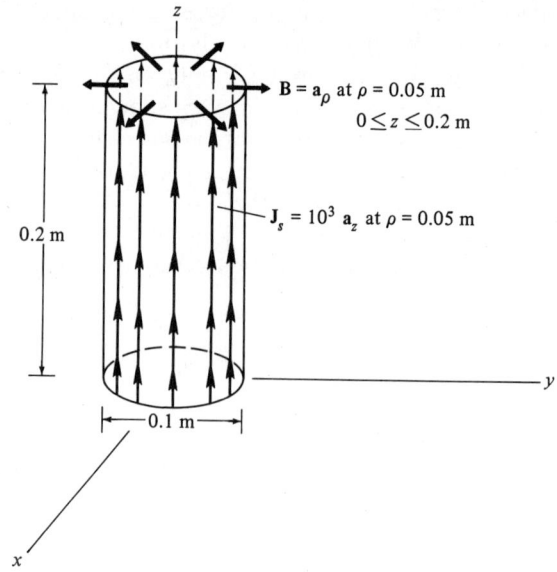

Figure 2.36. Idealized radial gap (dc) motor.

$$\mathbf{T} = \pi \mathbf{a}_z \quad (\text{N} \cdot \text{m})$$

for the numbers in Figure 2.36.

The power that is provided can be found from the energy relation $dW = T\, d\phi$ or $(dW/dt) = T(d\phi/dt) = T\omega = P$, where ω is the angular velocity of the armature. If this is $2000\pi/60$ rad/s = 1000 rpm, then $P = \pi(2000\pi/60) = 329$ W. This is a reasonable value for the numbers used in the example. ∎

2.15 CAPACITANCE, RESISTANCE, AND INDUCTANCE

Consider two conductors of arbitrary shape surrounded by a dielectric as shown in Figure 2.37. The dielectric is assumed to be lossless so that there is no conduction current in the dielectric, and static conditions exist. A battery has been connected so that the conductors carry equal magnitude but opposite sign charges. These charges will appear as surface charges, and the electric field will be normal outward from the positively charged conductor and normal inward toward the negatively charged conductor (Why?). The capacitance of the system is defined as the ratio of the (total) charge on the positively charged conductor to the potential difference $\Phi_{ab} = \Phi_a - \Phi_b$. Point a is taken to be on the positively charged conductor so that Φ_{ab} is positive, making the capacitance C a positive quantity. Thus, we have

$$\boxed{C = \frac{Q_a}{\Phi_{ab}} = \frac{\oiint_s \mathbf{D} \cdot d\mathbf{s}}{\Phi_{ab}} = \frac{\oiint_s \varepsilon \mathbf{E} \cdot d\mathbf{s}}{\Phi_{ab}} = \frac{-\oiint_s \varepsilon \nabla \Phi \cdot d\mathbf{s}}{-\int_b^a \mathbf{E} \cdot d\mathbf{l}}} \quad (\text{F}). \quad (2.138)$$

2.15 CAPACITANCE, RESISTANCE, AND INDUCTANCE

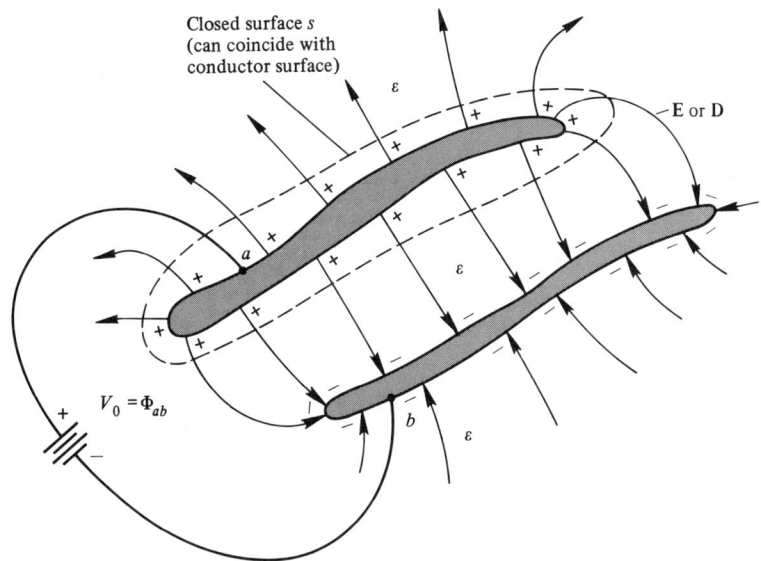

Figure 2.37. Two arbitrarily shaped conductors embedded in a lossless dielectric for calculating capacitance.

In a linear capacitor the capacitance is independent of Q_a and Φ_{ab}, and in order to find a formula for the capacitance, we must find Q_a in terms of Φ_{ab} or vice versa.

Generally speaking, when solving for capacitance, it is a good idea to try to determine what, if any, quantities ($\mathbf{D}$, $\mathbf{E}$, ρ_s) are uniform before using the above equations.

Example 2.23

It is very easy to find the capacitance of the parallel-plate capacitor of Figure 2.12 (Example 2.10) if we assume that the width and depth are large compared to the spacing (d) so that fringing of the electric field at the four sides can be ignored. In other words, we assume that the field is uniform in the capacitor. Since the charge density must then also be uniform, the charge on the upper plate is $\rho_s s$. It was shown (Example 2.10) that $\rho_s = \varepsilon V_0/d$ or $V_0 = \rho_s d/\varepsilon$, so

$$C = \frac{Q}{V_0} = \frac{\rho_s s}{\rho_s d/\varepsilon} = \frac{\varepsilon s}{d} \quad \text{(F)} \quad \textit{parallel-plate.} \tag{2.139}$$

■

Example 2.24

A coaxial cable (cross section) is shown in Figure 2.38. The total charge on the inner conductor (l is the length) is $+Q = 2\pi a l \rho_{sa}$, and the total charge on the outer conductor is $-Q = 2\pi b l \rho_{sb}$ ($\rho_{sa} = -\rho_{sb} b/a$). It is easy to show by means

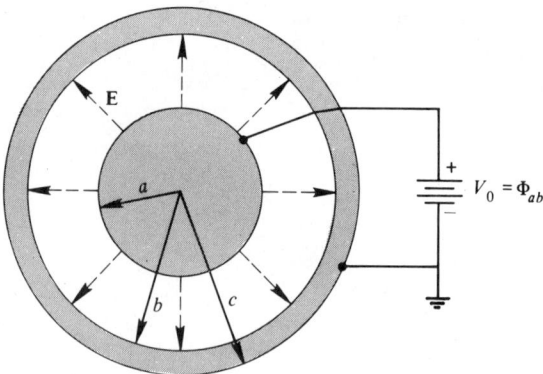

Figure 2.38. Coaxial conducting cylinders (coaxial capacitor).

of Gauss's law that $\mathbf{E} = \mathbf{a}_\rho a \rho_{sa}/(\varepsilon \rho)$ for $a \le \rho \le b$ and $\mathbf{E} = 0$ otherwise. Note that this field is identical to that produced by an infinite line charge density on the z axis if $\rho_l = 2\pi a \rho_{sa}$ (C/m). It is easy to find Φ:

$$\Phi = -\int E_\rho \, d\rho + C_1 = -\frac{a\rho_{sa}}{\varepsilon} \ln \rho + C_1;$$

and using the boundary conditions $\Phi(b) = 0$ and $\Phi(a) = V_0$, we obtain $\Phi_{ab} = (a\rho_{sa}/\varepsilon) \ln b/a = V_0$. So, we get

$$C = \frac{2\pi a l \rho_{sa} \varepsilon}{a \rho_{sa} \ln(b/a)} = \frac{2\pi \varepsilon l}{\ln(b/a)} \quad \text{(F)} \quad \text{coaxial cable} \quad (2.140)$$

■

Example 2.25

Another practical geometry for transmission lines is the two-wire line consisting of a pair of parallel conductors of equal radii (Figure 2.39a). These conductors are equipotential surfaces, $\Phi = \pm V_2$, but these surfaces do not represent a constant coordinate in any of our three standard systems (they do in *bicylindrical* coordinates). A pair of parallel (opposite) line charge densities, properly placed, will provide the correct equipotential surfaces to correspond to $\Phi = \pm V_2$. The potential of a single line charge density ρ_l on the z axis is $\Phi = [-\rho_l/(2\pi\varepsilon)] \ln \rho + C_2$. Using superposition for the two line charges of Figure 2.39(b), and choosing the added constants to make the potential vanish in the $x = 0$ plane, gives a *balanced* line whose potential is

$$\Phi = -\frac{\rho_l}{2\pi\varepsilon} \ln \left(\frac{(x-a)^2 + y^2}{(x+a)^2 + y^2} \right)^{1/2} = \frac{\rho_l}{4\pi\varepsilon} \ln \frac{(x+a)^2 + y^2}{(x-a)^2 + y^2}$$

2.15 CAPACITANCE, RESISTANCE, AND INDUCTANCE

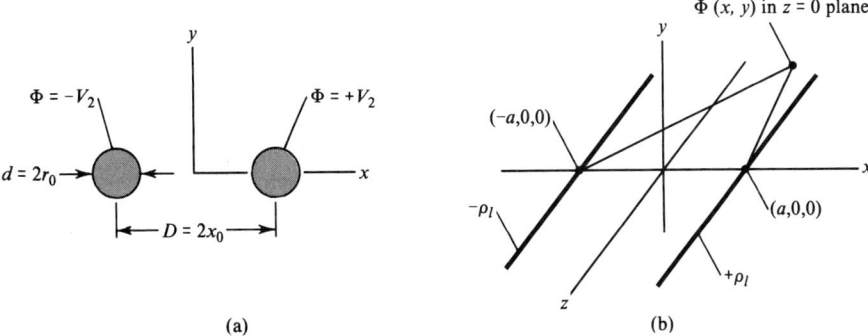

Figure 2.39. (a) Geometry for the two-wire line. (b) Geometry for finding $\Phi(x,y)$ in the $z=0$ plane.

for a point in the $z = 0$ plane. When the argument of the logarithm is constant (C_1) the potential is constant, $\Phi = V_2$ (for example):

$$V_2 = \frac{\rho_l}{4\pi\varepsilon} \ln C_1 = \frac{\rho_l}{2\pi\varepsilon} \ln \sqrt{C_1}, \qquad C_1 = \frac{(x+a)^2 + y^2}{(x-a)^2 + y^2},$$

which can be arranged to the equation of a circle:

$$\left(x - a\frac{C_1 + 1}{C_1 - 1}\right)^2 + y^2 = \left(\frac{2a\sqrt{C_1}}{C_1 - 1}\right)^2.$$

Therefore, the equipotentials are circular cylinders whose axes are located at $D/2 = x = [a(C_1 + 1)/(C_1 - 1)], 0, z$ for $x > 0$ and whose radii are $d/2 = 2a\sqrt{C_1}/(C_1 - 1)$. Equipotentials and electric field lines are shown in Figure 2.40. Note that equipotential surfaces are *not* concentric with the line charges. Solving for $\sqrt{C_1}$ and a gives $\sqrt{C_1} = [D + (D^2 - d^2)^{1/2}]/d$ and $a = [(D/2)^2 - (d/2)^2]^{1/2}$. Thus, V_2 becomes

$$V_2 = \frac{\rho_l}{2\pi\varepsilon} \ln \frac{D + (D^2 - d^2)^{1/2}}{d} = \frac{\rho_l}{2\pi\varepsilon} \cosh^{-1}(D/d).$$

The *potential difference* is $\Phi_{ab} = V_2 - (-V_2) = 2V_2$, and the charge on the positively charged conductor for a length l is $Q_a = \rho_l l$. Therefore, the capacitance is

$$C = \frac{Q_a}{2V_2} = \frac{\rho_l l}{\rho_l \cosh^{-1}(D/d)/(\pi\varepsilon)}$$

or

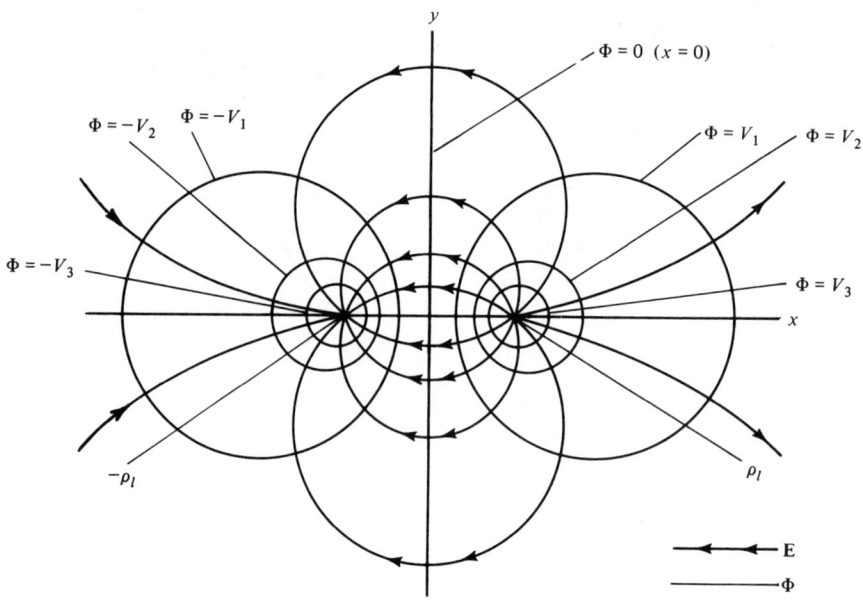

Figure 2.40. Equipotentials and electric field intensity lines about parallel (opposite sign) infinite, uniform line charge densities.

$$C = \frac{\pi \varepsilon l}{\cosh^{-1}(D/d)} = \frac{\pi \varepsilon l}{\ln\{[D + (D^2 - d^2)^{1/2}]/d\}} \quad \text{(F)} \quad \text{two-wire line.} \quad (2.141)$$

If $D \gg d$, $C \approx \pi \varepsilon l / \ln(2D/d)$.

Using the method of images,[12] it is easy to show that the capacitance of a single wire of radius b and located at a height h above and parallel to a grounded conducting plane is

$$C = \frac{2\pi \varepsilon l}{\cosh^{-1}(h/b)} \approx \frac{2\pi \varepsilon l}{\ln(2h/b)} \quad \text{(F)}. \quad (2.142)$$

■

We now return to Figure 2.37, and imagine that the medium surrounding the electrodes has conductivity σ as well as permittivity ε. We further assume that the conductivity of the electrodes and connecting wires is very much larger than that of the surrounding medium. There will then be a current I supplied by the battery that enters the upper electrode at a and leaves the lower electrode at b. The current that enters from the wire at a will leave the upper electrode in the form of a current density $\mathbf{J} = \sigma \mathbf{E}$ and terminate on the lower electrode. As a matter of fact $\mathbf{J}$ lines would look exactly like the $\mathbf{E}$ lines of Figure 2.37 (as long as σ and ε have the same spatial distribution). This occurs because it is still possible to use

---

[12]Refer to the last paragraph of Section 2.7.

2.15 CAPACITANCE, RESISTANCE, AND INDUCTANCE

$\nabla \times \mathbf{E} = 0$ and $\mathbf{E} = -\nabla\Phi$ even though the charges in the surrounding medium are in *uniform* motion. The charge distribution is the same (macroscopically) at any point at any instant of time. The flux of the current density, obtained by $\oint_s \mathbf{J} \cdot d\mathbf{s}$, for any closed surface that encloses the upper electrode will certainly be zero as demanded by conservation of charge, but in what follows we are only interested in the current *leaving* the upper electrode (I) so we exclude the current entering at a.

Resistance is defined as

$$\boxed{R = \Phi_{ab}/I = 1/G} \quad (\Omega), \tag{2.143}$$

where, as explained above, I is the current leaving the upper electrode. Thus, the conductance G (℧) is given by

$$\boxed{G = \frac{I}{\Phi_{ab}} = \frac{\oint_s \mathbf{J} \cdot d\mathbf{s}}{\Phi_{ab}} = \frac{\oint_s \sigma \mathbf{E} \cdot d\mathbf{s}}{\Phi_{ab}} = \frac{-\oint_s \sigma \nabla\Phi \cdot d\mathbf{s}}{-\int_b^a \mathbf{E} \cdot d\mathbf{l}}} \quad (℧) \tag{2.144}$$

Equation (2.144) is dual to equation (2.138), and so for the same geometry (including the same spatial distribution for σ and ε), we have

$$\boxed{\frac{C}{G} = \frac{\varepsilon}{\sigma}}. \tag{2.145}$$

The conductance of a conductor of length d, cross-sectional area s, and conductivity σ is given by combining Equations (2.145) and (2.139): $G = \sigma s/d$. Its resistance is $R = 1/G = d/(\sigma s)$. The *shunt* conductance of a coaxial cable with uniform σ (in the dielectric) is given by Equations (2.140) and (2.145) as

$$G = \frac{2\pi \sigma l}{\ln(b/a)} \quad (℧) \quad \text{coaxial cable.} \tag{2.146}$$

In a *linear* capacitor the capacitance is independent of Q_a and Φ_{ab}, while in a *linear* resistor the resistance is independent of I and Φ_{ab}. The *varactor* diode (semiconductor junction diode) is a nonlinear capacitor since it can be shown that C is proportional to $(\Phi_{ab})^{-1/2} = (V_0)^{-1/2}$. It is often used in automatic frequency control (AFC) and voltage controlled oscillator (VCO) applications.

Inductance L is the third parameter to be considered that is used in circuit theory. An inductor stores magnetic energy in the region surrounding it. Recall that resistance R depended on conductivity and the geometry of the conductor. Capacitance C depended on permittivity and the capacitor geometry. In the same way, we expect that inductance will depend on permeability and the inductor

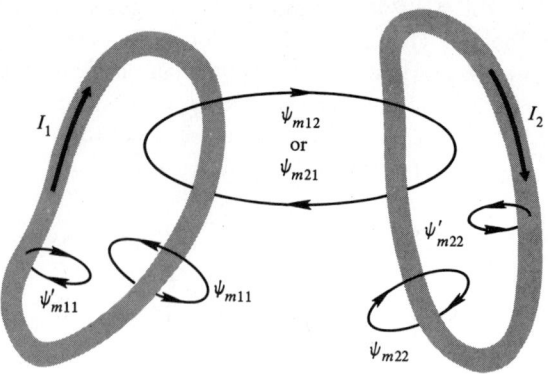

Figure 2.41. Magnetically coupled circuits showing the various flux linkages.

geometry. Consider Figure 2.41 which shows two circuits magnetically coupled. ψ'_{m11} is a flux line due to current I_1 linking partially with I_1 by passing through the conductor. This type of *flux linkage* gives rise to an *internal self-inductance* L'_{11}. Those flux lines like ψ_{m11}, due to I_1, which link I_1 externally, give rise to an *external* self-inductance L_{11}. The same things can be said about circuit 2. Flux line ψ_{m12} is produced by I_1 and links I_2, giving rise to a *mutual inductance* L_{12}. ψ_{m12} could just as well be called ψ_{m21} because it could also be considered to be due to I_2 and linking I_1. We suspect that $L_{12} = L_{21}$.

Inductance will be defined as the ratio of magnetic flux linkage to the current producing the flux linkage or

$$L = \frac{\psi_{ml}}{I} \quad \text{(H)}. \tag{2.147}$$

This definition can accommodate any of the various types of inductance mentioned in the previous paragraph. It is obvious from Equation (2.147) that the flux linkage must be proportional to the current if L is to be independent of I. This, in turn, requires that the medium be linear so that μ is constant. In this way, L, as defined in Equation (2.147), will depend only on permeability and the geometry. Inductance is a characteristic possessed by one or more *closed* circuits.

Flux linkage internal to a conductor leads to internal inductance and its value depends on the exact distribution of the current density. We must exercise care in finding the flux linkage, and this can best be done by finding the flux linkage on a differential basis and then integrating (summing) over the interior of the conductor to find the total flux linkage. The differential flux linkage is the differential flux times the fraction of the total current that is actually enclosed:

$$L = \frac{1}{I} \iint_s d\Psi_{ml} = \frac{1}{I} \iint_s \left[\frac{I_{enc}}{I}\right] d\Psi_m \quad \text{(H)}. \tag{2.148}$$

Example 2.26

The magnetic field intensity for the coaxial cable was found in Example 2.16 (Figure 2.25). Using Equation (2.148) and the results of Example 2.16 (l is the cable length), we get

$$L = \frac{1}{I}\int_0^a \int_0^l \left(\frac{I\rho/a^2}{I}\right)\left(\frac{\mu I \rho}{2\pi a^2}\right) d\rho\, dz$$

$$+ \frac{1}{I}\int_a^b \int_0^l \left(\frac{I}{I}\right)\left(\frac{\mu I}{2\pi \rho}\right) d\rho\, dz$$

$$+ \frac{1}{I}\int_b^c \int_0^l \left(\frac{I(c^2-\rho^2)/(c^2-b^2)}{I}\right)\left(\frac{\mu I}{2\pi\rho}\frac{c^2-\rho^2}{c^2-b^2}\right) d\rho\, dz,$$

assuming that μ is the same everywhere. Thus,

$$L = \frac{\mu l}{8\pi} + \frac{\mu l}{2\pi}\ln\frac{b}{a} + \frac{\mu l}{2\pi}\int_b^c \left(\frac{c^2-\rho^2}{c^2-b^2}\right)^2 \frac{d\rho}{\rho},$$

$$L = \frac{\mu l}{8\pi} + \frac{\mu l}{2\pi}\ln\frac{b}{a} + \frac{\mu l}{2\pi}\left(\frac{c^2}{c^2-b^2}\right)^2 \ln\frac{c}{b} \qquad (2.149)$$

$$- \frac{\mu l}{2\pi}\left(\frac{c^2}{c^2-b^2}\right) + \frac{\mu l}{8\pi}\left(\frac{c^2+b^2}{c^2-b^2}\right) \quad \text{(H)}.$$

∎

A nonlinear inductance will result when the medium is nonlinear. In Example 2.21, the flux $\Psi_m = 245.4 \times 10^{-6}$ links the current $I = 0.539$ N $= 10^3$ times so the inductance is $L = 10^3(245.4 \times 10^{-6})/0.539 = 455$ mH, but if the current is changed to 0.3 A, then L will change to 409 mH because the operating point on the nonlinear B-H curve will change. The flux linkage does not depend linearly on the current.

It will be shown later that at high frequencies the "skin-effect" occurs where the current density becomes largest at the surface of the conductor and decays rapidly in the interior of the conductor. The effect can be accounted for very accurately if the current density is assumed to be uniform over the "skin depth" δ. In the case of the coaxial cable, for example, the uniform current density will exist for $a - \delta < \rho < a$ for the inner conductance and for $b < \rho < b + \delta$ for the outer conductor. The reader is encouraged to show that for this case (and for $\delta \ll a$, $\delta \ll c - b$) the coaxial cable inductance is (l is the cable length)

$$L = \frac{\mu l}{2\pi}\left[\ln\left(\frac{b}{a}\right) + \frac{\delta}{2}\left(\frac{1}{a} + \frac{1}{b}\right)\right] \quad \text{coaxial cable, high frequency.} \qquad (2.150)$$

If the conductivity of the conductors is allowed to approach infinity, then δ approaches zero, the currents become *surface* currents, and the inductance is entirely external:

$$L_{\text{ext}} = \frac{\mu l}{2\pi} \ln(b/a) \quad \text{coaxial cable, perfect conductors.} \tag{2.151}$$

This is called the *external* inductance because there is no internal flux.

What is the *series* resistance of the coaxial cable under the conditions outlined for Equations (2.149), (2.150), and (2.151)?

Comparing Equations (2.151) and (2.140) shows that (μ and ε are independent of position) for a 1-m length ($l = 1$ m), we have

$$\boxed{L_{\text{ext}} C = \mu \varepsilon}. \tag{2.152}$$

It can be shown that under the stated conditions this result is *independent of geometry*. This is a result of the duality in the fields between the capacitance and inductance problem. **D** or **E** lines in the capacitance problem fall on $\Phi_m =$ constant lines in the external inductance problem, and **B** and **H** lines in the external inductance problem fall on $\Phi =$ constant lines in capacitance problem. Stated more basically, **E** must satisfy $\nabla^2 \mathbf{E} = 0$ in the surrounding medium, and **H** must satisfy the dual equation $\nabla^2 \mathbf{H} = 0$ in the surrounding medium.

Mutual inductance is calculated in the same manner as self-inductance, but we are concerned with the flux produced by the current I_1 (for example) that links the path of the current I_2 or vice versa. That is, from Equation (2.148), we get

$$L_{12} = \frac{\Psi_{ml,12}}{I_1} = \frac{1}{I_1} \iint_s \frac{I_{2\,\text{enc}}}{I_2} d\Psi_{ml,12}, \tag{2.153}$$

where $\psi_{ml,12}$ is the magnetic flux produced by I_1 that links I_2. It can be shown that when μ is a scalar constant, then $L_{12} = L_{21}$. L_{21} is obtained from Equation (2.153) by interchanging subscripts.

■ Example 2.27

Consider the magnetically coupled circuits shown in Figure 2.42. The currents are filamentary, so the labor required to find the mutual inductance should be much less than for the more general case. Using Equation (2.153), we have

$$L_{12} = \frac{1}{I_1} \int_1^2 \int_0^1 \frac{I_2}{I_2} \frac{\mu_0 I_1}{2\pi\rho} \mathbf{a}_\phi \cdot \mathbf{a}_\phi \, d\rho \, dz$$

or

$$L_{12} = \frac{\mu_0}{2\pi} \ln 2 = 0.139 \quad (\mu\text{H}).$$

■

2.17 CONCLUDING REMARKS

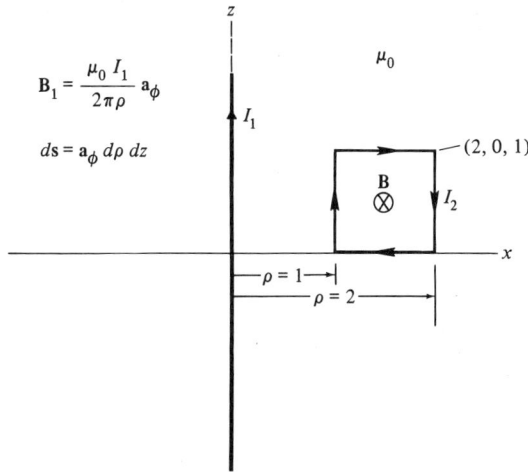

Figure 2.42. Infinite filamentary current coupled to a square filamentary loop.

Capacitance and inductance will be reconsidered in terms of energy when the general Poynting theorem is introduced in Chapter 3.

2.17 CONCLUDING REMARKS

This chapter was primarily concerned with developing a set of equations for electrostatics and magnetostatics called Maxwell's equations. These equations, in point form and integral form were developed from experimental laws, and are summarized:

$$\nabla \times \mathbf{E} = 0, \quad \oint \mathbf{E} \cdot d\mathbf{l} = 0; \tag{2.154}$$

$$\nabla \times \mathbf{H} = \mathbf{J}, \quad \oint \mathbf{H} \cdot d\mathbf{l} = I; \tag{2.155}$$

$$\nabla \cdot \mathbf{D} = \rho_v, \quad \oint_s \mathbf{D} \cdot d\mathbf{s} = \iiint_{vol} \rho_v \, dv; \tag{2.156}$$

$$\nabla \cdot \mathbf{B} = 0, \quad \oint_s \mathbf{B} \cdot d\mathbf{s} = 0; \tag{2.157}$$

$$\nabla \cdot \mathbf{J} = 0, \quad \oint_s \mathbf{J} \cdot d\mathbf{s} = 0. \tag{2.158}$$

Also, other fundamental results were

$$\mathbf{D} = \varepsilon\mathbf{E}, \qquad (2.159)$$

$$\mathbf{B} = \mu\mathbf{H}, \qquad (2.160)$$

$$\mathbf{J} = \sigma\mathbf{E}, \qquad (2.161)$$

$$\mathbf{F} = Q(\mathbf{E} + \mathbf{u} \times \mathbf{B}). \qquad (2.162)$$

The experience gained by working with these equations will be of much help in the chapters that follow.

Materials and boundary conditions were also briefly introduced and discussed in order to be able to treat several examples, including those involving the calculation of capacitance, conductance, and inductance. We will need this information when transmission lines are considered later.

Potential functions were introduced as auxiliary functions in that they offer alternate methods for calculating the (usually) more important field quantities. We will meet them again.

The relation between *energy* and the field quantities is investigated in Section 3.6 when Poynting's theorem is introduced. Capacitance and inductance are then reconsidered.

REFERENCES

Boast, W. B. *Vector Fields*. New York: Harper & Row, 1964. Sketches of the fields for many of the commonly used examples are given.

Cheng, D. K. *Field and Wave Electromagnetics*. Reading, MA: Addison-Wesley, 1983.

Dekker, A. J. *Electrical Engineering Materials*. Englewood Cliffs, NJ: Prentice-Hall, 1959. A short and readable book on the electrical properties of materials, including dielectrics.

Edminister, Joseph A. *Electromagnetics*, 4th ed. New York: McGraw-Hill, 1981. Many of our topics are covered at the same level in this textbook.

Fink, D. G. and Carroll, J. M. *Standard Handbook for Electrical Engineers*, 10th ed. New York: McGraw-Hill, 1968.

Hayt, Jr., W. H. *Engineering Electromagnetics*, 5th ed. New York: McGraw-Hill, 1989.

Kraus, J. D. and Carver, K. R. *Electromagnetics*, 2nd ed. New York: McGraw-Hill, 1973.

Lorrain, P. and Corson, D. R. *Electromagnetic Fields and Waves*. San Francisco: Freeman, 1970. Relatively is discussed in Chapter 5.

Moon, P. and Spencer, D. E. *Field Theory for Engineers*. New York: Van Nostrand, 1961. A highly mathematical book. Excellent for boundary value problems in coordinate systems other than the three usual ones.

Neff, Jr., H. P. *Basic Electromagnetic Fields*, 1st ed. New York: Harper & Row, 1981. The electromagnetic field is derived from Coulomb's law in Appendix F.

Page, L. "A Derivation of the Fundamental Relations of Electrodynamics from Those of Electrostatics," *Am. J. Sci.* **34** (1912):57–68. This is apparently the original work whereby the magnetostatic field is obtained from the electrostatic field of moving charges.

Paul, C. R. and Nasar, S. A. *Introduction to Electromagnetic Fields*, 2nd ed. New York: McGraw-Hill, 1987.

Plonsey, R. and Collin, R. E. *Principles and Applications of Electromagnetic Fields*. New York: McGraw-Hill, 1961. A classical approach to electromagnetics on a slightly higher level.

Plonus, M. A. *Applied Electromagnetics*. New York: McGraw-Hill, 1978. Relativity is discussed in Chapter 12.

Ramo, S., Whinnery, J. R., and Van Duzer, T. *Fields and Waves in Communications Electronics*, 2nd ed. New York: Wiley, 1984. Primarily a graduate level textbook, a discussion of polarization paralleling that herein is included in Chapter 2.

Shedd, P. C. *Fundamentals of Electric Waves*. New York: Prentice-Hall, 1954. The electromagnetic field is derived from Coulomb's law in Chapter 18.

Zahn, M. *Electromagnetic Field Theory*. New York: Wiley, 1979.

PROBLEMS

1. Point charge $Q_1 = 10^{-9}$ C is located at the origin, while point charge Q_2 is located at (0,0,1). If $E_z = 0$ at (2,2,2), find Q_2.

2. The $z = 0$ plane is covered with the uniform surface charge density $\rho_s = 10^{-9}$ C/m^2, while at the same time the point charge $Q = 10^{-9}$ C is located at (0,0,1). Where can an electron be place so that, if released, it will not move?

*3. The $z = 0$ plane contains the concentric circular line charge densities $\rho_l = \rho_s \Delta\rho$ located at radii $\Delta\rho/2$, $3\Delta\rho/2$, $5\Delta\rho/2$, ..., $(2n-1)\Delta\rho/2$, ...

   (a) Show that, for N such filamentary loops

   $$E_z(0,0,z) = \frac{\rho_s}{2\varepsilon_0} \sum_{n=1}^{N} \frac{z(2n-1)(\Delta\rho)^2/2}{\{z^2 + [(2n-1)\Delta\rho/2]^2\}^{3/2}}.$$

   (b) Let $\Delta\rho = 0.1$ m and $N = 100$ so that $(2N-1)\Delta\rho/2 = 9.95$. Plot $E_z(0,0,z)/(\rho_s/2\varepsilon_0)$ versus z for $0 \leq z \leq 1$.

   (c) Comment on the results.

4. A uniform surface charge density $\rho_s = 10^{-9}$ C/m^2 lies in the $z = 0$ plane for $-50 \leq x \leq 50$ and $-0.05 \leq y \leq 0.05$. Using reasonable approximations find $E(0,0,z)$ for (a) $z = 10^{-3}$, (b) $z = 1$, (c) $z = 10^4$.

*5. Figure 2.43 shows a two-dimensional *electrostatic deflection system*. The upper deflection plate is located $x = d/2$, $0 \leq z \leq l$, while the lower plate is located at $x = -d/2$, $0 \leq z \leq l$. The (assumed) uniform field is given by

---

*The more difficult problems are marked with an asterisk.

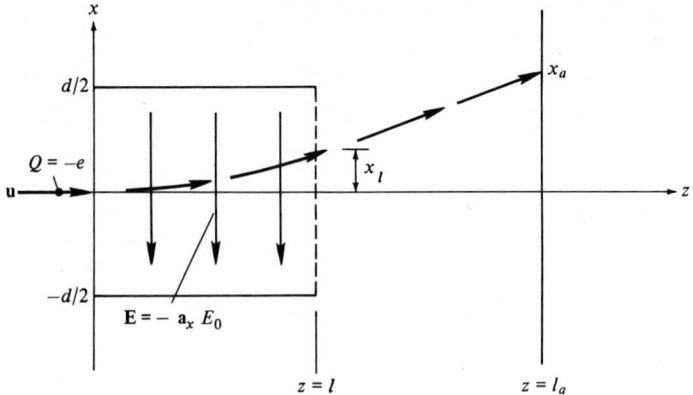

Figure 2.43. Two-dimensional electrostatic deflection system.

$\mathbf{E} = -\mathbf{a}_x E_0$ for the region between the plates only. An electron is accelerated by a cathode-accelerating grid arrangement (not shown) so that it enters at the origin with a velocity $\mathbf{u} = \mathbf{a}_z u_0$. Find x_1 at $z = l$ and x_a at $z = l_a$.

6. Find the work required to transport an electron ($Q = -1.602 \times 10^{-19}$ C) from (1,1,1) to (2,2,2) (choosing any path) in the field of
   (a) A point charge $Q = 10^{-9}$ C at the origin.
   (b) An infinite line charge density $\rho_l = 10^{-9}$ C/m on the $z = 0$ axis.
   (c) An infinite surface charge density $\rho_s = 10^{-9}$ C/m² in the $z = 0$ plane.

7. The $z = 0$ plane contains the uniform surface charge density 10^{-9} C/m², and this plane is also the reference for zero potential. Find $\Phi(z)$.

8. Newton's law of gravity is dual to Coulomb's law and given by

$$\mathbf{F}_1 = G \frac{m_1 m_2 (\mathbf{r}_2 - \mathbf{r}_1)}{|\mathbf{r}_1 - \mathbf{r}_2|^3},$$

where

$$G = 6.664 \times 10^{-11} \quad (\text{m}^3/\text{kg s}^2).$$

If the masses of earth and moon are 5.98×10^{24} kg and 7.35×10^{22} kg, and their centers are separated by 3.848×10^8 m (on the average), find
   (a) The force of attraction between the earth and moon. Let the center of the earth be the origin, and let the radius of the earth be 6.371×10^6 m (average).
   (b) The force on a point mass m at the earth's surface. Ignore the perturbing effect of the moon.
   (c) The acceleration due to gravity at the earth's surface.

(d) The earth's gravitational potential field Φ_g (zero reference at infinity).
(e) The energy required to move a point mass from the earth's surface to infinity.
(f) The escape velocity if the supplied energy is kinetic energy.

*9. The *electrostatic dipole* consists of $+Q$ at $(0,0,d/2)$, and $-Q$ at $(0,0,-d/2)$.
   (a) Show that the $z = 0$ plane is the equipotential surface $\Phi = 0$.
   (b) Show that equipotential surfaces are given by $r = K\sqrt{\cos\theta}$ for $r \gg d$.
   (c) If a uniform external field $\mathbf{E} = \mathbf{a}_x E_0$ is applied to the dipole in (a), find the energy required to rotate the dipole to a stable position.

10. The potential at a point z between the heated cathode $[\Phi(0) = 0]$ and the anode $[\Phi(d) = V_0]$ for a planar *vacuum diode* is given by $\Phi(z) = V_0(z/d)^{4/3}$. What is the electric field intensity midway between the cathode and anode?

11. Prove that the electric field intensities of (a) a point charge at the origin, (b) an infinite uniform line charge density on the z axis, and (c) a uniform surface charge density on the $z = 0$ plane are conservative.

12. What charge distributions produce the fields below?
   (a) $\mathbf{D} = (1/r^2)\mathbf{a}_r$.
   (b) $\mathbf{D} = (1/\rho)\mathbf{a}_\rho$.
   (c) $\mathbf{E} = (10r)^{-2}\mathbf{a}_r$, $r \geq a$ only.

13. The charge per unit length (normal to the cross section) on the outside of the inner conductor is the negative of the charge per unit length on the inside of the outer conductor of the coaxial cable shown in Figure 2.25:
   (a) Show that $\mathbf{E} = \mathbf{a}_\rho(a\rho_{sa}/\varepsilon\rho)$.
   (b) What is the relation between V_0 and ρ_{sa}?

14. The region $z > 0$ is free space, while the region $z < 0$ has $\varepsilon_R = 4$. The uniform electric field for $z > 0$ is 10 V/m and in a radial direction for which $\theta = 30°$ and $\phi = 45°$. Find $\mathbf{D}$ and $\mathbf{E}$ everywhere in Cartesian coordinates.

*15. A dielectric sphere $(\varepsilon = \varepsilon_R \varepsilon_0)$ of radius a is centered at the origin. A uniform electric field $\mathbf{E} = E_0 \mathbf{a}_z$ (without the dielectric sphere) is applied. The potential (with the sphere) is given by

$$\Phi(r,\theta) = \begin{cases} -\dfrac{3rE_0\cos\theta}{\varepsilon_R + 2}, & r \leq a; \\ -rE_0\cos\theta + \dfrac{a^3 E_0}{r^2}\dfrac{\varepsilon_R - 1}{\varepsilon_R + 2}\cos\theta, & r \geq a. \end{cases}$$

   (a) Find $\mathbf{E}$ for $r < a$.
   (b) Find $\mathbf{E}$ for $r > a$.
   (c) Find $\mathbf{E}$ for $r \gg a$.
   (d) Show that all boundary conditions are satisfied at $r = a$.

*16. For the two-wire line of Figure 2.39 show that

(a) $$\Phi = \frac{V_2}{2 \cosh^{-1}(D/d)} \ln \frac{(x+a)^2 + y^2}{(x-a)^2 + y^2},$$

(b) $$E_x = \frac{2V_2 a}{\cosh^{-1}(D/d)} \frac{x^2 - a^2 - y^2}{[(x+a)^2 + y^2][(x-a)^2 + y^2]},$$

(c) $$E_y = \frac{4V_2 a}{\cosh^{-1}(D/d)} \frac{xy}{[(x+a)^2 + y^2][(x-a)^2 + y^2]}.$$

(d) If $\varepsilon_R = 1$, $D = 2$ cm, and $d = 0.25$ cm, what is the maximum voltage that can be used?

17. As a crude model of the earth and a horizontal layer of charged clouds above, consider a pair of large parallel conducting plates with a lower plate (earth) at zero potential and the upper plate (cloud) at a negative potential.
   (a) Sketch equipotentials and **E** lines.
   (b) Place a conducting cone with a small apex angle resting (base down) on the lower plate with its tip about midway between the plates and repeat (a).
   (c) Explain how a lightning rod works.

18. Two semi-infinite conducting planes are inclined at an angle of ϕ_0 with respect to each other. At the apex, the two planes do not quite touch so that a battery of potential V_0 can be connected as shown in Figure 2.44.
   (a) Find $\Phi(\phi)$ between the planes.
   (b) Find ρ_s on the plane at $\phi = 0$.
   (c) Is the capacitance finite for a unit length in the z direction?

19. The remaining one-dimensional boundary value problem for Cartesian, cylindrical, and spherical coordinates is that for which equipotentials occur for $\theta = $ constant. Its solution is

$$\Phi = V_0 \frac{\ln(\tan \theta/2)}{\ln(\tan \theta_0/2)}.$$

Describe two geometries for which this result applies.

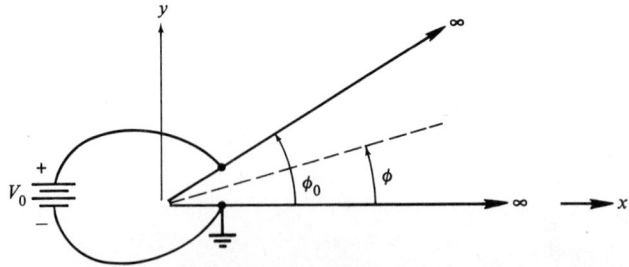

Figure 2.44. Geometry for Problem 18 (inclined, semi-infinite conducting planes).

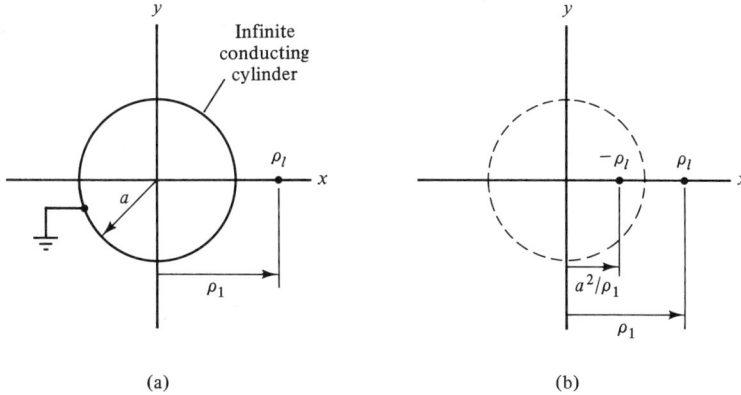

Figure 2.45. (a) Original problem: infinite, uniform line charge density outside and parallel to an infinite, grounded conducting cylinder. (b) Image problem equivalent to that shown in (a) for $\rho > a$.

*20. (a) Show that the problem in Figure 2.45(a) is equivalent ($\rho > a$) to the problem in Figure 2.45(b).
  (b) Find $\Phi(\rho, \phi)$, $\rho > a$.

*21. (a) Show that the problem in Figure 2.46(a) is equivalent ($r > a$) to the problem in Figure 2.46(b)
  (b) Find $\Phi(r, \theta, \phi)$, $r > a$.

*22. The differential equations and boundary conditions for Figure 2.47(a) are

$$\nabla^2 \Phi = 0, \text{ except at } (0, 0, d/2);$$

$$E_{\rho 1} = E_{\rho 2}, \quad z = 0; \qquad \varepsilon_1 E_{z1} = \varepsilon_2 E_{z2}, \quad z = 0.$$

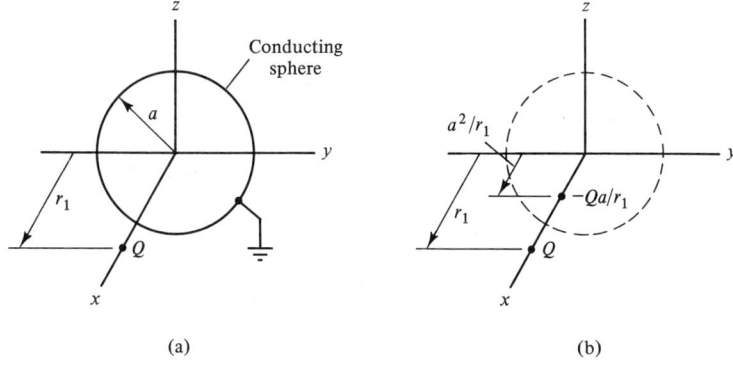

Figure 2.46. (a) Original problem: point charge outside a grounded conducting sphere. (b) Image problem equivalent to that shown in (a) for $r > a$.

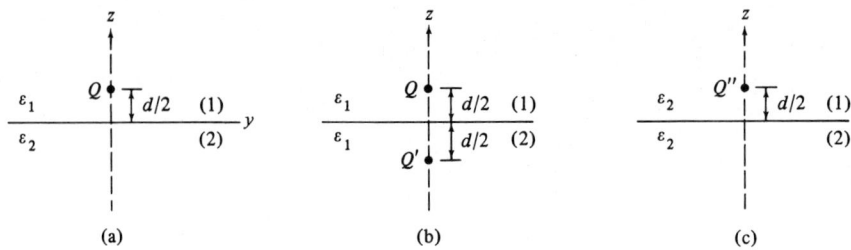

Figure 2.47. Another image problem (Problem 22).

Show that this problem is equivalent to that in Figure 2.47(b) for $z > 0$ (only) if $Q' = Q(\varepsilon_1 - \varepsilon_2)/(\varepsilon_1 + \varepsilon_2)$ and that this problem is equivalent to that in Figure 2.47(c) for $z < 0$ (only) if $Q'' = Q(2\varepsilon_2)/(\varepsilon_1 + \varepsilon_2)$.

*23. Find the surface charge density on the grounded plane in Example 2.11. Plot this charge density as a function of ρ for $\phi = 0$.

*24. Show that the total charge on the grounded plane of Figure 2.15 is $-Q$. The use of Gauss's law will avoid integration.

25. If $\mathbf{J} = \mathbf{a}_z/\rho$, $0 \le \rho \le a$,
   (a) Find $\nabla \cdot \mathbf{J}$ everywhere.
   (b) Find the total current in the $+\mathbf{a}_z$ direction.
   (c) Find the total current out of any closed surface.

*26. A long cylindrical conductor of conductivity σ_c and radius a is buried horizontally at a depth d ($d \gg a$) in earth whose conductivity is σ_e ($\sigma_c \gg \sigma_e$) as shown in Figure 2.48(a). A current I is being supplied to the conductor at a remote point. Insofar as $\mathbf{E}$ and $\mathbf{J}$ in the earth are concerned this problem can be replaced with the image problem of Figure 2.48(b) because boundary conditions ($J_n = E_n = 0$) are satisfied on the plane interface. Note that this

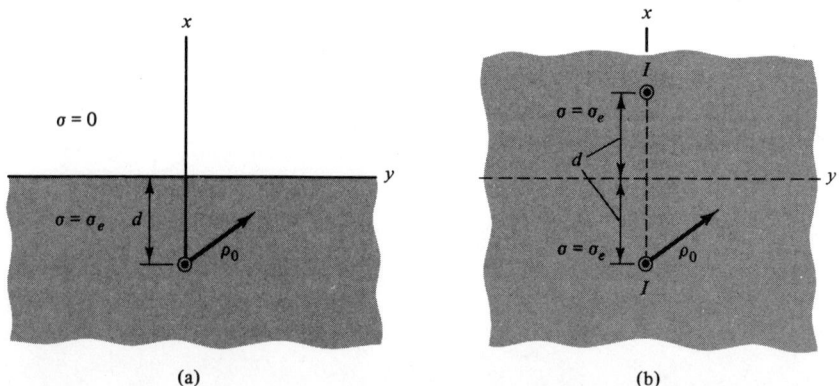

Figure 2.48. (a) Conductor buried in the ground. (b) Image problem for (a) (Problem 26).

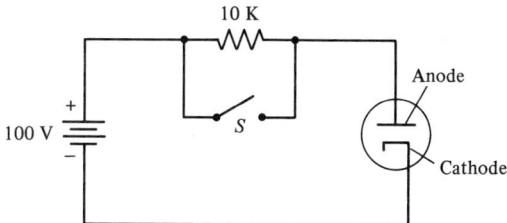

Figure 2.49. Nonlinear circuit (Problem 27).

requires that the current in the image conductor be in the same direction as that in the buried conductor. Assuming that the current density in the earth is $J = I/(2\pi\rho_0 l)$, find the maximum value of E_y at the earth's surface.

*27. A planar vacuum diode has a heated cathode at $z = 0$ [$\Phi(0) = 0$] and an anode at $z = d$ [$\Phi(d) = V_0$]. If an electron escapes from the cathode with zero initial velocity, then the total initial energy is zero, and the total energy is constant and zero ($-e\Phi + mu^2/2 = 0$). It can then be shown that $\Phi(z) = V_0(z/d)^{4/3}$.
  (a) Find $\rho_v(z)$.
  (b) Find $J_z(z)$ (A/m^2).
  (c) Show that $I = K V_0^{3/2}$ (*Child–Langmuir* or *three-halves power law*).
  (d) Find the time required for an electron to leave the cathode and reach the anode (*transit time*) if $V_0 = 100$ V and $d = 1$ mm.
  (e) If the current in Figure 2.49 is 10 mA when the switch is closed, what is the current when the switch is open?

28. An *idealized toroid* can be thought of as a finite length solenoid bent around to close on itself to form a doughnut shape as shown in Figure 2.50. The surface current density at $\rho = \rho_a - a$ is J_{sz}. It can be shown that

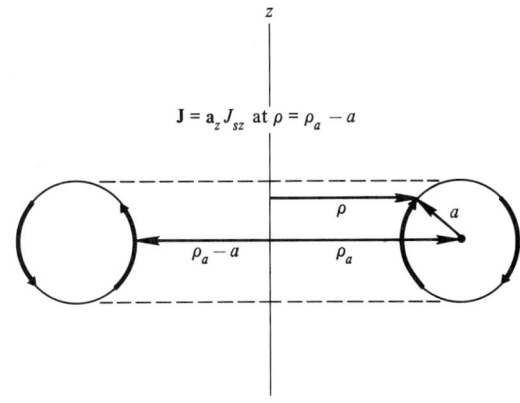

Figure 2.50. Idealized toroid (Problem 28).

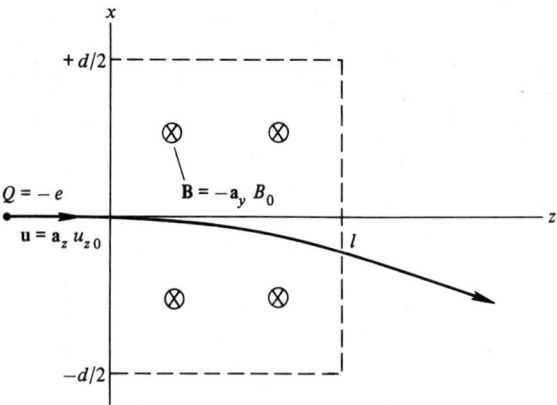

Figure 2.51. Magnetostatic deflection system (Problem 29).

$$\mathbf{H} = \begin{cases} J_{sz}\dfrac{\rho_a - a}{\rho}\mathbf{a}_\phi, & \text{inside toroid;} \\ 0, & \text{outside toroid.} \end{cases}$$

Find $\oint \mathbf{H} \cdot d\mathbf{l}$ for a circular path of radius b in the $z = 0$ plane if
(a) $0 < b < \rho_a - a$,
(b) $\rho_a - a < b < \rho_a + a$, and
(c) $b > \rho_a + a$.

*29. A uniform magnetic flux density $\mathbf{B} = -B_0 \mathbf{a}_y$ exists in the region, $-d/2 \leq x \leq d/2$, $0 \leq z \leq l$. Assume that there are no variations with y. An electron enters this field at $(0,0,0)$ with an initial velocity $u_{z0}\mathbf{a}_z$ as shown in Figure 2.51. Find the equations of motion for the electron while in the applied field (*magnetostatic deflection system*).

30. Find the force of repulsion per unit length between the two conductors of a planar transmission line. The two conductors are parallel plane strips, of width b and separation d, carrying equal and opposite surface currents. Assume $b \gg d$, and ignore fringing.

31. An idealized current density is given by $\mathbf{J} = \mathbf{a}_\rho J_{s\rho}/\rho$, $a \leq \rho \leq b$, $z = 0$, when a uniform external magnetic flux density $\mathbf{B} = -B_0 \mathbf{a}_z$ (Wb/m²) is applied. This is an *idealized axial gap motor*. See Figure 2.52.
   (a) Find the vector torque on the current if $J_{s\rho} = 10^3$ (A/m), $a = 1$ cm, $b = 5$ cm, and $B_0 = 1$ Wb/m².
   (b) If the armature rotates at 500 rpm, what power is provided?

32. A coaxial vacuum diode has a filamentary cathode on the z axis and its anode is located at $\rho = a$. If $\Phi = 0$ at $\rho = 0$, and $\Phi = V_0$ at $\rho = a$, then it can be shown that $\Phi(\rho) = A\rho^{2/3}$.
   (a) Find A.
   (b) Find $\rho_v(\rho)$.

PROBLEMS

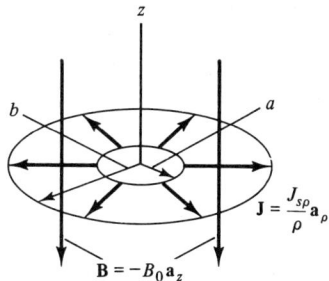

Figure 2.52. Idealized axial gap (dc) motor.

33. Region $z > 0$ has $\mu_R = 4$, while region $z < 0$ has $\mu_R = 1$. **B** is uniform for $z > 0$ with a magnitude of 1 Wb/m^2 and in a radial direction for which $\theta = 60°$ and $\phi = 45°$. Find **B** and **H** for $z < 0$.

*34. An infinitely long cylinder of relative permeability μ_R and a radius a is placed so that its axis is the z axis in a magnetic field that was (in free space) previously uniform $\mathbf{H} = H_0 \mathbf{a}_x$.
   (a) List the boundary conditions on **H** in terms of Φ_m.
   (b) Find Φ_m. Use Laplace's equation.
   (c) Show that the field inside the cylinder is uniform.

*35. The differential equations and boundary conditions for Figure 2.53(a) are

$$\nabla^2 A_x = 0, \text{ except at } (x, 0, d/2);$$

$$H_{y1} = H_{y2}, \quad z = 0; \qquad \mu_1 H_{z1} = \mu_2 H_{z2}, \quad z = 0.$$

Show that this problem is equivalent to that in Figure 2.53(b) for $z > 0$ (only) if $I' = I(\mu_2 - \mu_1)/(\mu_2 + \mu_1)$, and that this problem is equivalent to that in Figure 2.53(c) for $z < 0$ (only) if $I'' = I(2\mu_1)/(\mu_2 + \mu_1)$.

36. A magnetic core is shown in Figure 2.54. The mean lengths are as shown and the cross-sectional area is 4 cm^2 everywhere. If $H = 500B$, and a 1000-turn coil carrying 50 mA is placed on the left leg, find

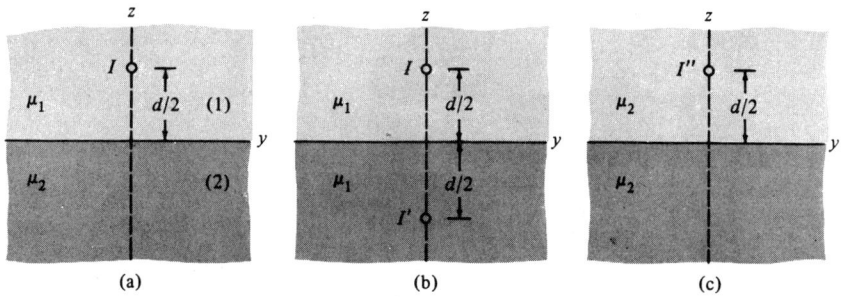

Figure 2.53. Filamentary current above a plane interface between two magnetic materials. (b) Equivalent problem for $z > 0$. (c) Equivalent problem for $z < 0$.

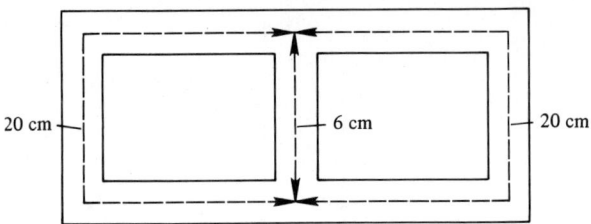

Figure 2.54. Magnetic core (Problem 36).

(a) B in each leg.
(b) The inductance of the coil.
(c) Repeat (a) and (b) if a 0.1 mm air gap is cut in the center leg.

37. If a convention is adopted whereby a dot is placed at a terminal of each of the windings where an *entering* current produces a flux that is *adding* to the flux being produced by the other winding, where should the dots be placed for the transformer in Figure 2.55?

38. A solenoid has N_1 turns, length l, and area s. A second solenoid has N_2 turns, length l, and area s. The physical arrangement is shown in Figure 2.56(a). Assume that a current I enters terminal a. If terminals b and c are connected, make reasonable approximations and find L for (a) an iron core, $\mu \gg \mu_0$; (b) an air core, $\mu = \mu_0$; (c) repeat (a) if terminals b and d are connected instead of b and c; (d) repeat (b) for terminals b and d connected; (e) repeat for the arrangement shown in Figure 2.56(b).

39. Refer to the coaxial capacitor of Figure 2.25.
   (a) Find E_ρ at $\rho = a$ in terms of V_0.
   (b) Find the minimum value of E_ρ at $\rho = a$ when b is fixed but a is variable.
   (c) What is the capacitance in (b)?

40. A transmission line is often fabricated as stripline. Assume that it consists of a thin strip of width 2 cm and spaced 0.25 cm from a large ground plane with a solid dielectric ($\varepsilon_R = 4$) between the two. Ignoring fringing of the field find the capacitance per unit length.

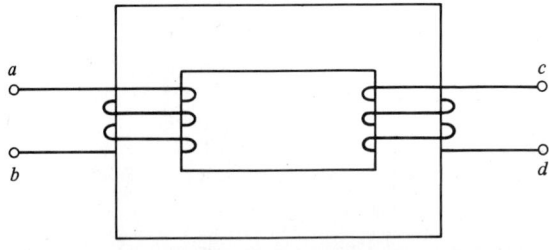

Figure 2.55. Geometry for Problem 37.

PROBLEMS

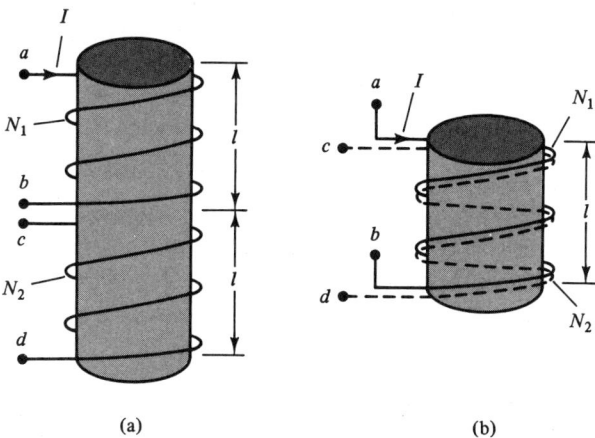

Figure 2.56. Two windings on a magnetic core. (a) Cascaded windings. (b) Bifilar windings.

*41. A cylindrical washer has inner and outer radii a and b, respectively. Its conductivity is σ and its thickness is t. Find the resistance between:
   (a) Inner and outer radii.
   (b) The flat sides.
   (c) The sides of a very thin radial cut all the way through the material.

42. A parallel-plate capacitor is charged to V_0 volts and the battery is disconnected. The solid dielectric is then removed. What is the new potential difference between the plates.

*43. A parallel-plate capacitor has plates of area 10^{-2} m^2 spaced by 10^{-2} m. The relative permittivity varies as $\varepsilon_R(z) = 1 + (z/d)^2$ when the lower plate is located at $z = 0$ and the upper plate is located at $z = d$. Find the capacitance.

44. Find the capacitance per unit length of the two-dielectric coaxial capacitor shown in Figure 2.57.

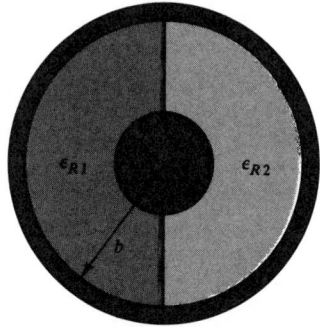

Figure 2.57. Two-dielectic coaxial capacitor.

45. What is the resistance per 100 m for a circular conductor that is steel for $0 \le \rho \le 10^{-2}$ and aluminum for $10^{-2} \le \rho \le 2 \times 10^{-2}$? Assume uniform current densities. Use $\sigma = 0.2 \times 10^7$ for steel. What is the "effective" conductivity of this conductor?

46. It is possible to construct an electric circuit with a pencil and a piece of paper. Assuming that graphite has a conductivity of 7×10^4 ℧/m, how "thick" would a 1-kΩ resistor be if it is 2 cm long and 1 mm wide?

*47. Certain junction diodes, called *varactor diodes*, behave as voltage dependent capacitors:

$$C = K(V_b + V + \Delta V)^{-1/2} = C_0[1 + \Delta V/(V_b + V)]^{-1/2},$$

where V_b is the unbiased barrier voltage, V is the external bias voltage, ΔV is the incremental bias voltage, and C_0 is the capacitance when $\Delta V = 0$. This diode is to be used to produce *frequency modulation*. If $V_b + V = 4$, what frequency deviation is produced for 1-mV modulating source (ΔV) when the carrier frequency is 100 MHz?

*48. Find the mutual inductance between an infinite filamentary wire on the z axis and a filamentary triangular loop with corners at $(0.5,0,0)$, $(1,0,0.5)$, and $(1,0,-0.5)$.

49. A filamentary current loop described by $\mathbf{m} = \mathbf{a}_z$ is centered at $(0,0,0.5)$ and an identical loop is centered at $(0,0,-0.5)$. Using reasonable approximations, find the mutual inductance if $\mu = \mu_0$ and the loop area is 0.05 m^2.

50. Find the external inductance of the stripline of Problem 40.

51. A uniform current density $\mathbf{J} = -J_0 \mathbf{a}_z$ exists in a conducting slab: $-\infty < x < \infty$, $-t/2 < y < t/2$, $-\infty < z < \infty$.
    (a) Show that

    $$\frac{d^2 A_z}{dy^2} = \mu J_0, \quad -t/2 < y < t/2.$$

    (b) Show that $\mathbf{H} = J_0 y \mathbf{a}_x$, $-t/2 < y < t/2$, if $H_x(0) = 0$.
    (c) Remove all but a finite width w $(-w/2 < x < w/2)$ of the slab, assume that $\mathbf{H}$ does not change (no fringing), and use Equation (3.65) to show that the internal inductance per unit length is

    $$L_{int} = \frac{\mu t}{12w} \quad \text{(H/m)}$$

52. Starting with Equation (2.147) show that inductance can be determined by

$$L = \frac{1}{I} \oint \mathbf{A} \cdot d\mathbf{l} \quad \text{(H)}.$$

PROBLEMS

53. An electron in the uniform field $\mathbf{B} = \mathbf{a}_z B_0$ experiences a force $\mathbf{F} = -e\mathbf{u} \times \mathbf{B} = m\mathbf{a}$. Express this relation in cylindrical coordinates, equate the ρ components and the ϕ components, and obtain a pair of coupled equations. Let $\rho = \rho_0$ (constant) to reduce the equations, and show that $\omega_c = eB_0/m$ (the *cyclotron frequency*) is the angular velocity of an electron in a circular orbit with radius $\rho_0 = u_\phi/\omega_c$.

54. Use Gauss's law to verify Equations (2.16) and (2.17).

3
MAXWELL'S EQUATIONS

This chapter begins by considering the consequences of conservation of charge in the general time-varying case. Next, those equations that retain their static point forms and apply to the general case are presented and discussed. Faraday's law and Ampere's (generalized) law are then introduced. We then have a complete set of equations that govern electromagnetics. The consolidation of this set of equations, including the concept of displacement current density, was primarily due to J. C. Maxwell. Therefore, this set of equations bears his name. One may accurately state that a study of electromagnetic field theory is a study of Maxwell's equations.

Auxiliary (scalar and vector) potential functions will be derived. Whether or not these potential functions are used to find electric or magnetic fields, or whether Maxwell's equations are used in a direct manner without the potentials, will depend to some extent on the problem itself. There are certain problems, which we will encounter in later chapters, that can be very easily solved using the potential functions, but other problems can be easily solved without them.

A general conservation of energy relationship, called Poynting's theorem, will be derived. This theorem, when applied to familiar practical situations, gives results that agree entirely with all of our prior experience, with both "field" theory and "circuit" theory.

Instantaneous field quantities (field quantities that are explicitly time-dependent) are denoted by script letters. Phasor field quantities are denoted by the same symbols that have been used for static (or steady) field quantities, although it will be explicitly indicated that these phasor quantities are functions of ω when it is not obvious that this is the case.

3.1 CONSERVATION OF CHARGE

The vector field quantity that relates to electric current is the time-varying current density, $\mathcal{J}$, measured in amperes per square meter. The scalar electric current,

3.1 CONSERVATION OF CHARGE

i, is given by the integral of the normal component of the current density over whatever *open* surface we wish to consider

$$i = \iint_s \mathcal{J} \cdot d\mathbf{s} \quad (A). \tag{3.1}$$

The current out of a *closed* surface is

$$i = \oiint_s \mathcal{J} \cdot d\mathbf{s}. \tag{3.2}$$

The *outward* flow of this positive charge must, in general, be accompanied by a *decrease* in time of positive charge in the volume represented by the closed surface s:

$$i = \oiint_s \mathcal{J} \cdot d\mathbf{s} = -\frac{dq}{dt}. \tag{3.3}$$

The charge q in the volume is obtained as the integral of the volume charge density ρ_v throughout the volume:

$$q = \iiint_{vol} \rho_v \, dv \quad (C). \tag{3.4}$$

Therefore, we have

$$i = \oiint_{s \leftarrow} \mathcal{J} \cdot d\mathbf{s} = -\frac{d}{dt} \iiint_{\rightarrow vol} \rho_v \, dv, \tag{3.5}$$

where the long double-headed arrow means that the volume on the right-hand side is that within the surface on the left-hand side. If this surface is fixed in time (not moving), then we may differentiate partially with time *before* integrating:

$$\oiint_{s \leftarrow} \mathcal{J} \cdot d\mathbf{s} = -\iiint_{\rightarrow vol} \frac{\partial \rho_v}{\partial t} \, dv. \tag{3.6}$$

The divergence theorem (Chapter One) equates the surface integral on the left-hand side of Equation (3.6) to the *divergence* of $\mathcal{J}$ throughout the volume defined by the closed surface:

$$\oiint_{s \leftarrow} \mathcal{J} \cdot d\mathbf{s} = \iiint_{\rightarrow vol} \operatorname{div} \mathcal{J} \, dv. \tag{3.7}$$

The divergence of a vector, such as $\mathcal{J}$, is the limit of the *flux* of the vector (electric current here) per unit volume out of a small volume as that volume is reduced to zero. In terms of the vector operator ∇ it is symbolized by div $\mathcal{J} = \nabla \cdot \mathcal{J}$. Thus, combining Equations (3.6) and (3.7), we have

$$\iiint_{\text{vol}} \nabla \cdot \mathcal{J}\, dv = -\iiint_{\text{vol}} \frac{\partial \rho_v}{\partial t}\, dv. \tag{3.8}$$

Since the volume is completely general, the integrands in Equation (3.8) must be identical so that

$$\boxed{\nabla \cdot \mathcal{J} = -\frac{\partial \rho_v}{\partial t}} \quad \text{(conservation of charge)}. \tag{3.9}$$

This result is called the continuity of current or conservation of charge equation, because if it does not hold, then charges are being created (or destroyed). It states that the limit of the electric current diverging from a small volume per unit volume is equal to the time rate of decrease of charge per unit volume at every point in space.

In the steady current case ($\partial/\partial t \equiv 0$), we obtain

$$\nabla \cdot \mathbf{J} = 0 \quad \text{(conservation of charge for steady current)} \tag{3.10}$$

as in Equation (2.71).

3.2 OTHER FIELD QUANTITIES

Current density is related to charge density with velocity $\mathbf{u}$ by

$$\mathcal{J} = \rho_v \mathbf{u} \quad \text{(convection current density)}. \tag{3.11}$$

This result applies for convection current densities (absence of conductors) as well as for conduction current densities if $\mathbf{u} = \mathbf{u}_d$ = drift velocity, although for conduction current densities it is more common to use

$$\mathcal{J} = \sigma \mathcal{E} \quad \text{(Ohm's law for field theory)}, \tag{3.12}$$

where $\mathcal{E}$ is the *electric field intensity* (V/m) and σ is the *conductivity* (℧/m).
The *constitutive* relations are

$$\mathcal{D} = \varepsilon \mathcal{E}, \tag{3.13}$$

where $\mathcal{D}$ is the *electric flux density* (C/m²) and ε is the *permittivity* (F/m), and

$$\mathcal{B} = \mu \mathcal{H}, \tag{3.14}$$

where $\mathcal{B}$ is the *magnetic flux density* (Wb/m²), μ is the *permeability* (H/m), and $\mathcal{H}$ is the *magnetic field intensity* (A/m). The electric field intensity is defined as the force per unit charge, and the magnetic field intensity is the force per unit current element.[1] The definitions are evident in the Lorentz force equation

$$\mathcal{F} = q(\mathcal{E} + \mathbf{u} \times \mathcal{B}) \quad \text{(N)} \quad \textit{(Lorentz force)}. \tag{3.15}$$

Gauss's law is fundamental, and in point form it is given by

$$\nabla \cdot \mathcal{D} = \rho_v \quad \textit{(Gauss's law)}, \tag{3.16}$$

which states that the electric flux diverging from an infinitesimal volume per unit volume is equal to the charge per unit volume at every point in space. Since no isolated *magnetic* charges have been discovered in nature it follows that

$$\nabla \cdot \mathcal{B} = 0 \tag{3.17}$$

and $\mathcal{B}$ is a divergenceless or solenoidal field. Put more simply, magnetic field lines close on themselves.

Note that, except for the fact that we are using script letters here to indicate instantaneous field quantities, Equations (3.11) through (3.17) are the same as they would be when dealing with electrostatic or steady magnetic fields.

■ Example 3.1

Assume that some *excess* charge is placed in a region internal to a large conductor that is otherwise charge neutral. As was mentioned in Chapter 2, this charge will ultimately appear as a surface charge density on the conductor surface. How is this accomplished? Equations (3.9), (3.12), and (3.13) give

$$\nabla \cdot \mathcal{J} = \nabla \cdot (\sigma \mathcal{E}) = \nabla \cdot (\sigma \mathcal{D}/\varepsilon) = -\frac{\partial \rho_v}{\partial t}.$$

If the conductor is homogeneous, then [also using Equation (3.16)]

$$\nabla \cdot \left(\frac{\sigma \mathcal{D}}{\varepsilon}\right) = \left(\frac{\sigma}{\varepsilon}\right) \nabla \cdot \mathcal{D} = \left(\frac{\sigma}{\varepsilon}\right) \rho_v = -\frac{\partial \rho_v}{\partial t}$$

or

$$\frac{\partial \rho_v}{\partial t} + \left(\frac{\sigma}{\varepsilon}\right) \rho_v = 0.$$

If σ is truly constant (independent of ρ_v), then the preceding equation has the simple solution

---

[1] The force is given by the vector product of the current element and the magnetic flux density.

$$\rho_v = \rho_{v0} e^{-(\sigma/\varepsilon)t},$$

where ρ_{v0} is the initial volume charge density at $t = 0$. Thus, the charge density at any point decays exponentially with a time constant ε/σ, called the *relaxation time*.

Copper is normally considered to be a good conductor, and it is for commonly encountered frequencies. The relaxation time for copper, however, is

$$\varepsilon/\sigma = 1.52 \times 10^{-19} \quad (s),$$

which is extremely short, and at ordinary frequencies where the period is large compared with this relaxation time, a field cannot be established in the conductor. At x-ray frequencies (on the other hand) the period is comparable to the relaxation time, and a field can be established in the conductor. In other words, at these frequencies copper acts more like a dielectric than a conductor. ∎

3.3 FARADAY'S LAW

In 1831, Faraday was successful in demonstrating that a *time changing* magnetic field could produce an electric current. It would perhaps be more accurate to say that what Faraday discovered was the following. When the magnetic flux linking a closed circuit is *altered*, a voltage, or *electromotive force* (emf), is induced which may produce a current in this circuit. Faraday's law is usually written mathematically

$$\boxed{\text{emf} = -\frac{d\psi_m}{dt}} \quad \text{(V or Wb/s)} \quad (\textit{Faraday's law}) \qquad (3.18)$$

where ψ_m is the magnetic flux passing through any open surface bounded by the circuit (closed path l). The flux that would be produced by the resultant current *opposes* the original variation of the flux. The last sentence is a statement of *Lenz's law* and accounts for the minus sign in Equation (3.18).

Electromotive force, emf, is a voltage due to some form of energy other than electric (e.g., batteries and generators produce emf). We shall define emf as

$$\boxed{\text{emf} = \oint_l \mathscr{E} \cdot d\mathbf{l}} \quad (V), \qquad (3.19)$$

which implies a *particular* closed path l and if some other closed path is chosen, the emf will, in general, change! In electrostatics, we spoke of voltage and potential difference interchangeably, and this is permissible. In the present situation, we will refrain from mentioning scalar potential, or potential difference, until it is necessary or desirable to do so. As we shall eventually see, voltage and potential difference

3.3 FARADAY'S LAW

are usually not equivalent except in electrostatics. One fact concerning Equation (3.19) is obvious: $\mathscr{E}$ cannot (in the general dynamic case) be a conservative field since its circulation [the right-hand side of Equation (3.19)] is not identically zero.

It is still true that magnetic flux is given by

$$\psi_m = \iint_{s_1} \mathscr{B} \cdot d\mathbf{s} \quad \text{(Wb)}, \tag{3.20}$$

where s_1 is any open surface bounded by the closed path in Equation (3.19). Combining Equations (3.18), (3.19), and (3.20), we have

$$\text{emf} = \oint_l \mathscr{E} \cdot d\mathbf{l} = -\frac{d}{dt} \iint_{s_1} \mathscr{B} \cdot d\mathbf{s} \quad \text{(V)}. \tag{3.21}$$

We follow the conventional right-hand rule for determining the positive sense of circulation with respect to positive flow through the surface. This is demonstrated in Figure 3.1. Applying the Stokes theorem[2] to the circulation of $\mathscr{E}$ in Equation (3.21) gives

$$\text{emf} = \iint_{s_2} \mathbf{\nabla} \times \mathscr{E} \cdot d\mathbf{s} = -\frac{d}{dt} \iint_{s_1} \mathscr{B} \cdot d\mathbf{s}, \tag{3.22}$$

where s_2, like s_1, is any open surface bounded by the closed path of Equation (3.19). Note that s_1 and s_2 are not necessarily the same surface, but their *limits* are the same.

If the closed path is fixed or stationary, then s_1 and s_2 are not time dependent. In this case, the limits of the integrals of Equation (3.22) are fixed, and we may differentiate $\mathscr{B}$ inside the integral sign *partially* with time. That is,

$$\text{emf} = \iint_{s_2} \mathbf{\nabla} \times \mathscr{E} \cdot d\mathbf{s} = -\frac{d}{dt} \iint_{s_1} \mathscr{B} \cdot d\mathbf{s} = -\iint_{s_1} \frac{\partial \mathscr{B}}{\partial t} \cdot d\mathbf{s}. \tag{3.23}$$

[2]Section 1.10.

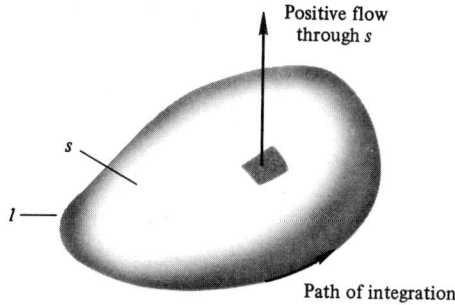

Figure 3.1. Right-hand rule applied to Faraday's law.

which is valid regardless of s_1 and s_2 (and the limits); so if s_1 and s_2 are identical, the left-hand and right-hand sides of Equation (3.23) can only be equal if the integrands are equal. Therefore,

$$\boxed{\nabla \times \mathscr{E} = -\frac{\partial \mathscr{B}}{\partial t}} \quad \text{(Maxwell's curl } \mathscr{E} \text{ equation)} \quad (3.24)$$

and $\mathscr{E}$ is indeed nonconservative ($\nabla \times \mathscr{E} \neq 0$) as predicted. This is the differential form of Maxwell's first equation. The integral form is obtained from Equation (3.21) with s fixed. It is

$$\boxed{\text{emf} = \oint_l \mathscr{E} \cdot d\mathbf{l} = -\int\int_s \frac{\partial \mathscr{B}}{\partial t} \cdot d\mathbf{s}} \quad \text{(transformer emf)}. \quad (3.25)$$

The mechanism that produces the emf here is called *transformer action*. Note particularly that if there are no time variations, we immediately obtain

$$\oint \mathbf{E} \cdot d\mathbf{l} = 0 \quad \left(\frac{\partial}{\partial t} = 0\right)$$

and

$$\nabla \times \mathbf{E} = 0 \quad \left(\frac{\partial}{\partial t} = 0\right),$$

which agree with our electrostatic results.

Example 3.2

Suppose that the magnetic flux density

$$\mathscr{B} = B_0 \cos(\omega t - x)\mathbf{a}_z$$

is being produced in free space by a source. For present purposes we are interested in a confined region near $x = 0$ where changes in x are small compared to changes in ωt. Later on we will learn that this is equivalent to requiring that changes in x are small compared to a *wavelength*. Thus, we can use the approximate form

$$\mathscr{B} = B_0 \cos(\omega t)\mathbf{a}_z.$$

The former form for $\mathscr{B}$ will satisfy all of Maxwell's equations (including the $\nabla \times \mathscr{H}$ equation to be seen later), but the latter form will not. It is important to recognize that the kind of approximation made here is always made when dealing

3.3 FARADAY'S LAW

with ac lumped circuit problems. The emf can be found for any closed path. Choosing a circular path of radius ρ lying in the $z = 0$ plane with equation (3.21) gives

$$\text{emf} = \int_0^{2\pi} \mathscr{E}_\phi \mathbf{a}_\phi \cdot \mathbf{a}_\phi \rho \, d\phi = -\frac{d}{dt} \int_0^\rho \int_0^{2\pi} B_0 \cos(\omega t) \mathbf{a}_z \cdot \mathbf{a}_z \rho \, d\rho \, d\phi.$$

The surface for the open surface integral was chosen to be the plane disk enclosed by the circular path as shown in Figure 3.2. Note that, because of symmetry, $\mathscr{E}_\phi$ is a constant on the chosen path and the right-hand rule has been followed. Thus

$$\text{emf} = 2\pi\rho\mathscr{E}_\phi = \pi\rho^2 \omega B_0 \sin(\omega t) \quad (\text{V}),$$

$$\mathscr{E}_\phi = \frac{\omega\rho}{2} B_0 \sin(\omega t).$$

Equation (3.24) gives

$$(\nabla \times \mathscr{E})_z = \frac{1}{\rho}\frac{\partial}{\partial \rho}(\rho\mathscr{E}_\phi) = -\frac{\partial}{\partial t}(B_0 \cos \omega t) = \omega B_0 \sin \omega t,$$

$$\frac{d}{d\rho}(\rho\mathscr{E}_\phi) = \omega\rho B_0 \cos(\omega t).$$

Integrating on ρ with t held constant, we get

$$\mathscr{E}_\phi = \frac{\omega\rho}{2} B_0 \sin(\omega t) + C_1/\rho,$$

where C_1 is independent of ρ. $\mathscr{E}_\phi$ must be zero if $\mathscr{B}$ is not changing in time, $\omega = 0$, and this requires that C_1 be zero. Thus, the two forms for $\mathscr{E}_\phi$ are the same.

It is interesting to verify *Lenz's law* for this example. For $0 \leq \omega t \leq \pi/2$ and $B_0 > 0$, we have $\mathscr{B}$ *decreasing* with t in the positive z direction while the emf

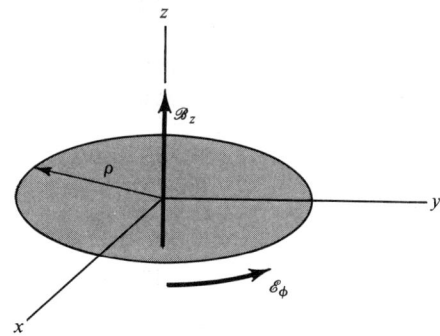

Figure 3.2. Circular path and plane (disk) surface for an emf calculation.

$[= \pi \rho^2 \omega B_0 \sin(\omega t)]$ is *increasing* in time. A filamentary circular current loop of finite conductivity, placed on the path where the emf calculation was made, would have induced in it a current in the $\mathbf{a}_\phi$ direction and *increasing* in time. The magnetic flux produced by this current would be *increasing* in time in the positive z direction, *opposing* the original applied flux in accordance with Lenz's law. ∎

We have seen how an emf can be produced for a closed *stationary* path by means of a time-changing magnetic flux. This has been called *transformer* action. It is also possible to produce an emf with a steady magnetic field if the path or circuit is changing in time. This is called *generator* action. Faraday's law [Equation (3.18)] clearly indicates that this is possible. An emf can be induced by moving a conductor through a steady magnetic field, or by moving a magnetic field past a fixed conductor (flux cutting in both cases), or by both of these schemes.

Generator action, or *motional* emf, is perhaps best introduced by means of a simple experiment. Consider a conductor moving through the flux lines of a steady magnetic field as shown in Figure 3.3. The force on a charge moving with velocity **u** in such a field is given by the second part of the Lorentz force (3.15): $\mathbf{F} = Q\mathbf{u} \times \mathbf{B}$. This force is experienced by both the positive and negative charges in the conductor and is capable of producing a current because we have a force per unit charge or motional electric field given by

$$\mathbf{E}_m = \frac{\mathbf{F}}{Q} = \mathbf{u} \times \mathbf{B} \quad (V/m). \tag{3.26}$$

The emf that can be produced by this field is called the motional emf and is given by

$$\text{emf} = \oint \mathbf{E}_m \cdot d\mathbf{l} = \oint \mathbf{u} \times \mathbf{B} \cdot d\mathbf{l} \quad (\textit{motional emf}). \tag{3.27}$$

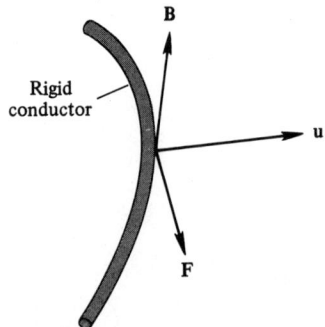

Figure 3.3. Force that is producing a motional emf (generator action) in a rigid conductor moving with a uniform velocity in a uniform **B** field.

3.3 FARADAY'S LAW

As indicated earlier, both types of emf can occur simultaneously giving a general result that can be rigorously derived[3]:

$$\text{emf} = \oint \mathcal{E} \cdot d\mathbf{l} = \underbrace{\oint \mathbf{u} \times \mathcal{B} \cdot d\mathbf{l}}_{l} - \underbrace{\iint \frac{\partial \mathcal{B}}{\partial t} \cdot d\mathbf{s}}_{s}. \qquad (3.28)$$

Example 3.3

Figure 3.4 shows a rectangular conducting loop having arms of negligible resistance with a high resistance volt meter inserted in one leg. Another leg (at $z = l$) has sliding contacts so that it can move with velocity $\mathbf{u} = \mathbf{a}_z u_0 = \mathbf{a}_z (dz)/(dt)$. Equation (3.21) gives

$$\text{emf} = -\frac{d\psi_m}{dt} = -\frac{d}{dt}\int_0^h \int_0^z B_0 \mathbf{a}_y \cdot \mathbf{a}_y \, dx \, dz = -\frac{d}{dt}(B_0 h z),$$

$$\text{emf} = -B_0 h \frac{dz}{dt} = -B_0 h u_0 \quad (\text{V}).$$

Equation (3.27) gives the motional emf:

$$\text{emf} = \int_0^h (u_0 \mathbf{a}_z \times \mathbf{a}_y B_0) \cdot \mathbf{a}_x \, dx = -B_0 h u_0,$$

which is the same result. Equation (3.28) also gives the same result.

Since the arms have negligible resistance, the entire emf appears as a *voltage* defined by

---

[3]See Owen or Neff in the references at the end of the chapter.

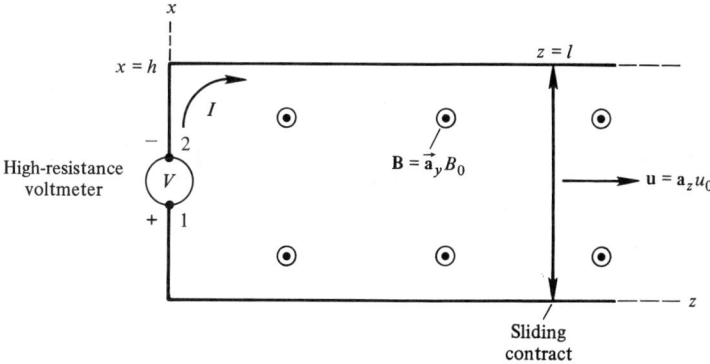

Figure 3.4. Geometry for a motional emf calculation.

$$V_{21} = -\int_1^2 \mathscr{E} \cdot d\mathbf{l} \quad (V).$$

Furthermore, the polarity at the voltmeter must be as shown, and the very small resultant current must be in the direction shown in order to provide an opposing flux (Lenz's law). Thus, the electric field at the voltmeter must be in the direction of a unit vector that points from the (+) to the (−) sign. Therefore, when integrating around the loop in the counterclockwise (positive) direction, we obtain a *negative* emf from the contribution at the voltmeter alone (emf $= -B_0 h u_0$). This has been a case of *generator action*.

If the flux density is changed to $\mathscr{B} = B_0 \cos(\omega t)\mathbf{a}_y$ (similar to that in Example 3.2), then both types of emf are present, and Equation (3.28) gives

$$\text{emf} = \int_0^h [u_0 \mathbf{a}_z \times \mathbf{a}_y B_0 \cos(\omega t)] \cdot \mathbf{a}_x \, dx - \int_0^h \int_0^z -\omega B_0 \sin(\omega t)\mathbf{a}_y \cdot \mathbf{a}_y \, dx \, dz,$$

$$\text{emf} = -B_0 h u_0 \cos(\omega t) + B_0 h z \omega \sin(\omega t) \quad (V),$$

and Equation (3.21) gives

$$\text{emf} = -\frac{d}{dt}\int_0^h \int_0^z B_0 \cos(\omega t)\mathbf{a}_y \cdot \mathbf{a}_y \, dx \, dz = -B_0 h \frac{d}{dt}(z \cos \omega t),$$

$$\text{emf} = +B_0 h z \omega \sin(\omega t) - B_0 h u_0 \cos(\omega t)$$

once again. ∎

Example 3.4

A more practical example, the ac generator, is considered next. Figure 3.5 shows a rectangular loop rotating with constant angular velocity $\omega = \alpha/t$. The emf ultimately appears at the terminals connected to the slip rings. This again is a case of motion only if $\mathscr{B} = \mathbf{a}_x B_0$ (constant). Two legs of the loop enter the calculation, the quantity $\mathbf{u} \times \mathscr{B} \cdot d\mathbf{l}$ being zero for the legs perpendicular to the axis of rotation. Then, we get

$$\text{emf} = \int_{-l/2}^{l/2} u B_0 \sin\alpha (+\mathbf{a}_z) \cdot \mathbf{a}_z \, dz + \int_{l/2}^{-l/2} u B_0 \sin\alpha (-\mathbf{a}_z) \cdot \mathbf{a}_z \, dz.$$

But $\alpha = \omega t$ and $u = \omega a$, so

$$\text{emf} = 2\omega B_0 la \sin \omega t = \omega B_0 s \sin \omega t \quad (V).$$

Application of Equation (3.21) to this problem gives the same result almost immediately. It is left as an exercise to verify that Lenz's law is obeyed.

3.4 MAXWELL'S SECOND EQUATION

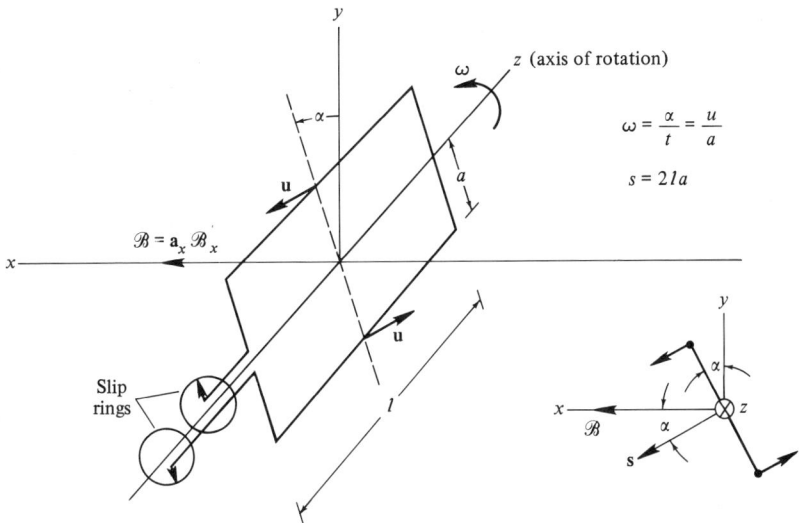

Figure 3.5. Simple ac generator.

Next suppose that $\mathcal{B} = (B_0 \cos \omega t) \mathbf{a}_x$ so that both types of emf are present. Using Equation (3.28), we have

$$\text{emf} = \int_{-l/2}^{l/2} u B_0 \cos \omega t \sin \alpha \mathbf{a}_z \cdot \mathbf{a}_z \, dz$$

$$+ \int_{+l/2}^{-l/2} u B_0 \cos \omega t \sin \alpha (-\mathbf{a}_z) \cdot \mathbf{a}_z \, dz$$

$$+ \omega B_0 \int_{-l/2}^{l/2} \int_{-a}^{a} \sin \omega t \, \mathbf{a}_x \cdot (\mathbf{a}_x \cos \alpha - \mathbf{a}_y \sin \alpha) dz \, dv,$$

where $-l/2 \leq z \leq l/2$ and $-a \leq v \leq a$. That is, $d\mathbf{s} = (\mathbf{a}_x \cos \alpha - \mathbf{a}_y \sin \alpha) dz \, dv$. The integration is straightforward and gives

$$\text{emf} = \omega B_0 s \sin 2\omega t \quad (\text{V}),$$

resulting in a *second harmonic* generator. ∎

3.4 MAXWELL'S SECOND EQUATION

Maxwell's first Equation (3.24) gives us a relation between electric and magnetic fields. Ampere's law, or Maxwell's second equation for magnetostatics, is Equation (2.99):

$$\nabla \times \mathbf{H} = \mathbf{J}. \tag{2.99}$$

Since the divergence of the curl of any vector is identically zero

$$\nabla \cdot (\nabla \times \mathbf{H}) \equiv 0 = \nabla \cdot \mathbf{J}, \tag{3.29}$$

Equation (3.29) is correct for magnetostatics, but, in light of Equation (3.9), Equation (3.29) cannot be correct for the time-varying case. This implies, in turn, that Equation (2.99) is incorrect for the general case also. A *correct* equation for the general case is Equation (3.29) with $\partial \rho_v / \partial t$ added to the right-hand side, since this would give a result that is consistent with conservation of charge:

$$\nabla \cdot (\nabla \times \mathcal{H}) \equiv 0 = \nabla \cdot \mathcal{J} + \frac{\partial \rho_v}{\partial t} \quad \left(\nabla \cdot \mathcal{J} = -\frac{\partial \rho_v}{\partial t}\right). \tag{3.30}$$

Since it is always true that $\rho_v = \nabla \cdot \mathcal{D}$, we have

$$\nabla \cdot (\nabla \times \mathcal{H}) \equiv 0 = \nabla \cdot \mathcal{J} + \frac{\partial}{\partial t}(\nabla \cdot \mathcal{D}). \tag{3.31}$$

If $\mathcal{D}$ and its spatial and time derivatives are continuous, then

$$\frac{\partial}{\partial t}(\nabla \cdot \mathcal{D}) = \nabla \cdot \frac{\partial \mathcal{D}}{\partial t}$$

and

$$\nabla \cdot \nabla \times \mathcal{H} \equiv 0 = \nabla \cdot \left(\mathcal{J} + \frac{\partial \mathcal{D}}{\partial t}\right). \tag{3.32}$$

Equation (3.32) suggests that Maxwell's second equation is, in general,

$$\boxed{\nabla \times \mathcal{H} = \mathcal{J} + \frac{\partial \mathcal{D}}{\partial t}} \quad \text{(Maxwell's curl } \mathcal{H} \text{ equation)}. \tag{3.33}$$

This result is consistent with all previous equations [including Equation (2.99)].

The added term $\partial \mathcal{D}/\partial t$ was Maxwell's primary contribution, and because of this contribution, his name is associated with the whole set of equations. This term is obviously a current density (A/m²), and so Maxwell named it *displacement current density* (time derivative of electric flux density). It is probably easier for us to recognize the need for this added term than it was for Maxwell. Hindsight always seems easy. We know that the ammeter in Figure 3.6 will give a continuous reading, indicating a continuous current. For closed surface s_1 it is obvious that the divergence of the *total* current density is zero, because only a conduction current density is involved. For closed surface s_2, however, we suddenly lose our

3.4 MAXWELL'S SECOND EQUATION

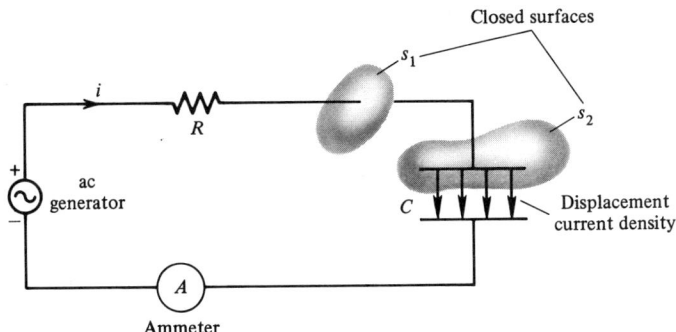

Figure 3.6. Simple ac circuit that shows the necessity of displacement current density.

conducting path. It is still true that the divergence of the *total* current density is zero, or

$$\nabla \cdot \left(\mathscr{J} + \frac{\partial \mathscr{D}}{\partial t} \right) = 0$$

as in Equation (3.32). In other words, the current density is continuous because of the added time derivative of electric flux density between the capacitor plates. In this simple (to us) example, the necessity of displacement current is obvious.

In many low-frequency applications, the displacement current term is not important and is usually neglected. This is one reason why its presence was not easy to verify or detect. With the advent of higher-frequency sources, the displacement current term became more important in relation to conduction current or convection current. This statement will be verified in the chapters that follow and in Example 3.5.

■ Example 3.5

We would like to demonstrate that the magnetic field used initially in Example 3.2, namely,

$$\mathscr{B} = B_0 \cos(\omega t - x)\mathbf{a}_z \quad \text{or} \quad \mathscr{H}_z = (B_0/\mu_0) \cos(\omega t - x),$$

satisfies both of Maxwell's curl equations as originally stated. Using Cartesian coordinates and the inside front cover we have for the $\nabla \times \mathscr{H}$ equation (free space)

$$-\frac{\partial \mathscr{H}_z}{\partial x} = \varepsilon_0 \frac{\partial \mathscr{E}_y}{\partial t} \quad \text{or} \quad -\frac{B_0}{\mu_0} \sin(\omega t - x) = \varepsilon_0 \frac{\partial \mathscr{E}_y}{\partial t},$$

$$\frac{\partial \mathscr{E}_y}{\partial t} = -\frac{B_0}{\mu_0 \varepsilon_0} \sin(\omega t - x).$$

Integrating, we have

$$\mathscr{E}_y = \frac{B_0}{\omega\mu_0\varepsilon_0}\cos(\omega t - x) + C_2(x,y,z),$$

where C_2 is independent of t. Using the $\nabla \times \mathscr{E}$ equation, we get

$$\frac{\partial \mathscr{E}_y}{\partial x} = -\mu_0 \frac{\partial \mathscr{H}_z}{\partial t} \quad \text{or} \quad \frac{B_0}{\omega\mu_0\varepsilon_0}\sin(\omega t - x) + \frac{\partial C_2}{\partial x} = -\mu_0 \frac{\partial \mathscr{H}_z}{\partial t},$$

$$\frac{\partial \mathscr{H}_z}{\partial t} = -\frac{B_0}{\omega\mu_0^2\varepsilon_0}\sin(\omega t - x) - \frac{1}{\mu_0}\frac{\partial C_2}{\partial x}.$$

Integrating, we get

$$\mathscr{H}_z = \frac{B_0}{\omega^2 \mu_0^2 \varepsilon_0}\cos(\omega t - x) - \frac{1}{\mu_0}\frac{\partial C_2}{\partial x}t + C_3(x,y,z).$$

Comparing this result for $\mathscr{H}_z$ with that with which we started, we see that *both* of Maxwell's equations are satisfied if $C_2 = C_3 = 0$ and $\omega^2 \mu_0 \varepsilon_0 = 1$; that is,

$$\mathscr{E}_y(x,t) = \frac{B_0}{\omega\mu_0\varepsilon_0}\cos(\omega t - x) \quad \text{and} \quad \mathscr{H}_z(x,t) = \frac{B_0}{\mu_0}\cos(\omega t - x)$$

coexist and represent an electromagnetic field in free space when $\omega^2 \mu_0 \varepsilon_0 = 1$. We will look closely into this in Chapter 4. ∎

3.5 A SUMMARY OF MAXWELL'S EQUATIONS

The story has been told about the professor who, at the first meeting of a class for a course in electromagnetics, wrote the set of equations called Maxwell's equations on the board and left the room. The implication is that these equations tell the whole story of electromagnetic theory. This is essentially true. Here we will do (more or less) the same thing.

We have found that

$$\boxed{\nabla \times \mathscr{E} = -\frac{\partial \mathscr{B}}{\partial t}} \quad (3.34)$$

and

$$\boxed{\nabla \times \mathscr{H} = \mathscr{J} + \frac{\partial \mathscr{D}}{\partial t}}. \quad (3.35)$$

3.5 A SUMMARY OF MAXWELL'S EQUATIONS

Two other equations give us the set of four equations most frequently called Maxwell's equations. These two equations are unchanged from their static forms:

$$\boxed{\nabla \cdot \mathcal{D} = \rho_v}, \tag{3.36}$$

$$\boxed{\nabla \cdot \mathcal{B} = 0}. \tag{3.37}$$

These last two equations relate flux lines to sources or sinks. It is emphasized, once again, that the magnetic flux density is solenoidal; having no source or sink. The *constitutive* relations are

$$\boxed{\mathcal{D} = \varepsilon \mathcal{E}} \tag{3.38}$$

and

$$\boxed{\mathcal{B} = \mu \mathcal{H}}, \tag{3.39}$$

where the parameters μ and ε depend on the material present. The force on a charge is of fundamental importance because we have seen that this is one way to define the electric and magnetic field. The Lorentz force is

$$\boxed{\mathcal{F} = q(\mathcal{E} + \boldsymbol{u} \times \mathcal{B})}. \tag{3.40}$$

Conduction current density and convection current density have been defined by

$$\boxed{\mathcal{J} = \sigma \mathcal{E}} \tag{3.41}$$

and

$$\boxed{\mathcal{J} = \rho_v \boldsymbol{u}}, \tag{3.42}$$

respectively. Conservation of charge is stated by

$$\boxed{\nabla \cdot \mathcal{J} = -\frac{\partial \rho_v}{\partial t}}. \tag{3.43}$$

Maxwell's equations, as given in Equations (3.34) through (3.37), are in point form. The integral forms of Maxwell's equations are related directly to the fundamental laws we accepted and used to "derive" Equations (3.34) through (3.37). Faraday's law is obtained by integrating both sides of Equation (3.34) over a fixed open surface and then applying the Stokes' theorem to the left-hand side. We have

$$\oint_l \mathscr{E} \cdot d\mathbf{l} = -\iint_s \frac{\partial \mathscr{B}}{\partial t} \cdot d\mathbf{s} \quad \text{(Faraday's law)}. \tag{3.44}$$

The same process applied to Equation (3.35) gives

$$\oint_l \mathscr{H} \cdot d\mathbf{l} = i + \iint_s \frac{\partial \mathscr{D}}{\partial t} \cdot d\mathbf{s} \quad \text{(Ampere's law)}. \tag{3.45}$$

If we integrate both sides of Equation (3.36) throughout a volume and apply the divergence theorem to the left side, we have

$$\oiint_s \mathscr{D} \cdot d\mathbf{s} = \iiint_{\text{vol}} \rho_v \, dv \quad \text{(Gauss's law)}. \tag{3.46}$$

The same process applied to Equation (3.37) gives

$$\oiint_s \mathscr{B} \cdot d\mathbf{s} = 0. \tag{3.47}$$

One special case needs to be given special attention because of its importance. The most important time-varying case, from an engineering applications view point, is the time-harmonic case. For this case, we need to assume sinusoidal time variations and steady-state conditions. If this is done, all the methods of Fourier analysis (superposition of different sinusoids) are available to us. In particular, all the advantages of *phasor* forms (complex algebra) can be utilized. If we merely replace $\partial/\partial t$ by $j\omega$, then Maxwell's equations become

$$\boxed{\nabla \times \mathbf{E} = -j\omega \mathbf{B}}, \tag{3.48}$$

$$\boxed{\nabla \times \mathbf{H} = \mathbf{J} + j\omega \mathbf{D}}, \tag{3.49}$$

$$\boxed{\nabla \cdot \mathbf{D} = \rho_v}, \tag{3.50}$$

and

$$\boxed{\nabla \cdot \mathbf{B} = 0}. \tag{3.51}$$

If we wish to examine the behavior of a function versus time, and we are employing phasor notation, we simply multiply the phasor by $e^{j\omega t}$ and take the real part of the resultant. As an example, suppose we have the phasor electric field

$$\mathbf{E}(\omega) = \mathbf{E}_r + j\mathbf{E}_i,$$

where $\mathbf{E}_r$ and $\mathbf{E}_i$ are *real* vectors. Then, in the real-time domain

$$\mathcal{E}(t) = \mathrm{Re}[(\mathbf{E}_r + j\mathbf{E}_i)e^{j\omega t}]$$

or

$$\mathcal{E}(t) = \mathbf{E}_r \cos \omega t - \mathbf{E}_i \sin \omega t.$$

Since $\cos \omega t \to 1$ and $\sin \omega t \to -j$, when going from the time domain to the frequency domain (phasor), the phasor form for $\mathcal{E}(t)$ above is

$$\mathbf{E}(\omega) = \mathbf{E}_r - (-j)\mathbf{E}_i = \mathbf{E}_r + j\mathbf{E}_i.$$

Note that in the preceding equations ω was inserted to show that the phasor $\mathbf{E}(\omega)$ was a function of radian frequency. This is not necessary, strictly speaking, because the use of phasors implies *single* frequency excitation. It is advantageous, however, to employ this notation because quite often we want to find a time-domain response when the excitation is not sinusoidal. That is, it is easy (formally, at least) to find the response to any excitation if the phasor response is known. This is accomplished by means of the inverse Fourier (or Laplace) transform.

3.6 POYNTING'S THEOREM

An identity of vector analysis from the inside back cover is

$$\nabla \cdot \mathcal{E} \times \mathcal{H} \equiv \mathcal{H} \cdot \nabla \times \mathcal{E} - \mathcal{E} \cdot \nabla \times \mathcal{H} \tag{3.52}$$

for *any* $\mathcal{E}$ and $\mathcal{H}$. Substituting Maxwell's equations [Equations (3.34) and (3.35)], into the right-hand side of Equation (3.52) gives

$$\nabla \cdot \mathscr{E} \times \mathscr{H} = \mathscr{H} \cdot \left(-\frac{\partial \mathscr{B}}{\partial t}\right) - \mathscr{E} \cdot \left(\mathscr{J} + \frac{\partial \mathscr{D}}{\partial t}\right)$$

or (3.53)

$$\nabla \cdot \mathscr{E} \times \mathscr{H} = -\mathscr{H} \cdot \frac{\partial \mathscr{B}}{\partial t} - \mathscr{E} \cdot \mathscr{J} - \mathscr{E} \cdot \frac{\partial \mathscr{D}}{\partial t}.$$

Equations (3.53) may be integrated throughout some region of interest, giving

$$-\iiint_{\text{vol}} \nabla \cdot (\mathscr{E} \times \mathscr{H}) \, dv = \iiint_{\text{vol}} \left(\mathscr{H} \cdot \frac{\partial \mathscr{B}}{\partial t} + \mathscr{E} \cdot \frac{\partial \mathscr{D}}{\partial t} + \mathscr{J} \cdot \mathscr{E}\right) dv$$

or, with the divergence theorem,

$$-\oiint_{s} (\mathscr{E} \times \mathscr{H}) \cdot d\mathbf{s} = \iiint_{\text{vol}} \left(\mathscr{H} \cdot \frac{\partial \mathscr{B}}{\partial t} + \mathscr{E} \cdot \frac{\partial \mathscr{D}}{\partial t} + \mathscr{J} \cdot \mathscr{E}\right) dv. \quad (3.54)$$

Now, if μ and ε are time independent scalars, then

$$\mathscr{E} \cdot \frac{\partial \mathscr{D}}{\partial t} = \frac{1}{2}\frac{\partial}{\partial t}(\mathscr{D} \cdot \mathscr{E}),$$

$$\mathscr{H} \cdot \frac{\partial \mathscr{B}}{\partial t} = \frac{1}{2}\frac{\partial}{\partial t}(\mathscr{B} \cdot \mathscr{H}),$$

and

$$-\oiint_{s} (\mathscr{E} \times \mathscr{H}) \cdot d\mathbf{s} = \iiint_{\text{vol}} \left[\frac{\partial}{\partial t}\left(\frac{\mathscr{D} \cdot \mathscr{E}}{2} + \frac{\mathscr{B} \cdot \mathscr{H}}{2}\right) + \mathscr{J} \cdot \mathscr{E}\right] dv. \quad (3.55)$$

Equation (3.55) may be rewritten

$$-\oiint_{s} (\mathscr{E} \times \mathscr{H}) \cdot d\mathbf{s} = \frac{\partial}{\partial t}\iiint_{\text{vol}} \left(\frac{\mathscr{D} \cdot \mathscr{E}}{2} + \frac{\mathscr{B} \cdot \mathscr{H}}{2}\right) dv + \iiint_{\text{vol}} \mathscr{J} \cdot \mathscr{E} \, dv \quad (3.56)$$

if the volume limits are fixed in time. The first term on the right-hand side of Equation (3.56) is the time rate of increase of the stored energy in the electric field, while the second term is the time rate of increase of the stored energy in the magnetic field. The last term represents energy dissipated (Joule's law) in heat per unit time or energy to accelerate isolated charges per unit time, depending on

3.6 POYNTING'S THEOREM

whether $\mathcal{J}$ is $\sigma\mathcal{E}$ or $\rho_v \mathbf{u}$, respectively. Normally we will use this term to calculate power lost or dissipated ($\mathcal{P}_d$) in a material with conductivity. If there are *sources* inside the volume, $\mathcal{J} \cdot \mathcal{E}$ will be of opposite sign, and will represent the power density added to the system by the sources. The left-hand side of Equation (3.56) must then represent the energy flow *into* the volume per unit time. The energy flow *out* of the volume per unit time is therefore

$$\mathcal{P}_f = \oint_s (\mathcal{E} \times \mathcal{H}) \cdot d\mathbf{s}$$
$$= -\iiint_{\text{vol}} \frac{\partial}{\partial t}\left(\frac{\mathcal{D} \cdot \mathcal{E}}{2} + \frac{\mathcal{B} \cdot \mathcal{H}}{2}\right) dv - \iiint_{\text{vol}} \mathcal{J} \cdot \mathcal{E}\, dv. \quad \text{(W)} \quad (3.57)$$

We *interpret* the vector $\mathcal{E} \times \mathcal{H}$ as the vector giving the direction and magnitude of power density at any point. This interpretation is a matter of convenience and does not follow directly from Poynting's theorem. The Poynting vector is

$$\boxed{\mathcal{S} = \mathcal{E} \times \mathcal{H}} \quad (\text{W/m}^2) \quad (\textit{Poynting vector}). \quad (3.58)$$

In the time-harmonic case, the time-average Poynting vector is (in phasor notation)

$$\boxed{<\mathcal{S}> = \tfrac{1}{2}\,\text{Re}\{\mathbf{E} \times \mathbf{H}^*\}} \quad (\text{W/m}^2) \quad (3.59)$$

analogous to $\tfrac{1}{2}\,\text{Re}\{VI^*\}$ in circuit theory. $\mathbf{H}^*$ is the complex conjugate of $\mathbf{H}$.

■ Example 3.6

The Poynting vector for the electromagnetic field of Example 3.5 is

$$\mathcal{S} = \mathcal{E} \times \mathcal{H} = \mathbf{a}_y \mathcal{E}_y \times \mathbf{a}_z \mathcal{H}_z = \mathbf{a}_x \mathcal{E}_y \mathcal{H}_z,$$
$$\mathcal{S} = \frac{B_0^2}{\omega \mu_0^2 \varepsilon_0} \cos^2(\omega t - x)\mathbf{a}_x \quad (\text{W/m}^2),$$

whose average value is obtained directly by integrating. Since the average value of $\cos^2(\omega t - x)$ over one period in time is $\tfrac{1}{2}$, we obtain

$$<\mathcal{S}> = \frac{B_0^2}{2\omega \mu_0^2 \varepsilon_0}\mathbf{a}_x \quad (\text{W/m}^2).$$

The phasor forms for $\mathcal{E}$ and $\mathcal{H}$ are $\mathbf{E} = [B_0/(\omega\mu_0\varepsilon_0)]e^{-jx}\mathbf{a}_y$ and $\mathbf{H} = (B_0/\mu_0)e^{-jx}\mathbf{a}_z$ (both *peak* values), so Equation (3.59) gives

$$<\mathcal{S}> = \frac{1}{2} \text{Re}\left(\frac{B_0}{\omega\mu_0\varepsilon_0}e^{-jx}\mathbf{a}_y \times \frac{B_0}{\mu_0}e^{+jx}\mathbf{a}_z\right),$$

$$<\mathcal{S}> = \frac{B_0}{2\omega\mu_0^2\varepsilon_0}\mathbf{a}_x,$$

as before.

The reader can easily verify by means of Equation (3.57) that the net power (W) out of a cube 1 m on a side, with sides parallel to the coordinate axes, is zero; that is, the same power flows in the cube as flows out! ∎

■ Example 3.7

Suppose we have a circular cylindrical conductor (resistor) of length l, radius a, and conductivity σ supporting a steady current I with uniform density $J_z = I/(\pi a^2)$ when the cylinder axis is the z axis. As we saw earlier, $H_\phi = I/(2\pi a)$ on the cylinder surface, and $E_z = J_z/\sigma$ on the cylinder surface. Thus, the power flow *out* of the cylindrical volume is given by Equation (3.57):

$$P_f = \iint_s \mathbf{E} \times \mathbf{H} \cdot d\mathbf{s} = \int_0^{2\pi}\int_0^l \left(\frac{J_z}{\sigma}\mathbf{a}_z \times \mathbf{a}_\phi \frac{I}{2\pi a}\right) \cdot (\mathbf{a}_\rho a \, d\phi \, dz),$$

$$P_f = -\frac{1}{2\sigma}\left(\frac{I}{\pi a}\right)^2 \int_0^{2\pi}\int_0^l d\phi \, dz = -I^2\left(\frac{l}{\sigma\pi a^2}\right) = -I^2 R \quad (W),$$

where the integration has been carried out over the lateral surface of the cylinder and R is the *dc* resistance. The power flow *into* the volume is the well-known result $I^2 R$.

If we accept $\mathcal{S}$ as correctly giving the power density flow at every point, then this concept tells us here that a battery (or some other source of emf) sets up fields such that energy flows through the fields and into the conductor through its surface. A circuit theory interpretation may differ in detail, but will correctly give $I^2 R$ as the power being dissipated. Regardless of the interpretation, the Poynting theorem, Equation (3.57), will always give the correct total power balance for a given region. ∎

Based on what was stated below Equation (3.56), the energy stored in the electric and magnetic fields must be

$$\mathcal{W}_E = \frac{1}{2}\iiint_{\text{vol}} \mathcal{D} \cdot \mathcal{E} \, dv \quad (J), \tag{3.60}$$

$$\mathcal{W}_H = \frac{1}{2}\iiint_{\text{vol}} \mathcal{B} \cdot \mathcal{H} \, dv \quad (J), \tag{3.61}$$

3.6 POYNTING'S THEOREM

respectively. These results apply to static electric and steady magnetic fields as well:

$$W_E = \frac{1}{2} \iiint_{\text{vol}} \mathbf{D} \cdot \mathbf{E} \, dv \quad \text{(J)}, \tag{3.62}$$

$$W_H = \frac{1}{2} \iiint_{\text{vol}} \mathbf{B} \cdot \mathbf{H} \, dv \quad \text{(J)}, \tag{3.63}$$

If we equate these energies to those in circuit theory for the energy *stored in a capacitor* and *inductor* ($CV_0^2/2$ and $LI^2/2$), we obtain alternate formulas for calculating capacitance and inductance:

$$C = \frac{1}{V_0^2} \iiint_{\text{vol}} \mathbf{D} \cdot \mathbf{E} \, dv \quad \text{(F)}, \tag{3.64}$$

$$L = \frac{1}{I^2} \iiint_{\text{vol}} \mathbf{B} \cdot \mathbf{H} \, dv \quad \text{(H)}, \tag{3.65}$$

■ Example 3.8

The electric field and potential difference for a coaxial cable were calculated in Example 2.24 (Figure 2.38). They are $\mathbf{E} = \mathbf{a}_\rho a \rho_{sa}/(\varepsilon \rho)$ and $V_0 = (a\rho_{sa}/\varepsilon) \ln b/a$, or, eliminating ρ_{sa}, we get

$$\mathbf{E} = \mathbf{a}_\rho \frac{V_0}{\rho \ln b/a}, \quad a \leq \rho \leq b.$$

Thus, for a length l, Equation (3.64) gives

$$C = \frac{\varepsilon}{V_0^2} \iiint_{\text{vol}} |\mathbf{E}|^2 \, dv = \frac{\varepsilon}{(\ln b/a)^2} \int_a^b \int_0^{2\pi} \int_0^l \frac{1}{\rho} \, d\rho \, d\phi \, dz,$$

$$C = \frac{2\pi \varepsilon l}{\ln b/a} \quad \text{(F)},$$

which agrees with Equation (2.140).

The magnetic field intensity for the cable was found in Example 2.16 (Figure 2.25) for uniform current densities. It consisted of a different form for each of the three regions. Using these in Equation (3.65) for a length l, we get

$$L = \frac{\mu}{I^2}\iiint_{\text{vol}} |H|^2 \, dv = \frac{\mu}{I^2}\int_0^a \int_0^{2\pi}\int_0^l \left(\frac{I\rho}{2\pi a^2}\right)^2 \rho \, d\rho \, d\phi \, dz$$

$$+ \frac{\mu}{I^2}\int_a^b \int_0^{2\pi}\int_0^l \left(\frac{I}{2\pi\rho}\right)^2 \rho \, d\rho \, d\phi \, dz$$

$$+ \frac{\mu}{I^2}\int_b^c \int_0^{2\pi}\int_0^l \left(\frac{I}{2\pi\rho}\frac{c^2-\rho^2}{c^2-b^2}\right)^2 \rho \, d\rho \, d\phi \, dz,$$

when the integration is carried out, the result is identical to Equation (2.149). ∎

We now return to Equation (3.63). Suppose that we have an ideal N-turn solenoid whose ferromagnetic core consists of two identical halves that are touching. We apply a mechanical force to separate the two halves, creating an air gap. The required work appears as magnetostatic energy stored in the (linear medium) air gap. This is a practical problem which must be answered in designing solenoid operated devices (e.g., relays). Equation (3.63) may be written

$$W_H = \frac{1}{2}\iiint_{\text{vol}} \frac{B^2}{\mu_0} \, dv.$$

Since B is the same in the ferromagnetic core and the air gap (and is uniform by the assumption of an ideal solenoid), we have in the air gap

$$W_H = \frac{B^2 s l_{\text{ag}}}{2\mu_0}$$

or

$$dW_H = \frac{B^2 s}{2\mu_0} \, dl_{\text{ag}} = F \, dl_{\text{ag}},$$

where F is the mechanical force required to create the air gap of length l_{ag}. Therefore, we have

$$F = \frac{B^2 s}{\mu_0} \quad (\text{N}).$$

The following example demonstrates the practical use of this result:

■ Example 3.9

A horseshoe electromagnet (Figure 3.7) is holding a smooth steel bar that weighs W pounds (1 pound = 0.454 kg). Assume that the magnet and bar are both cast steel, have the same cross-sectional area (0.05 m²), and that the mean magnetic

3.6 POYNTING'S THEOREM

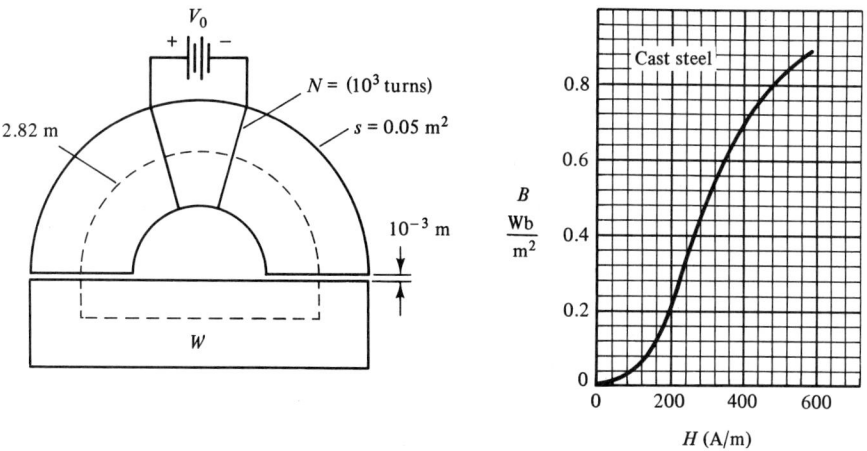

Figure 3.7. Horseshoe electromagnet and its magnetization curve (cast steel).

path length is 2.82 m. The mean air gaps are *assumed* to be 10^{-3} m. The winding is 12-gauge copper wire with a total length of 800 m and resistance of 4.168 Ω. The energy for holding the bar is supplied by flashlight batteries (1.5 V, D cells). There are 30 batteries (total) with 5 units in parallel. Each unit has six batteries in series. We want to find the weight that is being held. The coil current is

$$I = \frac{V}{R} = \frac{6(1.5)}{4.168} = 2.159 \quad (A).$$

So, we get

$$NI = 2159 = (NI)_{ag} + (NI)_{fc} = \psi_m \mathcal{R}_{ag} + Hl$$

$$NI = Bs\mathcal{R}_{ag} + Hl$$

The air-gap reluctance is

$$\mathcal{R}_{ag} = \frac{2l_{ag}}{s\mu_0}$$

or

$$s\mathcal{R}_{ag} = \frac{2l_{ag}}{\mu_0} = \frac{2 \times 10^{-3}}{4\pi \times 10^{-7}} = 1592.$$

Therefore, we have

$$2159 = 1592B + 2.82H.$$

Using trial and error with the given magnetization curve, we find $B = 0.67$ and $H = 390$. Thus, for two air gaps, we get

$$F = B^2 s/\mu_0 = (0.67)^2(0.05)/(4\pi \times 10^{-7}) = 17.86 \times 10^3 \quad (N);$$

converting to pounds, we get

$$W = \frac{17.86 \times 10^3}{9.8(0.454)} = 4014 \quad (pounds).$$

∎

3.7 POTENTIALS

In electrostatics and magnetostatics, we found that certain problems could be solved by first solving for auxiliary potential functions Φ or $\mathbf{A}$. The desired field quantity $\mathbf{E}$ or $\mathbf{B}$ could then be found from the potential by differentiation.[4] We would like now to find a set of potentials from which time-dependent fields may be derived and which are also consistent with those potential functions already found for statics.

From Equation (3.37), there is an $\mathcal{A}$ such that

$$\mathcal{B} = \nabla \times \mathcal{A}, \qquad (3.66)$$

since the divergence of the curl of *any* vector is identically zero. Equation (3.66), however, does not completely define $\mathcal{A}$, for if $\nabla \alpha$ (where α is any scalar field) is added to $\mathcal{A}$, Equations (3.37) and (3.66) still hold because $\nabla \times (\nabla \alpha) = 0$. α is called a *gauge function*. Since Equation (3.66) is the same as Equation (2.104) for magnetostatics, let us try to utilize $\mathcal{A}$ here. Substituting Equation (3.66) into Equation (3.34) gives

$$\nabla \times \mathcal{E} = -\frac{\partial}{\partial t}(\nabla \times \mathcal{A}).$$

Now, if $\mathcal{A}$ and its derivatives are continuous, we have

$$\nabla \times \mathcal{E} = -\nabla \times \frac{\partial \mathcal{A}}{\partial t}$$

or (3.67)

$$\nabla \times \left(\mathcal{E} + \frac{\partial \mathcal{A}}{\partial t}\right) = 0.$$

---

[4] $\mathbf{E} = -\nabla \Phi$ and $\mathbf{B} = \nabla \times \mathbf{A}$.

3.7 POTENTIALS

This means that $\mathscr{E} + \partial \mathscr{A}/\partial t$ is a conservative (zero curl) field and therefore a scalar Φ exists such that

$$\mathscr{E} + \frac{\partial \mathscr{A}}{\partial t} = -\nabla \Phi \tag{3.68}$$

Note also that

$$\int_a^b \left(\mathscr{E} + \frac{\partial \mathscr{A}}{\partial t} \right) \cdot d\mathbf{l}$$

is independent of the path. $\mathscr{E}$ is (in general) nonconservative, but $\mathscr{E} + \partial \mathscr{A}/\partial t$ is conservative. Equation (3.68) may be written

$$\boxed{\mathscr{E} = -\nabla \Phi - \frac{\partial \mathscr{A}}{\partial t}}, \tag{3.69}$$

and agrees with our electrostatic result ($\mathbf{E} = -\nabla \Phi$). Equation (3.69) is not exactly inviting for determining $\mathscr{E}$ because it contains both a scalar and a vector potential.

A pair of coupled equations for Φ and $\mathscr{A}$ may be obtained in the following way. Substituting Equation (3.38) into Equation (3.36), and then, in turn, substituting Equations (3.39) and (3.38) into (3.35) (μ, ε scalars) gives

$$\nabla \cdot \mathscr{E} = \frac{\rho_v}{\varepsilon} \tag{3.70}$$

and

$$\nabla \times \mathscr{B} - \mu\varepsilon \frac{\partial \mathscr{E}}{\partial t} = \mu \mathscr{J}. \tag{3.71}$$

Equations (3.70) and (3.71) are inhomogeneous in that they have *sources* on the right-hand sides. We next eliminate $\mathscr{B}$ with Equation (3.66) and $\mathscr{E}$ with Equation (3.69). We then have

$$\nabla \cdot \left(-\nabla \Phi - \frac{\partial \mathscr{A}}{\partial t} \right) = \frac{\rho_v}{\varepsilon} \tag{3.72}$$

and

$$\nabla \times (\nabla \times \mathscr{A}) - \mu\varepsilon \frac{\partial}{\partial t}\left(-\nabla \Phi - \frac{\partial \mathscr{A}}{\partial t} \right) = \mu \mathscr{J}, \tag{3.73}$$

a pair of coupled equations. If Φ and its derivatives are continuous, Equations (3.72) and (3.73) may be written

$$\nabla^2 \Phi + \frac{\partial}{\partial t}(\nabla \cdot \mathcal{A}) = -\frac{\rho_v}{\varepsilon} \tag{3.74}$$

and

$$\nabla \times (\nabla \times \mathcal{A}) + \mu\varepsilon \frac{\partial^2 \mathcal{A}}{\partial t^2} + \mu\varepsilon \nabla \frac{\partial \Phi}{\partial t} = \mu \mathcal{J}. \tag{3.75}$$

Now, as we have seen before, we have

$$\nabla \times (\nabla \times \mathcal{A}) \equiv \nabla(\nabla \cdot \mathcal{A}) - \nabla^2 \mathcal{A}$$

by vector identity (see the inside back cover); so the pair of coupled equations becomes

$$\nabla^2 \Phi + \frac{\partial}{\partial t}(\nabla \cdot \mathcal{A}) = -\frac{\rho_v}{\varepsilon} \tag{3.76}$$

and

$$\nabla^2 \mathcal{A} - \nabla\left(\nabla \cdot \mathcal{A} + \mu\varepsilon \frac{\partial \Phi}{\partial t}\right) - \mu\varepsilon \frac{\partial^2 \mathcal{A}}{\partial t^2} = -\mu \mathcal{J}. \tag{3.77}$$

We have already specified the curl of $\mathcal{A}$ in Equation (3.66). In order to completely define $\mathcal{A}$, we must also specify its divergence. Equations (3.76) and (3.77) suggest very strongly that we choose

$$\boxed{\nabla \cdot \mathcal{A} = -\mu\varepsilon \frac{\partial \Phi}{\partial t}} \quad \text{(Lorentz condition)} \tag{3.78}$$

If this is done, then $\mathcal{A}$ is said to be unique in the Lorentz gauge and Equations (3.76) and (3.77) are uncoupled, for then we have

$$\boxed{\nabla^2 \Phi - \mu\varepsilon \frac{\partial^2 \Phi}{\partial t^2} = -\frac{\rho_v}{\varepsilon}} \tag{3.79}$$

and

$$\boxed{\nabla^2 \mathcal{A} - \mu\varepsilon \frac{\partial^2 \mathcal{A}}{\partial t^2} = -\mu \mathcal{J}}. \tag{3.80}$$

Note that Equations (3.78), (3.79), and (3.80) are consistent with their static counterparts

3.7 POTENTIALS

$$\nabla \cdot \mathbf{A} = 0, \qquad (2.107)$$

$$\nabla^2 \Phi = -\frac{\rho_v}{\varepsilon}, \qquad (2.61)$$

and

$$\nabla^2 \mathbf{A} = -\mu \mathbf{J}, \qquad (2.108)$$

respectively. Equations (3.79) and (3.80) are called the inhomogeneous scalar and vector Helmholtz *wave equations*, respectively.

The static solutions for Φ and $\mathbf{A}$ were the Helmholtz integrals,

$$\Phi(\mathbf{r}) = \frac{1}{4\pi\varepsilon} \iiint\limits_{\text{vol}'} \frac{\rho_v(\mathbf{r}')}{R} \, dv' \qquad (\varepsilon_0 \to \varepsilon) \qquad (2.65)$$

and

$$\mathbf{A}(\mathbf{r}) = \frac{\mu}{4\pi} \iiint\limits_{\text{vol}'} \frac{\mathbf{J}(\mathbf{r}')}{R} \, dv' \qquad (\mu_0 \to \mu). \qquad (2.110)$$

Solutions to Equations (3.79) and (3.80) are similar as might be expected. Without proof,[5] the solutions are

$$\boxed{\Phi(\mathbf{r}, t) = \frac{1}{4\pi\varepsilon} \iiint\limits_{\text{vol}'} \frac{\rho_v(\mathbf{r}', t - R\sqrt{\mu\varepsilon})}{R} \, dv'} \qquad (3.81)$$

and

$$\boxed{\mathcal{A}(\mathbf{r}, t) = \frac{\mu}{4\pi} \iiint\limits_{\text{vol}'} \frac{\mathcal{J}(\mathbf{r}', t - R\sqrt{\mu\varepsilon})}{R} \, dv'}. \qquad (3.82)$$

That is, to evaluate Φ at $\mathbf{r}$ and time t, the value of ρ_v at $\mathbf{r}'$ and time $t' = t - R\sqrt{\mu\varepsilon}$, or *retarded* time, should be used in the integrand. In the same way, to evaluate $\mathcal{A}$ at $\mathbf{r}$ and time t, the value of $\mathcal{J}$ at $\mathbf{r}'$ and $t - R\sqrt{\mu\varepsilon}$ should be used. The potentials given by Equations (3.81) and (3.82) are thus called *retarded* potentials. In other words, a change in the source cannot be observed at the field point until a later time. Apparently, the effect propagates at a velocity given by $(\mu\varepsilon)^{-1/2}$ (the speed of light). Figures 3.8 and 3.9 demonstrate the geometry. If the region of

---

[5] See Neff, 1981 in the references at the end of this chapter.

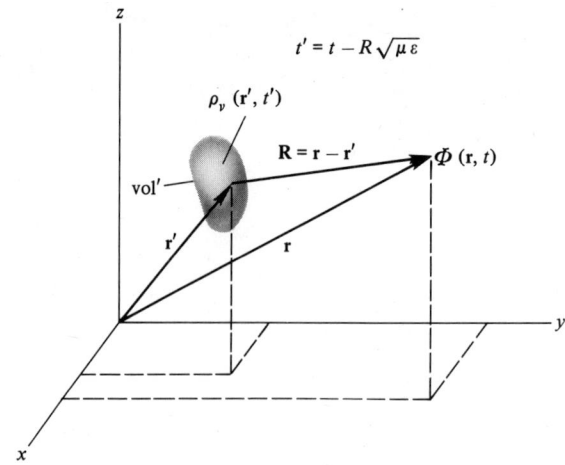

Figure 3.8. Geometry for the calculation of $\Phi(\mathbf{r},t)$ due to $\rho_v(\mathbf{r}',t')$.

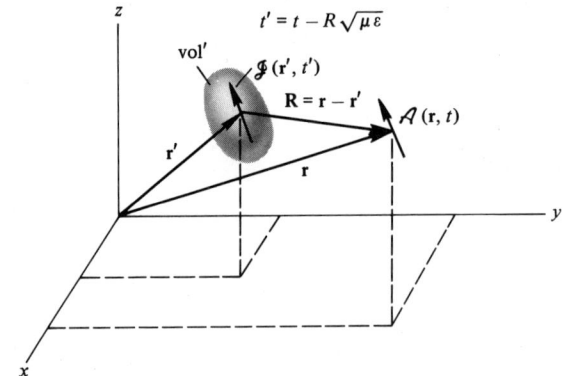

Figure 3.9. Geometry for the calculation of $\mathcal{A}(\mathbf{r},t)$ due to $\mathcal{J}(\mathbf{r}',t')$.

interest does not include the source (x,y,z not inside the region labeled vol′), then Equations (3.81) and (3.82) (still) are solutions to the *homogeneous* differential equations (wave equations)

$$\nabla^2 \Phi - \mu\varepsilon \frac{\partial^2 \Phi}{\partial t^2} = 0 \tag{3.83}$$

and

$$\nabla^2 \mathcal{A} - \mu\varepsilon \frac{\partial^2 \mathcal{A}}{\partial t^2} = 0. \tag{3.84}$$

3.7 POTENTIALS

For the special case of time dependent surface charge densities and surface current densities, superposition gives

$$\Phi(\mathbf{r},t) = \frac{1}{4\pi\varepsilon} \int\int_{s'} \frac{\rho_s(\mathbf{r}',t')}{R} \, ds' \qquad (3.85)$$

and

$$\mathcal{A}(\mathbf{r},t) = \frac{\mu}{4\pi} \int\int_{s'} \frac{\mathbf{J}_s(\mathbf{r}',t')}{R} \, ds'. \qquad (3.86)$$

In the same way, for filamentary charges and currents, we get

$$\Phi(\mathbf{r},t) = \frac{1}{4\pi\varepsilon} \int_{l'} \frac{\rho_l(\mathbf{r}',t')}{R} \, dl' \qquad (3.87)$$

and

$$\mathcal{A}(\mathbf{r},t) = \frac{\mu}{4\pi} \int_{l'} \frac{i(\mathbf{r}',t')}{R} \, d\mathbf{l}'. \qquad (3.88)$$

It is usually not necessary to work with both potentials. Consider the phasor form of Equation (3.80) ($\partial/\partial t \to j\omega$, $\partial^2/\partial t^2 \to -\omega^2$):

$$\nabla^2 \mathbf{A} + \omega^2 \mu\varepsilon \mathbf{A} = -\mu \mathbf{J}, \qquad (3.89)$$

whose solution is[6]

$$\boxed{\mathbf{A}(\mathbf{r},\omega) = \frac{\mu}{4\pi} \int\int\int_{vol'} \frac{\mathbf{J}(\mathbf{r}',\omega)}{R} e^{-j\omega\sqrt{\mu\varepsilon}R} dv'} \qquad (3.90)$$

In terms of this **A**, we get

$$\boxed{\mathbf{B} = \nabla \times \mathbf{A}}, \qquad (3.91)$$

and, from Equations (3.69) and (3.78), we get

$$\boxed{\mathbf{E} = \frac{1}{j\omega\mu\varepsilon} \nabla(\nabla \cdot \mathbf{A}) - j\omega \mathbf{A}}, \qquad (3.92)$$

---

[6]See Neff, 1981 in the references at the end of this chapter.

where

$$\Phi = -\frac{1}{j\omega\mu\varepsilon}(\nabla \cdot \mathbf{A}).$$ (3.93)

Equations (3.91) and (3.92) give the field in terms of **A**.

Example 3.10

An electromagnetic field in free space that satisfies all of Maxwell's equations was found in Example 3.5. The phasor form of this field is $E_y = [B_0/(\omega\mu_0\varepsilon_0)]e^{-jx}$, and the differential equation for **A** [Equation (3.89)] in free space is $\nabla^2\mathbf{A} + \omega^2\mu_0\varepsilon_0\mathbf{A} = 0$. That is, in free space $\mathbf{J} = 0$, but keep in mind that **J** must be nonzero somewhere[7] to act as a source for the field. The curl of **A** must produce a z-component of **B** or **H** that depends on x according to Equation (3.91). That is, $H_z = (1/\mu_0)\partial A_y/\partial x$, or

$$\frac{\partial A_y}{\partial x} = \mu_0 H_z = B_0 e^{-jx}, \qquad H_z = (B_0/\mu_0)e^{-jx};$$

or, integrating with x, and dropping the constant of integration, we have

$$A_y(x,\omega) = jB_0 e^{-jx}.$$

The differential equation for **A** becomes

$$\frac{d^2 A_y}{dx^2} + \omega^2\mu_0\varepsilon_0 A_y = \frac{d^2 A_y}{dx^2} + A_y = 0,$$

since it was shown in Example 3.5 that $\omega^2\mu_0\varepsilon_0 = 1$. Thus, the differential equation for **A** becomes

$$(-j)(-j)jB_0 e^{-jx} + jB_0 e^{-jx} = 0,$$

$$0 = 0,$$

and $A_y = jB_0 e^{-jx}$ is a solution. **E** is given by Equation (3.92):

$$\mathbf{E} = \frac{1}{j\omega\mu_0\varepsilon_0}\nabla(\nabla \cdot \mathbf{A}) - j\omega\mathbf{A}.$$

---

[7] We will be more explicit in Section 4.1.

3.7 POTENTIALS

This reduces to $\mathbf{E} = -j\omega\mathbf{A}$ since $\nabla \cdot \mathbf{A} \equiv 0$. We then have left $E_y = -j\omega A_y$ or $E_y = \omega B_0 e^{-jx} = [B_0/(\omega\mu_0\varepsilon_0)]e^{-jx}$ (since $\omega^2\mu_0\varepsilon_0 = 1$) in agreement with the result from Example 3.5. We did not use Equation (3.90) to find $\mathbf{A} = jB_0e^{-jx}\mathbf{a}_y$. ∎

Considerable labor is often saved by using the potential functions to obtain the fields, rather than obtaining the fields directly. The use of potential functions is, however, not necessary, and in order to show this, we would next like to solve Maxwell's equations simultaneously in a region where μ and ε are scalar constants and current density sources ($\mathcal{J}$) are present. It is advantageous to use the phasor forms for this. Taking the curl of both sides of Equation (3.48), and then substituting Equation (3.49) gives

$$\nabla \times (\nabla \times \mathbf{E}) = -j\omega\mu(\mathbf{J} + j\omega\varepsilon\mathbf{E}). \tag{3.94}$$

Taking the curl of both sides of Equation (3.49), and then substituting Equation (3.48) gives

$$\nabla \times (\nabla \times \mathbf{H}) = \nabla \times \mathbf{J} + j\omega\varepsilon(-j\omega\mu\mathbf{H}). \tag{3.95}$$

The left-hand sides of Equations (3.94) and (3.95) can be expanded by vector identity (inside back cover):

$$\nabla(\nabla \cdot \mathbf{E}) - \nabla^2\mathbf{E} = -j\omega\mu\mathbf{J} + \omega^2\mu\varepsilon\mathbf{E}, \tag{3.96}$$

$$\nabla(\nabla \cdot \mathbf{H}) - \nabla^2\mathbf{H} = \nabla \times \mathbf{J} + \omega^2\mu\varepsilon\mathbf{H}. \tag{3.97}$$

However, we have $\nabla \cdot \mathbf{E} = \rho_v/\varepsilon$ and $\nabla \cdot \mathbf{H} \equiv 0$, so we get

$$\nabla^2\mathbf{E} + \omega^2\mu\varepsilon\mathbf{E} = j\omega\mu\mathbf{J} + \nabla(\rho_v/\varepsilon), \tag{3.98}$$

$$\nabla^2\mathbf{H} + \omega^2\mu\varepsilon\mathbf{H} = -\nabla \times \mathbf{J}.$$

The volume charge density can be eliminated by means of the continuity equation, which for phasors is $\nabla \cdot \mathbf{J} = -j\omega\rho_v$ or $\rho_v = -(1/j\omega)\nabla \cdot \mathbf{J}$. Thus, in terms of the current density $\mathbf{J}$, we have the uncoupled partial differential equations that are called the phasor *inhomogeneous vector Helmholtz wave equations*:

$$\nabla^2\mathbf{E} + \omega^2\mu\varepsilon\mathbf{E} = j\omega\mu\mathbf{J} - \frac{1}{j\omega\varepsilon}\nabla(\nabla \cdot \mathbf{J}) \equiv -\mathbf{S}_E, \tag{3.99}$$

$$\nabla^2\mathbf{H} + \omega^2\mu\varepsilon\mathbf{H} = -\nabla \times \mathbf{J} \equiv -\mathbf{S}_H. \tag{3.100}$$

Note that the *sources* ($\mathbf{S}_E$ and $\mathbf{S}_H$) for these equations involve $\mathbf{J}$, but are not simply $\mathbf{J}$ alone. When the last two equations are compared with Equation (3.89) and its solution, Equation (3.90), we may immediately write the phasor solutions

$$\mathbf{E}(\mathbf{r}, \omega) = \frac{1}{4\pi} \iiint_{\text{vol}} \frac{\mathbf{S}_E(\mathbf{r}', \omega)}{R} e^{-j\omega\sqrt{\mu\varepsilon}\,R}\, dv', \qquad (3.101)$$

$$\mathbf{H}(\mathbf{r}, \omega) = \frac{1}{4\pi} \iiint_{\text{vol}} \frac{\mathbf{S}_H(\mathbf{r}', \omega)}{R} e^{-j\omega\sqrt{\mu\varepsilon}\,R}\, dv', \qquad (3.102)$$

Using the reverse of the process that was used to go from Equation (3.82) to (3.90), we obtain the time-domain solutions

$$\boldsymbol{\mathcal{E}}(\mathbf{r}, t) = \frac{1}{4\pi} \iiint_{\text{vol}} \frac{\boldsymbol{\mathcal{S}}_E(\mathbf{r}', t')}{R}\, dv', \qquad (3.103)$$

$$\boldsymbol{\mathcal{H}}(\mathbf{r}, t) = \frac{1}{4\pi} \iiint_{\text{vol}} \frac{\boldsymbol{\mathcal{S}}_H(\mathbf{r}', t')}{R}\, dv', \qquad (3.104)$$

where, as before, $t' = t - \sqrt{\mu\varepsilon}\,R$ is the retarded time, and a change at the source is observed at the field point at a later time, and the effect propagates at the speed of light. Figure 3.8 applies to the present situation if the source and field quantities are appropriately changed. Note that the integrands in Equations (3.103) and (3.104) are more complicated than that in Equation (3.82), and this is one reason why the use of $\mathcal{A}$ is often preferred. The wave character of $\boldsymbol{\mathcal{E}}$ and $\boldsymbol{\mathcal{H}}$ will be investigated in detail in Chapter 4.

■ Example 3.11

In free space, Equations (3.99) and (3.100) reduce to the homogeneous forms $\nabla^2 \mathbf{E} + \omega^2 \mu_0 \varepsilon_0 \mathbf{E} = 0$ and $\nabla^2 \mathbf{H} + \omega^2 \mu_0 \varepsilon_0 \mathbf{H} = 0$, respectively. For the electromagnetic field of Example 3.10, they become

$$\frac{d^2 E_y}{dx^2} + \omega^2 \mu_0 \varepsilon_0 E_y = 0 = \frac{d^2 E_y}{dx^2} + E_y = 0,$$

$$\frac{d^2 H_z}{dx^2} + \omega^2 \mu_0 \varepsilon_0 H_z = 0 = \frac{d^2 H_z}{dx^2} + H_z = 0$$

or

$$-\omega B_0 e^{-jx} + \omega B_0 e^{-jx} = 0, \qquad 0 = 0;$$

$$-\frac{B_0}{\mu_0} e^{-jx} + \frac{B_0}{\mu_0} e^{-jx} = 0, \qquad 0 = 0,$$

3.9 BOUNDARY CONDITIONS

and our electromagnetic field that we have been using is apparently a wave, since **E** and **H** satisfy homogeneous wave equations. ∎

3.8 VOLTAGE AND POTENTIAL DIFFERENCE

The *voltage* between two points may be defined *in general* as the negative of the line integral of the electric field taken along a *specific* path from point 1 to point 2. That is, the voltage between point 2 and point 1 is

$$v_{21} = -\int_1^2 \mathscr{E} \cdot d\mathbf{l} = \int_2^1 \mathscr{E} \cdot d\mathbf{l} = -v_{12} \quad (V). \tag{3.105}$$

Substituting Equation (3.69) into Equation (3.105) gives

$$v_{21} = +\int_1^2 \left(\nabla \Phi + \frac{\partial \mathscr{A}}{\partial t} \right) \cdot d\mathbf{l},$$

or

$$v_{21} = \int_1^2 \left(\frac{\partial \Phi}{\partial x} dx + \frac{\partial \Phi}{\partial y} dy + \frac{\partial \Phi}{\partial z} dz \right) + \int_1^2 \frac{\partial \mathscr{A}}{\partial t} \cdot d\mathbf{l},$$

or

$$v_{21} = \int_1^2 d\Phi + \int_1^2 \frac{\partial \mathscr{A}}{\partial t} \cdot d\mathbf{l},$$

$$\boxed{v_{21} = \Phi_2 - \Phi_1 + \int_1^2 \frac{\partial \mathscr{A}}{\partial t} \cdot d\mathbf{l}}. \tag{3.106}$$

Thus, voltage and potential difference are generally not identical. Note that they are identical when $\partial \mathscr{A}/\partial t = 0$ (the static case) or when $\mathscr{A} \cdot d\mathbf{l} = 0$. It is also true, according to Equation (3.69), that when $\mathscr{E} \cdot \mathscr{A} = 0$, then $\mathscr{E}$ can be obtained from Φ alone. See Problem 3.24.

3.9 BOUNDARY CONDITIONS

We have already seen in electrostatics and magnetostatics that it is necessary to have a complete set of boundary conditions in order to solve many problems, especially when dealing with boundary-value problems explicitly. The same thing is true in the more general dynamic case. In fact, the boundary conditions are almost the same for statics and dynamics. The required boundary conditions are most easily determined from Maxwell's equations in integral form, Equations (3.44) through (3.47).

Applying Equation (3.44) to the rectangular path in Figure 3.10(a) leads to

$$\mathscr{E}_{t1} = \mathscr{E}_{t2} \quad \text{or} \quad \boxed{\mathbf{a}_n \times (\mathscr{E}_1 - \mathscr{E}_2) = 0}, \tag{3.107}$$

as long as $\partial \mathscr{B}/\partial t$ is finite, and we allow Δw to approach zero. Equation (3.107) simply states that between any two *physically realizable media* the tangential components of $\mathscr{E}$ are continuous. The unrealizable (but useful) perfect conductor is an important special case and will be considered shortly.

Applying Equation (3.45) to the same path leads (no surface currents) to

$$\mathscr{H}_{t1} = \mathscr{H}_{t2} \quad \text{or} \quad \boxed{\mathbf{a}_n \times (\mathscr{H}_1 - \mathscr{H}_2) = 0}, \tag{3.108}$$

as long as $\mathscr{J}$ and $\partial \mathscr{D}/\partial t$ are finite. (Perfect conductors are again an exception.) Equation (3.108) states in words that the tangential components of $\mathscr{H}$ are continuous at the interface between two physically realizable media.

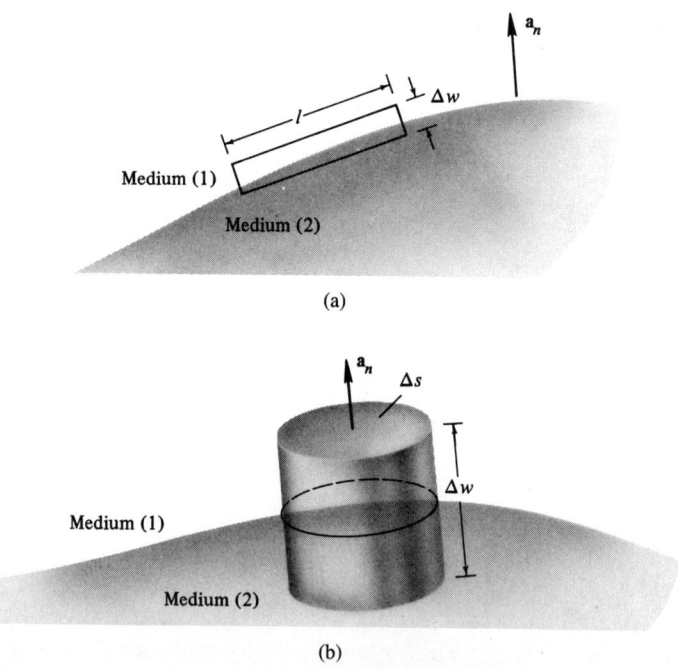

Figure 3.10. (a) Rectangular path for evaluating $\oint \mathscr{E} \cdot d\mathbf{l}$ or $\oint \mathscr{H} \cdot d\mathbf{l}$. (b) Cylindrical surface for evaluating $\oint \mathscr{D} \cdot d\mathbf{s}$ or $\oint \mathscr{B} \cdot d\mathbf{s}$.

3.9 BOUNDARY CONDITIONS

Applying Equation (3.46) to the cylindrical can of Figure 3.10(b) (for $\Delta w \to 0$) gives the same result as in electrostatics, namely,

$$\mathscr{D}_{n1} - \mathscr{D}_{n2} = \rho_s \quad \text{or} \quad \boxed{\mathbf{a}_n \cdot (\mathscr{D}_1 - \mathscr{D}_2) = \rho_s}. \tag{3.109}$$

Equation (3.47) applied to this surface gives

$$\mathscr{B}_{n1} = \mathscr{B}_{n2} \quad \text{or} \quad \boxed{\mathbf{a}_n \cdot (\mathscr{B}_1 - \mathscr{B}_2) = 0}. \tag{3.110}$$

Next, suppose that region 2 in Figure 3.10(a) and Figure 3.10(b) is a perfect conductor. In a perfect conductor the conductivity σ is infinite. Insofar as external fields are concerned, it is often convenient to treat good conductors, like copper or silver, as if they were perfect conductors, neglecting the small errors introduced. Ohm's law $\mathscr{J} = \sigma \mathscr{E}$ requires that $\mathscr{J}$ be infinite unless $\mathscr{E}$ is zero. In other words, to keep $\mathscr{J}$ finite in the perfect conductor $\mathscr{E}$ must be identically zero. Then from Equation (3.34), $\mathscr{H}$ must be identically zero inside the perfect conductor. It then follows from Equation (3.35) that $\mathscr{J} = 0$ inside the perfect conductor. The only way left for a current to exist is in the form of a surface current density $\mathscr{J}_s$. In this case, Equation (3.107) becomes

$$\mathscr{E}_{t1} = \mathscr{E}_{t2} = 0 \quad \text{or} \quad \boxed{\mathbf{a}_n \times \mathscr{E}_1 = 0} \quad (\sigma_2 \to \infty), \tag{3.111}$$

while Equation (3.109) becomes

$$\mathscr{D}_{n1} = \rho_s \quad \text{or} \quad \boxed{\mathbf{a}_n \cdot \mathscr{D}_1 = \rho_s} \quad (\sigma_2 \to \infty), \tag{3.112}$$

and Equation (3.110) becomes

$$\mathscr{B}_{n1} = 0 \quad \text{or} \quad \boxed{\mathbf{a}_n \cdot \mathscr{B}_1 = 0} \quad (\sigma_2 \to \infty). \tag{3.113}$$

In light of the existence of surface current (zero thickness layer), Equation (3.45) is reapplied to Figure 3.10(a) giving

$$\mathscr{H}_{t1} = \mathscr{J}_s \quad \text{or} \quad \boxed{\mathbf{a}_n \times \mathscr{H}_1 = \mathscr{J}_s} \quad (\sigma_2 \to \infty), \tag{3.114}$$

where the current flows perpendicular to $\mathscr{H}_{t1}$. Examples of the use of these boundary conditions will occur frequently in the chapters to follow.

3.10 CIRCUIT THEORY FROM FIELD THEORY

It is informative to show that circuit theory is a very special case of field theory. Put another way, the equations of circuit theory are Maxwell's equations recast into more applicable forms. This does not imply that one should start with Maxwell's equations to solve a circuit problem. What we intend to show here is where circuit equations come from and, in particular, what assumptions must be made for these equations to be valid.

■ Example 3.12

Consider first the rather innocent looking geometry of Figure 3.11. A time-dependent source is applied at the terminals in the form of a time-dependent electric field. Some kind of current would be expected on and inside the conductor of finite conductivity. An internal electric field would be necessary to support this current. Some kind of nonuniform distributed self-inductance, both internal and external, would be associated with this geometry. Also, a nonuniform resistance (in the dc sense) would be expected. If the linear dimensions of the geometry are appreciable compared with $(f\sqrt{\mu\varepsilon})^{-1}$ (the *wavelength*),[8] *radiation* would occur introducing further complications. We would certainly hesitate to call this geometry a circuit. About the best we could do, and this only for sinusoidal time variation, would be to associate a resistance and reactance (inductive or capacitive?) or impedance, with the driven or "input" terminals. Any theoretical development past the starting equations would depend on having a nice simple geometry.

The preceding paragraph paints a rather bleak picture. Actually, many of the difficulties encountered there may be overcome if certain assumptions can be justified. First of all, and fundamentally, let us consider only those cases for which the linear dimensions of the geometry are small compared with $(f\sqrt{\mu\varepsilon})^{-1}$. In other words, the geometry is small physically, or the frequency is low, or both. This essentially eliminates radiation. Next, let us investigate the geometry of Figure 3.12, which certainly looks like a "circuit." Let the interconnecting conductors be filamentary everywhere, except for the small lossy cylinder of conductivity σ and

---

[8] This was mentioned first in Example 2.

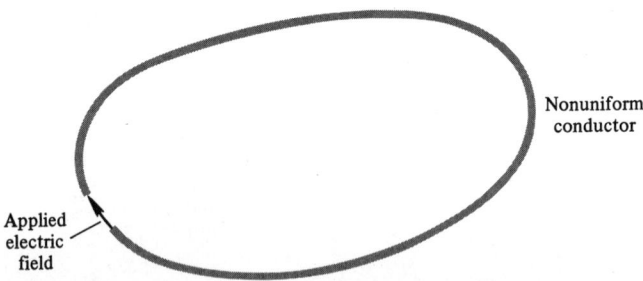

Figure 3.11. Electric field applied to an arbitrary conducting loop.

3.10 CIRCUIT THEORY FROM FIELD THEORY

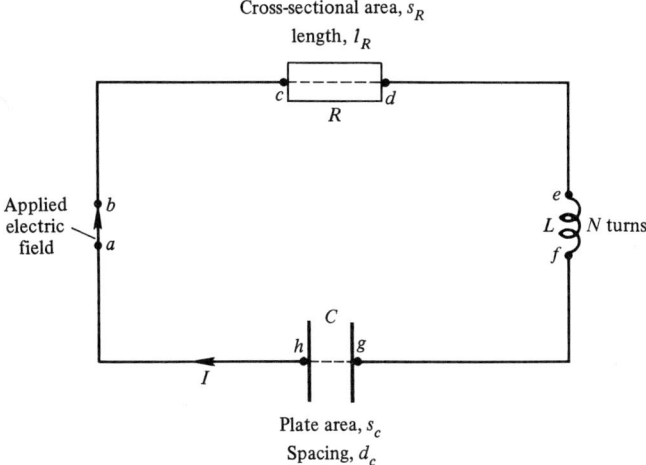

Figure 3.12. Geometry for developing the circuit equation (*RLC* series circuit) from Faraday's law.

the plates of areas s_c. This last requirement eliminates internal effects except for the lossy cylinder and the region between the plates. The dimensions are indicated in Figure 3.12. Let a time-dependent electric field from some completely independent source be applied at the terminals *a-b*. This is the "input."

Let us evaluate the circulation of the electric field around the closed path *l*, defined by the filamentary conductors and the dashed lines. Faraday's law is

$$\oint_l \mathscr{E} \cdot d\mathbf{l} = -\frac{d}{dt} \int\int_s \mathscr{B} \cdot d\mathbf{s}.$$

Let us consider the left-hand side of the above equation and make additional assumptions as they seem necessary. The terminals *a-b* are very close so that the only effect here is that produced by the applied field, and

$$\int_a^b \mathscr{E} \cdot d\mathbf{l} = -v_{ba}.$$

according to Equation (3.105). v_{ba} is called the applied voltage (between points *b* and *a*) in circuit theory.

Let us next assume that the filamentary conductor is perfectly conducting or that the conductivity is large enough so that $\mathscr{E}$ is negligible on the conducting path. In this case, we have

$$\int_b^c \mathscr{E} \cdot d\mathbf{l} = \int_d^e \mathscr{E} \cdot d\mathbf{l} = \int_e^f \mathscr{E} \cdot d\mathbf{l} = \int_f^g \mathscr{E} \cdot d\mathbf{l} = \int_h^a \mathscr{E} \cdot d\mathbf{l} = 0.$$

Note particularly that the line integral, *e* to *f*, around the helical path is zero. Now, we have

$$\int_c^d \mathscr{E} \cdot d\mathbf{l} = \int_c^d \frac{\mathscr{J}}{\sigma} \cdot d\mathbf{l} = \frac{\mathscr{J}}{\sigma} \int_c^d dl = \frac{il_R}{\sigma s_R} = iR,$$

where uniform σ and $\mathscr{E}$ are assumed in the lossy material. R is the dc resistance of the lossy cylinder. We have assumed no displacement currents in the lossy material. In the region between the plates, we assume no conduction current. The current that flows between the plates is entirely displacement current, but it must equal (as we have seen before) the conduction current in the lossy material. Then

$$\int_g^h \mathscr{E} \cdot d\mathbf{l} = \int_g^h \frac{\mathscr{D}}{\varepsilon} \cdot d\mathbf{l} = \frac{\mathscr{D}}{\varepsilon} \int_g^h dl = \frac{\rho_s s_c}{\varepsilon s_c} d_c = \frac{q}{C},$$

where C is the "static" capacitance (geometry dependent). The current can be related to the charge, because this current, flowing to the left *away from terminal h*, must equal the time rate of *increase* of charge at h (outside the plate). Thus,

$$i = \frac{dq}{dt} \quad \text{or} \quad q = \int i\, dt + Q_0$$

Note that the same current (dq/dt) is flowing *into terminal g* (outside the plate). This last result should be compared to Equation (3.5). Finally, we have

$$\int_g^h \mathscr{E} \cdot d\mathbf{l} = \frac{1}{C} \int i\, dt + \frac{Q_0}{C}$$

and

$$\oint_l \mathscr{E} \cdot d\mathbf{l} = -v_{ba} + iR + \frac{1}{C} \int i\, dt + \frac{Q_0}{C} = -\frac{d}{dt} \int\int_s \mathscr{B} \cdot d\mathbf{s}.$$

There remains the right-hand side which contains a surface integral. The surface s is very complicated since it is that surface bounded by the path l, but at least it is fixed in time. It is reasonable to neglect the flux linking the open surface everywhere *except* through the N-turn coil, for the magnetic field threading through the coil is much larger than anywhere else. Then,

$$-\frac{d}{dt}\int\int_s \mathscr{B} \cdot d\mathbf{s} = -\frac{d\psi_m}{dt} = -\frac{d}{dt}(Li) = -L\frac{di}{dt},$$

since L is the external inductance, determined only by the fixed coil geometry. The final result is

$$-v_{ba} + iR + \frac{1}{C}\int i\, dt + \frac{Q_0}{C} = -L\frac{di}{dt}$$

or

$$v_{ba} = iR + L\frac{di}{dt} + \frac{1}{C}\int i\,dt + \frac{Q_0}{C},$$

a well-known result. It is obvious from the development that we have a lumped R, L, and C forming our circuit, and Faraday's law has led to *Kirchhoff's voltage law*.

In a similar manner, it can be shown that conservation of charge, Equation (3.9), leads to *Kirchhoff's current law*.

It is extremely important to remember the large number of assumptions made in this development. To summarize, they are listed.

1. All linear dimensions are much less than $(f\sqrt{\mu\varepsilon})^{-1}$.
2. A filamentary closed path is employed.
3. Perfect conductors exist everywhere in the circuit except at the input gap, between capacitor plates, and between the *resistor* terminals.
4. Displacement current is confined to the *capacitor*.
5. Magnetic flux is confined to the *inductor*.
6. The geometry is fixed in time.

These restrictions can be relaxed somewhat only after considerable experience has been gained with the simplest cases. ∎

3.11 CONCLUDING REMARKS

The material presented in this chapter is extremely important. It is the essence of electromagnetic field theory. Maxwell's equations have been developed from fundamental laws starting with Faraday's law. These equations (with perhaps auxiliary potential functions derived from them) with the appropriate boundary conditions enable us, formally at least, to solve the problems of electromagnetics in the general case. They reduce identically to the electrostatic and magnetostatic results of Chapter 2. We will examine some important practical applications in the following chapters, beginning with a detailed look at the uniform plane wave in Chapter 4.

REFERENCES

Bradshaw, M. D. and Byatt, W. J. *Introductory Engineering Field Theory*. Englewood Cliffs, NJ: Prentice-Hall, 1967. It is interesting for the reader to examine a textbook that begins with the dynamic equations.

Johnk, C. T. A. *Engineering Electromagnetic Fields and Waves*. New York: Wiley, 1975. The text begins with the dynamic equations.

Liao, S. Y. *Engineering Applications of Electromagnetic Theory*, St. Paul, MN: West, 1988.

Maxwell, J. C. *A Treatise on Electricity and Magnetism*, 3rd ed. New York: Oxford University Press, 1964; or New York: Dover, 1954.

Neff, Jr., H. P. *Basic Electromagnetic Fields*. New York: Harper & Row, 1981.

Owen, G. E. *Introduction to Electromagnetic Theory*. Boston, Mass.: Allyn and Bacon, 1963. Faraday's law is derived in a moving frame.

Rao, N. N. *Elements of Engineering Electromagnetics*. Englewood Cliffs, NJ: Prentice-Hall, 1977. The text begins with the dynamic equations.

Shen, L. C. and Kong, J. A. *Applied Electromagnetism*. Boston, Mass.: PWS Engineering, 1987.

Skitek, G. G. and Marshall, S. V. *Electromagnetic Concepts and Applications*. Englewood Cliffs, NJ: Prentice-Hall, 1982.

PROBLEMS

1. If $\mathbf{J}(\omega) = \mathbf{a}_x x^2 + \mathbf{a}_y yz + \mathbf{a}_z xyz$ (A/m^2) in free space, find the current leaving the unit cube that occupies the first octant with a corner at the origin. Use $\oint_s \mathbf{J} \cdot d\mathbf{s}$. Apply the divergence theorem and repeat.

2. Starting with Ampere's law in point form (Maxwell's equation), derive the continuity of current (conservation of charge) equation.

3. Refer to Figure 3.4. Let the legs at $z = 0$ and $z = l$ (both fixed) be replaced with resistors R. If $\mathcal{B} = \mathbf{a}_y B_0 \cos \omega t$, find the resultant current, and indicate its direction.

4. Refer to Figure 2.42 and Example 2.27. Find the induced emf in the square loop if $i_1 = I_0 \cos \omega t$. Compare the result to that obtained from the (phasor circuit) open-circuit voltage $V_2 = j\omega L_{12} I_1$.

*5. Repeat Problem 4 (find the induced emf) for the geometry of Chapter 2, Problem 48.

6. A solenoid consists of 10 closely spaced turns of filamentary wire where the radius of a turn is 1 cm. A uniform axially directed field $10^{-3} \cos 10^3 t$ (Wb/m^2) is impressed. What is the induced open-circuit voltage between the two ends of the wire?

*7. Given the uniform magnetic field $\mathbf{B} = \mathbf{a}_z 10^{-2}$ Wb/m^2 and the circular path of radius ρ lying in the $z = 0$ plane, find the induced emf if $\rho = \rho(t) = -10t + 0.1$ m.

8. A thin conducting strip is located in the $z = 0$ plane, $0 \le x \le l$, $-\infty < y < \infty$ (Figure 3.13). The strip is moving with velocity $\mathbf{u} = \mathbf{a}_y u_0$. A steady magnetic field $\mathbf{B} = \mathbf{a}_z B_0$ is applied. Compare this situation to that of a straight wire moving at the right angles to steady magnetic field. The emf may be picked off by a fixed voltmeter connected by brushes at the edge of the strip.

PROBLEMS

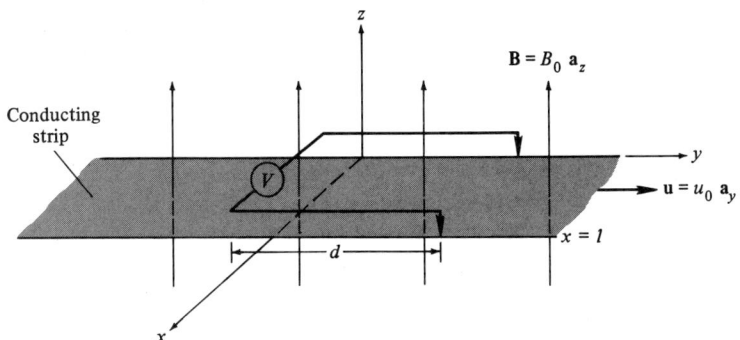

Figure 3.13. Moving conducting strip in a uniform magnetic field (flux cutting).

(a) Calculate the emf.
(b) Calculate the emf if $\mathcal{B} = \mathbf{a}_z B_m \cos \omega t$ and the plane containing the voltmeter leads is normal to the strip.
(c) Calculate the emf if $\mathcal{B} = \mathbf{a}_z B_m \cos \omega t$ and the plane containing the voltmeter leads is parallel to the strip as in Figure 3.13.

9. The Faraday disk generator is a more practical version of Problem 8. It consists of a thin circular conducting disk rotating about its axis with angular velocity ω, while the magnetic field is applied normal to the disk. The emf is picked off by brushes, one on the shaft and one on the periphery of the disk. The radius of the disk is a, and the voltmeter leads define three sides of a rectangle ($b \times a$) parallel to the plane of the disk. Let $\boldsymbol{\omega} = -\mathbf{a}_z \omega_0$. See Figure 3.14.
(a) Calculate the emf.
(b) Calculate the emf if $\mathcal{B} = \mathbf{a}_z B_m \sin \omega_b t$.

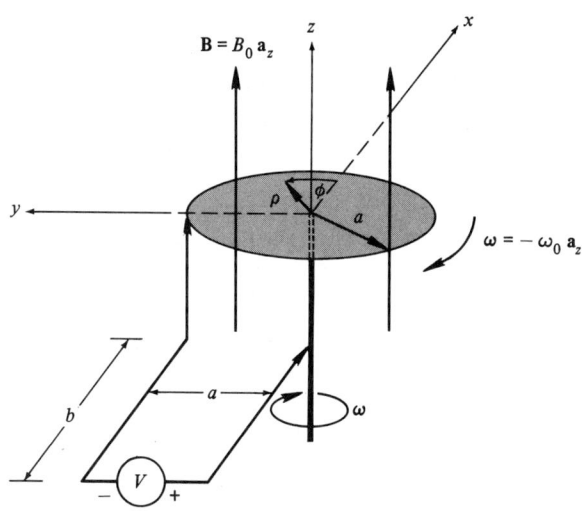

Figure 3.14. Faraday disk generator.

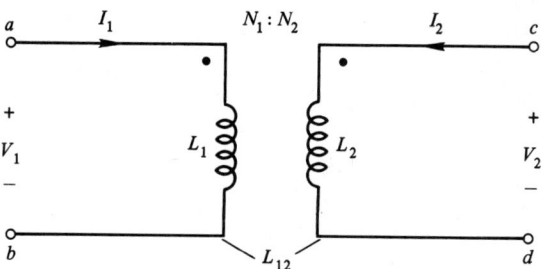

Figure 3.15. Schematic diagram of the transformer of Figure 2.55.

10. The transformer of Figure 2.55 is shown schematically in Figure 3.15 using the dot convention of Chapter 2, Problem 37. Using the definitions of inductance and voltage with Faraday's law:
    (a) Derive the circuit equations for the transformer in terms of L_1, L_2, and L_{12}. Assume that the transformer is linear. Use phasors.
    (b) Assume that the transformer has negligible leakage flux ($k_{mc} \equiv 1$) so that it is ideal. Show that $I_1/I_2 = V_2/V_1 = -N_2/N_1$.

11. A lossy parallel-plate capacitor has a dielectric with conductivity σ_d and permittivity ε. Ignore fringing and show that the ratio of conduction current density to displacement current density is $\sigma_d/(\omega\varepsilon)$ in magnitude.

12. The capacitor in Problem 11 is charged to V_0 (V), and the battery is disconnected at $t = 0$. Show that the voltage across the capacitor decays exponentially with a time constant that equals the relaxation time.

13. The electric field for the dominant mode in a coaxial cable with perfect conductors (Figure 2.38) is given in phasor form by

$$E_\rho(\rho, z) = \frac{E_0}{\rho} e^{-j\omega\sqrt{\mu\varepsilon}\,z}, \quad a \leq \rho \leq b.$$

(a) Using $\nabla \times \mathbf{E} = -j\omega\mu\mathbf{H}$, find $\mathbf{H}$.
(b) Find $\mathbf{J}_s$, the surface current density that supports the electromagnetic field.

14. (a) Find the surface charge density in Problem 13.
    (b) Find the displacement current per unit length in Problem 13.

15. Show that

$$\mathbf{A} = -\mathbf{a}_z \sqrt{\mu\varepsilon} E_0 e^{-j\omega\sqrt{\mu\varepsilon}\,z} \cdot \begin{cases} \ln(a/b), & 0 \leq \rho \leq a, \\ \ln(\rho/b), & a \leq \rho \leq b, \\ 0, & \rho \leq b \end{cases}$$

gives the correct electric field for the coaxial cable (Problem 13). Use Equation (3.92).

16. The wavelength for time-harmonic fields is given by $\lambda = (f\sqrt{\mu\varepsilon})^{-1}$. We arbitrarily decide that "much less than a wavelength" means less than 0.1λ. What does this mean for the usual power line frequency, 60 Hz?

17. How far is the surface of the earth from the surface of the moon if a radar pulse requires 2.56 s for a round trip?

*18. The radiated field at large distances from a *half-wave dipole* antenna with z-directed current is given in phasor form by

$$E_\theta(r,\theta) = j60I_t \frac{e^{-j\omega\sqrt{\mu_0\varepsilon_0}\,r}}{r} \frac{\cos(\pi/2\cos\theta)}{\sin\theta},$$

$$H_\phi(r,\theta) = \sqrt{\varepsilon_0/\mu_0}\,E_\theta.$$

Show that Maxwell's first two equations hold when (only) the terms that vary as $1/r$ are retained.

19. If $\Phi = t - x\sqrt{\mu\varepsilon}$ and $\mathcal{A} = \mathbf{a}_x(t - x\sqrt{\mu\varepsilon})\sqrt{\mu\varepsilon}$, show that
   (a) $\nabla \cdot \mathcal{A} = -\mu\varepsilon\,\partial\Phi/\partial t$
   (b) Find $\mathcal{E}$ and $\mathcal{H}$.

20. Find an equation for the transformer induced emf in a stationary path in terms of $\mathcal{A}$.

21. Using Equations (3.34), (3.35), and (3.43) derive Equations (3.36) and (3.37).

22. Given the phasor magnetic field intensity

$$\mathbf{H} = \mathbf{a}_y \cos \pi x\, e^{-j\beta z},$$

find $\mathbf{E}$.

23. (a) Find the phasor forms for $\mathbf{E}$ and $\mathbf{H}$ for the electromagnetic field:

$$\mathcal{E}_x = E_0 \cos\omega(t - z\sqrt{\mu_0\varepsilon_0}), \quad \mathcal{H}_y = \sqrt{\varepsilon_0/\mu_0}\,E_0 \cos\omega(t - z\sqrt{\mu_0\varepsilon_0})$$

   (b) Find the Poynting vector.
   (c) Find the average power passing through a square that is 1 m by 1 m lying on a plane parallel to the $z = 0$ plane.

24. (a) Given $\mathbf{A}$ for the coaxial cable (Problem 15), find Φ using Equation (3.93).
   (b) Find $\Phi_{ab} = \Phi_a - \Phi_b$.
   (c) Find $V_{ab} = \int_b^a \mathbf{E} \cdot d\mathbf{l}$.
   (d) Why is $V_{ab} = \Phi_{ab}$?

25. Based on Problem 16, estimate the highest usable frequency for a planar vacuum triode if the cathode–anode spacing is 2 mm.

26. A filamentary current loop carries the current $i = I_0 \cos\phi \cos\omega t$ in the $\mathbf{a}_\phi$ direction, find $\mathscr{A}(0,0,z)$. The loop has a radius $\rho = a$ and lies in the $z = 0$ plane.

27. Assuming that $\mathscr{A} = \mathbf{a}_x \cos\omega t \cos kz$ $(k = \omega\sqrt{\mu\varepsilon})$ and $\sigma = 0$, (a) find $\mathscr{H}$, (b) find $\mathscr{E}$, and (c) find Φ.

*28. Find the power radiated by the half-wave dipole of Problem 18 by integrating the Poynting vector over a (large) sphere of radius r.

29. Given

$$\mathscr{E} = \mathbf{a}_y E_0 \sin(\pi x/a)\cos(\omega t - k_z z), \quad k_z = [\omega^2\mu\varepsilon - (\pi/a)^2]^{1/2}, \quad \sigma = 0.$$

(a) Find **E** (phasor form).
(b) Find $\mathscr{H}$.

4
UNIFORM PLANE WAVE PROPAGATION

The uniform plane wave represents a very simple example of the simultaneous use of Maxwell's equations in solving an electromagnetic field problem. A good understanding of the uniform plane wave is also necessary if one is to successfully master the more complex types of electromagnetic phenomena. Furthermore, it is very easy to move from a study of uniform plane waves into a study of transmission lines by way of the parallel plane system. Proceeding in this fashion gives one a better understanding of the relation between field quantities and circuit quantities as applied to the transmission line. The alternate approach, as far as the transmission line is concerned, is to assume an equivalent circuit and derive the voltage and current relationships. This seems rather incomplete and not related directly to previous work.

The uniform undamped plane wave in unbounded space is considered first in this chapter. Next, the damped uniform plane wave in unbounded (but lossy) space is considered. We then have the necessary background to consider the case of reflection (first normal, then oblique incidence) from a plane interface. The material on either side of the interface may be quite general, but the only interface considered here is a plane one where the boundary conditions are easy to observe and apply. The treatment of plane wave reflection (scattering) from other geometries (an infinite cylinder, for example) is beyond the scope of this material and will not be pursued. With a good basic understanding of uniform plane waves, many avenues of investigation are open to us, and we will in Chapter 5 explore one: the transmission line.

The scalar and vector inhomogeneous *wave* equations were derived in the preceding chapter. Formal solutions (the Helmholtz integrals) to these equations were presented. We are now ready to proceed toward the simplest practical cases of wave phenomena. Sinusoidal excitation in time and steady-state conditions will usually be assumed so that the advantageous use of phasors is permissible.

A plane wave may be defined as a wave for which the electric and magnetic field vectors lie in a plane. This plane is perpendicular to the direction of wave motion or propagation. In the special case of time-harmonic excitation (phasors), we may also say that contours of *constant phase* are these same planes. In a *cylindrical wave*, contours of constant phase are *cylinders*, and for a *spherical wave*, contours of constant phase are *spheres*. Plane waves are difficult to generate in practice, but may be well approximated by cylindrical waves or spherical waves at large distances from their respective sources. The field at large distances from a radio antenna is essentially a plane wave over a limited region. These electromagnetic waves are examples of *transverse* waves, as opposed to an acoustic wave, which is an example of a *longitudinal* wave.

In the special case of a *uniform* plane wave, the electric and magnetic field vectors are *uniform* in a plane that is perpendicular to the direction of propagation. In the case of phasor quantities, the electric and magnetic fields are uniform in a plane of constant phase.

4.1 THE UNDAMPED UNIFORM PLANE WAVE

The phasor differential equations for **E** and **H** were obtained as Equations (3.99) and (3.100) for a lossless region, where μ and ε are scalar constants. The integral solutions were Equations (3.101) and (3.102), but, for present purposes, we will simply work with the differential equations. Consider the case where $H_x = H_z \equiv 0$, and we are seeking a plane wave solution involving H_y. If the plane wave to be found is to be *uniform* in planes parallel to the $z = 0$ plane, for example, then H_y cannot depend on x or y. That is, the magnetic field intensity is $\mathbf{H} = \mathbf{a}_y H_y(z, \omega)$. In this case, Equation (3.100) reduces to

$$\frac{\partial^2 H_y(z, \omega)}{\partial z^2} + k^2 H_y(z, \omega) = - (\nabla \times \mathbf{J})_y = -\frac{\partial J_x}{\partial z}, \qquad (4.1)$$

where $k^2 \equiv \omega^2 \mu \varepsilon$ or

$$\boxed{k = \omega \sqrt{\mu \varepsilon}}, \qquad (4.2)$$

and k is called the *wave number* or *phase constant*.

Suppose that the only source present is the uniform x-directed surface current density on the $z = 0$ plane, then

$$\mathbf{J}_x = -J_{s0} \mathbf{a}_x \quad (\text{A/m}). \qquad (4.3)$$

In this case, Equation (4.1) simplifies so long as we require that $z \neq 0$:

$$\frac{d^2 H_y(z, \omega)}{dz^2} + k^2 H_y(z, \omega) = 0, \quad z \neq 0. \qquad (4.4)$$

4.1 THE UNDAMPED UNIFORM PLANE WAVE

The behavior at $z = 0$ can be determined by Ampere's law (3.45) in a manner similar to that used several times in prior work. Consider a small rectangle lying in the $x = 0$ plane with its center at the origin and with sides having a length l in the y direction and Δz in the z direction as shown in Figure 4.1. We will let $\Delta z \to 0$ so that the area of the rectangle becomes zero, and there will be no contribution to the circulation from displacement current (**E** will be zero on the path). The current enclosed by the path is $J_{s0}l$ (A), so Ampere's law gives $lH_y|_{z=0^+} - lH_y|_{z=0^-} = lJ_{s0}$ for $\Delta z \to 0$. Thus, we have

$$H_y = \begin{cases} -\tfrac{1}{2}J_{s0}, & z = 0^-, \\ +\tfrac{1}{2}J_{s0}, & z = 0^+, \end{cases} \quad (4.5)$$

and H_y is discontinuous by an amount that equals the surface current density. It is easy to show that a solution to Equation (4.4) that satisfies Equation (4.5) is

$$H_y = \begin{cases} -\dfrac{J_{s0}}{2}e^{+jkz}, & z < 0, \\ +\dfrac{J_{s0}}{2}e^{-jkz}, & z > 0. \end{cases} \quad (4.6)$$

The choice of signs on the exponents is such that waves traveling *away* from the source will be produced.

The electric field is best found from Maxwell's equation $\nabla \times \mathbf{H} = j\omega\varepsilon\mathbf{E}$, or $\mathbf{E} = [1/(j\omega\varepsilon)][\nabla \times \mathbf{H}]$, $z \neq 0$:

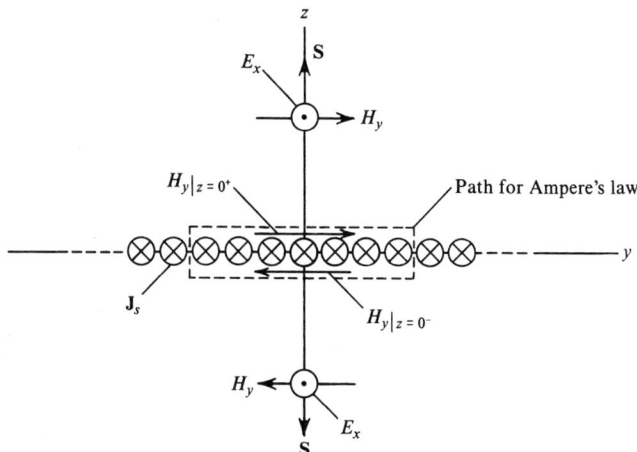

Figure 4.1. Geometry for finding **E** and **H** due to an infinite sheet of surface current density over the $z = 0$ plane.

$$E_x = -\frac{1}{j\omega\varepsilon}\frac{\partial H_y}{\partial z} = \begin{cases} +\dfrac{k}{\omega\varepsilon}\dfrac{J_{s0}}{2}e^{+jkz}, & z<0, \\ +\dfrac{k}{\omega\varepsilon}\dfrac{J_{s0}}{2}e^{-jkz}, & z>0. \end{cases} \quad (4.7)$$

Note that E_x is continuous at $z = 0$ and is normal to the path used for finding H_y. Since $k = \omega\sqrt{\mu\varepsilon}$,

$$E_x = \begin{cases} +\dfrac{1}{2}\sqrt{\dfrac{\mu}{\varepsilon}}J_{s0}e^{+jkz}, & z<0, \\ +\dfrac{1}{2}\sqrt{\dfrac{\mu}{\varepsilon}}J_{s0}e^{-jkz}, & z>0. \end{cases} \quad (4.8)$$

Collecting results and letting $\sqrt{\mu/\varepsilon}\, J_{s0}(\omega)/2 \equiv E_0(\omega)$, we get

$$E_x(z,\omega) = \begin{cases} +E_0(\omega)e^{+jkz}, & z<0, \\ +E_0(\omega)e^{-jkz}, & z>0, \end{cases} \quad (4.9)$$

$$H_y(z,\omega) = \begin{cases} -\dfrac{E_0(\omega)}{\sqrt{\mu/\varepsilon}}e^{+jkz}, & z<0, \\ +\dfrac{E_0(\omega)}{\sqrt{\mu/\varepsilon}}e^{-jkz}, & z>0, \end{cases} \quad (4.10)$$

$$\mathbf{J}_s(\omega) = -\left[2E_0(\omega)/\sqrt{\mu/\varepsilon}\right]\mathbf{a}_x. \quad (4.11)$$

This surface current density over the entire $z = 0$ plane launches uniform plane waves:

$$E_x = E_0 e^{+jkz}, \quad H_y = -\frac{E_0}{\sqrt{\mu/\varepsilon}}e^{+jkz}, \quad z<0 \quad (4.12)$$

in the $-z$ direction, and

$$E_x = E_0 e^{-jkz}, \quad H_y = +\frac{E_0}{\sqrt{\mu/\varepsilon}}e^{-jkz}, \quad z>0 \quad (4.13)$$

in the $+z$ direction. The wave character will be apparent shortly, but it can be stated now that these are *transverse* waves since **E** and **H** lie in planes *transverse* to the direction of propagation, which in the case above is the $-z$ direction for Equation (4.12) and the $+z$ direction for Equation (4.13). In a similar manner, a large, but thin vibrating plate launches acoustic waves in both directions perpendicular to the plate, but these are *longitudinal* waves consisting of expansions and compressions of the medium *in* the direction of propagation.

4.1 THE UNDAMPED UNIFORM PLANE WAVE

It should be pointed out that our source [Equation (4.11)] is impossible to duplicate in practice, and uniform plane waves are difficult to generate over a large region of space.

In a region where the source is not present E_x and H_y must satisfy the homogeneous equations

$$\frac{d^2 E_x(z,\omega)}{dz^2} + k^2 E_x(z,\omega) = 0, \tag{4.14}$$

$$\frac{d^2 H_y(z,\omega)}{dz^2} + k^2 H_y(z,\omega) = 0. \tag{4.15}$$

It is easy to show that *general* solutions to these second-order ordinary differential equations are

$$E_x(z,\omega) = f(\omega)e^{-jkz} + g(\omega)e^{+jkz} \tag{4.16}$$

and

$$H_y(z,\omega) = \frac{f(\omega)}{\sqrt{\mu/\varepsilon}} e^{-jkz} - \frac{g(\omega)}{\sqrt{\mu/\varepsilon}} e^{+jkz}, \tag{4.17}$$

where k is given by Equation (4.2). The "constants" $f(\omega)$ and $g(\omega)$ are independent of x, y, and z, but, as indicated, may be functions of ω.

Corresponding to Equations (4.14) and (4.15), the time-domain forms are

$$\frac{\partial^2 \mathscr{E}_x(z,t)}{\partial z^2} - \mu\varepsilon \frac{\partial^2 \mathscr{E}_x(z,t)}{\partial t^2} = 0, \tag{4.18}$$

$$\frac{\partial^2 \mathscr{H}_y(z,t)}{\partial z^2} - \mu\varepsilon \frac{\partial^2 \mathscr{H}_y(z,t)}{\partial t^2} = 0. \tag{4.19}$$

The reader can easily verify that solutions to these equations are

$$\mathscr{E}_x(z,t) = f(t - z\sqrt{\mu\varepsilon}) + g(t + z\sqrt{\mu\varepsilon}), \tag{4.20}$$

$$\mathscr{H}_y(z,t) = \frac{1}{\sqrt{\mu/\varepsilon}} f(t - z\sqrt{\mu\varepsilon}) - \frac{1}{\sqrt{\mu/\varepsilon}} g(t + z\sqrt{\mu\varepsilon}), \tag{4.21}$$

where f and g are any functions whose first and second derivatives exist. It is worth mentioning that Equations (4.16) and (4.17) lead directly to Equations (4.20) and (4.21) via Fourier methods.

Let us examine, as promised, the wave or propagating nature of our solutions. Consider the rectangular pulse $p(t)$ of height A and width t_0 that starts at $t = 0$. With $f(t) = p(t)$ and $g(t) = 0$ in Equations (4.20) and (4.21), an electromagnetic wave could exist in the form

$$\mathscr{E}_x(z,t) = p(t - z\sqrt{\mu\varepsilon}), \quad \mathscr{H}_y = \frac{1}{\sqrt{\mu/\varepsilon}} p(t - z\sqrt{\mu\varepsilon}). \tag{4.22}$$

Figure 4.2(a) shows $\mathscr{E}_x$ (or $\sqrt{\mu/\varepsilon}\,\mathscr{H}_y$) as a function of z at several instants of time (like snapshots): $t_1 = -2t_0$, $t_2 = 0$, $t_3 = 2t_0$, $t_4 = 4t_0$, ... That is, $p(z, t_3) = p(t_3 - z\sqrt{\mu\varepsilon}) = p(2t_0 - z\sqrt{\mu\varepsilon})$, for example. The leading edge of the pulse appears at $t - z\sqrt{\mu\varepsilon} = 0$ or $z = t/\sqrt{\mu\varepsilon}$, and moves in the $+z$ direction with velocity $dz/dt = 1/\sqrt{\mu\varepsilon}$, which is the speed of light for the medium:

$$\boxed{u = 1/\sqrt{\mu\varepsilon}} \quad \text{(m/s)}. \qquad (4.23)$$

Thus, $\mathscr{E}_x(z, t)$ and $\mathscr{H}_y(z, t)$ are *traveling waves* with $\mathscr{E}_x$ polarized in the $+x$ direction. $\mathscr{E}_x$ and $\mathscr{H}_y$ are related in magnitude by $\mathscr{E}_x/\mathscr{H}_y = \sqrt{\mu/\varepsilon}$. Note that for free space $u = 1/\sqrt{\mu_0\varepsilon_0} \approx 3 \times 10^8$ m/s and $\sqrt{\mu_0/\varepsilon_0} \approx 120\pi$ (Ω). It should be clear that the field

$$\mathscr{E}_x(z, t) = p(t + z\sqrt{\mu\varepsilon}), \quad \mathscr{H}_y(z, t) = -\frac{1}{\sqrt{\mu/\varepsilon}} p(t + z\sqrt{\mu\varepsilon}) \qquad (4.24)$$

consists of traveling waves in the $-z$ direction at the speed of light. Note that in this case $\mathscr{E}$ is still polarized in the $+x$ direction, but the sign of $\mathscr{H}$ has reversed.

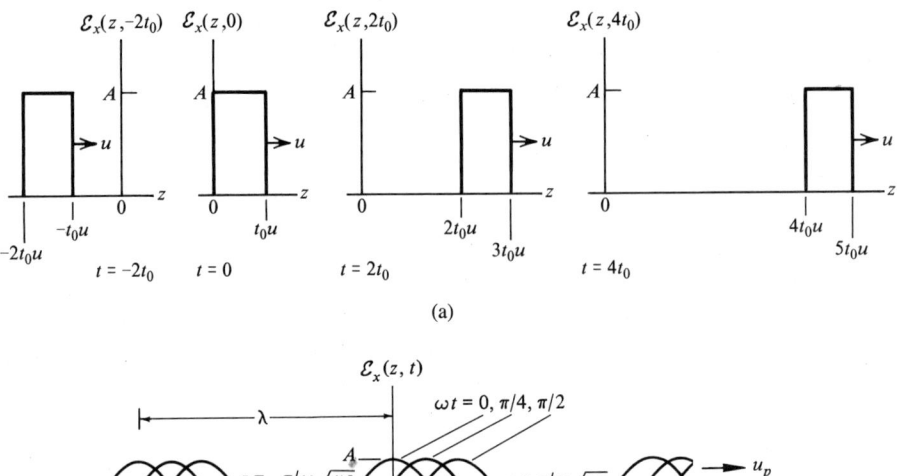

Figure 4.2. (a) Rectangular pulse electric field as a function of z for several values of t (propagating in the $+z$ direction). (b) Sinusoidal electric field as a function of z for three values of t (propagating in the $+z$ direction).

4.1 THE UNDAMPED UNIFORM PLANE WAVE

Finally, note that pulse shape is preserved for these waves. This behavior is a direct result of the fact that the medium has no losses, and pulse shape is usually not preserved when the medium is lossy.

Suppose that $f(\omega) = E^+$ (a *real* constant) and $g(\omega) = 0$ in Equations (4.16) and (4.17). Then, the phasor forms are $E_x(z, \omega) = E^+ e^{-jkz}$, $H_y(z, \omega) = [E^+/\sqrt{\mu/\varepsilon}]e^{-jkz}$. The time-domain forms corresponding to these phasors are obtained by multiplying by $e^{j\omega t}$ and taking the real part of the resultant

$$\mathscr{E}_x(z,t) = \text{Re}(E^+ e^{j(\omega t - kz)}), \quad \mathscr{H}_y(z,t) = \text{Re}\left[\left(E^+/\sqrt{\mu/\varepsilon}\right)e^{j(\omega t - kz)}\right]$$

$$\mathscr{E}_x(z,t) = E^+ \cos(\omega t - kz), \quad \mathscr{H}_y(z,t) = [E^+/\sqrt{\mu/\varepsilon}] \cos(\omega t - kz) \quad (4.25)$$

and this result is just a special case of Equations (4.20) and (4.21). Note that $\mathscr{E}_x/\mathscr{H}_y = \sqrt{\mu/\varepsilon}$, and $\mathscr{E}_x$ and $\mathscr{H}_y$ are in time phase. $\mathscr{E}_x$ (or $\sqrt{\mu/\varepsilon}\,\mathscr{H}_y$) is shown in Figure 4.2(b), and again represents a wave traveling in the $+z$ direction.

■ Example 4.1

The electromagnetic field of Example 3.5 was

$$\mathscr{E}_y(x,t) = \frac{B_0}{\omega\mu_0\varepsilon_0} \cos(\omega t - x), \quad \mathscr{H}_z(x,t) = \frac{B_0}{\mu_0} \cos(\omega t - x),$$

with $\omega^2 \mu_0 \varepsilon_0 = 1 = k^2$ or $k = 1$. We can now see that this is a uniform plane wave that is propagating in the $+x$ direction with its electric field polarized in the $+y$ direction. $\mathscr{E}_y/\mathscr{H}_z = 1/(\omega\varepsilon_0) = \sqrt{\mu_0/\varepsilon_0}$ since $\omega\sqrt{\mu_0\varepsilon_0} = 1$. $\mathscr{E}_y$ and $\mathscr{H}_z$ are in time phase. ■

If $f(\omega) = 0$ and $g(\omega) = E^-$ (a real constant) in Equations (4.16) and (4.17), then following the same procedure as before, we get

$$\mathscr{E}_x(x,t) = E^- \cos(\omega t + kz),$$
$$\mathscr{H}_y(z,t) = -[E^-/\sqrt{\mu/\varepsilon}] \cos(\omega t + kz), \quad (4.26)$$

which represents traveling waves in the $-z$ direction with $\mathscr{E}_x/\mathscr{H}_y = -\sqrt{\mu/\varepsilon}$.

We can now draw the following general conclusions for uniform plane waves propagating at the speed of light in a lossless medium. $\mathscr{E}$ and $\mathscr{H}$ at right angles to each other travel in a direction at right angles to both $\mathscr{E}$ and $\mathscr{H}$. To be more specific, the direction of $\mathscr{E}$, $\mathscr{H}$, and the direction of propagation form a right-handed orthogonal set in that order. For the time-harmonic case, $|\mathbf{E}|/|\mathbf{H}| = \sqrt{\mu/\varepsilon}$. There is no loss in magnitude with propagation due to losses or due to a *spreading* effect. There is a spreading effect with *cylindrical* waves and *spherical* waves in a lossless medium.

4.2 WAVELENGTH AND PHASE VELOCITY

The distance between two successive points having a phase difference of 2π radians, as shown in Figure 4.2(b), is a wavelength λ in meters. Thus, with $k = \omega\sqrt{\mu\varepsilon}$, Equation (4.25) gives

$$(\omega t - \omega\sqrt{\mu\varepsilon}\,z_1) - (\omega t - \omega\sqrt{\mu\varepsilon}\,z_2) = 2\pi$$

or

$$\omega\sqrt{\mu\varepsilon}(z_2 - z_1) = 2\pi = \omega\sqrt{\mu\varepsilon}\,\lambda.$$

Therefore,

$$\boxed{\lambda = \frac{2\pi}{\omega\sqrt{\mu\varepsilon}} = \frac{1}{f\sqrt{\mu\varepsilon}} = \frac{2\pi}{k}} \quad \text{(m)} \quad \textit{(wavelength)}. \qquad (4.27)$$

Note that λ depends on the material parameters μ and ε, as well as the frequency f, and the concept of wavelength only applies to the single frequency sinusoidal case (where the phasor concept applies).

The velocity of a point of constant phase in the z direction is the phase velocity u_p in the z direction. Setting the phase equal to a constant and differentiating with time will give the phase velocity. Thus, Equation (4.25) gives

$$\omega t - \omega\sqrt{\mu\varepsilon}\,z = \text{constant}$$

and

$$\frac{d}{dt}(\omega t - \omega\sqrt{\mu\varepsilon}\,z) = 0$$

or

$$\omega - \omega\sqrt{\mu\varepsilon}\,\frac{dz}{dt} = 0, \quad \frac{dz}{dt} = u_p.$$

Then, we have

$$\boxed{u_p = \frac{1}{\sqrt{\mu\varepsilon}} = \frac{\omega}{k} = f\lambda} \quad \text{(m/s)} \quad \textit{(phase velocity)}. \qquad (4.28)$$

Note that u_p is also the velocity of a point of constant $\mathscr{E}$ field (or $\mathscr{H}$ field) amplitude in Figure 4.2(b).

In a vacuum $\varepsilon = \varepsilon_0 = 8.854 \times 10^{-12} \approx 10^{-9}/36\pi$, $\mu = \mu_0 \equiv 4\pi \times 10^{-7}$, and $u_p = 2.998 \times 10^8$ m/s $\approx 3 \times 10^8$ m/s, which has previously been recognized as the speed of light in a vacuum (or free space).

4.3 POWER DENSITY AND VELOCITY OF ENERGY FLOW

The time-average Poynting vector is given by Equation (3.59) as

$$<\mathcal{S}> = \frac{1}{2}\text{Re}\{\mathbf{H} \times \mathbf{H}^*\} \quad (\text{W/m}^2). \tag{3.59}$$

For the wave traveling in the positive z direction, Equation (3.59) and the phasor forms of Equation (4.25) give

$$<\mathcal{S}> = \mathbf{a}_z \frac{(E^+)^2}{2\sqrt{\mu/\varepsilon}}$$

or simply

$$<\mathcal{S}_z> = \frac{(E^+)^2}{2\sqrt{\mu/\varepsilon}}. \tag{4.29}$$

The interpretation of this result is quite simple. The power density flow is in the positive z direction (with the wave velocity). For the wave traveling in the negative z direction, Equation (3.59) and the phasor forms of Equation (4.26) give

$$<\mathcal{S}_z> = -\frac{(E^-)^2}{2\sqrt{\mu/\varepsilon}}, \tag{4.30}$$

indicating a power density flow in the negative z direction.

For *any* uniform plane wave the direction of power density flow may be obtained by the "right-hand rule" as indicated in Equation (3.59). Rotate the fingers of the right hand from $\mathbf{E}$ to $\mathbf{H}$ and the thumb points in the direction of $<\mathcal{S}>$ (which is also the direction of wave propagation).

The time-average power flow in watts $<\mathcal{P}_f>$ is the integral of the normal component of $<\mathcal{S}>$ over whatever surface one wishes to investigate. It is the net power flowing through that surface. The time-average total energy stored per unit length $<\mathcal{W}>$ can be determined as the volume integral (for a length of unity in the propagation direction) of the time-average energy densities in the electric and magnetic fields. These densities are discussed in Section 3.6. From the preceding definitions, we get

$$<\mathcal{P}_f> = \iint \frac{1}{2}\text{Re}\{\mathbf{E} \times \mathbf{H}^*\} \cdot d\mathbf{s} \quad (\text{W}) \tag{4.31}$$

and

$$<\mathcal{W}> = \iiint \frac{1}{2}\text{Re}\left(\frac{\mathbf{D}}{2} \cdot \mathbf{E}^* + \frac{\mathbf{B}}{2} \cdot \mathbf{H}^*\right) dx\, dy\, dz \quad (\text{J/m}). \tag{4.32}$$

It is implied in Equation (4.32) that the limits of integration are to have a difference of unity for whichever dimension represents the direction of propagation. This ensures a per unit length basis so that $<\mathcal{W}>$ will have the units J/m.

We can now quite easily *define* the velocity of energy flow as

$$\boxed{u_e = \frac{<\mathcal{P}_f>}{<\mathcal{W}>}} \quad \text{(W/J/m or m/s)} \quad \textit{(velocity of energy flow)}. \quad (4.33)$$

For the wave traveling in the positive z direction, we may take a cube 1 m on a side and apply Equation (4.33). This gives [from Equation (4.25)]

$$u_e = \frac{\int_0^1 \int_0^1 \frac{1}{2}\mathrm{Re}\{\mathbf{E} \times \mathbf{H}^*\} \cdot \mathbf{a}_z \, dx \, dy}{\int_0^1 \int_0^1 \int_0^1 \frac{1}{2}\mathrm{Re}\left[(\mathbf{D}/2) \cdot \mathbf{E}^* + (\mathbf{B}/2) \cdot \mathbf{H}^*\right] dx \, dy \, dz}$$

or

$$u_e = \frac{(E^+)^2/2(\sqrt{\mu/\varepsilon})}{(\varepsilon/4)(E^+)^2 + (\mu/4)(\varepsilon/\mu)(E^+)^2}$$

or

$$\boxed{u_e = \frac{1}{\sqrt{\mu\varepsilon}}} \quad \text{(m/s)}. \quad (4.34)$$

Note that, for the case being considered, we have $u_e = u_p$. It is *not true in general* that the velocity of energy flow is the same as the phase velocity. This fact will be demonstrated in later sections dealing with the hollow waveguide. The concept of velocity of energy flow is usually reserved for lossless systems.

4.4 WAVE IMPEDANCE

Inspection of the phasor form of Equation (4.25) reveals that E_x and H_y are very simply related for the special case that has been treated. In fact, we have

$$\frac{E_x}{H_y} = \sqrt{\frac{\mu}{\varepsilon}} \quad (\Omega)$$

for the wave traveling in the positive z direction, while the phasor form of Equation (4.26) gives

$$\frac{E_x}{H_y} = -\sqrt{\frac{\mu}{\varepsilon}} \quad (\Omega)$$

4.4 WAVE IMPEDANCE

for the wave traveling in the negative z direction. Moreover, in general, we have

$$\mathbf{E} = \sqrt{\frac{\mu}{\varepsilon}} \mathbf{H} \times \mathbf{a}_n \tag{4.35}$$

for a uniform plane wave traveling in the direction of the general unit vector $\mathbf{a}_n$ in a perfect dielectric region. The parameters μ and ε have been assumed to be scalar constants to provide a perfect, or *lossless*, dielectric. These parameters are *intrinsic* to the medium through which the wave is propagating. Hence, we have

$$\boxed{\eta \equiv \sqrt{\frac{\mu}{\varepsilon}}} \quad (\Omega) \quad (\textit{intrinsic impedance}), \tag{4.36}$$

which is called the *intrinsic impedance* of the medium. Here, the intrinsic impedance is purely real, but in general (with any loss mechanism in the material) it may be complex, as we shall shortly see.

Another concept that will prove useful to us, particularly for more general waves, is that of *wave impedance*. The *transverse wave impedance* is defined in terms of a particular direction, usually the direction of power density flow. It is given by a ratio of $|\mathbf{E}|$ to $|\mathbf{H}|$ in a manner such that this ratio is greater than 0. $\mathbf{E}$ and $\mathbf{H}$ are *transverse* to the particular direction of power density flow. For the wave traveling in the $+z$ direction, we have

$$\eta_z^+ = \frac{E_x}{H_y} = \sqrt{\frac{\mu}{\varepsilon}} = \eta,$$

while for the wave traveling in the $-z$ direction we have

$$\eta_z^- = -\frac{E_x}{H_y} = \sqrt{\frac{\mu}{\varepsilon}} = \eta.$$

Results as simple as these will not always occur. Another view of the two waves we have been describing is given in Figure 4.3.

The uniform plane wave is an example of a *transverse electromagnetic mode* or simply a TEM mode. Our particular example (*not* Example 4.1) has been a TEM to z mode. This designation implies that both the electric and magnetic fields lie in planes normal to the direction of propagation. Or, in other words, a TEM to z mode has $E_z = H_z = 0$. Of course, we could just as well have described a TEM to x mode (Example 4.1), a TEM to y mode, or a mode TEM to any direction.

■ Example 4.2

Suppose, for example, we have a uniform plane wave propagating in the $+z$ direction (in a lossless medium), and we know that the peak value of the electric field is 10 V/m and is aligned or *polarized* in the positive y direction. Then, we have

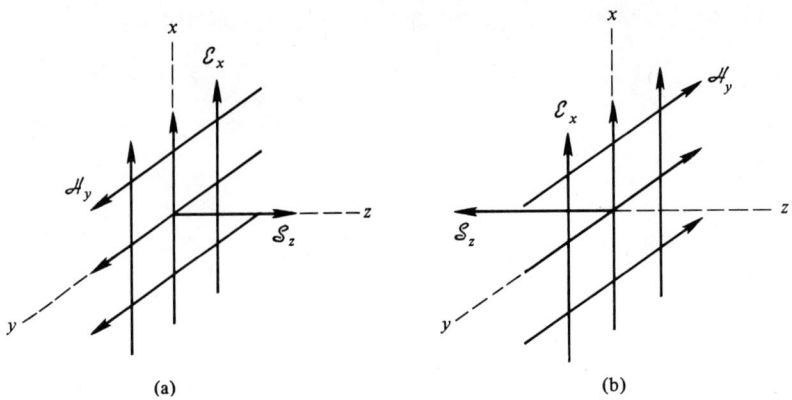

Figure 4.3. Polarization of uniform plane waves with t fixed showing $\mathscr{E}$ and $\mathscr{H}$ at $z = 0$ only for clarity. (a) $+z$ traveling wave. (b) $-z$ traveling wave.

$$E_y = 10e^{-jkz}.$$

The right-hand rule gives us the direction of the magnetic field. Thus,

$$H_x = -\frac{10}{\eta}e^{-jkz}.$$

If the medium is a dielectric for which $\varepsilon_R = 4$, $\mu = \mu_0$, and $f = 300$ MHz, then from Equation (4.27) we get

$$\lambda = \frac{1}{f\sqrt{\mu\varepsilon}} \approx 0.5 \quad \text{m},$$

while from Equation (4.35), we get

$$\eta = \sqrt{\frac{\mu}{\varepsilon}} \approx 60\pi \quad \Omega.$$

Also, we have

$$u_p \approx 1.5 \times 10^8 \quad \text{m/s}$$

and

$$u_e \approx 1.5 \times 10^8 \quad \text{m/s}$$

from Equations (4.28) and (4.34), respectively. From Equation (3.59) the time-average power density is

$$<\mathscr{S}_z> = \frac{100}{120\pi} = 0.265 \quad \text{W/m}^2.$$

The wave impedance is

$$\eta_z^+ = -\frac{E_y}{H_x} = \sqrt{\frac{\mu}{\varepsilon}} = \eta \approx 60\pi \quad \Omega.$$

∎

The equations for the electromagnetic field of a uniform plane wave propagating in *any* direction will be given when oblique incidence on a plane interface is considered in Section 4.7.

4.5 UNIFORM DAMPED PLANE WAVES

We now consider the effects of a lossy medium. This loss may be due to several mechanisms, but the general effect may be observed in a simple manner if we simply allow σ to be nonzero. Loss is considered in more detail later. The medium is still linear, homogeneous, isotropic, and ρ_v is still zero. We are still seeking uniform plane waves with $E_y = E_z = 0$ and propagation in the $\pm z$ direction. Maxwell's equations may be written

$$\nabla \times \mathbf{E} = -j\omega\mu\mathbf{H}, \tag{4.37}$$

$$\nabla \times \mathbf{H} = \mathbf{J} + j\omega\varepsilon\mathbf{E} = (\sigma + j\omega\varepsilon)\mathbf{E} = j\omega\varepsilon'\mathbf{E}, \tag{4.38}$$

$$\nabla \cdot \mathbf{E} = 0,$$

$$\nabla \cdot \mathbf{H} = 0,$$

where

$$\varepsilon' = \varepsilon[1 - j\sigma/(\omega\varepsilon)]. \tag{4.39}$$

The only difference between these equations and those for the lossless case is that ε has been replaced by a complex permittivity ε'. Note also that $\mathbf{J} = \sigma\mathbf{E}$ is not a source current density, but is simply the current density in a conducting medium. We now merely replace $jk = j\omega\sqrt{\mu\varepsilon}$ with $j\omega\sqrt{\mu\varepsilon'}$ in our lossless equations: $j\omega\sqrt{\mu\varepsilon'} = \sqrt{-\omega^2\mu\varepsilon'} = \sqrt{-\omega^2\mu\varepsilon[1 - j\sigma/(\omega\varepsilon)]}$, or

$$\gamma = \alpha + j\beta \equiv \sqrt{-\omega^2\mu\varepsilon[1 - j\sigma/(\omega\varepsilon)]} \tag{4.40}$$

replaces jk. Then, we have

$$\alpha = \text{Re}\{\gamma\} = \text{Re}\left\{\sqrt{-\omega^2\mu\varepsilon[1 - j\sigma/(\omega\varepsilon)]}\right\} \quad (\text{Np/m}), \tag{4.41}$$

$$\beta = \text{Im}\{\gamma\} = \text{Im}\left\{\sqrt{-\omega^2\mu\varepsilon[1 - j\sigma/(\omega\varepsilon)]}\right\} \quad (\text{rad/m}). \tag{4.42}$$

α is the *attenuation constant*, β is the *phase constant*, and γ is the *complex propagation constant*. One neper (Np) equals 8.69 dB and is actually dimensionless, as is the radian. After some complex algebra, the last two equations reduce to

$$\alpha = \omega\sqrt{\mu\varepsilon}\left\{\frac{1}{2}\left[\sqrt{1+\left(\frac{\sigma}{\omega\varepsilon}\right)^2}-1\right]\right\}^{1/2} \geq 0 \qquad (4.43)$$

and

$$\beta = \omega\sqrt{\mu\varepsilon}\left\{\frac{1}{2}\left[\sqrt{1+\left(\frac{\sigma}{\omega\varepsilon}\right)^2}+1\right]\right\}^{1/2} > 0. \qquad (4.44)$$

The intrinsic impedance becomes complex:

$$\eta = \sqrt{\mu/\varepsilon'} = \frac{j\omega\mu}{j\omega\sqrt{\mu\varepsilon'}} = \frac{j\omega\mu}{\gamma} = \frac{\omega\mu}{\alpha^2+\beta^2}(\beta+j\alpha) \qquad (4.45)$$

Equations (4.14) and (4.15) become ($k^2 \to -\gamma^2$)

$$\frac{d^2E_x(z,\omega)}{dz^2} - \gamma^2 E_x(z,\omega) = 0,$$

$$\frac{d^2H_y(z,\omega)}{dz^2} - \gamma^2 H_y(z,\omega) = 0.$$

It is easy to show by direct substitution that solution sets are

$$E_x(z,\omega) = E^+ e^{-\alpha z}e^{-j\beta z}, \quad H_y(z,\omega) = (E^+/\eta)e^{-\alpha z}e^{-j\beta z} \qquad (4.46)$$

and (or)

$$E_x(z,\omega) = E^- e^{+\alpha z}e^{+j\beta z}, \quad H_y(z,\omega) = -(E^-/\eta)e^{+\alpha z}e^{+j\beta z}. \qquad (4.47)$$

The interpretation of these equations is easy. The waves are traveling waves, but they are *damped* exponentially as they progress (*in either direction*). This damping is, of course, due to the finite conductivity of the medium which is the *only* loss mechanism (heating) present. The loss shows up in another way also: **E** and **H** are no longer in-phase for either the $+z$ traveling wave or the $-z$ traveling wave since η is complex.

The damped electromagnetic wave that is propagating in the positive z direction is given by Equation (4.46), whose time-domain form (when E^+ is real) is

$$\mathscr{E}_x(z,t) = E^+ e^{-\alpha z}\cos(\omega t - \beta z),$$

$$\mathscr{H}_y(z,t) = E^+ e^{-\alpha z}\text{Re}\left(\frac{1}{\eta}e^{j(\omega t-\beta z)}\right) = E^+ e^{-\alpha z}\frac{1}{|\eta|}\cos\left(\omega t - \beta z - \tan^{-1}\frac{\alpha}{\beta}\right). \qquad (4.48)$$

4.5 UNIFORM DAMPED PLANE WAVES

The phase velocity can be found by setting the phase of $\mathscr{E}_x(z,t)$ equal to a constant and differentiating (as before). This results in

$$u_p = \frac{\omega}{\beta},\qquad(4.49)$$

a *general result*, which reduces to Equation (4.28) for the lossless case ($\alpha = 0$). The wavelength can be found (as before) to be

$$\lambda = \frac{2\pi}{\beta},\qquad(4.50)$$

a general result, which reduces to Equation (4.27) for the lossless case.

Good conductors and good dielectrics are distinguished by whether conduction current densities dominate or whether displacement current densities dominate, respectively. We can now be even more explicit for the damped uniform plane wave.

Since $(1+x)^{1/2} \approx 1 + x/2$ for $x \ll 1$, we find that for good dielectrics where $\sigma/(\omega\varepsilon) \ll 1$:

$$\alpha \approx \frac{\omega}{2}\sqrt{\mu\varepsilon}\left(\frac{\sigma}{\omega\varepsilon}\right)\quad\text{(good dielectrics)},\qquad(4.51)$$

$$\beta \approx \omega\sqrt{\mu\varepsilon}\left[1 + \frac{1}{8}\left(\frac{\sigma}{\omega\varepsilon}\right)^2\right]\quad\text{(good dielectrics)},\qquad(4.52)$$

$$\eta \approx \sqrt{\mu/\varepsilon}\left\{\left[1 - \frac{3}{8}\left(\frac{\sigma}{\omega\varepsilon}\right)^2\right] + j\frac{1}{2}\left(\frac{\sigma}{\omega\varepsilon}\right)\right\}\quad\text{(good dielectrics)}\qquad(4.53)$$

from Equations (4.43), (4.44), and (4.45), respectively. The ratio $\sigma/(\omega\varepsilon)$ is called the *loss tangent*, and $\hat{y} = \sigma + j\omega\varepsilon$ is the *admittivity*. The phase velocity and wavelength are easily determined from Equations (4.49) and (4.50), respectively.

When $\sigma/(\omega\varepsilon) \gg 1$, we have the good conductor case, and it is easy to show that

$$\alpha = \beta = \sqrt{\omega\mu\sigma/2}\quad\text{(good conductors)}\qquad(4.54)$$

$$u_p = \sqrt{\frac{2\omega}{\mu\sigma}}\quad\text{(good conductors)}\qquad(4.55)$$

$$\eta = \sqrt{\frac{\omega\mu}{2\sigma}} + j\sqrt{\frac{\omega\mu}{2\sigma}}\quad\text{(good conductors)}\qquad(4.56)$$

from Equations (4.41), (4.42), (4.49), and (4.45), respectively.

■ **Example 4.3**

Consider the possibility of propagating a plane wave through seawater ($\sigma = 4$, $\varepsilon_r = 81$). We want to find α, β, u_p, and λ at 1 MHz. Direct calculation gives $\sigma/\omega\varepsilon = 889$, while Equation (4.41) gives

$$\alpha = 3.972 \quad \text{Np/m} \quad \text{or} \quad 34.5 \quad \text{dB/m}.$$

From Equation (4.42), we get

$$\beta = 3.976 \quad \text{rad/m}.$$

From Equation (4.49), we get

$$u_p = 1.58 \times 10^6 \quad \text{m/s}.$$

From Equation (4.50), we get

$$\lambda = 1.58 \quad \text{m}.$$

At 100 kHz the same calculations give

$$\alpha = 1.257 \quad \text{Np/m}, \qquad \beta = 1.257 \quad \text{rad/m},$$
$$u_p = 0.5 \times 10^6 \quad \text{m/s} \qquad \lambda = 5.0 \quad \text{m}. \qquad ■$$

This example should make the reader wonder about some things. Which frequency would a submariner choose? Another question that may not have occurred to all of us, is: what is the effect of the different phase velocities for the two frequencies in the example? An obvious answer is that the two plane waves encounter very different time delays. The medium is said to be *dispersive* because of the loss.

Suppose a wave is propagating in a lossy or dispersive medium such as that in the preceding example. Further, suppose that this wave is the result of a modulated signal or perhaps it is a repetitive pulse. Using the methods of Fourier analysis we can decompose the signal into its sinusoidal components and each component (a phasor) produces a wave. Together, all the waves produce the composite wave, which is propagating. Note particularly that the various parts do not propagate with the same velocity, and the wave will be dispersed. The result of this after propagation over some distance is that if the modulating signal is extracted or demodulated, it will no longer be the original modulating signal. It is distorted. In the case of a pulse, its shape will be changed. This distortion is due to the dispersion, but, more particularly, it is a direct result of the fact that the phase velocity is frequency dependent, or, since $u_p = \omega/\beta$, it is a result of the fact that β, as in Equation (4.42), is not a linear function of frequency.

These ideas can often be expressed more concisely if a quantity called the *group velocity* u_g is introduced. In order to obtain an equation for the group velocity, consider a wave in a lossless medium traveling in the $+z$ direction and polarized in the x direction and consisting of two parts with equal amplitudes. Let the two frequencies be different (as in the discussion above), and let β be some general function of ω (as in the discussion above). One frequency is $\omega_0 - \Delta\omega$, and the other is $\omega_0 + \Delta\omega$, while the two phase constants are $\beta_0 - \Delta\beta$ and $\beta_0 + \Delta\beta$, respectively. Thus, we have

$$\mathscr{E}_x = E_0 \cos\left[(\omega_0 - \Delta\omega)t - (\beta_0 - \Delta\beta)z\right]$$
$$+ E_0 \cos\left[(\omega_0 + \Delta\omega)t - (\beta_0 + \Delta\beta)z\right].$$

Using some trigonometry and rearranging, we get

$$\mathscr{E}_x = 2E_0 \cos\left[\Delta\omega\left(t - \frac{\Delta\beta}{\Delta\omega}z\right)\right] \cos\left[\omega_0\left(t - \frac{\beta_0}{\omega_0}z\right)\right].$$

In this form, the wave is the product of two traveling waves, the second cosine term being a wave traveling with velocity ω_0/β_0, while the first cosine term is a wave traveling with velocity $(\Delta\omega)/(\Delta\beta)$. Thinking of the first term as being like a modulation envelope, the "envelope" of the overall wave moves with velocity $(\Delta\omega)/(\Delta\beta)$ and the "carrier" moves with velocity ω_0/β_0. The envelope velocity is defined as the group velocity for *small* excursions. That is,

$$u_g \equiv \lim_{\Delta\omega \to 0} \frac{\Delta\omega}{\Delta\beta} = \frac{d\omega}{d\beta} = \left(\frac{d\beta}{d\omega}\right)^{-1} \quad (group\ velocity). \tag{4.57}$$

The last form is usually most useful, as Equation (4.44) shows.

Thus, if the dispersion in frequency is small, a group of waves will have an envelope which travels at the group velocity. The "signal" travels at the group velocity, not the phase velocity. If the dispersion in frequency is large, the concept is rather meaningless.[1] The *group time delay* over a distance l is

$$\tau_D = \frac{l}{u_g} = l\frac{d\beta}{d\omega} \quad (group\ time\ delay). \tag{4.58}$$

A somewhat analogous situation to that above occurs when a caterpillar is propagating itself along. The caterpillar is moving at group velocity, while the waves which ripple along its back from tail to head move at the phase velocity.[2]

4.6 REFLECTION OF PLANE WAVES, NORMAL INCIDENCE

The source for an electromagnetic wave is a time-dependent electric current that must exist somewhere. The wave that is produced inevitably strikes material bodies of some shape, causing reflections and, at least, partial transmission into the body. Much insight is gained if we examine what happens when a uniform plane wave strikes a plane interface that separates two different media. The analysis of this

---

[1] This will be discussed again.
[2] See Skilling, 1948 in the references at the end of Chapter 1.

situation is also very helpful in viewing the propagation of waves supported by transmission lines and waveguides. The simplest case, that of normal incidence, is sufficient to not only utilize previously established boundary conditions, but to introduce new parameters that will be very useful. For the time being, at least, we will only consider nonmagnetic media ($\mu = \mu_0$), permittivities that are simple scalar constants, and conductivities that are scalar constants. These assumptions still allow sufficient freedom to investigate the most important cases. Oblique incidence is considered in Section 4.7.

Dielectric-Conductor Interface

Consider the situation shown in Figure 4.4. It is ultimately more convenient, but certainly not necessary, to choose the positive or increasing z direction as shown. For $z > 0$ the medium is a loss-free dielectric having $\mu = \mu_0$, $\varepsilon = \varepsilon_1$, and $\sigma_1 = 0$, while for $z < 0$ the medium is a lossy dielectric having parameters μ_0, ε_2, and σ_2. A source for uniform plane waves exists somewhere to the left (large z) and produces a wave that is *incident* on the interface at $z = 0$. The incident wave is *reflected* (at least partially) back toward the source. Thus, we need both $-z$ traveling waves and $+z$ traveling waves for $z > 0$, and a suitable form for the sum of these is

$$E_x = E^- e^{+jk_1 z} + E^+ e^{-jk_1 z}, \quad z > 0 \tag{4.59}$$

and

$$H_y = -\frac{E^-}{\eta_1} e^{+jk_1 z} + \frac{E^+}{\eta_1} e^{-jk_1 z}, \quad z > 0, \tag{4.60}$$

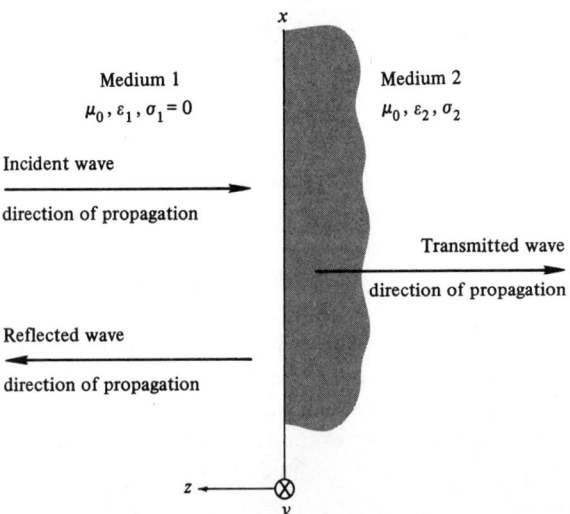

Figure 4.4. Dielectric-conductor plane interface.

4.6 REFLECTION OF PLANE WAVES, NORMAL INCIDENCE

where $k_1 = \omega\sqrt{\mu_0 \varepsilon_1}$ and $\eta_1 = \sqrt{\mu_0/\varepsilon_1}$. We have assumed that $E_y = E_z \equiv 0$, as before. Note that subscripts refer to medium 1. The interpretation of Equations (4.59) and (4.60) is as follows. The first terms of E_x and H_y represent an undamped electromagnetic wave traveling in the *negative* z direction and which is incident on the interface. This wave is called the *incident* wave. The second terms are an undamped electromagnetic wave traveling in the *positive* z direction, representing the *reflected wave*. Since at least part of the incident wave is apparently reflected, we suspect that part is also transmitted into medium 2. A suitable representation for the field in medium 2 must be a damped electromagnetic wave traveling in the negative z direction since $\sigma_2 \neq 0$. The *transmitted wave* must have the same polarization as the incident and reflected waves in order to preserve continuity of tangential components at the interface, as required by Equations (3.107) and (3.108). That is, if **E** (for $z < 0$) has a component other than E_x at $z = 0$, it cannot be canceled by the field from the other side at $z = 0$. It must be of the form $e^{+\gamma_2 z}$ (e.g., not $e^{+\gamma_2 y}$) in order to reduce to a constant at $z = 0$ and allow the cancellation to occur. No new frequencies are created by linear media, so the transmitted wave will be at the same frequency (in γ_2) as the incident and reflected waves. Thus, suitable forms for $z < 0$ are

$$E_x = E_2 e^{+\alpha_2 z} e^{+j\beta_2 z}, \quad z < 0 \tag{4.61}$$

and

$$H_y = -\frac{E_2}{\eta_2} e^{+\alpha_2 z} e^{+j\beta_2 z}, \quad z < 0, \tag{4.62}$$

where α_2, β_2, and η_2 are determined from appropriate forms depending on $\sigma_2/\omega\varepsilon_2$. Setting $z = 0$ and equating tangential components gives

$$E_x\bigg|_{z=0^+} = E_x\bigg|_{z=0^-}$$

and

$$H_y\bigg|_{z=0^+} = H_y\bigg|_{z=0^-}$$

or

$$E^- + E^+ = E_2$$

and

$$-\frac{E^-}{\eta_1} + \frac{E^+}{\eta_1} = -\frac{E_2}{\eta_2}.$$

In the last two equations there are apparently three unknowns E^-, E^+, and E_2. E^- is actually not an unknown because the peak value of the incident wave is usually a given quantity. In other words, the incident wave must have come from some source to the left ($z \gg 0$), and the incident wave assumes the role of forcing function. In any case, the logical procedure is to solve for E^+ and E_2 in terms of E^-. This results in

$$E^+ = E^- \frac{\eta_2 - \eta_1}{\eta_2 + \eta_1}$$

and

$$E_2 = E^- \frac{2\eta_2}{\eta_2 + \eta_1}.$$

The *coefficient of reflection* is defined as the ratio of the reflected to incident electric fields at $z = 0$ or

$$\boxed{\Gamma \equiv \frac{E^+}{E^-} = \frac{\eta_2 - \eta_1}{\eta_2 + \eta_1} = |\Gamma| e^{j\phi_k}} \qquad (4.63)$$

and the *coefficient of transmission* is defined as the ratio of the transmitted to incident electric field at $z = 0$ or

$$\boxed{T \equiv \frac{E_2}{E^-} = \frac{2\eta_2}{\eta_2 + \eta_1} = 1 + \Gamma}. \qquad (4.64)$$

The field equations may be written in a more convenient form with these definitions:

$$\begin{aligned} E_x &= E^- e^{jk_1 z}(1 + |\Gamma| e^{j(\phi_k - 2k_1 z)}), \quad z > 0, \\ H_y &= -\frac{E^-}{\eta_1} e^{jk_1 z}(1 - |\Gamma| e^{j(\phi_k - 2k_1 z)}), \quad z > 0 \end{aligned} \qquad (4.65)$$

and

$$\begin{aligned} E_x &= TE^- e^{\alpha_2 z} e^{j\beta_2 z}, \quad z < 0, \\ H_y &= -\frac{TE^-}{\eta_2} e^{\alpha_2 z} e^{j\beta_2 z}, \quad z < 0. \end{aligned} \qquad (4.66)$$

The current density in medium 2 is $J_x = \sigma_2 E_x$:

$$J_x = \sigma_2 TE^- e^{\alpha_2 z} e^{j\beta_2 z}, \quad z < 0. \qquad (4.67)$$

Inspection of the terms in parentheses in Equations (4.65) reveals that $|E_x|$ (or $|H_y|$) varies between maxima and minima ($z > 0$) in what is called a *standing wave pattern*. In this regard, it is convenient to define the *standing wave ratio* as

4.6 REFLECTION OF PLANE WAVES, NORMAL INCIDENCE

$$\text{SWR} \equiv \frac{|E_x|_{\max}}{|E_x|_{\min}} = \frac{|1 + |\Gamma|e^{j(\phi_k - 2k_1 z)}|_{\max}}{|1 + |\Gamma|e^{j(\phi_k - 2k_1 z)}|_{\min}}. \quad (4.68)$$

$|E_x|_{\max}$ occurs when the angle $\phi_k - 2k_1 z = -2n\pi$, $n = 0, 1, 2, \ldots$, or when $z = z_{\max} = (\phi_k + 2n\pi)/(2k_1)$, $z_{\max} > 0$. $|E_x|_{\min}$ occurs when the angle $\phi_k - 2k_1 z = -(2n-1)\pi$, $n = 0, 1, 2, \ldots$, or when $z = z_{\min} = [\phi_k + (2n-1)\pi]/(2k_1)$, $z_{\min} > 0$. We agree to use positive values for ϕ_k since z must be positive. Thus $|E_x|_{\max} = 1 + |\Gamma|$, and $|E_x|_{\min} = 1 - |\Gamma|$, so that

$$\boxed{\text{SWR} = \frac{1 + |\Gamma|}{1 - |\Gamma|}} \quad \text{(standing wave ratio)}. \quad (4.69)$$

The distance between successive peaks ($|E_x|_{\max}$) or between successive valleys ($|E_x|_{\min}$) corresponds to a change in phase of 2π radians of the numerator or denominator, respectively, of Equation (4.68). Then, we have

$$2k_1(z_2 - z_1) = 2\pi$$

or

$$z_2 - z_1 = \frac{\pi}{k_1} = \frac{\lambda_1}{2},$$

where we have used Equation (4.27). Thus, the spacing between adjacent maxima (or minima) is one-half wavelength, and either a maximum or minimum will always lie less than, or at most equal to, one-quarter wavelength from the interface at $z = 0$. The maxima and minima are $\lambda_1/4$ apart.

■ **Example 4.4**

Suppose that medium 1 (Figure 4.4) is air, while medium 2 is a material for which $\mu_2 = \mu_0$, $\varepsilon_2 = \varepsilon_0$, and $\sigma_2/(\omega\varepsilon_2) = 1$ at a frequency $\omega = 3 \times 10^8$ (rad/s). These parameters give

$$\gamma_2 = 0.455 + j1.099 \text{ m}^{-1},$$

$$\eta_2 = 292.8 + j121.3 = 316.9 \underline{|22.5°} \ \Omega,$$

$$\Gamma = -89.96 \times 10^{-3} + j197.4 \times 10^{-3} = 216.9 \times 10^{-3} \underline{|114.5°},$$

$$T = 910.0 \times 10^{-3} + j197.4 \times 10^{-3} = 931.2 \times 10^{-3} \underline{|12.24°}$$

$$\text{SWR} = 1.55,$$

$$z_{\max} = 1.0 \text{ m} \quad (\text{or } 0.159\lambda_1).$$

Figure 4.5 shows $|E_x|$ plotted as a function of z, demonstrating the (partial) standing wave that results from the successive incomplete cancellations and reinforcements

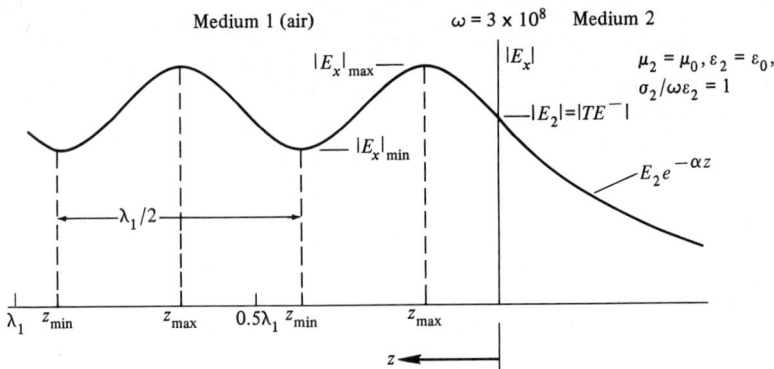

Figure 4.5. Reflection and transmission at a dielectric-conductor interface $|E_x|$ versus z. Medium 1, air: Medium 2, $\sigma_2/\omega\varepsilon_2 = 1$.

of the incident and reflected waves traveling in opposite directions. The damped transmitted wave decays as exp $(0.455z)$, $z < 0$. ∎

Assuming that E^- is real, the time-average power density for the *incident wave* (alone) is

$$<\mathcal{S}_{\text{inc}}> = \frac{1}{2}\text{Re}[\mathbf{a}_x E^- e^{jk_1 z} \times (-\mathbf{a}_y)(E^-/\eta_1)e^{-jk_1 z}],$$

$$<\mathcal{S}_{\text{inc}}> = -\mathbf{a}_z (E^-)^2/(2\eta_1) \quad (\text{W/m}^2).$$

The time-average power density for the *reflected wave* (alone) is

$$<\mathcal{S}_{\text{ref}}> = \frac{1}{2}\text{Re}[\mathbf{a}_x E^-|\Gamma|e^{j(\phi_k - k_1 z)} \times \mathbf{a}_y (E^-|\Gamma|/\eta_1^*)e^{-j(\phi_k - k_1 z)}],$$

$$<\mathcal{S}_{\text{ref}}> = \mathbf{a}_z (E^-|\Gamma|)^2/2\eta_1 \quad (\text{W/m}^2).$$

The time-average power density for the *transmitted wave* (alone) is

$$<\mathcal{S}_{\text{tr}}> = \frac{1}{2}\text{Re}[\mathbf{a}_x E^- T e^{(\alpha_2 + j\beta_2)z} \times (-\mathbf{a}_y) E^- (T/\eta_2)^* e^{(\alpha_2 - j\beta_2)z}],$$

$$<\mathcal{S}_{\text{tr}}> = -\mathbf{a}_z (E^-|T|)^2 e^{2\alpha_2 z} \frac{1}{2}\text{Re}(1/\eta_2^*) \quad (\text{W/m}^2).$$

It is intuitively obvious that for $z = 0$, we have

$$\boxed{<\mathcal{S}_{\text{inc}}> + <\mathcal{S}_{\text{ref}}> = <\mathcal{S}_{\text{tr}}>}. \tag{4.70}$$

4.6 REFLECTION OF PLANE WAVES, NORMAL INCIDENCE

The reader can verify that this is the case, and, with $E^- = 1$ V/m, the numbers in Example 4.4 give $<\mathcal{S}_{tr}> = -1.246\mathbf{a}_z$ (mW/m²). This power density that enters medium 2 at $z = 0$ is ultimately dissipated as heat in medium 2. That is, the time-average power dissipated *per unit area* $<\mathcal{P}_d>$ is the power flow *into* a volume that is a uniform column: $0 \le x \le 1, 0 \le y \le 1, -\infty \le z \le 0$:

$$<\mathcal{P}_d> = -\int_0^1 \int_0^1 \frac{1}{2} \text{Re}\{\mathbf{E} \times \mathbf{H}^*\}\bigg|_{z=0} \cdot \mathbf{a}_z \, dx \, dy,$$

where **E** and **H** are transmitted forms. Substituting Equations (4.66), the last equation can be put in the form:

$$\boxed{<\mathcal{P}_d> = \frac{1}{2}|H_y|^2_{z=0} \text{Re}\{\eta_2\}} \quad (\text{W/m}^2). \tag{4.71}$$

The reader can verify that the last two equations give $<\mathcal{P}_d> = 1.246$ mW/m² for the numbers in Example 4.4 ($E^- = 1$). Equation (4.71) is very similar in form to equations for average power that arise in circuit theory.

Returning to Equation (3.57), we see that the time-average power dissipated per unit area can also be calculated from

$$<\mathcal{P}_d> = \frac{1}{2}\text{Re}\left(\iiint_{\text{vol}} \mathbf{E} \cdot \mathbf{J} \, dv\right)$$

or

$$\boxed{<\mathcal{P}_d> = \frac{\sigma_2}{2} \iiint_{\text{vol}} |\mathbf{E}|^2 \, dv = \frac{1}{2\sigma_2} \iiint_{\text{vol}} |\mathbf{J}|^2 \, dv}. \tag{4.72}$$

It is implied in Equation (4.72) that $0 \le x \le 1$ and $0 \le y \le 1$. The reader should once again verify that $<\mathcal{P}_d> = 1.246$ mW/m² for the numbers in Example 4.4 ($E^- = 1$).

The Skin Effect

We have seen that in the steady-current case (dc) the current density **J** is uniformly distributed in a circular cylindrical conductor of uniform cross section. Time-varying current densities, on the other hand, have a tendency to congregate near the surface of good conductors. If the frequency is high, the current density becomes very nearly a surface current density, being concentrated in a thin layer at the conductor surface. This is known as the *skin effect*. Outdoor television antennas, for example, do not require solid conductors. Hollow tubular conductors serve better because they are lighter, stronger, and there would be essentially no current in the interior anyway. Silver plating is often used in waveguides for reducing losses.

Time-varying magnetic fields in conductors behave similarly, and, because of this, solid magnetic cores are not used in electric power equipment (motors, generators, transformers, etc.). These cores are laminated (layered), being made up of thin sheets pressed together (but insulated from each other). For higher frequencies, magnetic cores are made of a ferromagnetic powder to reduce losses (called *eddy* current losses). The skin effect, in either the electric or magnetic case, is the result of induction. A time-varying magnetic field is accompanied by a time-varying electric field which produces time-varying currents and *secondary* fields. The total fields are the sums of the original and secondary fields.

Equation (3.96), with $\mathbf{J} = \sigma \mathbf{E}$ and $\sigma \gg \omega\varepsilon$ for good conductors, becomes

$$\boxed{\nabla(\nabla \cdot \mathbf{J}) - \nabla^2 \mathbf{J} = -j\omega\mu\sigma\mathbf{J}}, \quad \sigma \gg \omega\varepsilon. \tag{4.73}$$

This is the partial differential equation that the conduction current density must satisfy in a good conductor *for any coordinate system*. Returning to Equation (3.100):

$$\nabla^2 \mathbf{H} + \omega^2 \mu\varepsilon \mathbf{H} = -\nabla \times \mathbf{J} = \nabla^2 \mathbf{H} + \omega\mu(\omega\varepsilon)\mathbf{H} = -\sigma\nabla \times \mathbf{E},$$

or, with $\nabla \times \mathbf{E} = -j\omega\mu\mathbf{H}$, we get

$$\nabla^2 \mathbf{H} = -\omega\mu(\omega\varepsilon)\mathbf{H} + j\omega\mu(\sigma)\mathbf{H};$$

and, for $\sigma \gg \omega\varepsilon$ and $\mathbf{B} = \mu\mathbf{H}$, we get

$$\boxed{\nabla^2 \mathbf{B} = j\omega\mu\sigma\mathbf{B}}, \quad \sigma \gg \omega\varepsilon. \tag{4.74}$$

This is the partial differential equation that $\mathbf{B}$ must satisfy in a good conductor.

Equation (4.73) reduces to

$$-\mathbf{a}_x \frac{d^2 J_x}{dz^2} = -j\omega\mu_2\sigma_2\mathbf{a}_x J_x \quad [\mathbf{J} = \mathbf{a}_x J_x(z)],$$

$$\frac{d^2 J_x}{dz^2} - j\omega\mu_2\sigma_2 J_x = 0$$

for the geometry of Figure 4.4. The solution to this *ordinary* differential equation must be Equation (4.67), where α_2 and β_2 are given by Equation (4.54) for a good conductor:

$$J_x(z,\omega) = \sigma T E^- e^{\sqrt{\omega\mu\sigma/2}\,z} e^{j\sqrt{\omega\mu\sigma/2}\,z}, \quad z < 0.$$

4.6 REFLECTION OF PLANE WAVES, NORMAL INCIDENCE

Subscripts (2) have been dropped for simplicity. The other possible solution involving *negative* exponents cannot be allowed for $z < 0$. The amplitude term $\sigma T E^-$ is the current density at $z = 0$, so

$$J_x(z, \omega) = J_x(0, \omega) e^{\sqrt{\omega\mu\sigma/2}\, z} e^{j\sqrt{\omega\mu\sigma/2}\, z}.$$

It is customary to define the *skin depth* as

$$\boxed{\delta = \sqrt{2/(\omega\mu\sigma)}} \quad \text{(m)} \quad \text{(skin depth)}, \tag{4.75}$$

so that

$$J_x(z, \omega) = J_x(0, \omega) e^{z/\delta} e^{jz/\delta}, \quad z < 0.$$

Thus, for a distance $z = -\delta$ into the conductor, the magnitude of the current density is $e^{-1} = 0.368$ times its value at the surface.

The power per unit area dissipated in the conductor can be found from Equations (4.70), (4.71), or (4.72). Equation (4.72) gives

$$<\mathcal{P}_d> \;=\; \frac{1}{2\sigma} \int_0^1 \int_0^1 \int_{-\infty}^0 |J_x|^2 \, dx \, dy \, dz = \frac{1}{2\sigma} \int_{-\infty}^0 |J_x|^2 \, dz,$$

$$<\mathcal{P}_d> \;=\; \frac{1}{2\sigma}|J_x(0, \omega)|^2 \int_{-\infty}^0 e^{2z/\delta} \, dz = \frac{\delta}{4\sigma}|J_x(0, \omega)|^2 \quad (\text{W/m}^2). \tag{4.76}$$

An alternate and very useful interpretation of this skin effect is obtained if the *total current* is assumed to flow *uniformly* (rather than as $e^{z/\delta}$) over *one skin depth* ($-\delta \le z \le 0$) only. The total current is given by Equation (2.69), where we integrate the actual current density over the infinite depth ($-\infty < z \le 0$) in the negative z direction and 1 m in the y direction:

$$I = \int_0^1 \int_{-\infty}^0 J_x(0, \omega) e^{z/\delta} e^{jz/\delta} \, dy \, dz = \frac{\delta J_x(0, \omega)}{1 + j}.$$

The density of this current is uniform over the area $(\delta)(1) = \delta$ (m^2), so the uniform current density is $J_x(0, \omega)/(1 + j)$ whose magnitude is $|J_x(0, \omega)|/\sqrt{2}$ as shown in Figure 4.6. The dc resistance (uniform current density) of the strip whose cross section is δ meters by 1 m and whose length is 1 m is $R = 1/[\sigma(\delta)1] = 1/(\sigma\delta)$. The time-average power dissipated in the strip is

$$<\mathcal{P}_d> \;=\; |I_{\text{eff}}|^2 R = \left|\frac{\delta J_x(0, \omega)}{\sqrt{2}(1 + j)}\right|^2 \frac{1}{\sigma\delta} = \frac{\delta}{4\sigma}|J_x(0, \omega)|^2,$$

which is identical to the result given by Equation (4.76).

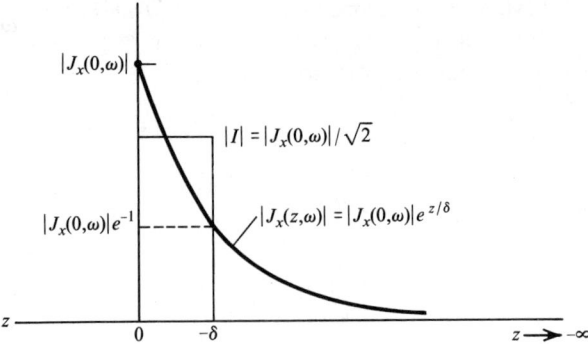

Figure 4.6. Actual distribution of $|J_x(z, \omega)|$ and the equivalent uniform current density (skin effect).

We now have a way of treating the skin effect in a simple manner. We merely need to use a uniform current density in a depth δ, and proceed on this basis. Although the result that we have derived applies only to plane surfaces, the concept can be applied to *curved surfaces* so long as the skin depth, as calculated by Equation (4.75), is small compared to the radius of curvature. The skin depth for copper at 60 Hz is 8.63 mm, so Equation (4.75) would not apply for a calculation of the resistance of a circular copper wire of 2 mm radius. On the other hand, at a frequency of 1 MHz the skin depth is only 0.068 mm, so Equation (4.75) applies, and we have

$$R = \frac{l}{\sigma \pi [a^2 - (a - \delta)^2]} \approx \frac{l}{2\pi a \sigma \delta} \quad (\Omega), \quad \delta \ll a$$

for a length l, radius a, and where the current density is confined to the radius ρ, $a - \delta \leq \rho \leq a$.

The skin effect also modifies *internal* inductance since the nonuniform current density alters internal magnetic flux linkage. This also can be treated in a relatively simple manner for good conductors. The wave impedance for a good conductor is given by Equation (4.56) where it is seen that the real and imaginary parts are equal. In light of the skin effect concept, it would certainly be appropriate to call this impedance a *surface* impedance. That is

$$\eta = R_s + jX_s = \sqrt{\frac{\omega \mu}{2\sigma}} + j\sqrt{\frac{\omega \mu}{2\sigma}}$$

or

$$R_s + jX_s = \frac{1}{\sigma \delta} + j\frac{1}{\sigma \delta} \quad (\Omega), \quad \sigma \gg \omega \varepsilon. \quad (4.77)$$

Equation (4.71) then becomes

4.6 REFLECTION OF PLANE WAVES, NORMAL INCIDENCE

$$<\mathcal{P}_d> = \frac{1}{2}|H_y|^2_{z=0} R_s \quad (\text{W/m}^2), \quad \sigma \gg \omega\varepsilon. \tag{4.78}$$

Thus, for good conductors where the skin effect applies, the inductive reactance $X_s = \omega L_{\text{int}}$ must equal the resistance R_s. These results can be written in the following forms for other geometries (particularly useful for guiding systems):

$$\boxed{R = \frac{l}{\sigma \delta w}} \quad (\Omega), \quad \sigma \gg \omega\varepsilon, \tag{4.79}$$

where l is the length, and w is the *width* through which the skin current flows for a depth δ, and

$$\boxed{L_{\text{int}} = R/\omega} \quad (\text{H}), \quad \sigma \gg \omega\varepsilon. \tag{4.80}$$

For example, a coaxial cable has *widths* that are approximately $2\pi a$ and $2\pi b$ for the inner and outer conductors, respectively, so the *series resistance per unit length* is approximately

$$R \approx \frac{1}{2\pi\sigma\delta}\left(\frac{1}{a} + \frac{1}{b}\right) \quad (\Omega/\text{m}), \quad \sigma \gg \omega\varepsilon. \tag{4.81}$$

If $\omega = 10^8$, $a = 2 \times 10^{-3}$, and $b = 6 \times 10^{-3}$, then $\delta = 16.57 \times 10^{-6}$ m for copper conductors, and $R = 110.4$ mΩ/m and $L_{\text{int}} = 1.104$ nH/m.

The time-domain fields, appropriate to Figure 4.4 when $J_x(0, \omega)$ is real and $\sigma \gg \omega\varepsilon$ are

$$\mathcal{J}_x(z, t) = J_x(0, \omega) e^{z/\delta} \cos(\omega t + z/\delta), \tag{4.82}$$

$$\mathcal{E}_x(z, t) = \mathcal{J}_x(z, t)/\sigma, \tag{4.83}$$

$$\mathcal{H}_y(z, t) = -\frac{\delta}{\sqrt{2}} J_x(0, \omega) e^{z/\delta} \cos(\omega t + z/\delta - \pi/4). \tag{4.84}$$

Note that the magnetic field intensity lags the electric field intensity by 45° (the magnetic field intensity vector is in the *negative y* direction), further supporting our conclusions in the preceding paragraph.

While on the subject of the skin effect it would be worthwhile to digress and investigate the skin effect in a thin sheet used in the laminated magnetic core of a transformer (for example).

Example 4.5

Consider the thin conducting sheet shown in Figure 4.7, where it is assumed that the magnetic flux density is $\mathbf{B} = B_0 \mathbf{a}_x$ at the surface of the sheet. The induced current will be in the $-y$ direction (to produce an opposing magnetic field). As

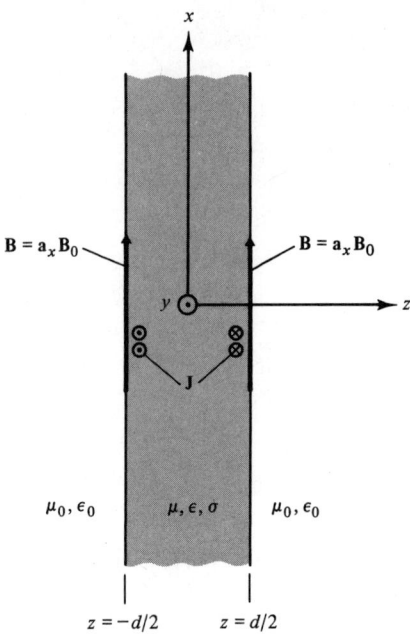

Figure 4.7. Thin conducting sheet in free space to demonstrate the skin effect on the magnetic flux density.

before, we assume that both **B** and **J** depend only on z and ω, in which case Equation (4.74) reduces to

$$\frac{d^2 B_x}{dz^2} = j\omega\mu\sigma B_x$$

whose solution can be expressed in terms of hyperbolic functions:

$$B_x(z,\omega) = A_1 \cosh\left(\sqrt{j\omega\mu\sigma}\,z\right) + A_2 \sinh\left(\sqrt{j\omega\mu\sigma}\,z\right).$$

The second term is not needed because of the even symmetry in the problem. B_x must be an even function of z, and $\sinh(\sqrt{j\omega\mu\sigma}\,z)$ is an odd function of z. Furthermore, $B_z(\pm d/2, \omega) = B_0$, so $A_2 = 0$ and

$$B_0 = A_1 \cosh\left(\sqrt{j\omega\mu\sigma}\,d/2\right),$$

$$A_1 = \frac{B_0}{\cosh\left(\sqrt{j\omega\mu\sigma}\,d/2\right)}.$$

Therefore, we get

$$B_x(z,\omega) = B_0 \frac{\cosh\left(\sqrt{j\omega\mu\sigma}\,z\right)}{\cosh\left(\sqrt{j\omega\mu\sigma}\,d/2\right)}$$

4.6 REFLECTION OF PLANE WAVES, NORMAL INCIDENCE

or

$$B_x(z, \omega) = B_0 \frac{\cosh(z/\delta + jz/\delta)}{\cosh(d/2\delta + jd/2\delta)}. \tag{4.85}$$

It is easy to calculate $|B_x(z, \omega)|/B_0$ if a calculator that has a built-in complex mode is available. Otherwise, it is necessary to use the double angle formulas for the hyperbolic cosine. The latter gives

$$\frac{|B_x(z, \omega)|}{B_0} = \left(\frac{\cosh^2 z/\delta - \sin^2 z/\delta}{\cosh^2 d/2\delta - \sin^2 d/2\delta} \right)^{1/2}.$$

Figure 4.8 shows $|B_x(z, \omega)|/B_0$ plotted versus z for $d/\delta = 1, 2,$ and 3.

Silicon is often added to steel for the magnetic material making up the lamination. This reduces σ, increases δ, and thus reduces d/δ, making B_x more nearly uniform. If this is done, then it can be assumed that $d/\delta \ll 1$. Now, $\mathbf{J} = \nabla \times \mathbf{H}$, or with Equation (4.85), we get

$$J_y = \frac{\partial H_x}{\partial z} = \frac{1+j}{\mu\delta} B_0 \frac{\sinh(z/\delta + jz/\delta)}{\cosh(d/2\delta + jd/2\delta)}.$$

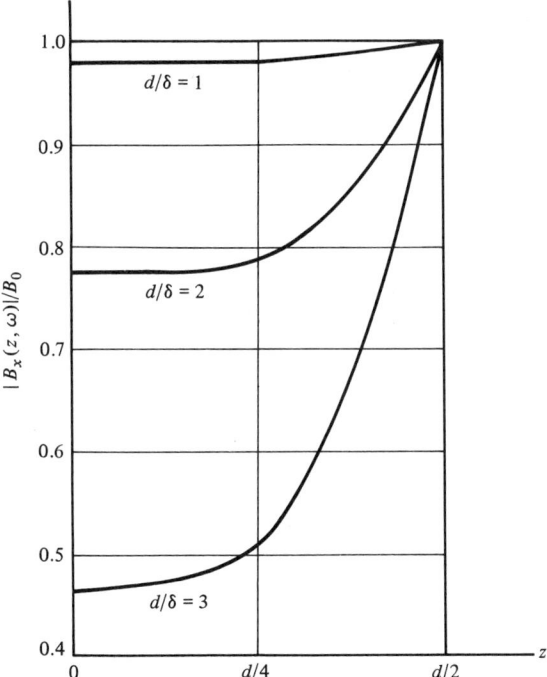

Figure 4.8. $|B_x(z, \omega)|/B_0$ versus z $(d/\delta = 1, 2, 3)$ for the conducting sheet of Figure 4.7.

Next, sinh $(z/\delta + jz/\delta) \approx z/\delta + jz/\delta$ and cosh $(d/2\delta + jd/2\delta) \approx 1$ for $z/\delta \ll 1$ and $d/\delta \ll 1$, respectively, so

$$J_y \approx \frac{1}{\mu} B_0 \left(\frac{1+j}{\delta}\right)^2 z.$$

According to Equation (4.72), the time-average power dissipated is

$$<\mathcal{P}_d> = \frac{1}{2\sigma} \iiint_{\text{vol}} |J_y|^2 \, dv.$$

For a lamination with unit area and thickness d, we have

$$<\mathcal{P}_d> = \frac{1}{2\sigma} \int_{-d/2}^{d/2} |J_y|^2 \, dz = \frac{1}{2\sigma} \int_{-d/2}^{d/2} \left(\frac{B_0}{\mu}\right)^2 \frac{4}{\delta^4} z^2 \, dz,$$

$$<\mathcal{P}_d> = \frac{2 B_0^2}{\sigma \mu^2} \frac{1}{\delta^4} \int_{-d/2}^{d/2} z^2 \, dz = \frac{1}{\sigma} \frac{B_0^2}{\mu^2} \frac{1}{\delta^4} \frac{d^3}{6}.$$

Eliminating δ^4 with Equation (4.75), we get

$$<\mathcal{P}_d> = \frac{1}{24} B_0^2 \omega^2 \sigma d^3 \quad (\text{W/m}^2)$$

for a thickness d and $d/\delta \ll 1$. Note that this loss due to *eddy currents* is proportional to the *square* of the frequency and the cube of the sheet thickness. ∎

Dielectric-Dielectric Interface

A dielectric-dielectric interface results in Figure 4.4 if we allow σ_2 to go to zero. The preceding results apply, but it is obvious that the most important effect of this is that there is no attenuation in medium 2, and the transmitted wave proceeds without damping. Since $\eta_1/\eta_2 = \sqrt{\varepsilon_2/\varepsilon_1}$, it is easy to show that

$$\Gamma = \frac{1 - \sqrt{\varepsilon_2/\varepsilon_1}}{1 + \sqrt{\varepsilon_2/\varepsilon_1}}, \quad z > 0, \quad \sigma_2 = 0$$

and

$$\text{SWR} = \begin{cases} \sqrt{\varepsilon_2/\varepsilon_1}, & \varepsilon_2 > \varepsilon_1, \\ \sqrt{\varepsilon_1/\varepsilon_2}, & \varepsilon_1 > \varepsilon_2. \end{cases}$$

The electric field intensity at the interface $z = 0$ is

$$E_2 = TE^- = \frac{2E^-}{1 + \sqrt{\varepsilon_2/\varepsilon_1}}.$$

Example 4.6

Suppose that $\varepsilon_2 = 4\varepsilon_1$. Then,

$$\Gamma = \frac{1-2}{1+2} = -\frac{1}{3} = \frac{1}{3}e^{j\pi},$$

$$T = \frac{2}{3},$$

and

$$\text{SWR} = 2.$$

From Equations (4.65) and (4.66), we have

$$E_x = E^- e^{jk_1 z}[1 + \tfrac{1}{3}e^{j(\pi - 2k_1 z)}], \quad z > 0,$$

$$H_y = -\frac{E^-}{\eta_1} e^{jk_1 z}[1 - \tfrac{1}{3}e^{j(\pi - 2k_1 z)}], \quad z > 0,$$

$$E_x = \frac{2E^-}{3} e^{jk_2 z}, \quad z < 0,$$

$$H_y = -\frac{2E^-}{3\eta_2} e^{jk_2 z}, \quad z < 0,$$

and

$$k_2 = \omega\sqrt{\mu_0 \varepsilon_2} = 2k_1 = \frac{4\pi}{\lambda_1} = \frac{2\pi}{\lambda_2}$$

and

$$\eta_2 = \sqrt{\frac{\mu_0}{\varepsilon_2}} = \frac{\eta_1}{2}.$$

A plot of $|E_x|$ and $|H_y|$ is shown in Figure 4.9. ∎

Dielectric-Perfect Conductor Interface

Referring to Figure 4.4 again, if we let $\sigma_2 \to \infty$ (perfect conductor) then $\eta_2 \to 0$ [see Equations (4.40) and (4.45)]. In this case, $\Gamma = -1$, $T = 0$, and SWR $\to \infty$. This is an example of total reflection, and E_x and H_y become *pure standing waves*, giving no net transmitted power density. The field ($z > 0$) is [see Equation (4.65)]

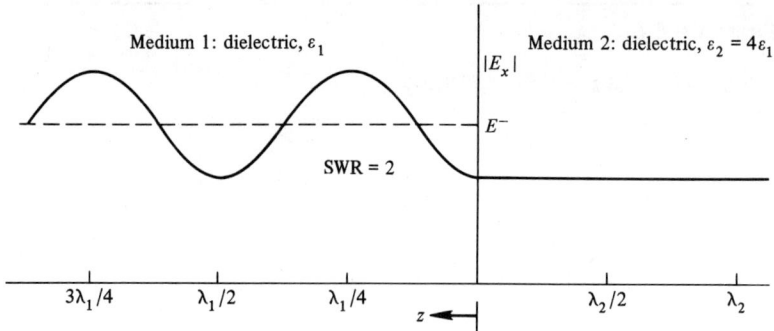

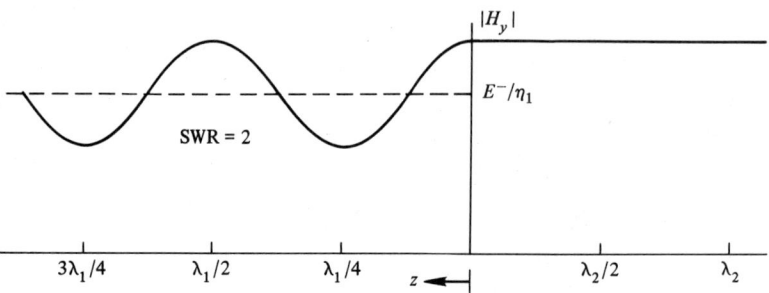

Figure 4.9. $|E_x|$ and $|H_y|$ as functions of z for a dielectric-dielectric interface $\varepsilon_2 = 4\varepsilon_1$, SWR = 2.

$$|E_x| = 2E^-|\sin k_1 z|,$$

$$|H_y| = \frac{2E^-}{\eta_1}|\cos k_1 z|$$

and is plotted in Figure 4.10(a).

It is worthwhile at this point to examine the behavior of a pure standing wave (SWR $\to \infty$) as a function of time. Consider E_x. Using Equation (4.65) with $\Gamma = -1 = e^{j\pi}$,

$$E_x = E^- e^{jk_1 z}(1 - e^{-j2k_1 z})$$

or

$$E_x = E^-(e^{jk_1 z} - e^{-jk_1 z}).$$

We obtain the time-domain form by multiplying the $e^{j\omega t}$ and then taking the real part. Thus,

$$\mathscr{E}_x(z,t) = E^- \operatorname{Re}\left(e^{j(\omega t + k_1 z)} - e^{j(\omega t - k_1 z)}\right)$$
$$= E^-[\cos(\omega t + k_1 z) - \cos(\omega t - k_1 z)]$$

4.6 REFLECTION OF PLANE WAVES, NORMAL INCIDENCE

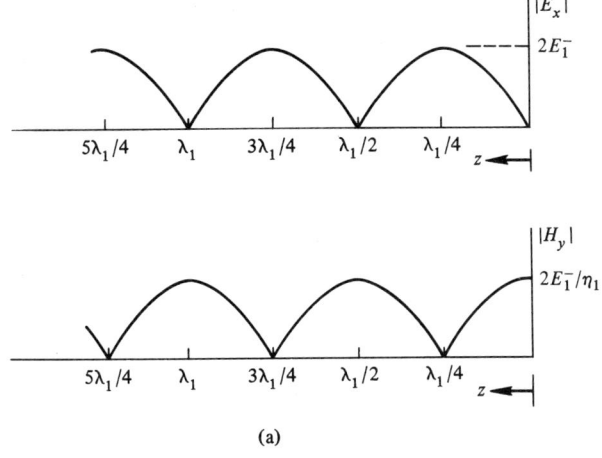

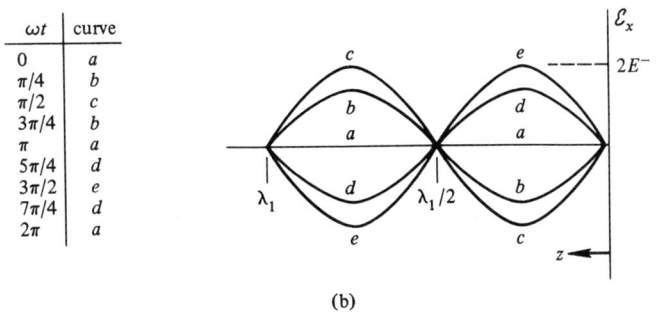

Figure 4.10. Pure standing waves, SWR $\to \infty$. (a) $|E_x|$ and $|H_y|$ versus z, dielectric-perfect conductor interface. (b) $\mathcal{E}_x$ versus z for various values of t, $\mathcal{E}_x(z,t) = -2E^- \sin(\omega t) \sin(k_1 z)$.

or

$$\mathcal{E}_x(z,t) = -2E^- \sin \omega t \sin k_1 z.$$

An observer with an electric field strength meter would measure nothing at $z = 0$, while at $z = \lambda/8$, he or she would measure $-\sqrt{2} E^- \sin \omega t$, and at $z = \lambda/4$, he or she would measure $-2E^- \sin \omega t$. Figure 4.10(b) shows this particular pure standing wave. Compare this standing wave to the traveling wave of Figure 4.2(b). Does this standing wave behavior resemble that of a vibrating string?

In the case of a perfect conductor, the skin depth is zero, and the current flow is in the form of a surface current. Multiplying this current density times the *width* across which it flows gives the *total* current. Formally, the surface current density is written [Equation (3.114)]

$$\mathbf{J}_s = \mathbf{a}_n \times \mathbf{H} \Big|_{\text{on } s} \quad \text{(A/m)},$$

where $\mathbf{a}_n$ is a unit vector normal to the perfect conductor surface s and pointing *into* the region of interest. Thus, as we saw in Chapter 3, the magnetic field is *discontinuous* at the surface of a perfect conductor. For Figure 4.10(a), we have

$$\mathbf{J}_s = \mathbf{a}_z \times \mathbf{a}_y \left(-\frac{2E^-}{\eta_1} \right) = \mathbf{a}_x \frac{2E^-}{\eta_1}$$

or

$$J_{sx} = \frac{2E^-}{\eta_1}.$$

Conductor-Perfect Conductor Interface

There will again be no field for $z < 0$ in this case because total reflection again occurs. For $z > 0$, the situation is now more complicated. Since $\eta_2 = 0$, it follows that $\Gamma = -1$ and $T = 0$. The field quantities for $z > 0$ become

$$E_x = E^- e^{+\alpha_1 z} e^{+j\beta_1 z} + E^+ e^{-\alpha_1 z} e^{-j\beta_1 z}$$

and

$$H_y = -\frac{E^-}{\eta_1} e^{+\alpha_1 z} e^{+j\beta_1 z} + \frac{E^+}{\eta_1} e^{-\alpha_1 z} e^{-j\beta_1 z}$$

or

$$E_x = E^- e^{(\alpha_1 + j\beta_1)z} (1 - e^{-2(\alpha_1 + j\beta_1)z})$$

and

$$H_y = -\frac{E^-}{\eta_1} e^{(\alpha_1 + j\beta_1)z} (1 + e^{-2(\alpha_1 + j\beta_1)z}).$$

The *magnitudes* of the two quantities are

$$|E_x| = E^- e^{\alpha_1 z} |1 - e^{-2(\alpha_1 + j\beta_1)z}|$$

and

$$|H_y| = \frac{E^-}{|\eta_1|} e^{\alpha_1 z} |1 + e^{-2(\alpha_1 + j\beta_1)z}|.$$

$|E_x|$ and $|H_y|$ are no longer periodic with z ($z > 0$) and the SWR concept becomes rather meaningless. A plot of $|E_x|$ against z is shown in Figure 4.11 for $\beta_1 \approx 10\alpha_1$. Note that the effect of the reflected wave is rapidly disappearing as the source of the incident wave is approached (large z).

4.7 OBLIQUE INCIDENCE 205

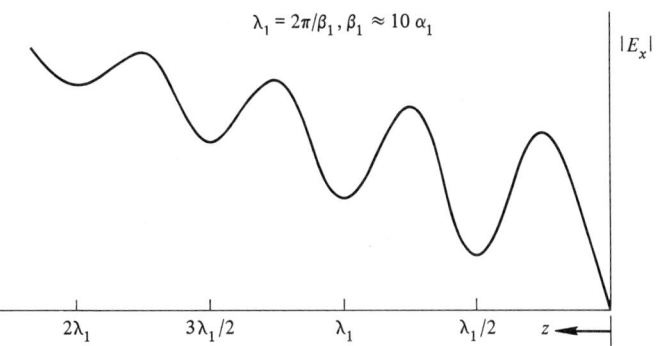

Figure 4.11. $|E_x|$ versus z conductor-perfect conductor interface $\beta_1 \approx 10\alpha_1$.

4.7 OBLIQUE INCIDENCE

It is not too difficult to show from the equation of a plane that a general expression for a uniform plane wave is (phasor form)

$$\mathbf{E} = \mathbf{E}_0 e^{-\gamma \mathbf{a}_n \cdot \mathbf{r}} \qquad (4.86)$$

for a wave propagating in the $\mathbf{a}_n$ direction. In Equation (4.86), $\mathbf{a}_n$ is a unit vector normal to an equiphase plane and making angles θ_x, θ_y, and $\theta_z = \theta$ with x, y, and z axes, respectively; $\mathbf{r}$ is a radial vector to any point on the equiphase plane. Then, we get

$$\mathbf{a}_n = \mathbf{a}_x \cos\theta_x + \mathbf{a}_y \cos\theta_y + \mathbf{a}_z \cos\theta_z, \qquad (4.87)$$

$$\mathbf{r} = \mathbf{a}_x x + \mathbf{a}_y y + \mathbf{a}_z z, \qquad (4.88)$$

and

$$\mathbf{a}_n \cdot \mathbf{r} = x \cos\theta_x + y \cos\theta_y + z \cos\theta_z. \qquad (4.89)$$

The vector $\mathbf{E}_0$ may be complex, but in any case both it and $\mathbf{H}$ lie in the equiphase plane, so that

$$\mathbf{a}_n \cdot \mathbf{E} = 0 \quad \text{and} \quad \frac{\mathbf{a}_n \times \mathbf{E}}{\eta} = \mathbf{H} \qquad (4.90)$$

This is shown in Figure 4.12.

Let us now consider the case of oblique incidence on a plane interface. We will explicitly consider the cases of **E** *parallel* to the interface and **H** *parallel* to the interface,[3] arbitrary polarization being a superposition of these two cases. We

---

[3]These are commonly called *horizontal* and *vertical* polarization, respectively, when the earth is the reflecting surface.

4 UNIFORM PLANE WAVE PROPAGATION

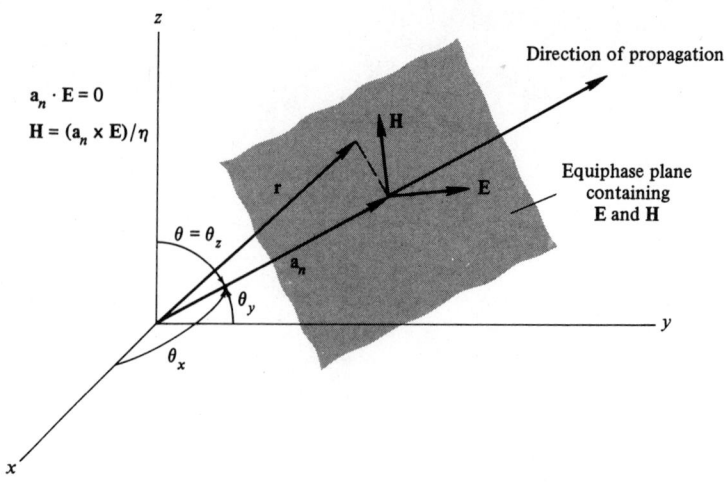

Figure 4.12. Geometry for an arbitrarily oriented uniform plane wave.

will follow the same procedure as that used for normal incidence in Section 4.6. Consider Figure 4.13.

(1) For **E** parallel to the $z = 0$ plane, **E** will be in the $\pm x$ direction. We choose the positive direction. Using Equations (4.87) through (4.90), we have, for the incident wave,

$$\theta_x = \frac{\pi}{2}, \quad \theta_y = \frac{\pi}{2} - \theta_i, \quad \theta_z = \pi - \theta_i,$$

$$E_x = E^- e^{-jk_1(y \sin \theta_i - z \cos \theta_i)}, \quad (4.91)$$

and

$$\mathbf{H} = (-\mathbf{a}_y \cos \theta_i - \mathbf{a}_z \sin \theta_i)\frac{E^-}{\eta_1} e^{-jk_1(y \sin \theta_i - z \cos \theta_i)} \quad (4.92)$$

if medium (1) is lossless ($\gamma = jk$). The *incident wave impedance in the* $-z$ *direction* is (see Section 4.4).

$$\eta_z^- \equiv -\frac{E_x}{H_y} \equiv \frac{\eta_1}{\cos \theta_i}. \quad (4.93)$$

For the reflected wave, we have

$$E_x = E^+ e^{-jk_1(y \sin \theta_r + z \cos \theta_r)}, \quad (4.94)$$

$$\mathbf{H} = (\mathbf{a}_y \cos \theta_r - \mathbf{a}_z \sin \theta_r)\frac{E^+}{\eta_1} e^{-jk_1(y \sin \theta_r + z \cos \theta_r)} \quad (4.95)$$

4.7 OBLIQUE INCIDENCE

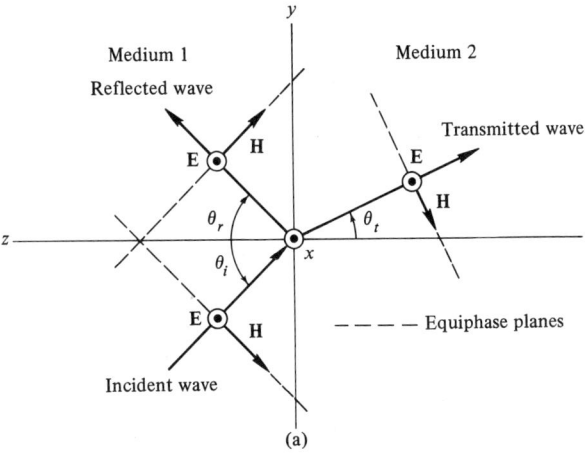

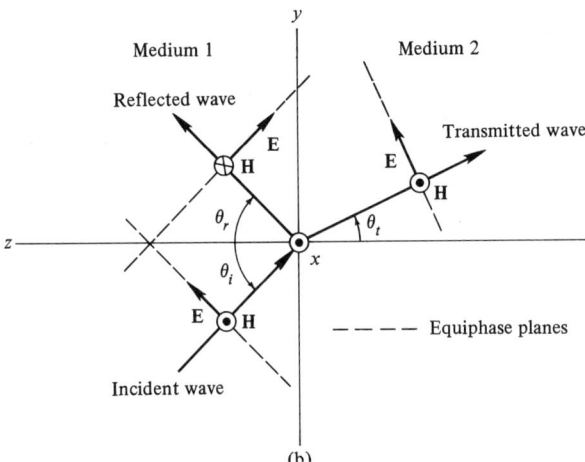

Figure 4.13. Geometry for oblique incidence calculations. (a) **E** parallel to the $z = 0$ plane (horizontal polarization). (b) **H** parallel to the $z = 0$ plane (vertical polarization).

and

$$\eta_z^+ \equiv \frac{E_x}{H_y} = \frac{\eta_1}{\cos \theta_r} \tag{4.96}$$

For the transmitted wave (if medium 2 is lossless), we have

$$E_x = E_2 e^{-jk_2(y \sin \theta_t - z \cos \theta_t)}, \tag{4.97}$$

$$\mathbf{H} = (-\mathbf{a}_y \cos \theta_t - \mathbf{a}_z \sin \theta_t)\frac{E_2}{\eta_2} e^{-jk_2(y \sin \theta_t - z \cos \theta_t)}, \tag{4.98}$$

and

$$\bar{\eta_z} \equiv -\frac{E_x}{H_y} = \frac{\eta_2}{\cos \theta_t}. \qquad (4.99)$$

(2) For **H** parallel to the interface, we again choose, or assume, positive tangential **E** components. Then, for the incident wave, we get

$$\mathbf{E} = (\mathbf{a}_y \cos \theta_i + \mathbf{a}_z \sin \theta_i) E^- e^{-jk_1(y \sin \theta_i - z \cos \theta_i)}, \qquad (4.100)$$

$$H_x = \frac{E^-}{\eta_1} e^{-jk_1(y \sin \theta_i - z \cos \theta_i)}, \qquad (4.101)$$

and

$$\bar{\eta_z} \equiv \frac{E_y}{H_x} = \eta_1 \cos \theta_i. \qquad (4.102)$$

For the reflected wave, we have

$$\mathbf{E} = (\mathbf{a}_y \cos \theta_r - \mathbf{a}_z \sin \theta_r) E^+ e^{-jk_1(y \sin \theta_r + z \cos \theta_r)}, \qquad (4.103)$$

$$H_x = -\frac{E^+}{\eta_1} e^{-jk_1(y \sin \theta_r + z \cos \theta_r)}, \qquad (4.104)$$

and

$$\eta_z^+ \equiv -\frac{E_y}{H_x} = \eta_1 \cos \theta_r. \qquad (4.105)$$

For the transmitted wave, we have

$$\mathbf{E} = (\mathbf{a}_y \cos \theta_t + \mathbf{a}_z \sin \theta_t) E_2 e^{-jk_2(y \sin \theta_t - z \cos \theta_t)}, \qquad (4.106)$$

$$H_x = \frac{E_2}{\eta_2} e^{-jk_2(y \sin \theta_t - z \cos \theta_t)}, \qquad (4.107)$$

and

$$\bar{\eta_z} \equiv \frac{E_y}{H_x} = \eta_2 \cos \theta_t. \qquad (4.108)$$

Note that the directions of **H** and **S** with respect to **E** are determined by $\mathbf{S} = \mathbf{E} \times \mathbf{H}$, where **S** is the power density vector, pointing in the direction of propagation, and *positive* tangential components of **E** were *assumed* in all cases. The transverse wave impedances were defined in Section 4.4.

4.7 OBLIQUE INCIDENCE

Boundary conditions at the interface simply require that the tangential components of **E** and **H** are continuous at $z = 0$. Thus, for (1) **E** parallel to the interface, we have

$$E^- e^{-jk_1 y \sin \theta_i} + E^+ e^{-jk_1 y \sin \theta_r} = E_2 e^{-jk_2 y \sin \theta_t}, \qquad (4.109)$$

$$-\cos\theta_1 \frac{E^-}{\eta_1} e^{-jk_1 y \sin \theta_i} + \cos\theta_r \frac{E^+}{\eta_1} e^{-jk_1 y \sin \theta_r} = -\cos\theta_t \frac{E_2}{\eta_2} e^{-jk_2 y \sin \theta_t}; \qquad (4.110)$$

and for (2) **H** parallel to the interface, we have

$$\cos\theta_i E^- e^{-jk_1 y \sin \theta_i} + \cos\theta_r E^+ e^{-jk_1 y \sin \theta_r} = \cos\theta E_2 e^{-jk_2 y \sin \theta_t}, \qquad (4.111)$$

$$\frac{E^-}{\eta_1} e^{-jk_1 y \sin \theta_i} - \frac{E^+}{\eta_1} e^{-jk_1 y \sin \theta_r} = \frac{E_2}{\eta_2} e^{-jk_2 y \sin \theta_t}. \qquad (4.112)$$

The phenomena we are observing at the interface must be independent of y, and this can only occur if all the phase angles of the phasor quantities are the same in the last four equations. Therefore, we have

$$k_1 \sin \theta_i = k_1 \sin \theta_r = k_2 \sin \theta_t. \qquad (4.113)$$

From the first of Equations (4.113), we have

$$\boxed{\theta_i = \theta_r}, \qquad (4.114)$$

or the angle of incidence is equal to the angle of reflection. From the second of Equations (4.113), we have

$$\boxed{\frac{\sin \theta_t}{\sin \theta_r} = \frac{\sin \theta_t}{\sin \theta_i} = \frac{k_1}{k_2} = \frac{u_2}{u_1} = \sqrt{\frac{\mu_1 \varepsilon_1}{\mu_2 \varepsilon_2}}}. \qquad (4.115)$$

This result is known as *Snell's law*. In the most common case, we have $\mu_1 = \mu_2 = \mu_0$, for which

$$\sin \theta_t = \sin \theta_i \sqrt{\frac{\varepsilon_1}{\varepsilon_2}}. \qquad (4.116)$$

We now have for (1) **E** parallel to the interface,

$$E^- + E^+ = E_2, \qquad (4.117)$$

$$\frac{\cos \theta_i}{\eta_1}(E^- - E^+) = \cos\theta_t \frac{E_2}{\eta_2}; \qquad (4.118)$$

and for (2) **H** parallel to the interface,

$$\cos \theta_i (E^- + E^+) = \cos \theta_t E_2, \quad (4.119)$$

$$\frac{E^-}{\eta_1} - \frac{E^+}{\eta_1} = \frac{E_2}{\eta_2}. \quad (4.120)$$

The coefficients of reflection (E^+/E^-) are obtained by solving the preceding pairs of equations simultaneously. They are [from Equations (4.117) and (4.118)]

$$\Gamma_E = \frac{\eta_2 \cos \theta_i - \eta_1 \cos \theta_t}{\eta_2 \cos \theta_i + \eta_1 \cos \theta_t} \quad (\textbf{E parallel}) \quad (4.121)$$

and [from Equations (4.119) and (4.120)]

$$\Gamma_H = \frac{\eta_2 \cos \theta_t - \eta_1 \cos \theta_i}{\eta_2 \cos \theta_t + \eta_1 \cos \theta_i} \quad (\textbf{H parallel}) \quad (4.122)$$

for **E** parallel to the interface and **H** parallel to the interface, respectively. Note that these equations are in the same form as Equation (4.63) (normal incidence) to which Equations (4.121) and (4.122) reduce when $\theta_i = \theta_t = 0$.

Total transmission occurs when $\Gamma = 0$ or

$$\sin \theta_i = \sqrt{\frac{\varepsilon_2/\varepsilon_1 - \mu_2/\mu_1}{\mu_1/\mu_2 - \mu_2/\mu_1}} \quad (\textbf{E parallel}) \quad (4.123)$$

and

$$\sin \theta_i = \sqrt{\frac{\varepsilon_2/\varepsilon_1 - \mu_2/\mu_1}{\varepsilon_2/\varepsilon_1 - \varepsilon_1/\varepsilon_2}} \quad (\textbf{H parallel}). \quad (4.124)$$

Equations (4.123) and (4.124) may not have solutions in all cases, but one special case is important. For **H** parallel to the interface and $\mu_1 = \mu_2 = \mu_0$, Equation (4.124) gives

$$\boxed{\theta_i = \theta_B = \sin^{-1} \sqrt{\frac{\varepsilon_2}{\varepsilon_1 + \varepsilon_2}} = \tan^{-1} \sqrt{\frac{\varepsilon_2}{\varepsilon_1}}} \quad (Brewster\ angle). \quad (4.125)$$

This angle is the *polarizing angle* or *Brewster angle* because an arbitrarily polarized wave incident at this angle will be reflected with **E** polarized parallel to the interface. That is, the other part of **E** is totally transmitted.

Example 4.7

A uniform plane wave is incident at the Brewster angle with **H** parallel to the interface at $z=0$ from air onto a plastic with $\varepsilon_R = 4$. Thus, from Equation (4.125), we get

$$\theta_i = \theta_B = \tan^{-1}\sqrt{\varepsilon_2/\varepsilon_1} = \tan^{-1} 2 = 63.43°,$$

$\Gamma_H = 0$, and from Equation (4.115), we get

$$\theta_t = \sin^{-1}\left(\sin\theta_i \sqrt{\varepsilon_1/\varepsilon_2}\right) = \sin^{-1} 0.447 = 26.57°$$

With $E^+ = 0$, Equation (4.119) gives

$$E_2 = E^- \cos\theta_i/\cos\theta_t = 0.5 E^-.$$

The incident field [from Equations (4.100) and (4.101)] is

$$\mathbf{E} = (0.447\mathbf{a}_y + 0.894\mathbf{a}_z)E^- e^{-jk_1(0.894y - 0.447z)},$$

$$H_x = (E^-/\eta_1)e^{-jk_1(0.894y - 0.447z)}.$$

The reflected field is zero, and the transmitted field [from Equations (4.106) and (4.107)] is

$$\mathbf{E} = (0.894\mathbf{a}_y + 0.447\mathbf{a}_z)(0.5 E^-)e^{-jk_2(0.447y - 0.894z)},$$

$$H_x = (0.5 E^-/\eta_2)e^{-jk_2(0.447y - 0.894z)}.$$

The time-average power density for the incident wave is

$$<\mathcal{S}_{\text{inc}}> = \tfrac{1}{2}\text{Re}(\mathbf{E} \times \mathbf{H}^*),$$

$$<\mathcal{S}_{\text{inc}}> = \tfrac{1}{2}\text{Re}\{-\mathbf{a}_z[0.447(E^-)^2/\eta_1] + \mathbf{a}_y[0.894(E^-)^2/\eta_1]\}.$$

Taking E^- to be real and equal to 1 V/m with $\eta_1 = 120\pi$, we get

$$<\mathcal{S}_{\text{inc}}> = -593\mathbf{a}_z + 1186\mathbf{a}_y \quad \mu\text{W/m}^2, \quad (z > 0),$$

while for the transmitted wave ($\eta_2 = 60\pi$), we have

$$<\mathcal{S}_{\text{tr}}> = \tfrac{1}{2}\text{Re}\{-\mathbf{a}_z[0.894(E^-)^2/4\eta_2] + \mathbf{a}_y[0.447(E^-)^2/4\eta_2]\},$$

$$<\mathcal{S}_{\text{tr}}> = -593\mathbf{a}_z + 297\mathbf{a}_y \quad \mu\text{W/m}^2 \quad (z < 0).$$

The results just obtained for $<\mathcal{S}_{\text{inc}}>$ and $<\mathcal{S}_{\text{tr}}>$ may, at first glance, appear to be incorrect, but they are not. The time-average power flow out of any finite volume must be zero in this example since there are no losses or sources in evidence. There must be sources somewhere, of course, to create the incident wave. Thus, using Equations (3.57) and (3.59), we have

$$<\mathcal{P}_f> = \iint_s <\mathcal{S}> \cdot ds \equiv 0.$$

Taking a cube that is 1 m on a side, centered at the origin, and with edges parallel to the coordinate axes, to be the closed surface, we have

$$<\mathcal{P}_f> = \int_{-1/2}^{1/2}\int_0^{1/2} <\mathcal{S}_{\text{inc}}> \cdot (-\mathbf{a}_y)\, dx\, dz + \int_{-1/2}^{1/2}\int_{-1/2}^{1/2} <\mathcal{S}_{\text{inc}}> \cdot \mathbf{a}_z\, dx\, dy$$

$$+ \int_{-1/2}^{1/2}\int_0^{1/2} <\mathcal{S}_{\text{inc}}> \cdot \mathbf{a}_y\, dx\, dz + \int_{-1/2}^{1/2}\int_{-1/2}^0 <\mathcal{S}_{\text{tr}}> \cdot (-\mathbf{a}_y)\, dx\, dz$$

$$+ \int_{-1/2}^{1/2}\int_{-1/2}^{1/2} <\mathcal{S}_{\text{tr}}> \cdot (-\mathbf{a}_y)\, dx\, dy + \int_{-1/2}^{1/2}\int_{-1/2}^0 <\mathcal{S}_{\text{tr}}> \cdot \mathbf{a}_y\, dx\, dz,$$

$$<\mathcal{P}_f> = (-1186/2 - 593 + 1186/2 - 297/2 + 593 + 297/2)\ \mu\text{W},$$

$$<\mathcal{P}_f> = 0.$$

The faces for $x = \pm\frac{1}{2}$ were ignored since $<\mathcal{S}_x> \equiv 0$. ∎

Total reflection[4] occurs for $|\Gamma| = 1$. Inspection of Equations (4.121) and (4.122) reveals that this cannot occur for real values of θ_i and θ_t. Suppose that θ_t is imaginary, or $\sin \theta_t > 1$. For convenience let

$$k_2 \sin \theta_t \equiv \beta$$

and

$$k_2 \cos \theta_t = k_2\sqrt{1 - \sin^2 \theta_t} \equiv \pm j\alpha.$$

Choosing the minus sign to give a damped or decreasing wave and using these definitions in Equations (4.97) and (4.98) for the transmitted wave (**E** parallel), we have

$$E_x = E_2 e^{+\alpha z} e^{-j\beta y}, \quad (4.126)$$

$$\mathbf{H} = \left(\mathbf{a}_y \frac{j\alpha}{k_2} - \mathbf{a}_z \frac{\beta}{k_2}\right)\frac{E_2}{\eta_2} e^{+\alpha z} e^{-j\beta y}. \quad (4.127)$$

---

[4]The operation of fiber optic cables is based on the principle of total reflection. A wave that exists internal to a dielectric rod can be totally reflected at the surface of the rod so that it remains within the rod, but has a net propagation in the direction of the axis of the rod. This will be examined more closely later.

4.7 OBLIQUE INCIDENCE

The wave impedance *into medium 2* is

$$\bar{\eta_z} \equiv -\frac{E_x}{H_y} = j\frac{\eta_2 k_2}{\alpha}, \qquad (4.128)$$

which is imaginary (η_2, k_2 real) indicating *no power flow* into medium 2!

It is a straightforward matter to show that Equations (4.126) and (4.127) simultaneously satisfy Maxwell's equations. These equations represent waves propagating in the $+y$ direction and are *evanescent* (exponential damping only) as far as the z direction is concerned. Now, $\sin\theta_t$ becomes greater than unity (for *both* types of polarization) when $\sin\theta_i$ becomes greater then $\sqrt{\mu_2\varepsilon_2/\mu_1\varepsilon_1}$, as Equation (4.115) shows. The *critical angle* then is

$$\theta_{ic} = \sin^{-1}\sqrt{\frac{\mu_2\varepsilon_2}{\mu_1\varepsilon_1}} \quad (\sin\theta_{ic} < 1) \qquad (4.129)$$

and a wave incident on a plane interface at an angle equal to or greater than this angle will be totally reflected. When $\sin\theta_t$ is greater than unity, Equation (4.121), for example, becomes

$$\Gamma_E = \frac{\eta_2\sqrt{1-(\mu_2\varepsilon_2/\mu_1\varepsilon_1)^2} + j(\eta_1\alpha/k_2)}{\eta_2\sqrt{1-(\mu_2\varepsilon_2/\mu_1\varepsilon_1)^2} - j(\eta_1\alpha/k_2)} = \frac{a+jb}{a-jb}$$

so $|\Gamma_E|$ is indeed unity when $\sqrt{(\mu_2\varepsilon_2)/(\mu_1\varepsilon_1)} < 1$.

■ Example 4.8

A uniform plane wave is incident at the angle $\theta_i = 60°$ (with **E** parallel to the interface) at $z = 0$ from a plastic with $\varepsilon_R = 4$ onto free space at $\omega = 3\times 10^8$ rad/s. Equation (4.129) gives the critical angle

$$\theta_{ic} = \sin^{-1}\sqrt{1/4} = 30°;$$

and since we have $\theta_i > \theta_{ic}$, total reflection occurs. From Equation (4.115), we get

$$\sin\theta_t = 2\sin\theta_i = \sqrt{3} > 1.$$

Thus, we have

$$\beta = k_2 \sin\theta_t = \omega\sqrt{\mu_0\varepsilon_0}\sin\theta_t = \sqrt{3}$$

and

$$\cos\theta_t = \sqrt{1-\sin^2\theta_t} = \sqrt{-2} = -j\sqrt{2},$$

so that

$$k_2 \cos \theta_t = -j\alpha = \omega\sqrt{\mu_0\varepsilon_0}(-j\sqrt{2})$$

or

$$\alpha = \sqrt{2}.$$

The coefficient of reflection ($\eta_1 = 60\pi$, $\eta_2 = 120\pi$) is given by Equation (4.121):

$$\Gamma_E = \frac{2\cos\theta_i - \cos\theta_t}{2\cos\theta_i + \cos\theta_t} = \frac{1 + j\sqrt{2}}{1 - j\sqrt{2}} = 1\underline{|109.48°};$$

and from Equation (4.120), we get

$$E_2 = \frac{\eta_2}{\eta_1}(E^- - E^+) = \frac{\eta_2}{\eta_1}(E^- + \Gamma_E E^-),$$

$$E_2 = 2E^-(1 + 1\underline{|109.48°}) = (2.309\underline{|54.74°})E^-$$

The incident field is

$$E_x = E^- e^{-jk_1(0.866y - 0.5z)} \quad (z > 0),$$

$$\mathbf{H} = (-0.5\mathbf{a}_y - 0.866\mathbf{a}_z)(E^-/\eta_1)e^{-jk_1(0.866y - 0.5z)} \quad (z > 0);$$

the reflected field is

$$E_x = \Gamma_E E^- e^{-jk_1(0.866y + 0.5z)} \quad (z > 0),$$

$$\mathbf{H} = (0.5\mathbf{a}_y - 0.866\mathbf{a}_z)(\Gamma_E E^-/\eta_1)e^{-jk_1(0.866y + 0.5z)} \quad (z > 0);$$

and the transmitted field is

$$E_x = (2.309\underline{|54.74°})E^- e^{\sqrt{2}z - j\sqrt{3}y} \quad (z < 0),$$

$$\mathbf{H} = (j\sqrt{2}\mathbf{a}_y - \sqrt{3}\mathbf{a}_z)(2.309\underline{|54.74°})(E^-/\eta_2)e^{\sqrt{2}z - j\sqrt{3}y} \quad (z < 0).$$

Taking E^- to be real and equal to 1 V/m, we have

$$\langle \mathscr{S}_{\text{inc}} \rangle = -1.326\mathbf{a}_z + 2.297\mathbf{a}_y \quad (\text{mW/m}^2) \quad (z > 0),$$
$$\langle \mathscr{S}_{\text{ref}} \rangle = +1.326\mathbf{a}_z + 2.297\mathbf{a}_y \quad (\text{mW/m}^2) \quad (z > 0),$$
$$\langle \mathscr{S}_{\text{tr}} \rangle = 24.49e^{2\sqrt{2}z}\mathbf{a}_y \quad (\text{mW/m}^2) \quad (z < 0).$$

If we take a cube that is 1 m on a side, centered at the origin, and with edges parallel to the coordinate axes, as in Example 4.3, we find that the time-average power flow out of the cube is

4.7 OBLIQUE INCIDENCE

$$\langle \mathcal{P}_f \rangle = -2.297 + 0 + 2.297 - 6.555 + 0 + 6.555 \quad \text{m/W},$$

$$\langle \mathcal{P}_f \rangle = 0.$$

This is the expected result, and can be obtained by inspection of the time-average power densities without integrating. Note the evanescent character of the transmitted field that is being *guided* along the interface. This field decays as $e^{\alpha z}$, $z < 0$, and suggests that a dielectric rod can guide electromagnetic fields. This phenomenon will be investigated in more detail later. ∎

All of the results of this section can be extended to the more general case where medium 2 is lossy by simply replacing ε with $\varepsilon' = \varepsilon + \sigma/(j\omega)$, as we did in Section 4.5. The case of plane wave incidence at an oblique angle from air ($\mu_1 = \mu_0$, $\varepsilon_1 = \varepsilon_0$, and $\sigma_1 = 0$) onto a plane earth ($\mu_2 = \mu_0$, $\varepsilon_2 = \varepsilon_e$, and $\sigma_2 = \sigma_e$ is important in propagation studies. Note that $\varepsilon_2 = \varepsilon' = \varepsilon_e + \sigma_e/(j\omega)$ appears in both η_2 and $\cos \theta_t$. For this case we obtain the following results:

$$\Gamma_E = \frac{\eta' \cos \theta_i - \eta_0 \cos \theta'}{\eta' \cos \theta_i + \eta_0 \cos \theta'} \quad \text{(\textbf{E} parallel to earth)} \tag{4.130}$$

and

$$\Gamma_H = \frac{\eta' \cos \theta' - \eta_0 \cos \theta_i}{\eta' \cos \theta' + \eta_0 \cos \theta_i} \quad \text{(\textbf{H} parallel to earth)}, \tag{4.31}$$

where

$$\eta_0 = \sqrt{\mu_0/\varepsilon_0} = 120\pi \quad (\Omega),$$
$$\eta' = \sqrt{\mu_0/\varepsilon'} = \sqrt{\mu_0/(\varepsilon_e + \sigma_e/j\omega)} \quad (\Omega),$$
$$\cos \theta' = \sqrt{1 - \sin^2 \theta_t} = \sqrt{1 - \sin^2 \theta_i \varepsilon_1/\varepsilon_2},$$
$$\cos \theta' = \sqrt{1 - \sin^2 \theta_i \varepsilon_0/(\varepsilon_e + \sigma_e/j\omega)} \quad \text{(complex)}.$$

We next consider a situation that will be very helpful when rectangular waveguides are studied. If medium 2 is perfectly conducting, then $\eta_2 = 0$, $\Gamma_E = \Gamma_H = -1$, and there is no field in region 2. For the case of **E** parallel to the interface the *total* field is obtained from Equations (4.91) through (4.96):

$$E_x = E^- e^{-jk_1 y \sin \theta_i}[2j \sin(k_1 z \cos \theta_i)], \quad z > 0 \tag{4.132}$$

and

$$\mathbf{H} = -\frac{E^-}{\eta_1} e^{-jk_1 y \sin \theta_i}[2\mathbf{a}_y \cos \theta_i \cos(k_1 z \cos \theta_i)$$
$$+ 2j\mathbf{a}_z \sin \theta_i \sin(k_1 z \cos \theta_i)], \quad z > 0 \tag{4.133}$$

The time-average power densities are

$$<\mathcal{S}_z> = \tfrac{1}{2}\mathrm{Re}(E_x H_y^*) = 0 \tag{4.134}$$

and

$$<\mathcal{S}_y> = -\tfrac{1}{2}\mathrm{Re}(E_x H_z^*) = \frac{2(E^-)^2}{\eta_1}\sin\theta_i \sin^2(k_1 z \cos\theta_i). \tag{4.135}$$

Inspection of Equations (4.132) and (4.133) reveals, first of all, through the term $\sin(k_1 z \cos\theta_i)$ that for E_x and H_z a *standing wave* pattern exists in the z direction with nulls where $k_1 z \cos\theta_i = p\pi$, $p = 0, 1, 2, \ldots$, or $z = p/(2f\sqrt{\mu_1\varepsilon_1}\cos\theta_i)$. One of these loci is indicated in Figure 4.14 by the horizontal (long) dashed line. Second, the time-average power density is in the y direction only. Last, the term $e^{-jk_1 y \sin\theta_i}$ reveals that a *traveling wave* exists in the y direction with a phase velocity given by setting the phase constant $k_1 \sin\theta_i$ equal to ω/u_p. That is, $u_p = \omega/(k_1 \sin\theta_i)$, or

$$u_p = \frac{1}{\sqrt{\mu_1\varepsilon_1}\sin\theta_i} \geq \frac{1}{\sqrt{\mu_1\varepsilon_1}}. \tag{4.136}$$

The most striking feature of Equation (4.136) is the fact that the phase velocity is greater ($\theta_i < \pi/2$) than the speed of light, or "intrinsic" velocity in the same medium, and may even approach infinity. This feature does not violate anything of a fundamental nature. Figure 4.14 reveals that the phase velocity in the y direction must be greater than the intrinsic velocity. The field we are considering here is the superposition of two simple plane waves, each having a phase velocity of $1/\sqrt{\mu_1\varepsilon_1}$.

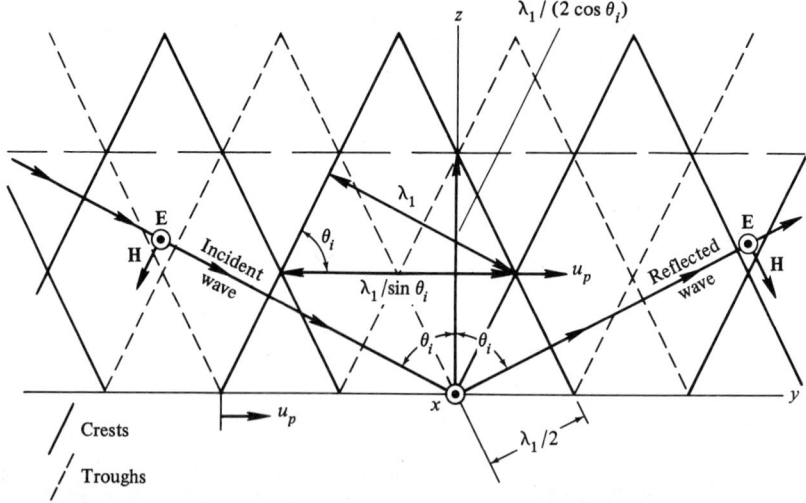

Figure 4.14. Oblique incidence (horizontal polarization), plane wave in a lossless dielectric onto a perfect conductor.

The *intersection* of their equiphase planes at a crest (for example) moves a distance $\lambda_1/\sin\theta_i$, while over the same time interval the same point of constant phase in the simple plane wave moves only a distance λ_1. Thus, the phase velocity in the composite wave is given by the plane wave phase velocity divided by $\sin\theta_i$. The intersection between one of the equiphase planes and the $z = 0$ plane also moves at this velocity. Many of us have observed an ocean wave striking a straight shoreline at an oblique angle such that the point of intersection between the crest of a wave and the shoreline moves with a greater velocity than the crest itself.

The concept of velocity of energy flow is ordinarily reserved for guiding systems, but if it were calculated according to Equation (4.33) for the system under present discussion, we would have

$$u_e = u_g = \frac{\sin\theta_i}{\sqrt{\mu_1\varepsilon_1}} \leq \frac{1}{\sqrt{\mu_1\varepsilon_1}},$$

which is comforting.

Note that since the electric field is zero where $z = p/(2f\sqrt{\mu_1\varepsilon_1}\cos\theta_i)$, $p = 1, 2, 3, \ldots$, a perfectly conducting plane parallel to the $z = 0$ plane could be placed there *without changing the field between the planes*. The dashed line in Figure 4.13 is the locus ($p = 1$) for such a plane. Thus, we could construct a parallel-plane waveguide with little more mathematical effort. Furthermore, since the fields in this case have no variation with x, we could also add perfectly conducting planes at $x = 0$ and $x = a$, forming a hollow one-conductor rectangular waveguide. E_x would, of course, terminate in surface charge densities on the planes at $x = 0$ and $x = a$. The mode that would be propagating is not a TEM mode since the magnetic field is not transverse to the direction of propagation (the y direction). **H** has a y component. The electric field does lie in planes that are transverse to the direction of propagation since it has only an x component. This field is designated as a TE (*transverse electric*) to y mode. These concepts will be examined more closely when the one-conductor waveguide is studied in detail, but it is convenient at this time to point out that when $\theta_i = 0$, $u_p \to \infty$, $u_e \to 0$, and there is *no net propagation in the y direction*. We then say that the waveguide is *cutoff* or nonpropagating in the y direction.

4.8 POLARIZATION

In several instances we have referred to *polarization*. In fact, we have considered several cases of *linear* polarization, where the lines of electric field **E** have been parallel to some fixed axis. Figure 4.3 is a good example of this behavior. In order to examine other types of polarization, consider the amplitudes A and B (which may be complex) and the field

$$\mathbf{E} = (\mathbf{a}_x A + \mathbf{a}_y B)e^{-jkz}, \tag{4.137}$$

$$\mathbf{H} = (\mathbf{a}_z \times \mathbf{E})/\eta. \tag{4.138}$$

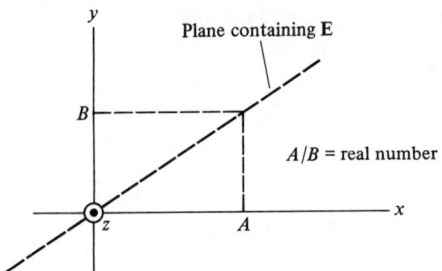

Figure 4.15. Linear polarization.

(1) If A and B have the same phase angle, the wave is *linearly polarized* and $\mathbf{E}$ will always lie in the plane containing the z axis, but inclined at an angle whose tangent is B/A from the $y = 0$ plane. This is seen in Figure 4.15. If $B = 0$, the wave is obviously polarized in the x direction, while if $A = 0$, the wave is obviously polarized in the y direction.

(2) If A and B are complex and have different phase angles, then $\mathbf{E}$ will no longer remain in one plane. Suppose $A = |A|e^{ja}$ and $B = |B|e^{jb}$. Then, we have

$$\mathbf{E} = \mathbf{a}_x |A| e^{j(a-kz)} + \mathbf{a}_y |B| e^{j(b-kz)} \tag{4.139}$$

and

$$\mathbf{H} = -\mathbf{a}_x \frac{|B|}{\eta} e^{j(b-kz)} + \mathbf{a}_y \frac{|A|}{\eta} e^{j(a-kz)} \tag{4.140}$$

In the time domain, we have

$$\mathscr{E}_x(z,t) = |A| \cos(\omega t + a - kz) \tag{4.141}$$

and

$$\mathscr{E}_y(z,t) = |B| \cos(\omega t + b - kz). \tag{4.142}$$

The locus of the endpoint of the vector $\mathscr{E}(z,t)$ will trace out an ellipse once each cycle, giving *elliptical polarization*. For example, suppose $|A| = 2|B| = 1$, $a = 0$ and $b = \pi/2$. Then, we have *left-handed* elliptic polarization where

$$\mathscr{E}_x(z,t) = \cos(\omega t - kz)$$

and

$$\mathscr{E}_y(z,t) = -\tfrac{1}{2} \sin(\omega t - kz).$$

A plot of $\mathscr{E}(z,t)$ in the $z = 0$ plane is given in Figure 4.16.

4.8 POLARIZATION

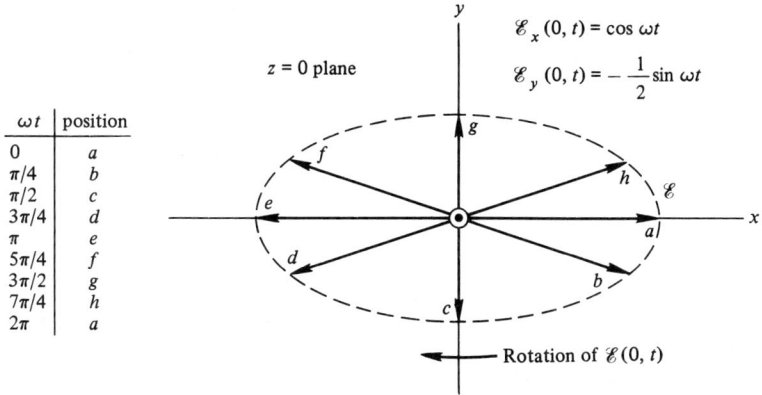

Figure 4.16. Left-handed elliptic polarization.

(3) If A and B are equal in amplitude and differ in phase angle by $\pi/2$, the ellipse becomes a circle, and we have the special case of circular polarization. If the thumb of the right hand points in the direction of propagation while the fingers point in the direction of rotation of $\mathscr{E}$, the polarization is said to be *right-handed circular*. For example, if $A = jB$, or $|A| = |B| = 1$, $a = 0$, $b = -\pi/2$, we have, from Equations (4.141) and (4.142),

$$\mathscr{E}_x(z,t) = \cos(\omega t - kz)$$

and

$$\mathscr{E}_y(z,t) = \sin(\omega t - kz).$$

It is easy to see that this is an example of right-handed circular polarization. If t were held constant and a plot of $\mathscr{E}$ versus z made, the endpoint of $\mathscr{E}$ would trace out a helix. A circularly polarized wave also has the interesting properties that both the power density and energy density are independent of time and space!

■ Example 4.9

The electric field of a right-handed circularly polarized plane wave is given by

$$\mathscr{E} = \mathbf{a}_x \cos(\omega t - kz) + \mathbf{a}_y \sin(\omega t - kz).$$

The magnetic field intensity can be found by any of several methods we have learned:

$$\mathscr{H} = [\mathbf{a}_y \cos(\omega t - kz) - \mathbf{a}_x \sin(\omega t - kz)]/\eta.$$

The Poynting vector is

$$\mathcal{S} = \mathcal{E} \times \mathcal{H} = \mathbf{a}_z[\cos^2(\omega t - kz) + \sin^2(\omega t - kz)]/\eta = \mathbf{a}_z/\eta \quad (\text{W/m}^2),$$

which is independent of time and space! ∎

Polarization is important in propagation between transmitting and receiving antennas. If the transmitting antenna is vertically polarized, or transmits a vertically polarized wave, then the receiving antenna must be compatible. That is, it must be capable of extracting a signal from a vertically polarized wave in free space. A linearly polarized wave that is produced by a transmitting antenna and enters the ionosphere may leave the ionosphere with elliptic polarization. This is an important consideration for long-distance communications via the ionosphere. Polarization is important in the operation of many devices such as microwave *ferrite isolators*.

4.9 DISPERSION

It was mentioned in Section 4.5, where group velocity was defined, that the frequency spread of the group must be small, or else the concept is meaningless. This needs further comment. A *normally* dispersive medium or system is one for which $du_p^{-1}/d\omega > 0$ and $u_g < u_p$, and an *anomalously* dispersive medium or system is one for which $du_p^{-1}/d\omega < 0$ and $u_g > u_p$. Systems that are anomalously dispersive can have $u_g > 1/\sqrt{\mu\varepsilon}$, which is clearly impossible. This occurs because of the way in which u_g was defined, but simply means that the derivation of the equation for u_g is not valid because the frequency spread is not small. Thus, the identity of group velocity is not valid for systems with anomalous dispersion. The uniform wave propagating in a lossy dielectric, or TEM waves on a lossy transmission line can have $u_g > 1/\sqrt{\mu\varepsilon}$ or $u_g > 1/\sqrt{LC}$, respectively. This is best seen in the $\omega - \beta$ diagram of Figure 4.17. Note that for oblique incidence on a perfectly conducting plane (Section 4.7), $u_g \leq 1/\sqrt{\mu\varepsilon}$. The same is true of the lossless waveguide to be investigated later.

Another way to reach the conclusions of the preceding paragraph occurs when we realize that a signal (or group) can never travel at a velocity greater than that of light, for if it did, some observers would detect some effects *before* their causes, thus violating the *principle of causality*.

4.10 CONCLUDING REMARKS

The main objective of this chapter was to develop a good understanding of uniform plane waves with normal and oblique incidence on plane interfaces. There is a close relationship between these phenomena and those that occur in guiding systems. More general problems, such as oblique incidence with arbitrary electric field polarization, or problems including more general media (such as magnetic materials) were not considered. The consideration of oblique incidence allowed us to predict that both dielectrics and conductors can guide waves.

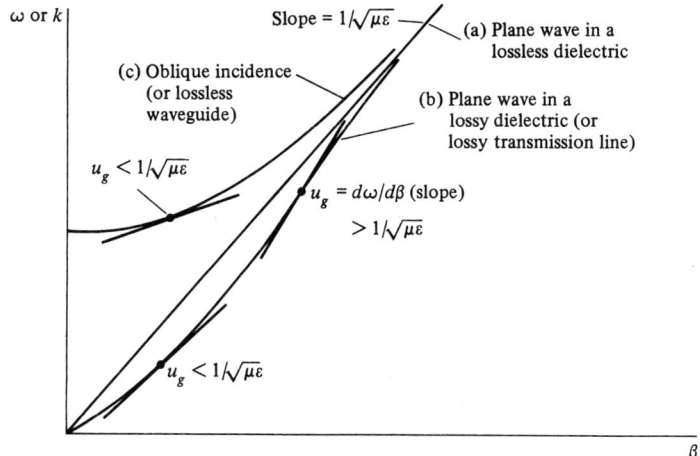

Figure 4.17. ω–β diagram for various systems. (a) Plane wave in a lossless dielectric, or a lossless transmission line, $u_p = u_e = u_g = 1/\sqrt{\mu\varepsilon}$. $\beta = \omega\sqrt{\mu\varepsilon}$. (b) Plane wave in a lossy dielectric, or a lossy transmission line, $u_g = d\omega/d\beta$. (c) Plane wave with oblique incidence on a perfectly conducting plane, or a lossless waveguide, or a plane wave in an ionized region, $u_g = (1/\sqrt{\mu\varepsilon}\sqrt{1 - \omega_c/\omega})^2$.

REFERENCES

Jordan, E. C. and Balmain, K. G. *Electromagnetic Waves and Radiating Systems*. 2nd ed. Englewood Cliffs, NJ: Prentice-Hall, 1968. A revised edition of a popular graduate level textbook.

Mott, H. *Polarization in Antennas and Radar*. New York: Wiley, 1986.

Pierce, J. R. *Almost All About Waves*. Cambridge, MA: The MIT Press, 1974. A very readable little book more general in nature than the material being presented herein.

Popovic, B. D. *Introductory Engineering Electromagnetics*. Reading, MA: Addison-Wesley, 1971.

Ramo, S., Whinnery, J. R., and Van Duzer, T. *Fields and Waves in Communications Electronics*. 2nd ed. New York: Wiley, 1984.

PROBLEMS

1. Verify that Equations (4.20) and (4.21) are solutions to Equations (4.18) and (4.19), respectively.

2. (a) What field is produced by the surface current density $\mathbf{J}_s = -\mathbf{a}_x E_0/\eta$ on the $z = 0$ plane plus the surface current density $\mathbf{J}_s = +j\mathbf{a}_x E_0/\eta$ on the $z = -\lambda/4$ plane. See the first part of Section 4.1.

   (b) What current density produces the field $E_x(z) = E_0 e^{-jkz}$, $z > 0$?

3. The magnetic field of a uniform plane wave is given by $\mathbf{H}(y, \omega) = \mathbf{a}_z 10 e^{-j10y}$.
   (a) Find phasor $\mathbf{E}$ and (b) find the time-domain forms for $\mathbf{E}$ and $\mathbf{H}$.

4. (a) If $\mathcal{E} = \mathbf{a}_y 10 \sin(\omega t - x)$ is propagating in free space, what is $\mathcal{H}$?
   (b) Repeat if the lossless medium has $\varepsilon_R = 4$.

5. A uniform plane wave is propagating in free space. It is described as having a time-average power density $<\mathcal{S}> = \mathbf{a}_y$ (W/m²) with the electric field polarized in the x direction. Find appropriate time-domain forms for $\mathcal{E}$ and $\mathcal{H}$.

6. (a) If microwave radiation is considered safe for humans when the power density is less than 10^{-2} W/m², what rms electric field intensity is represented (plane wave in air)?
   (b) If radiation from the sun at the earth's surface is given by 1.3×10^3 W/m² and is assumed to take the form of a monochromatic plane wave, what rms electric field intensity is represented?

7. A plane wave at 100 MHz is propagating in a lossy material. The phase of the electric field shifts 90° over a distance of 0.5 m, and its peak value is reduced by 25% for each meter traveled. (a) Find α, (b) find β, and (c) find u_p.

8. When the wave in Problem 7 travels 1 m, by how much is the power density reduced?

9. Show that 1 neper equals 8.69 decibels.

10. Derive Equations (4.41) and (4.42).

11. The uniform plane wave

$$\mathbf{E} = (4\mathbf{a}_x + 3\mathbf{a}_y)e^{-j\beta z}$$

is propagating in a lossless dielectric ($\varepsilon_R = 4$) at 100 MHz. (a) Find β, (b) λ, (c) $\mathcal{H}(x,y,z,t)$.

12. If seawater is described by $\sigma = 4$ ℧/m and $\varepsilon_R = 81$, calculate α and β for $\omega = 10^6$ rad/s to 10^{12} rad/s in decade steps.

13. When a uniform plane wave propagates in a certain material at 300 MHz, it is known that the wavelength is 0.472 m, the attenuation constant is 1 Np/m, and the intrinsic impedance has a magnitude of 195 Ω. Find μ_R, ε_R, and σ.

14. Surface resistance R_s can be written $R_s = K\sqrt{f}$. Find R_s in this form for the first six conductors listed in Table B.1 (Appendix B).

15. What percentage of the transmitted power density in a conductor is dissipated in heat over the distance δ (from normal incidence)?

16. It is known that $u_p = 0.005/\sqrt{\mu_0 \varepsilon_0}$ and $\lambda = 0.5$ mm when a uniform plane wave is propagating in a certain good conductor. Find the conductivity of the conductor and the frequency of the plane wave.

PROBLEMS

17. The induced current density at the surface of a large copper plate (at $z = 0$) from a normally incident uniform plane wave is 10 kA/m² if $f = 10$ kHz. Find **E**, **H**, and **J** everywhere if $\mathbf{E} = \mathbf{a}_x E_x$ and the incident wave is in air.

18. A plane wave is normally incident from air onto a material for which $\eta = 100\underline{/30°}$. Find (a) Γ, (b) T, and (c) SWR.

*19. Consider the dielectric-dielectric interface. Show that the "input" wave impedance $\eta_{in} = -E_x/H_y$ (where E_x and H_y represent the total fields) is given by

$$\eta_{in} = \eta_1 \frac{1 + \Gamma e^{-j2k_1 z}}{1 - \Gamma e^{-j2k_1 z}}$$

*20. A uniform plane wave at 300 MHz is normally incident from air onto a dielectric ($\varepsilon_R = 3$) window as in Fig. 4.18. See Problem 19.
   (a) What is the coefficient of reflection?
   (b) What is the SWR to the left of the interface at $z = d$?
   (c) What percentage of the incident power density is transmitted through the dielectric?
   (d) What is $|\mathbf{E}|$ at $z = 0$ and $z = d$ if $E^- = 1$ V/m?

*21. The region $z < 0$ in Figure 4.18 has the air replaced by a perfect conductor, but is otherwise unchanged.
   (a) Find $|\mathbf{E}|$ and $|\mathbf{H}|$ at $z = 0$ and $z = d$ if $E^- = 1$ V/m?
   (b) What values of d give a pure standing wave for $z > d$?

22. A 10 GHz uniform plane wave is normally incident on a fiberglass radome ($\sigma \approx 0$, $\varepsilon_R = 4.9$).
   (a) Specify the radome thickness if there is no reflection.
   (b) If the frequency is changed to 10.1 GHz, find the percentage of the incident power density that is transmitted through the radome.

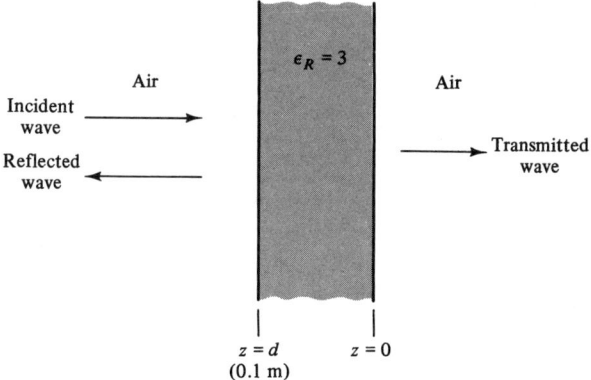

Figure 4.18. Normal incidence on a dielectric window (Problem 20).

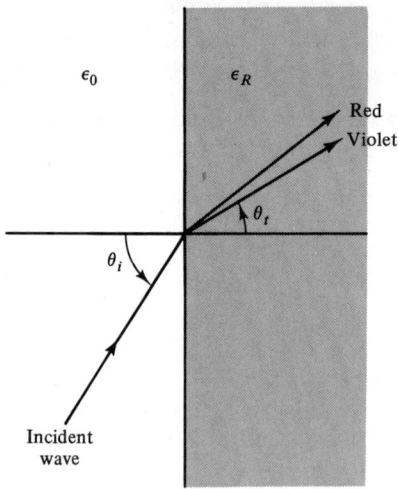

Figure 4.19. Prism effect in glass (Problem 24).

*23. A plate-glass window ($\sigma \approx 0$, $\varepsilon_R = 4$) is coated on the outside with a transparent dielectric such that red light ($f = 428.6 \times 10^{12}$ Hz) is not reflected.
  (a) Specify the thickness and dielectric constant of the coating.
  (b) What percentage of the incident power density is reflected for violet light ($f = 750 \times 10^{12}$ Hz?)

24. The dielectric constant for glass is actually not a constant, but depends slightly on frequency. Assume that this dependence is expressed by $\varepsilon_R = (2 + 2 \times 10^{-15}/\lambda^2)^2$ (*Cauchy's equation*). White light, containing the visible spectrum, is incident at $\theta_i = 65°$ onto a large plate of this glass. Calculate θ_t for the colors red ($f = 428.6 \times 10^{12}$ Hz) and violet ($f = 750 \times 10^{12}$ Hz). This dispersion gives the prism effect. See Figure 4.19.

*25. Calculate the percentage of the incident power density that is reflected back toward the light source for the glass ($\varepsilon_R = 4$) prism shown below in Figure 4.20.

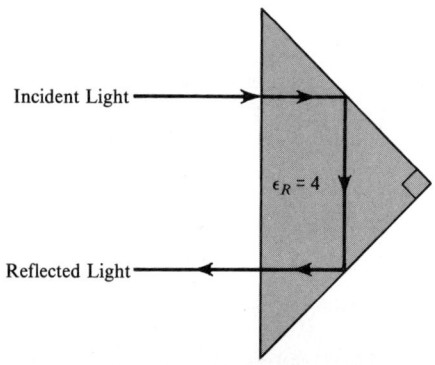

Figure 4.20. Reflection from a glass prism.

PROBLEMS

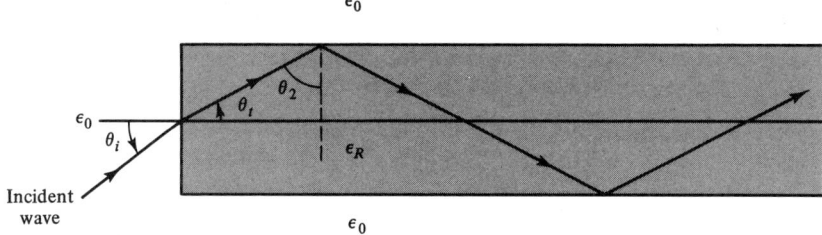

Figure 4.21. Total internal reflection in a dielectric slab.

26. What is the smallest value of ε_R such that total internal reflection occurs and the transparent dielectric guides light for any incident angle θ_i? See Figure 4.21.

27. Find the phasor form of a uniform plane wave propagating in the direction $\mathbf{a}_n = (\mathbf{a}_x + 2\mathbf{a}_y + 3\mathbf{a}_z)/\sqrt{14}$ in free space if $\mathbf{E}_0 = \mathbf{a}_x + \mathbf{a}_y - \mathbf{a}_z$ (V/m).

28. If $\mathscr{P} = 100 \cos^2(10^8 t - z/3) \mathbf{a}_z$ (W/m^2), find the average power per unit area crossing: (a) the $z = 0$ plane; (b) the plane defined by the points $(2,0,0)$, $(0,4,0)$, and $(0,0,8)$.

29. Calculate the polarizing angle for a plane wave from (a) air to water ($\varepsilon_R = 80$) and (b) water to air.

30. Calculate the critical angle for a plane wave from (a) water ($\varepsilon_R = 80$) to air and (b) polystyrene to air.

31. In viewing an object located under water, does it appear to be closer to, or more distant from, the surface than it actually is?

32. (a) Show that
$$u_p \approx \sqrt{2\omega/(\mu\sigma)}, \quad u_g \approx 2\sqrt{2\omega/(\mu\sigma)} \quad \text{(good conductors)};$$

(b) Show that
$$u_p \approx \frac{1}{\sqrt{\mu\varepsilon}}\left[1 + \frac{1}{8}\left(\frac{\sigma}{\omega\varepsilon}\right)^2\right]^{-1}, \quad u_g \approx \frac{1}{\sqrt{\mu\varepsilon}}\left[1 - \frac{1}{8}\left(\frac{\sigma}{\omega\varepsilon}\right)^2\right]^{-1} \quad \text{(good dielectrics)}.$$

Note that in (b), and possibly in (a), we have $u_g > 1/\sqrt{\mu\varepsilon}$, so that over much of the spectrum a lossy dielectric is anomalously dispersive.

33. If silver and brass shields are to have the same effectiveness in shielding, compare their thicknesses.

34. Plot the resistance of copper wire with 0.5 cm diameter versus frequency, and comment on the assumptions made.

35. A standing wave is the sum of two traveling waves. Show that a traveling wave is the sum of two standing waves.

36. (a) Find $\Gamma_H(\omega)$ [Equation (4.122)] for $\omega \to \infty$ when θ_i is less than the Brewster angle.
    (b) Repeat for θ_i equal to the Brewster angle.
    (c) Repeat for θ_i greater than the Brewster angle.

37. Show that a linearly polarized plane wave can be expressed as the sum of a left-handed circularly polarized wave and a right-handed circularly polarized plane wave.

38. An elliptically polarized plane wave is characterized by

$$\mathbf{E} = (\mathbf{a}_x 10 + \mathbf{a}_y 15\underline{|30°})e^{-jkz}$$

   (a) Find $\mathbf{H}$ and (b) find $<\mathscr{S}>$ if $\eta = \eta_0$.

39. (a) Using $\mathbf{J} = \rho_v \mathbf{u}$ and $\mathbf{F} = q\mathbf{E} = m\mathbf{a}$, show that

$$\mathbf{J} = -j\frac{Nq^2}{\omega m}\mathbf{E}$$

   in a region containing N electrons per cubic meter (charge q, mass m) if the excitation is sinusoidal in time and electronic collisions with molecules may be neglected.

   (b) Replacing conduction current density with that in part (a), show that a plane wave solution for the region is

$$E_x(z, \omega) = E_0(\omega)e^{-\gamma(\omega)z},$$

$$H_y(z, \omega) = \frac{E_0(\omega)}{\eta(\omega)}e^{-\gamma(\omega)z},$$

   where

$$\gamma(\omega) = \sqrt{\mu\varepsilon}\sqrt{\omega_p^2 - \omega^2},$$

$$\eta(\omega) = \sqrt{\mu/\varepsilon}\,[1 - (\omega_p/\omega)^2]^{-1/2},$$

$$\omega_p^2 \equiv \frac{Nq^2}{m\varepsilon}, \quad \text{plasma frequency.}$$

*40. Assume that the phasor current density in a long straight wire of radius $\rho = a$ is $\mathbf{J} = \mathbf{a}_z J_z(\rho)$. That is, axial and azimuthal variations in the current density are ignored.

   (a) Expand Equation (4.73) in cylindrical coordinates, and show that for a good conductor:

$$\frac{d^2 J_z}{d\rho^2} + \frac{1}{\rho}\frac{dJ_z}{d\rho} = j\omega\mu\sigma J_z.$$

(b) Assume (temporarily) that a solution to this equation is $J_z(\rho) = J_z(0)e^{-(1+j)\rho/\delta}$, which is identical in form to that for plane surfaces. The skin depth δ is given by Equation (4.75). Find

$$\left(\frac{1}{\rho}\frac{dJ_z}{d\rho}\bigg/\frac{d^2J_z}{d\rho^2}\right)$$

for $\rho = a$, and determine when it is small.

(c) Adjust the differential equation above accordingly, and show that the assumed solution is correct. This justifies the use of Equation (4.75) for conductors with curved surfaces.

41. Plot $|J_x(z,\omega)/J_x(0)|$ versus z for $0 \leq z \leq 0.25$ m when the material is seawater ($\mu = \mu_0$, $\sigma = 4$, $\varepsilon_R = 81$) and the frequency is 1, 10, and 100 MHz.

42. At what distance into a good conductor will the current density first be *oppositely directed* to the current density at the surface?

43. (a) Calculate the power loss per unit area for a magnetic material for which $B_0 = 0.5$ Wb/m², $\sigma = 10^6$ U/m, $f = 60$ Hz, and $d = 1$ mm.
(b) Repeat (a) if $\omega = 400$ Hz and $d = 0.5$ mm.

5
TRANSMISSION LINES: TEM MODES

The characteristics of uniform plane waves were developed in some detail in Chapter 4. Here we will use these results to help explore a two-conductor guiding system, commonly called a transmission line. Transmission lines are important in those applications where it is necessary to convey energy or information from one point to another at frequencies from 0 Hz to 3 or 4 GHz. The usual mode of operation is the TEM mode, where the electromagnetic field vectors **E** and **H** lie in planes that are *transverse* to the direction of propagation or axial direction. We will consider the entire line, including the generator at the sending end and the load at the receiving end. Many of the parameters that we will encounter will be similar to those that occur from plane waves with normal incidence on a plane interface.

Most of the important aspects of transmission line behavior can be studied when the line is lossless for this is the simplest case. It is introduced first, and continues through Section 5.13. Sections 5.14 through 5.18 deal with transmission line loss (material that can be omitted if time does not allow its inclusion).

5.1 THE PARALLEL PLANE GUIDING SYSTEM

Consider a (phasor) uniform plane wave propagating in a lossless dielectric, free of conductors, as shown in Figure 5.1. As defined in Chapter 4, this is a TEM to z mode of propagation since **E** and **H** lie in a plane (which extends to infinity in two dimensions) perpendicular to the z direction of propagation or power density flow. Is it possible to place perfectly conducting planes of infinite extent along the dashed lines in Figure 5.1 without disturbing the field distribution? A little thought should convince us that the answer is yes. We conclude that a positive surface charge density exists on the lower conductor, while a (equal magnitude)

5.1 THE PARALLEL PLANE GUIDING SYSTEM

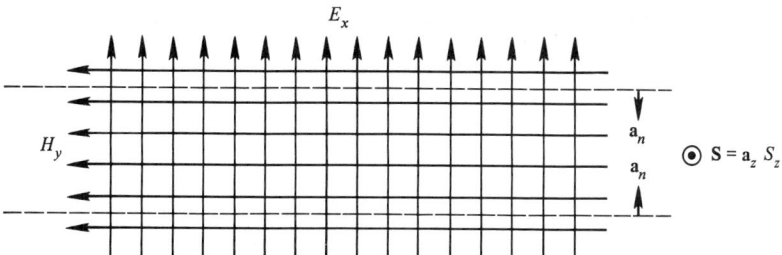

Figure 5.1. Uniform plane wave or TEM to z mode propagating in a lossless dielectric (time = constant).

negative surface charge density exists on the upper conductor. These charges fulfill the boundary condition [see Equations (3.111) through (3.114)]

$$\rho_s = \mathbf{a}_n \cdot \mathbf{D} \quad (\text{C/m}^2) \tag{5.1}$$

or simply

$$|\rho_s| = |D_n|, \tag{5.2}$$

where $\mathbf{D} = \varepsilon \mathbf{E}$ and the unit vector $\mathbf{a}_n$ is as shown in Figure 5.1. Also, these charges are changing in time (since $\mathscr{E}$ is changing in time) and therefore represent surface currents. The relation between the currents and charges is the *continuity of current* equation, or as it is often called, the *conservation of charge* equation, Equation (3.9), which for surface currents and charges becomes

$$\nabla \cdot \mathbf{J}_s = -j\omega \rho_s. \tag{5.3}$$

Simply stated, the surface currents obey the boundary condition

$$\mathbf{J}_s = \mathbf{a}_n \times \mathbf{H} \quad (\text{A/m}), \tag{5.4}$$

which was introduced as Equation (3.114). Thus, a surface current ($|J_s| = |H_{\tan}|$) will flow *out* of the page on the *inside* of the *lower* conductor and the "return" current will flow *into* the page on the *inside* of the *upper* conducting plane. The current flow is pictured in Figure 5.2.

As far as the *region between the conducting planes* is concerned, we must have a possible solution, because we have satisfied Maxwell's equations *in* the region and *boundary* conditions *on* the boundaries. The mode of propagation is still TEM. Other modes, called TE and TM modes, can exist for the system we have constructed. The possibility of the existence of a TE mode between parallel planes was pointed out at the end of Section 4.7.

Suppose next that we remove a *finite* width from this system and examine those changes that must occur. We expect that a TEM mode will still exist because we *still* essentially have a two-dimensional system. There must be some fringing or distortion of the fields near the edges of the finite planes as indicated in Figure 5.3.

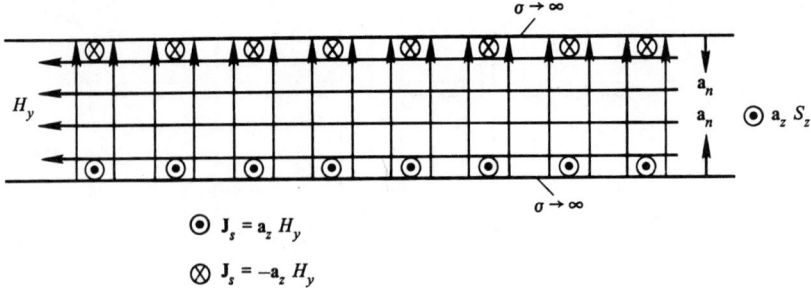

Figure 5.2. Surface current densities in an infinite parallel plane guiding system (TEM to z mode).

This *new* problem can be solved exactly, but at the present time that procedure would merely complicate the simple concept we are presenting. The final system that has evolved is a practical two-conductor transmission line with *no loss*.

5.2 THE GENERAL LOSSLESS LINE

Let us now consider two uniform perfect conductors of infinite length located in a perfect dielectric such that the conductors form cylinders with axes parallel to the z axis. The cross-sectional geometry is shown in Figure 5.4. Further, assume $E_z = H_z = 0$, and z dependence of the form $e^{-\gamma z}$ (a TEM to z mode). It is important to recognize that a similar geometry existed when we considered "static" capacitance and inductance in Chapter 2. We next apply Maxwell's equations in rectangular coordinates. Rewriting Maxwell's equations in components yields

$$\boldsymbol{\nabla} \times \mathbf{E} = -j\omega\mu\mathbf{H} \begin{cases} \gamma E_y = -j\omega\mu H_x, & (5.5) \\ \gamma E_x = j\omega\mu H_y, & (5.6) \\ \dfrac{\partial E_y}{\partial x} - \dfrac{\partial E_x}{\partial y} = 0 & (5.7) \end{cases}$$

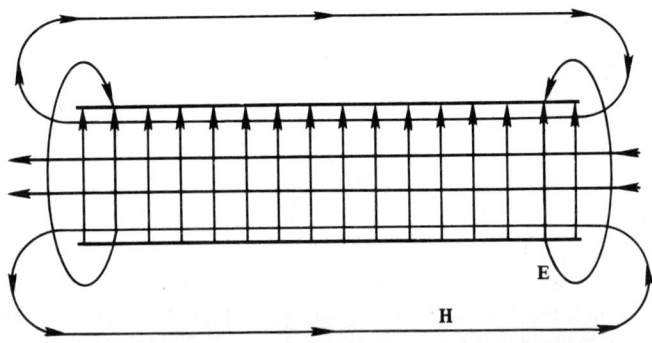

Figure 5.3. Field around a finite width parallel plane guiding system (TEM to z mode).

5.2 THE GENERAL LOSSLESS LINE

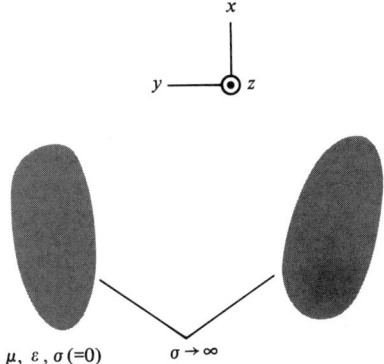

Figure 5.4. Arbitrary (uniform cross section) two-conductor guiding system or transmission line in a lossless medium.

and

$$\nabla \times \mathbf{H} = j\omega\varepsilon\mathbf{E} \begin{cases} \gamma H_y = j\omega\varepsilon E_x, & (5.8) \\ \gamma H_x = -j\omega\varepsilon E_y, & (5.9) \\ \dfrac{\partial H_y}{\partial x} - \dfrac{\partial H_x}{\partial y} = 0. & (5.10) \end{cases}$$

Substituting Equation (5.5) into Equation (5.9) shows that $\gamma = \pm j\omega\sqrt{\mu\varepsilon} = \pm jk$, where k is the intrinsic phase constant. Thus, the z variation is $e^{\mp jkz}$, giving undamped traveling waves in the negative or positive z directions, respectively. We choose e^{-jkz}, giving waves traveling in the positive z direction, and in this case $-j\omega\mu/\gamma = \omega\mu/k = \sqrt{\mu/\varepsilon} = \eta$. Equations (5.5) through (5.10) can then be described by

$$\boxed{\mathbf{E} = \eta \mathbf{H} \times \mathbf{a}_z} \tag{5.11}$$

and

$$\boxed{\mathbf{H} = \frac{1}{\eta}\mathbf{a}_z \times \mathbf{E}}. \tag{5.12}$$

It is then apparent that

$$\mathbf{E} \cdot \mathbf{H} = 0 \tag{5.13}$$

or $\mathbf{E}$ and $\mathbf{H}$ are normal everywhere. Solving Equations (5.5) through (5.10) simultaneously yields Laplace's equation in two dimensions for all four field components:

$$\frac{\partial^2 E_x}{\partial x^2} + \frac{\partial^2 E_x}{\partial y^2} = 0, \tag{5.14}$$

$$\frac{\partial^2 E_y}{\partial x^2} + \frac{\partial^2 E_y}{\partial y^2} = 0, \tag{5.15}$$

$$\frac{\partial^2 H_x}{\partial x^2} + \frac{\partial^2 H_x}{\partial y^2} = 0. \tag{5.16}$$

and

$$\frac{\partial^2 H_y}{\partial x^2} + \frac{\partial^2 H_y}{\partial y^2} = 0. \tag{5.17}$$

The field components must also satisfy the boundary conditions,

$$\begin{aligned} E_{\tan} &= 0, \\ H_{\text{norm}} &= 0 \end{aligned} \tag{5.18}$$

on the perfect conductor surfaces.

Equations (5.14), (5.15), and (5.18) represent exactly the same boundary value problem for **E** that we have already solved for the *electrostatic* case. Equations (5.16), (5.17), and (5.18) represent exactly the same boundary value problem for **H** that we have already solved for the *magnetostatic* case. Electric field lines in Figure 5.5 will be the same (for conductors with the same geometry, of course) as the electric field lines in the electrostatic capacitance problem. Likewise, magnetic field lines in Figure 5.5 will be the same as the magnetic field lines in the

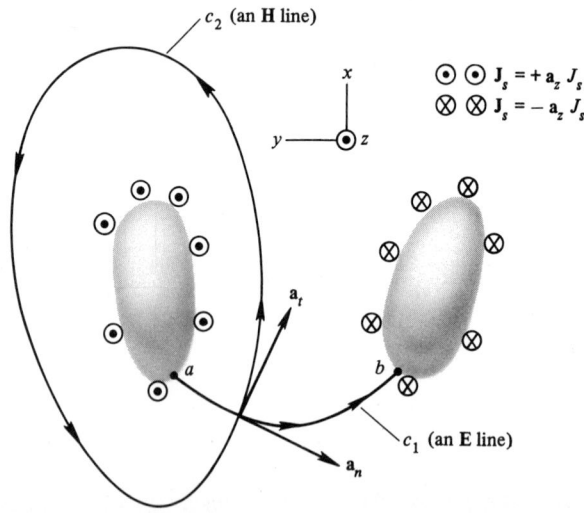

Figure 5.5. Integrating contours for determining V and I for the two-conductor system with surface currents (Figure 5.4).

5.2 THE GENERAL LOSSLESS LINE

magnetostatic (external) inductance problem. Therefore, we are permitted to use "static" capacitance and "static" external inductance, when the need for their use arises in this system, even though the fields are varying sinusoidally in time. This rather unusual result has occurred because the *mode* we are describing is TEM to z ($E_z = H_z \equiv 0$), whose z variation is simply $e^{\pm jkz}$.

A voltage, or emf, and current at any z for our infinite lossless two-wire line can be found by integrating along the contours shown in Figure 5.5. Note that z is constant on these contours. Note also that with z directed currents the fields are derivable from a vector potential $\mathbf{A}$ lying entirely in the z direction, and according to Equation (3.106) and the integrating contours of Figure 5.5, voltage and potential difference will be identical.[1] This is another property of TEM waves.

The appropriate defining equations are

$$V_{ab} = \int_{a,c_1}^{b} \mathbf{E} \cdot d\mathbf{l}_1 \quad \text{(V)}, \tag{5.19}$$

the voltage between the conductor carrying the $+z$ directed current and the conductor carrying the $-z$ directed current, and

$$I = \oint_{c_2} \mathbf{H} \cdot d\mathbf{l}_2 \tag{5.20}$$

from Ampere's law. Note that displacement current does not enter Equation (5.20) (Why not?). Since $\mathbf{E}$ and $\mathbf{H}$ represent traveling waves, it follows from Equations (5.19) and (5.20) that *V and I also represent traveling waves*! That is, V and I must have the same z variation as $\mathbf{E}$ and $\mathbf{H}$, respectively.

Substituting Equation (5.12) into Equation (5.20) gives

$$I = \frac{1}{\eta} \oint_{c_2} \mathbf{a}_z \times \mathbf{E} \cdot d\mathbf{l}_2,$$

but $\mathbf{E} = \mathbf{a}_n E_n$ at the intersection of c_1 and c_2, where $\mathbf{a}_n$ and $\mathbf{a}_t$ are unit vectors, normal and tangent to c_2, respectively, and are *not constant* vectors (in general). Also, $\mathbf{H} = \mathbf{a}_t H_t$ at the intersection of c_1 and c_2. Equation (5.20) can thus be written ($d\mathbf{l}_2 = \mathbf{a}_t \, dl_2$)

$$I = \frac{1}{\eta} \oint_{c_2} E_n \, dl_2. \tag{5.21}$$

From electrostatics, the definition of capacitance is $C = Q/V$, where, for a unit length in the z direction, $Q = \varepsilon \oint_{c_2} E_n \, dl_2(1)$. Therefore, the capacitance per unit length is

---

[1] $\mathbf{E} \cdot \mathbf{A} \equiv 0$

$$C = \frac{\varepsilon}{V} \oint E_n \, dl_2 \quad \text{(F/m)}. \tag{5.22}$$

Comparing Equations (5.21) and (5.22), it is easy to see that we may find the ratio of voltage to current as

$$\boxed{R_0 \equiv \frac{V}{I} = \eta \frac{\varepsilon}{C}} \quad (\Omega) \quad (\textit{characteristics resistance}) \tag{5.23}$$

and call it the *characteristic resistance* of the lossless line. In the same way ($d\mathbf{l}_1 = \mathbf{a}_n \, dl_1$),

$$V_{ab} = \eta \int_{a,c_1}^{b} \mathbf{H} \times \mathbf{a}_z \cdot d\mathbf{l}_1 = \eta \int_{c_1} H_t \, dl_1 \tag{5.24}$$

and inductance is $L \equiv \psi_m / I$, where ψ_m, the flux, for a unit length in the z direction, is $\psi_m = \mu \int_{c_1} H_t \, dl_1(1)$. Therefore, the inductance per unit length is

$$L = \frac{\mu}{I} \int_{c_1} H_t \, dl_1 \quad \text{(H/m)}. \tag{5.25}$$

Comparing Equations (5.24) and (5.25), it is easy to show that

$$\boxed{R_0 = \frac{V}{I} = \eta \frac{L}{\mu}} \quad (\Omega). \tag{5.26}$$

Comparing Equations (5.23) and (5.26), it follows, as we have already seen in equation (2.152), that

$$\boxed{LC = \mu \varepsilon} \quad (L = L_{\text{ext}}). \tag{5.27}$$

Then, if *either* L or C is known, the other may be found using Equation (5.27). Remember that the values of L and C used here are *per unit length* values. Some common geometries, with their characteristic resistances, are shown in Figure 5.6.

5.3 THE LOSSLESS LINE EQUIVALENT CIRCUIT

We have shown that the voltage and current at any point on the infinite lossless line are traveling waves, as long as the z dependence is $e^{\pm jkz}$. The ratio of V to I is R_0, where

5.3 THE LOSSLESS LINE EQUIVALENT CIRCUIT

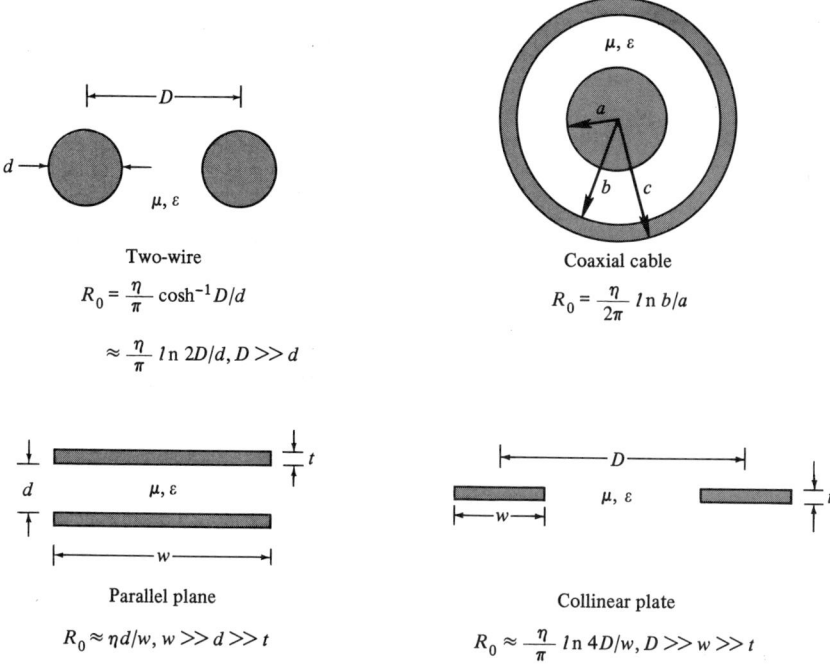

Figure 5.6. Some common lossless two-conductor transmission lines and their characteristic resistances (R_0).

$$R_0 = \eta \frac{\varepsilon}{C} = \eta \frac{L}{\mu}.$$

L and C are the static *external* inductance and static capacitance, respectively, per unit length. They are *distributed* parameters. Since the line is of infinite length, the ratio of V to I at any point z is also the *input impedance* at the point z. Apparently, then, V, I, and R_0 play roles similar to those of $\mathbf{E}$, $\mathbf{H}$, and η. We would like to show that this system represents the equivalent circuit of a lossless two-conductor line when supporting the TEM to z mode. It should be pointed out again that such an equivalent circuit is not unique to the geometry, for this geometry can support other modes. The TEM mode is, however, by far the most important mode. It is instructive to first calculate the input impedance to such a system assuming that it extends to infinity as indicated in Figure 5.7. If the system is broken at $B - B'$, the input impedance at $B - B'$ is *still* that being sought. Then, we have

$$Z_{in} = \frac{j\omega L}{2}\Delta z + \frac{(1/j\omega C\Delta z)(Z_{in} + j\omega L\Delta z/2)}{1/j\omega C\Delta z + Z_{in} + j\omega L\Delta z/2}$$

or, simplifying, we get

$$Z_{in}^2 = \frac{L}{C} - (\omega L\Delta z/2)^2.$$

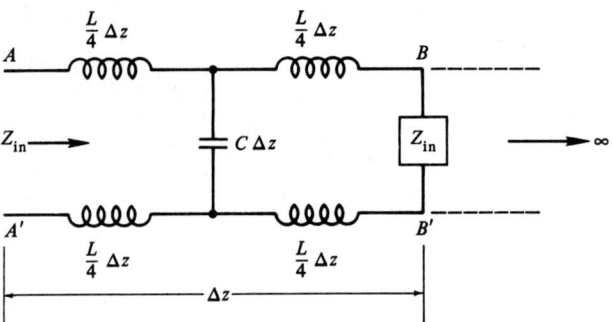

Figure 5.7. Equivalent circuit for determining the input impedance Z_{in} for the infinite lossless line.

The line does not consist of lumped parameters, but is a uniformly distributed system. Hence, we must let Δz approach zero; and in the limit

$$Z_{in} = \sqrt{\frac{L}{C}} \quad (\Omega). \tag{5.28}$$

Now, from Equations (5.23) and (5.26), we get

$$\boxed{R_0 = \sqrt{\frac{L}{C}}} \quad (\Omega) \tag{5.29}$$

and R_0 is indeed the input impedance to an infinite length of the distributed system, as was noted in the preceding paragraph.

The circuit voltage and current differential equations may also be derived from the distributed system of Figure 5.7, redrawn and labeled in Figure 5.8. Considering

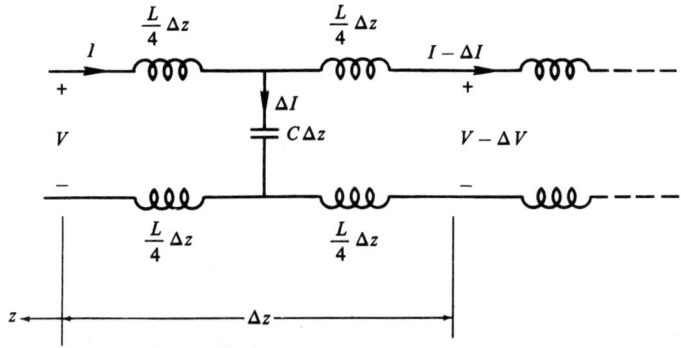

Figure 5.8. Equivalent circuit for determining the voltage and current differential equations for a lossless line.

5.3 THE LOSSLESS LINE EQUIVALENT CIRCUIT

a length Δz of the system and applying Kirchhoff's voltage and current laws, we get

$$V - j\omega \frac{L}{2}\Delta z I - j\omega \frac{L}{2}\Delta z (I - \Delta I) - (V - \Delta V) = 0 \tag{5.30}$$

and

$$I - j\omega C\Delta z \left(V - \frac{\Delta V}{2}\right) - (I - \Delta I) = 0 \tag{5.31}$$

or

$$\frac{\Delta V}{\Delta z} = j\omega L I - j\omega \frac{L}{2}\Delta I \tag{5.32}$$

and

$$\frac{\Delta I}{\Delta z} = j\omega C V - j\omega C \frac{\Delta V}{2}. \tag{5.33}$$

Now, in the limit as $\Delta z \to 0$ both ΔI and ΔV approach zero also. Recalling the definition of a derivative, we obtain, in the limit,

$$\frac{dV}{dz} = j\omega L I \tag{5.34}$$

and

$$\frac{dI}{dz} = j\omega C V. \tag{5.35}$$

Differentiating Equation (5.34) with z, and substituting for dI/dz from Equation (5.35) gives

$$\frac{d^2V}{dz^2} = -\omega^2 LCV$$

or

$$\frac{d^2V}{dz^2} + k^2 V = 0, \tag{5.36}$$

since $k = \omega\sqrt{LC} = \omega\sqrt{\mu\varepsilon}$, from Equation (5.27). In the same way, we have

$$\frac{d^2I}{dz^2} + k^2 I = 0. \tag{5.37}$$

Equations (5.36) and (5.37) are the *homogeneous wave equations* for V and I, respectively. We already know that V and I are traveling waves. We also know from Chapter 4 that solutions to Equations (5.36) and (5.37) will be traveling waves with phase velocity $u_p = 1/\sqrt{\mu\varepsilon} = 1/\sqrt{LC}$. The distributed system of Figure 5.8 thus leads to an accurate representation of V and I, and therefore, does indeed represent the equivalent circuit of the lossless two-wire line when it is supporting the TEM mode.

5.4 GENERAL SOLUTIONS FOR V AND I—LOSSLESS LINE

The *general* solutions for V and I [in Equations (5.36) and (5.37)] at any point z are

$$V = C_1 e^{+jkz} + C_2 e^{-jkz} \tag{5.38}$$

and

$$I = C_3 e^{+jkz} + C_4 e^{-jkz}, \tag{5.39}$$

where, according to Equation (5.27), we have

$$\boxed{k = \omega\sqrt{\mu\varepsilon} = \omega\sqrt{LC}}. \tag{5.40}$$

Substituting Equation (5.38) into (5.34) and solving for I eliminates C_3 and C_4:

$$I = \frac{C_1}{R_0} e^{+jkz} - \frac{C_2}{R_0} e^{-jkz}. \tag{5.41}$$

The first terms of V and I represent undamped waves traveling in the $-z$ direction, and the second terms represent undamped waves traveling in the $+z$ direction.

A complete transmission line is shown in Figure 5.9. Note that the load is located at $z = 0$, and the generator is located at $z = l$. The ratio V/I at $z = 0$ must be Z_l, so using this fact with Equations (5.38) and (5.41) leads to

$$\frac{C_2}{C_1} = \frac{Z_l - R_0}{Z_l + R_0}.$$

This ratio is quite logically called the *voltage coefficient of reflection at the load* Γ_l:

$$\boxed{\Gamma_l = |\Gamma_l| e^{j\phi_l} = \frac{Z_l - R_0}{Z_l + R_0}}. \tag{5.42}$$

5.4 GENERAL SOLUTIONS FOR V AND I—LOSSLESS LINE

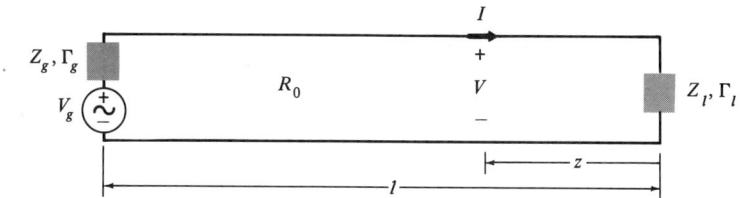

Figure 5.9. Complete two-wire transmission line system.

Compare this result to that given by Equation (4.63). The voltage and current at any point on the line may now be written ($C_2 = \Gamma_l C_1$) as

$$V(z, \omega) = V^-(\omega)(e^{+jkz} + \Gamma_l e^{-jkz}) \tag{5.43}$$

$$I(z, \omega) = V^-(\omega)(e^{+jkz} - \Gamma_l e^{-jkz})/R_0, \tag{5.44}$$

where $V^-(\omega) = C_1$ is the peak value of the $-z$ traveling wave, and may be a function of ω.

The *transmission line* voltage and current at the generator ($z = l$) are

$$V(l, \omega) = V^-(e^{+jkl} + \Gamma_l e^{-jkl}) \tag{5.45}$$

and

$$I(l, \omega) = V^-(e^{+jkl} - \Gamma_l e^{-jkl})/R_0. \tag{5.46}$$

Kirchhoff's voltage law, applied at the generator (Figure 5.9), gives $I(l, \omega) = [V_g(\omega) - V(l, \omega)]/Z_g$. Substituting this result into Equation (5.46), and solving for V^- gives

$$V^-(\omega) = \frac{V_g R_0}{Z_g + R_0} \frac{e^{-jkl}}{1 - \Gamma_l \Gamma_g e^{-j2kl}}, \tag{5.47}$$

where the *voltage coefficient of reflection at the generator* is

$$\Gamma_g = \frac{Z_g - R_0}{Z_g + R_0}. \tag{5.48}$$

Equation 5.47 can also be derived by considering the *first* incident wave that leaves the generator, travels to the load, is reflected, travels back to the generator, is reflected (Γ_g), travels back to the load, and so on. This leads to an infinite geometric series that has a closed form, resulting in Equation (5.47). It is suggested that the reader carry out this exercise to verify Equation (5.47).

As far as the *load* is concerned, we can find the Thévenin equivalent circuit in the usual way. Open circuiting the load ($z = 0$, $\Gamma_l = +1$) gives [from Equations (5.43) and (5.47)] the Thévenin voltage

$$V_T = V_g \frac{2R_0}{R_0 + Z_g} \frac{e^{-jkl}}{1 - \Gamma_g e^{-j2kl}} \tag{5.49}$$

or

$$V_T = \frac{V_g}{\cos(kl) + j(Z_g/R_0)\sin(kl)}. \tag{5.50}$$

Short circuiting the load ($z = 0$, $\Gamma_l = -1$) gives the Norton current

$$I_N = V_g \frac{2}{R_0 + Z_g} \frac{e^{-jkl}}{1 + \Gamma_g e^{-j2kl}}. \tag{5.51}$$

Therefore, the Thévenin impedance $Z_T = V_T/I_N$ is

$$Z_T = R_0 \frac{1 + \Gamma_g e^{-j2kl}}{1 - \Gamma_g e^{-j2kl}} \tag{5.52}$$

The load current in the Thévenin equivalent circuit of Figure 5.10 is $I_l = V_T/(Z_T + Z_l)$.

The input impedance to the line at the generator is given by the ratio of Equation (5.45) to Equation (5.46):

$$Z_{\text{in}} = R_0 \frac{1 + \Gamma_l e^{-j2kl}}{1 - \Gamma_l e^{-j2kl}} \qquad (z = l); \tag{5.53}$$

and as far as the generator is concerned, the equivalent circuit of Figure 5.11 applies.

In order to eliminate multiple reflections, it is highly desirable to make the impedance looking into the generator Z_g equal to the line impedance R_0. In this case, $\Gamma_g = 0$ so that Equations (5.43) and (5.44) with Equation (5.47) give

$$V(z, \omega) = \frac{V_g}{2} e^{-jkl}(e^{jkz} + \Gamma_l e^{-jkz}) \qquad (Z_g = R_0) \tag{5.54}$$

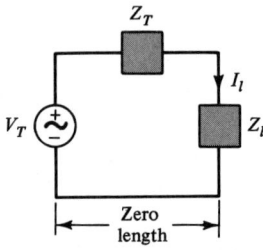

Figure 5.10. Equivalent circuit for the lossless transmission line as viewed from the load (Thévenin equivalent circuit).

5.4 GENERAL SOLUTIONS FOR V AND I—LOSSLESS LINE

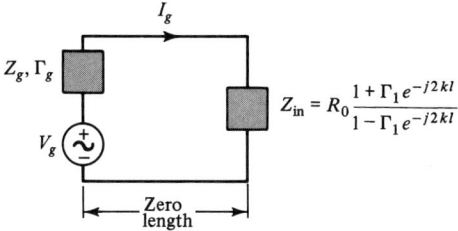

Figure 5.11. Equivalent circuit for the lossless transmission line as viewed from the generator.

and

$$I(z, \omega) = \frac{V_g}{2R_0} e^{-jkl}(e^{jkz} - \Gamma_l e^{-jkz}) \qquad (Z_g = R_0). \tag{5.55}$$

If, in addition to $Z_g = R_0$, we have a matched load ($Z_l = R_0$), then Figure 5.12 applies. The time-average load power is the time-average power in $Z_{in} = R_0$:

$$<\mathcal{P}> = \frac{V_g^2}{8R_0} \quad (W), \tag{5.56}$$

since V_g is a *peak value*. Note that the original time delay is not retained in our equivalent circuit.

For the case $Z_g = R_0$, $Z_l \neq R_0$, it is not difficult to show that the time-average *incident* power is

$$<\mathcal{P}_{inc}> = \frac{V_g^2}{8R_0}, \tag{5.57}$$

which is obviously the same as the power to a matched load [Equation (5.56)]. It is also easy to show that the reflected power is

$$<\mathcal{P}_{ref}> = \frac{|\Gamma_l|^2 V_g^2}{8R_0}. \tag{5.58}$$

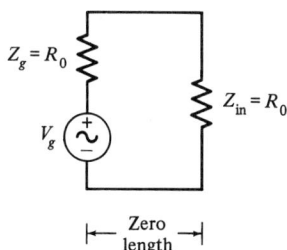

Figure 5.12. Equivalent circuit for the matched (lossless) two-wire line ($Z_g = R_0$, and omitting the time delay).

The *net* load power is then

$$<\mathcal{P}_l> = <\mathcal{P}_{\text{inc}}> - <\mathcal{P}_{\text{ref}}>$$

or

$$\boxed{<\mathcal{P}_l> = \frac{V_g^2}{8R_0}(1 - |\Gamma_l|^2)} \quad . \tag{5.59}$$

Note that Equation (5.59) reduces to Equation (5.56) if $\Gamma_l = 0$ (or $Z_l = R_0$).

■ Example 5.1

A 10-V (peak) sinusoidal source has an internal impedance of $Z_g = 100 + j50 \ \Omega$, and is connected to a load $Z_l = 75 + j100 \ \Omega$ through a 50-Ω lossless line that is 5λ in length. We want to find the power that is dissipated in the load using both Figures 5.10 and 5.11. Using Equation (5.53) for Figure 5.11, we get

$$Z_{\text{in}} = Z_l = 75 + j100 \quad (\Omega) \quad (kz = kl = 10\pi),$$

$$I_g = \frac{V_g}{Z_g + Z_{\text{in}}} = \frac{10\,\lfloor 0°}{175 + j150} = 43.39\,\lfloor -40.60° \quad \text{(mA)},$$

$$<\mathcal{P}_l> = |I_{g,\text{eff}}|^2 \text{Re}(Z_{\text{in}}) = (43.39/\sqrt{2})^2 (75) = 70.6 \quad \text{(mW)}.$$

Using Equations (5.50) and (5.52) for Figure 5.10, we get

$$V_T = V_g = 10\,\lfloor 0° \quad (\text{V}), \quad Z_T = Z_g = 100 + j50 \quad (\Omega),$$

$$<\mathcal{P}_l> = |I_{l,\text{eff}}|^2 R_l = \frac{1}{2}\left|\frac{10\,\lfloor 0°}{175 + j150}\right|^2 (75) = 70.6 \quad \text{(mW)},$$

which is the same result as before. Note carefully that, as far as the line is concerned, it is lossless.

Now, using Figure 5.11 once more, and recalling circuit theory results, maximum power can only occur when $Z_{\text{in}} = Z_g^* = 100 - j50$, in which case the maximum power that can be supplied by the given source voltage is

$$<\mathcal{P}> = \frac{1}{2}\left|\frac{10\,\lfloor 0°}{2(100)}\right|^2 (100) = 125 \quad \text{(mW)}.$$

We shall soon see how to ensure that the maximum power is dissipated in R_l regardless of the value of Z_l ($R_l \neq 0$, of course). ■

5.5 VOLTAGE STANDING WAVE RATIO

The voltage standing wave ratio (VSWR) is defined as the ratio $|V|_{max}/|V|_{min}$, and from Equation (5.43), it is

$$\boxed{\text{VSWR} = \frac{1 + |\Gamma_l|}{1 - |\Gamma_l|}} \quad \text{(voltage standing wave ratio).} \quad (5.60)$$

$|V|_{max}$ and $|V|_{min}$ are identified in Figure 5.13. $|V|_{max}$ occurs where

$$\phi_l - 2kz = -2n\pi, \quad n = 0, 1, 2, \ldots \quad (z > 0)$$

or

$$z_{max} = (\phi_l + 2n\pi)/2k. \quad (5.61)$$

$|V|_{min}$ occurs where

$$\phi_l - 2kz = -(2n - 1)\pi, \quad n = 0, 1, 2, \ldots \quad (z > 0)$$

or

$$z_{min} = [\phi_l + (2n - 1)\pi]/2k. \quad (5.62)$$

We will agree to use only positive values for ϕ_l, and, in any case, must use positive values for z_{max} or z_{min}. The spacing between adjacent crests ($|V|_{max}$) or adjacent troughs ($|V|_{min}$) will be $\lambda/2$. Either a crest or a trough will occur at a distance less than (or at most, equal to) $\lambda/4$ from the load; see Figure 5.13.

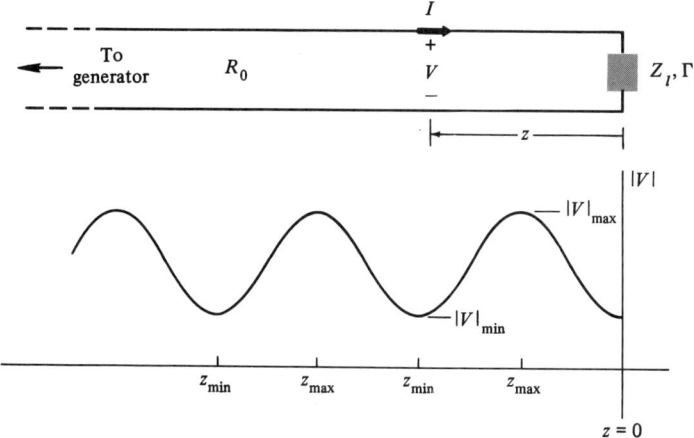

Figure 5.13. Location of an arbitrary load impedance on a lossless line and the resultant (incomplete) voltage standing wave.

Solving for ϕ_l in Equation (5.62), we get

$$\phi_l = 4\pi z_{min}/\lambda - (2n-1)\pi \quad \text{(rad)}, \tag{5.63}$$

since $k = 2\pi/\lambda$. Solving for $|\Gamma_l|$ in Equation (5.60) gives

$$|\Gamma_l| = \frac{\text{VSWR} - 1}{\text{VSWR} + 1}. \tag{5.64}$$

Therefore, using Equation (5.42), we get

$$\boxed{\Gamma_l = \left[\frac{\text{VSWR} - 1}{\text{VSWR} + 1}\right] e^{j[(4\pi z_{min}/\lambda) \pm \pi]}}. \tag{5.65}$$

Since both VSWR and z_{min} can be measured with a *slotted line*, as will be discussed later, Γ_l can be determined from these two pieces of information. Furthermore, if Γ_l is known, then Z_l can be found from Equation (5.42):

$$\boxed{Z_l = R_0 \frac{1 + \Gamma_l}{1 - \Gamma_l}}. \tag{5.66}$$

This is an important practical consideration, for in many cases the load impedance is not easy to identify. An antenna is a good example. It certainly represents a load on the end of the line, but the impedance of this kind of load is certainly not obvious by visual inspection. The similarity between the preceding equations (and the interpretation of them) and those of Chapter 4 for plane waves with normal incidence on an interface is obvious.

■ Example 5.2

The VSWR on a 50-Ω line with negligible loss is measured as 2.5, and z_{min} is measured at 0.1 m, 0.4 m, 0.7 m, etc. from the load. What is the load impedance? Since the spacing between the locations of $|V|_{min}$ is $\lambda/2$, we have $\lambda = 2(0.3) = 0.6$ m, and by equation (5.65) and (5.66), we have

$$\Gamma_l = 0.429 \underline{|300°|}, \quad Z_l = 54.05 - j49.15 \quad (\Omega).$$

■

5.6 INPUT IMPEDANCE

The input impedance at any point on the line (looking toward the load) can be measured with an *impedance bridge*. It is found as the ratio of Equation (5.43) to (5.44):

5.6 INPUT IMPEDANCE

$$Z_{in} = \frac{V}{I} = R_0 \frac{e^{jkz} + \Gamma_l e^{-jkz}}{e^{jkz} - \Gamma_l e^{-jkz}}$$

or

$$\boxed{Z_{in} = R_0 \frac{Z_l \cos kz + jR_0 \sin kz}{R_0 \cos kz + jZ_l \sin kz}} \quad (5.67)$$

If the line is terminated in a short circuit ($Z_l = 0$), Equation (5.67) gives

$$Z_{in}^{sc} = jR_0 \tan kz; \quad (5.68)$$

and if the line is terminated in an open circuit ($Z_l \to \infty$), Equation (5.67) gives

$$Z_{in}^{oc} = -jR_0 \cot kz. \quad (5.69)$$

Combining Equations (5.68) and (5.69) gives

$$\boxed{R_0 = \sqrt{Z_{in}^{sc} Z_{in}^{oc}}} \quad (5.70)$$

Substituting Equations (5.68) and (5.69) into Equation (5.67), and then solving for Z_l gives

$$\boxed{Z_l = Z_{in}^{oc} \frac{Z_{in}^{sc} - Z_{in}}{Z_{in} - Z_{in}^{oc}}} \quad (\Omega). \quad (5.71)$$

■ Example 5.3

A series of measurements is made on a line (with negligible loss) using an impedance bridge. All measurements are made at the same point on the line (z is constant). First, the input impedance with the unknown load impedance in place is measured as $Z_{in} = 30 + j60$ Ω. Since the line is not marked, its characteristic resistance is not known, so the load is replaced by a *short circuit*, and the input impedance is measured as $Z_{in}^{sc} = j53.1$ Ω. Repeating the procedure with the load replaced by an *open circuit* gives $Z_{in}^{oc} = -j48.3$ Ω. Equations (5.70) and (5.71) give

$$R_0 = 50.64 \ (\Omega), \quad Z_l = 11.63 + j6.30 \ (\Omega).$$

■

Some important practical questions arise in considering the preceding examples:

1. How is the position of $|V|_{\min}$ (that is, $z_{\min}$) actually measured?
2. How is the voltage standing wave ratio measured?
3. Is it possible to develop a simpler method to perform the preceding calculations?

These questions will all be answered as we proceed.

The two procedures that we have discussed for calculating Z_l are outlined below, and this outline can be used as a guide for writing simple computer programs from given or measured data.

Impedance Bridge	Slotted Line
Data $\rightarrow Z_{\text{in}}, Z_{\text{in}}^{\text{sc}}, Z_{\text{in}}^{\text{oc}}$ (5.71) $\downarrow$ Z_l	VSWR, $z_{\min}, \lambda, R_0$ (5.65) $\downarrow$ Γ_l (5.66) $\downarrow$ Z_l

5.7 SPECIAL LOAD IMPEDANCES

An examination of several special values of Z_l will shed additional light on the behavior of transmission lines. Characteristics are listed:

Open Circuit	Short Circuit
$Z_l \rightarrow \infty$	$Z_l = 0$
$\Gamma_l = +1$	$\Gamma_l = -1$
VSWR $\rightarrow \infty$	VSWR $\rightarrow \infty$
$Z_{\text{in}}^{\text{oc}} = -jR_0 \cot kz$	$Z_{\text{in}}^{\text{sc}} = +jR_0 \tan kz$
$|V| = 2|V^-\cos kz|$	$|V| = 2|V^-\sin kz|$
$|I| = \dfrac{2}{R_0}|V^-\sin kz|$	$|I| = \dfrac{2}{R_0}|V^-\cos kz|$

5.7 SPECIAL LOAD IMPEDANCES   247

Mismatched Case	Matched Case
$Z_l = 2R_0$	$Z_l = R_0$
$\Gamma_l = +\frac{1}{3}$	$\Gamma_l = 0$
VSWR = 2	VSWR = 1
$Z_{\text{in}} = R_0 \dfrac{2\cos kz + j\sin kz}{\cos kz + j2\sin kz}$	$Z_{\text{in}} = R_0$
$\|V\| = \|V^-\|\left\| 1 + \dfrac{1}{3}e^{-j2kz} \right\|$	$\|V\| = \|V^-\|$
$\|I\| = \dfrac{\|V^-\|}{R_0}\left\| 1 - \dfrac{1}{3}e^{-j2kz} \right\|$	$\|I\| = \|V^-\|/R_0$

A plot of $|V|$, $|I|$, and $Z_{\text{in}}^{\text{oc}}$ against z is shown in Figure 5.14. It is customary to plot magnitudes for V and I since normal detectors in an experimental setup do not reproduce phase. The infinite standing wave ratios are indicative of the lack of power transfer to the open circuit. That is, there is just as much power flowing

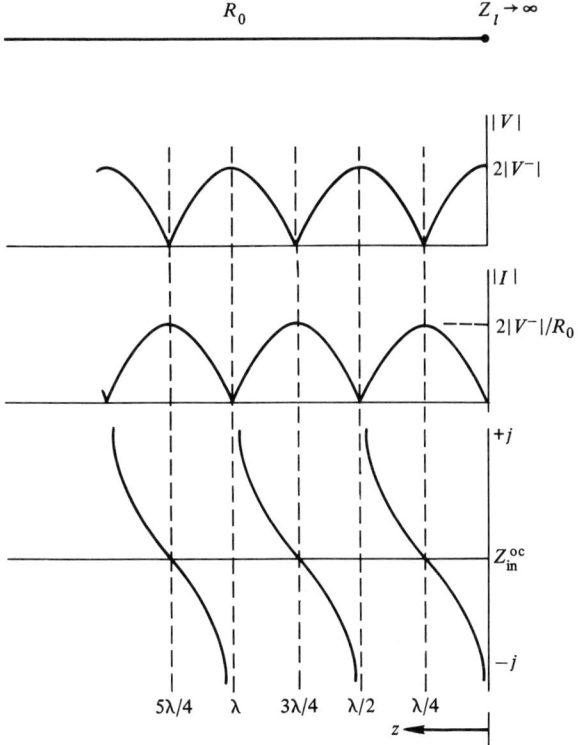

Figure 5.14. $|V|$, $|I|$, and $Z_{\text{in}}^{\text{oc}}$ versus z for $Z_l \to \infty$ (open-circuit load).

toward the open circuit as there is reflected back toward the generator. The input impedance indicates total reflection also, but in a slightly different manner. Z_{in}^{oc} is always a pure reactance $-\infty \le X \le +\infty$. Depending on the distance to the load z, the line can look like *any* capacitance or *any* inductance *at the terminals* where Z_{in}^{oc} is measured. At $z = 0, \lambda/2, \lambda, 3\lambda/2, 2\lambda, \ldots$, the line looks like a lossless parallel resonant circuit ($Z_{in}^{oc} \to \infty$), while for $z = \lambda/4, 3\lambda/4, 5\lambda/4, \ldots$, the line looks like a lossless series resonant circuit ($Z_{in}^{oc} = 0$).

The equations, as well as Figure 5.15, indicate that the short-circuited line is complementary to the open-circuited line. There is no power transfer to the load, resulting in infinite standing wave ratios. Z_{in}^{sc} is always a pure reactance, so the line may look like a capacitance or inductance at the terminals where Z_{in}^{sc} is measured. At $z = 0, \lambda/2, \lambda, 3\lambda/2, \ldots$, the line looks like a lossless series resonant circuit ($Z_{in}^{sc} = 0$), while for $z = \lambda/4, 3\lambda/4, 5\lambda/4, \ldots$, the line looks like a lossless parallel resonant circuit ($Z_{in}^{sc} \to \infty$).

Figure 5.16 shows plots of R_{in}, X_{in}, $|V|$, and $|I|$ against z when $Z_l = 2R_0$. Note that these functions are repetitive with a period of $\lambda/2$. It is easy to show that this is *always* the case when there is a reflection from the load (no line loss).

In the special and highly desirable matched ($Z_l = R_0$) case, the length of line is unimportant except in terms of phase shift or total time delay. To the incident

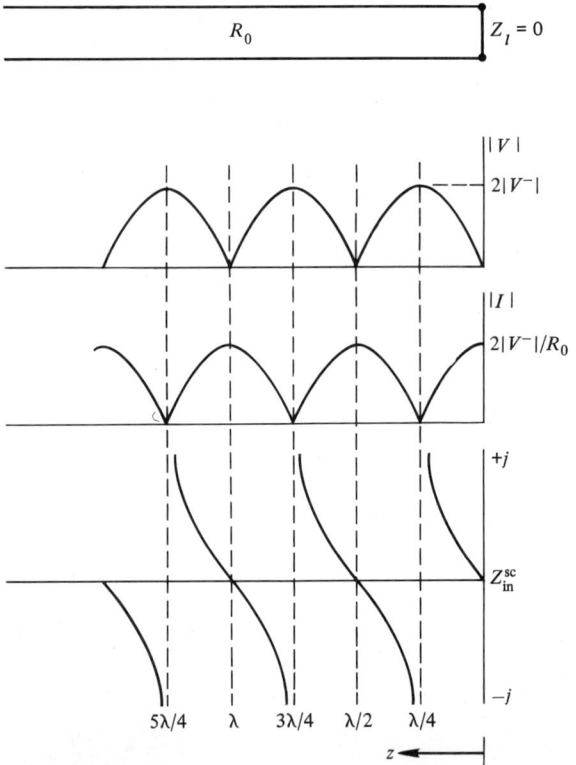

Figure 5.15. $|V|$, $|I|$, and Z_{in}^{sc} versus z for $Z_l \to 0$ (short-circuit load).

5.7 SPECIAL LOAD IMPEDANCES

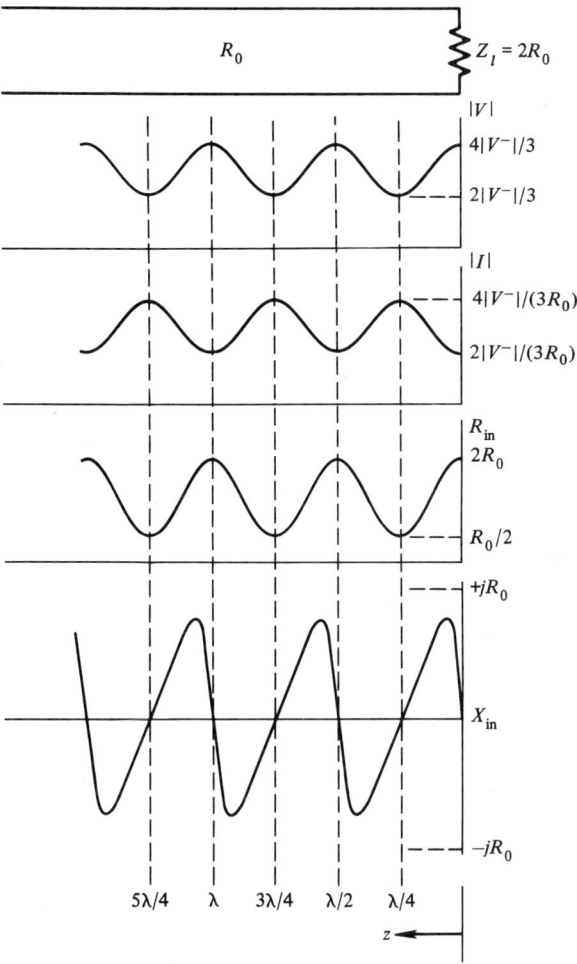

Figure 5.16. $|V|$, $|I|$, R_{in}, and X_{in} ($Z_{in} = R_{in} + jX_{in}$) versus z for $Z_l = 2R_0$.

wave, the load $Z_l = R_0$ looks just like more transmission line with R_0 characteristic resistance. The line is said to be *terminated in a match*. Results are shown in Figure 5.17.

Transmission lines can be used in place of "lumped" parameters in many applications at high frequencies. An engineer designing a UHF "tank" circuit for an oscillator or amplifier, for example, may find that any available inductor represents a reactance that is too large. A possible solution is the use of a section of silver-plated transmission line whose length z is less than $\lambda/4$, giving an input impedance that is equivalent to that of an ordinary inductor. This example, along with some others, is demonstrated in Figure 5.18.

In Figure 5.18(a) the length $z < \lambda/4$ is adjusted so that $X_L = X_C$, or [Equation (5.68)] $1/\omega C = R_0 \tan(2\pi z/\lambda)$. When this is done, parallel resonance occurs at the $x - x$ terminals. In Figure 5.18(b) the impedance at $x - x$ is the parallel

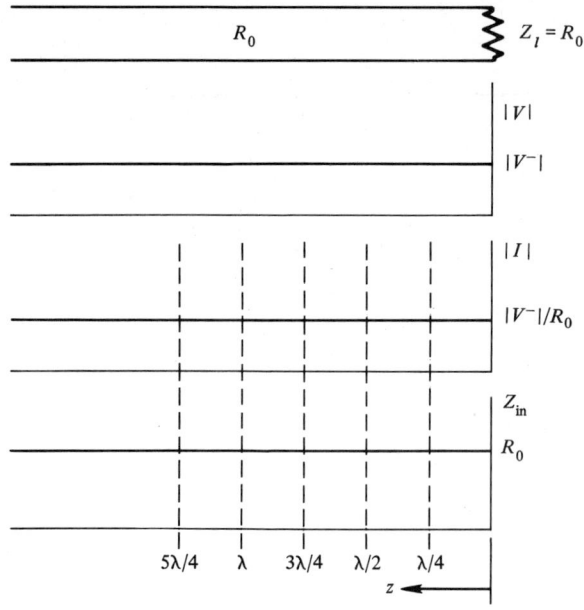

Figure 5.17. $|V|$, $|I|$, and $Z_{in} = R_0$ versus z for $Z_l = R_0$ (matched load case).

combination of that from a section $d < \lambda/4$ with $Z_l = 0$, and that from a section $(\lambda/4 - d) < \lambda/4$ with $Z_l \to \infty$. Equations listed at the beginning of this section are appropriate and give $Z_{in} \to \infty$ (parallel resonance) *for any value of d*. There are various ways in which this section may be used as a bandpass filter, where the center of the band will be $f = u_p/\lambda = 1/(\lambda\sqrt{LC})$.

Using Equation (5.67) with $z = \lambda/4$, it is easy to show that the input impedance at $x - x$ (looking to the right) in Figure 5.18(c) is R_{01}, and hence the system is matched, if $R_{02} = \sqrt{R_{01} R_l}$. Note that this technique only applies to purely resistive loads R_l. The impedance looking *into* the $\lambda/4$ section or "stub" of Figure 5.18(d) is infinite [Equation (5.68)]; hence the stub has no effect on the main line.

■ Example 5.4

Suppose that we want to match a resistive load $Z_l = 100\,\underline{|0°}\ \Omega$, to a 50-$\Omega$ line. As in the preceding paragraph, the system is matched [Figure 5.18(c)] if

$$R_{02} = \sqrt{R_{01} R_l} = \sqrt{50(100)} = 70.7\ \Omega.$$

Suppose further that we define the bandwidth as the "half-power bandwidth." That is, the bandwidth is to be measured between the two frequencies f_1 and f_2 where the time-average load power is one-half of its maximum value; that is, where $(<\mathcal{P}_l>/<\mathcal{P}_{inc}> = 0.5)$. Using Equations (5.57) and (5.59), we get

$$<\mathcal{P}_l>/<\mathcal{P}_{inc}> = 1 - |\Gamma_{xx}|^2 = 0.5$$

5.7 SPECIAL LOAD IMPEDANCES

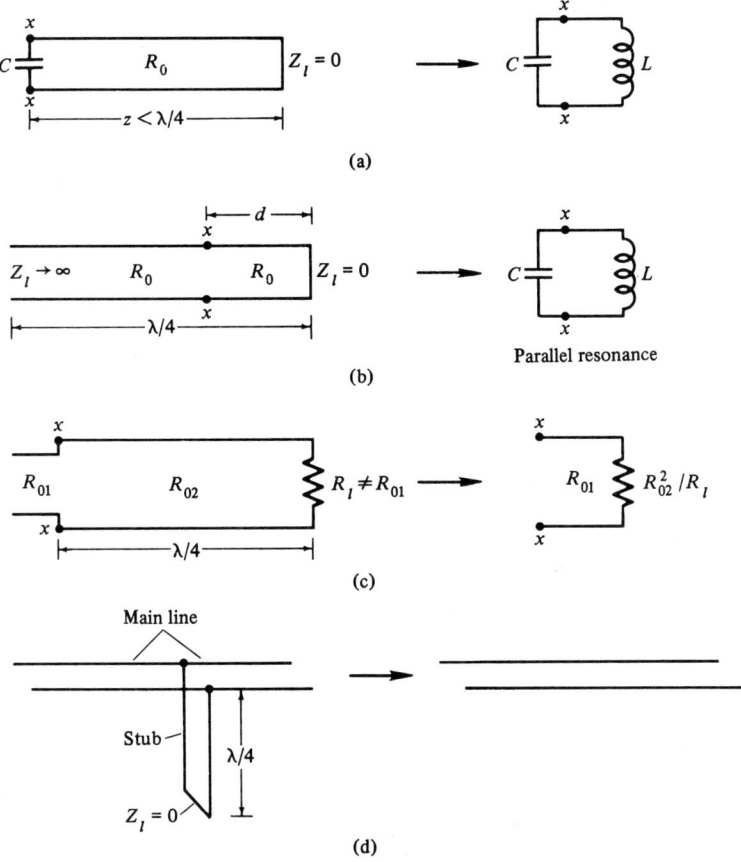

Figure 5.18. Some examples of the use of line sections to replace lumped parameters. (a) A shorted line acting as an inductor. (b) A tapped resonant $\lambda/4$ line. (c) A $\lambda/4$ transformer. (d) A "conducting" insulator.

or

$$|\Gamma_{xx}| = \left|\frac{Z_{xx} - R_{01}}{Z_{xx} + R_{01}}\right| = 0.707,$$

where the double subscript refers to the $x-x$ terminals in Figure 5.18(c). According to Equation (5.67), we have

$$Z_{xx} = R_{02}\frac{Z_l \cos kz + jR_{02} \sin kz}{R_{02} \cos kz + jZ_l \sin kz}$$

or

$$Z_{xx} = 70.7\frac{\sqrt{2} + j \tan[(\pi/2)(f/f_0)]}{1 + j\sqrt{2} \tan[(\pi/2)(f/f_0)]},$$

where f_0 is the design frequency, or

$$kz = \omega\sqrt{\mu\varepsilon}\,z = (\omega/\omega_0)(\omega_0\sqrt{\mu\varepsilon}\,\lambda/4) = (f/f_0)(2\pi/\lambda)(\lambda/4)$$
$$kz = (\pi/2)(f/f_0),$$

Calculation for *any* f/f_0 shows that $|\Gamma_{xx}| < 0.707$ and therefore, in this case, there are no half-power frequencies f_1 or f_2, and the bandwidth, as defined above, is infinite. Stated more explicitly, the load power is greater than one-half the incident power for any frequency! ∎

Example 5.5

A UHF transmitter operates at a carrier frequency of 300 MHz, but produces an unwanted second harmonic at 600 MHz. The tapped quarter-wave line of Figure 5.18(b) can be used to prevent the signal at 600 MHz from reaching the load if the length of its open-circuited section $(\lambda/4 - d)$ is one-quarter wavelength long *at the second-harmonic frequency*. If the dielectric is air, then $\lambda = (3 \times 10^8)/(3 \times 10^8) = 1$ m at 300 MHz. Thus, the *total* length of the line section is 0.25 m. At 600 MHz the wavelength is 0.5 m, so the length of the open-circuited section should be $0.5/4 = 0.125$ m. Thus, $d = 0.25 - 0.125 = 0.125$ m. The signal at 300 MHz sees the main line with an open-circuit across it at the $x - x$ terminals of Figure 5.18(b), and therefore is unaltered as it progresses toward the load. On the other hand, the second-harmonic signal at 600 MHz at $x - x$ sees three impedances in parallel: (i) the input impedance of the main line looking toward the load; (ii) the input impedance of the shorted section that is one-quarter wavelength long (d) at 600 MHz and therefore appears as an open circuit; (iii) the input impedance of the open section that is one-quarter wavelength long at 600 MHz and therefore appears as a short circuit. The parallel combination of these three impedances results in zero impedance at the $x - x$ terminals at 600 MHz, and therefore this second harmonic is totally reflected back toward the source, never reaching the load. ∎

5.8 TRANSMISSION PARAMETERS

It is sometimes helpful to consider a finite length of transmission line as a two-port network. This is particularly true when two or more different sections are *cascaded*. The equations for the voltage and current at a point $z = z_1$, on a lossless line [Equations (5.43) and (5.44), Figure 5.9)] are

$$V_1 = V^- e^{+jkz_1} + V^-\Gamma_l e^{-jkz_1}, \tag{5.72}$$

$$I_1 = (V^-/R_0)e^{+jkz_1} - (V^-/R_0)\Gamma_l e^{-jkz_1}, \tag{5.73}$$

and at $z = z_2$ $(z_2 < z_1)$ they are

5.8 TRANSMISSION PARAMETERS

$$V_2 = V^- e^{+jkz_2} + V^- \Gamma_l e^{-jkz_2}, \tag{5.74}$$

$$I_2 = (V^-/R_0) e^{+jkz_2} - (V^-/R_0) \Gamma_l e^{-jkz_2}. \tag{5.75}$$

V^- is given by Equation (5.47). Solving for $V^- e^{+jkz_1}$ and $V^- \Gamma_l e^{-jkz_1}$ in Equations (5.74) and (5.75), and then substituting into Equations (5.72) and (5.73) gives

$$V_1 = \cos[k(z_1 - z_2)] V_2 + jR_0 \sin[k(z_1 - z_2)] I_2, \tag{5.76}$$

$$I_1 = (j/R_0) \sin[k(z_1 - z_2)] V_2 + \cos[k(z_1 - z_2)] I_2. \tag{5.77}$$

This set of equations gives the line voltage and current at $z = z_1$ (the port nearest the generator) in terms of the line voltage and current at $z = z_2$ (the port nearest the load). Note that $z_2 < z_1$ and z_2 may be zero, in which case $V_2 = V_l$ and $I_2 = I_l$. The set can be put in the matrix form

$$\begin{bmatrix} V_1 \\ I_1 \end{bmatrix} \begin{bmatrix} t_{11} & t_{12} \\ t_{21} & t_{22} \end{bmatrix} \begin{bmatrix} V_2 \\ I_2 \end{bmatrix} = [t] \begin{bmatrix} V_2 \\ I_2 \end{bmatrix}. \tag{5.78}$$

The transmission (t) parameters are also called the *ABCD* parameters. Note that $\Delta_t = t_{11}t_{22} - t_{12}t_{21} \equiv 1$.

Figure 5.19(a) shows a pair of cascaded line sections. Applying Equation (5.78) gives

$$\begin{bmatrix} V_1 \\ I_1 \end{bmatrix} = [t_A] \begin{bmatrix} V_2 \\ I_2 \end{bmatrix} = [t_A] \begin{bmatrix} V_3 \\ I_3 \end{bmatrix},$$

$$\begin{bmatrix} V_3 \\ I_3 \end{bmatrix} = [t_B] \begin{bmatrix} V_4 \\ I_4 \end{bmatrix}$$

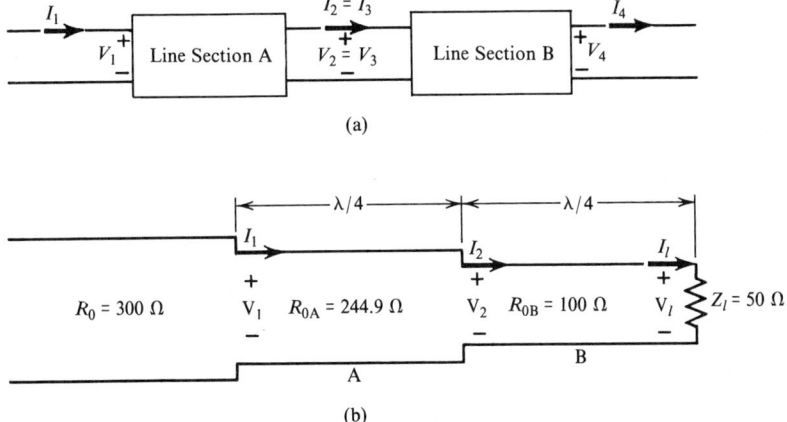

Figure 5.19. Pair of cascaded quarter-wave line sections.

or

$$\begin{bmatrix} V_1 \\ I_1 \end{bmatrix} = [t_A][t_B]\begin{bmatrix} V_4 \\ I_4 \end{bmatrix} = [t]\begin{bmatrix} V_4 \\ I_4 \end{bmatrix}, \qquad (5.79)$$

and the t parameters for the overall system are given in the matrix product $[t] = [t_A][t_B]$.

Example 5.6

Suppose that we want to match a resistive load $Z_l = 50\,\underline{/0°}\,\Omega$, to a 300-$\Omega$ line using two $\lambda/4$ transformers like in Example 5.4. We would expect a smoother transition and greater bandwidth than would be obtained with a single $\lambda/4$ transformer. The characteristic resistance of section B in Figure 5.19(b) is arbitrarily chosen to be $R_{0b} = 100\,\Omega$ so that the input impedance to this section is $(R_{0b})^2/R_l = 10^4/50 = 200\,\Omega$ [Equation (5.67)]. This impedance is the load for section A, so that $R_{0a} = \sqrt{(200)(300)} = 244.9\,\Omega$. Using Equation (5.78) ($z_1 - z_2 = \lambda/4$), we get

$$[t_A] = \begin{bmatrix} 0 & jR_{0a} \\ j/R_{0a} & 0 \end{bmatrix}, \qquad [t_B] = \begin{bmatrix} 0 & jR_{0b} \\ j/R_{0b} & 0 \end{bmatrix},$$

$$[t] = [t_A][t_B] = \begin{bmatrix} 0 & jR_{0a} \\ j/R_{0a} & 0 \end{bmatrix}\begin{bmatrix} 0 & jR_{0b} \\ j/R_{0b} & 0 \end{bmatrix} = \begin{bmatrix} -R_{0a}/R_{0b} & 0 \\ 0 & -R_{0b}/R_{0a} \end{bmatrix},$$

$$[t] = \begin{bmatrix} -2.449 & 0 \\ 0 & -0.408 \end{bmatrix},$$

$$\begin{bmatrix} V_1 \\ I_1 \end{bmatrix} = \begin{bmatrix} -2.449 & 0 \\ 0 & -0.408 \end{bmatrix}\begin{bmatrix} V_l \\ I_l \end{bmatrix} = \begin{bmatrix} -2.449\,V_l \\ -0.408\,I_l \end{bmatrix}$$

and

$$\frac{V_1}{I_1} = \frac{-2.449\,V_l}{-0.408\,I_l} = 6Z_l = 300\,\Omega$$

as expected. The reader is invited to compare the bandwidth of this matching scheme to that when only one $\lambda/4$ transformer is used with $R_{0c} = \sqrt{(50)(300)} = 122.5\,\Omega$. ∎

5.9 MULTIPLE LOADS

Lumped elements and active devices are often used with transmission lines. When these devices are connected across the line (parallel), it is advantageous to work with admittances, and when they are connected in series with the line, impedances are usually used.

5.9 MULTIPLE LOADS

It is often necessary to supply multiple loads from a single (main) line and source, and, furthermore, these loads may require *specified* voltages or currents. The linear antenna array is a good example because the individual elements (loads) require specified currents to produce a specified radiation pattern when transmitting.

Consider a $\lambda/4$ section of line like those used in Examples 5.4 and 5.6:

$$\begin{bmatrix} V_1 \\ I_1 \end{bmatrix} = \begin{bmatrix} V(\lambda/4) \\ I(\lambda/4) \end{bmatrix} = \begin{bmatrix} 0 & jR_0 \\ j/R_0 & 0 \end{bmatrix} \begin{bmatrix} V_l \\ I_l \end{bmatrix}$$

Thus, we have

$$V(\lambda/4) = jR_0 I_l, \qquad I_l = -jV(\lambda/4)/R_0, \tag{5.80}$$

$$I(\lambda/4) = jV_l/R_0, \qquad V_l = -jR_0 I(\lambda/4). \tag{5.81}$$

According to Equation (5.80), the load current is *independent of the load impedance*, and depends only on the voltage $V(\lambda/4)$ and the characteristic resistance of the line. Furthermore, the input admittance at the input to the $\lambda/4$ section is $I(\lambda/4)/V(\lambda/4)$ or

$$Y_{in} = Z_l/R_0^2. \tag{5.82}$$

Example 5.7

It is desired to supply three loads (antennas, for example) with the currents indicated in Figure 5.20 by way of three $\lambda/4$ feeders. $V(\lambda/4)$ is obviously the same for all three lines, and if we specify that $R_0 = 50 \ \Omega$ for the feeder that supplies the current I_2, then it follows from Equation (5.80) that the characteristic resistance of the other two feeders ($I_1 = I_3 = I_2/2$) must be $R_0 = 2(50) = 100 \ \Omega$. Equation (5.80)

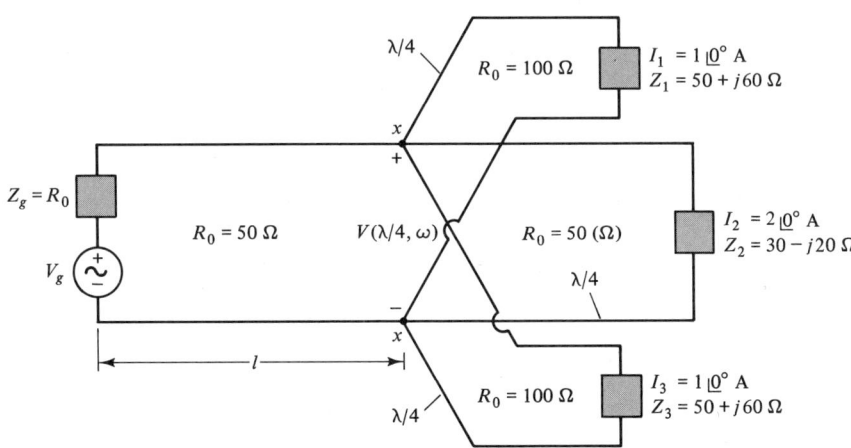

Figure 5.20. Transmission line supplying three load impedances with specified currents.

also gives $V(\lambda/4) = j(50)(2) = 100\underline{|90°}$. The admittance at $x - x$ is the *sum* of the input admittances for each feeder. These are given by Equation (5.82), so we have

$$Y_{xx} = \frac{Z_1}{10^4} + \frac{Z_2}{2.5 \times 10^3} + \frac{Z_3}{10^4}$$

Solving, we get

$$Y_{xx} = 22 + j4 \quad (m\mho), \qquad Z_{xx} = 44 - j8 \quad (\Omega).$$

Since the main line has $R_0 = 50\ \Omega$, there will be a small reflection

$$\Gamma_l = \frac{44 - j8 - 50}{44 - j8 + 50} = 0.106\underline{|-122°}.$$

We will see later how to correct for this with a matching scheme. Suppose that $l = 3.1\lambda$. Equation (5.54), with z being measured from the $x - x$ terminals, gives

$$V_{xx} = V(\lambda/4) = \frac{V_g}{2} e^{-jkl}(1 + \Gamma_l) \quad (z = 0)$$

or

$$V_g = \frac{2V(\lambda/4)}{1 + \Gamma_l} e^{+jkl} = 211\underline{|131.4°} \quad (V).$$

Thus, the phase of the generator with respect to the load currents is $+131.4°$.

Note that

$$I_{xx} = V_{xx}Y_{xx} = +j100(22 + j4) \times 10^{-3} = 2.236\underline{|100.3°} \quad (A),$$

so the power that is ultimately dissipated in the loads is

$$<\mathcal{P}_l> = \tfrac{1}{2}\text{Re}(V_{xx}I_{xx}^*) = 110 \quad (W).$$

This is correct since it is also true that

$$<\mathcal{P}_l> = \tfrac{1}{2}(1)^2(50) + \tfrac{1}{2}(2)^2(30) + \tfrac{1}{2}(1)^2(50) = 110 \quad (W). \qquad \blacksquare$$

5.10 TRANSMISSION LINE MEASUREMENTS

The success of any measurement scheme depends on the ability of the measurement equipment to be in place without appreciably disturbing the phenomenon being examined. The voltmeter, ammeter, and wattmeter (with minor modifications) can

5.11 THE SMITH CHART

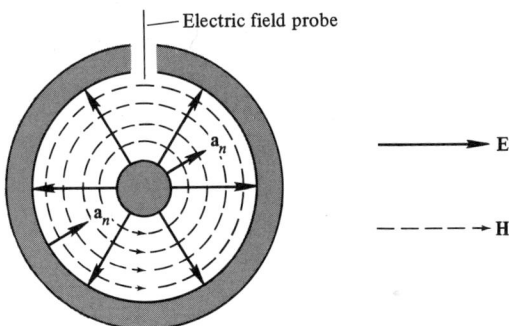

Figure 5.21. Coaxial slotted line with an electric field probe. The electromagnetic field (TEM mode) is also shown.

be used in transmission line measurements at low frequencies. The impedance (or admittance) bridge is also easy to use at low frequencies.

Other devices are usually used for transmission line measurements at high frequencies where the dimensions of the equipment may be appreciable compared to a wavelength. The impedance bridge, for example, can be used at high frequencies for impedance measurements, although it is usually more sophisticated than the low frequency bridge. See Example 5.3 in this regard. *Directional couplers* which can sample the incident wave and reflected wave separately can also be used (in-line) to continuously monitor VSWR or Γ_l. The directional coupler inherently takes a small amount of energy from the main line.

The *slotted line* is used in-line to measure $|V|$ (hence VSWR), f, and the location of $|V|_{\min}$ (or $|V|_{\max}$), and thus enables one to calculate Z_l. See Example 5.2 in this regard. It usually consists of a section of rigid coaxial line with an air dielectric and a narrow longitudinal slot cut in the outer conductor. The dominant TEM mode field distribution is shown over a cross section in Figure 5.21. The tangential electric field is (essentially) zero at the surface of the conductors, the normal magnetic field is (essentially) zero at the surface of the conductors, and **E** and **H** are everywhere perpendicular. **E** and **H** are distributed exactly as they were in the "static" cases considered earlier (for the coaxial cable). As seen in Figure 5.21, an electric field probe in the slot picks up a signal proportional to the line voltage. The slot has very little effect on the fields inside because it has almost no effect on the (axial) surface current on the outer conductor.

5.11 THE SMITH CHART

A graphical technique using a transmission line calculator, or Smith chart,[2] can be developed for making transmission line calculations. These calculations are usually accurate enough for engineering purposes. On the other hand, hand-held programmable calculators that handle complex numbers are readily available, and

---

[2] P. H. Smith, Transmission Line Calculator, in *Electronics* **12**:29–31, January (1939).

with these (or computers) and the equations of this chapter, it is very easy to simply calculate the desired quantities in a direct manner. One is then tempted to ignore the Smith chart entirely, and this can be done if desired, but what happens on the Smith chart correlates so well with what is actually happening on the line that it is still rewarding to devote some time to it.

The Smith chart results from a series of simple transformations. The starting point is the input impedance formula, given in Section 5.6,

$$Z_{in} = \frac{V}{I} = R_0 \frac{e^{jkz} + \Gamma_l e^{-jkz}}{e^{jkz} - \Gamma_l e^{-jkz}} \qquad (5.83)$$

which is valid regardless of Z_g! (note that the preceding statement is mathematically true, but from a practical point of view, a large reflection from the load may have undesirable effects on the generator.) The *normalized* input impedance is defined as

$$z_{in} \equiv \frac{Z_{in}}{R_0} = \frac{e^{jkz} + \Gamma_l e^{-jkz}}{e^{jkz} - \Gamma_l e^{-jkz}} \qquad (5.84)$$

or

$$z_{in} = \frac{1 + |\Gamma_l| e^{j(\phi_l - 2kz)}}{1 - |\Gamma_l| e^{j(\phi_l - 2kz)}} \equiv r_a + jx_a. \qquad (5.85)$$

Equation (5.85) may be written

$$|\Gamma_l| e^{j(\phi_l - 2kz)} = \frac{r_a - 1 + jx_a}{r_a + 1 + jx_a} \equiv U + jV. \qquad (5.86)$$

It is convenient at this point to recognize that

$$|U + jV| = (U^2 + V^2)^{1/2} \leq 1$$

since $|\Gamma_l| \leq 1$ for passive loads whose real parts are non-negative. Tunnel diodes, for example, are exceptions and offer the possibility for $|\Gamma_l| > 1$, and hence amplification!

Solving for U and V in Equation (5.86) gives

$$U = \frac{r_a^2 - 1 + x_a^2}{(r_a + 1)^2 + x_a^2} \qquad (5.87)$$

and

$$V = \frac{2x_a}{(r_a + 1)^2 + x_a^2}. \qquad (5.88)$$

5.11 THE SMITH CHART

Solving for r_a and x_a, then algebraically rearranging gives

$$\left(U - \frac{r_a}{r_a + 1}\right)^2 + (V)^2 = \left(\frac{1}{r_a + 1}\right)^2 \tag{5.89}$$

and

$$(U - 1)^2 + \left(V - \frac{1}{x_a}\right)^2 = \left(\frac{1}{x_a}\right)^2. \tag{5.90}$$

Equation (5.89) represents a family of constant r_a circles having centers on the U axis at $r_a/(r_a + 1)$ and radii $1/(r_a + 1)$ (the largest radius is *unity*). Equation (5.90) represents a family of constant x_a circles having centers at 1, $1/x_a$ in the U, V plane and having radii $1/x_a$. We also note that

$$|\Gamma_l|e^{j(\phi_l - 2kz)} = U + jV = (U^2 + V^2)^{1/2} e^{j \tan^{-1}(V/U)}$$

or

$$\tan(\phi_l - 2kz) = V/U. \tag{5.91}$$

The *normalized* input admittance y_{in} on the line at any point is the reciprocal of the normalized input impedance at the same point. It is a straightforward matter to begin with the reciprocal of Equation (5.84) to represent y_{in}, and then show that y_{in} at a distance z from the load is identical with z_{in} at a distance $z \pm \lambda/4$ from the load. Since $|\Gamma_l| = (U^2 + V^2)^{1/2} \leq 1$ for passive loads, all possible values of z_{in} lie inside, or on, the *unit* circle in the U, V plane! Also, from Equation (5.91), it can be seen that one revolution in the U, V plane (at constant radius) corresponds to a change in distance on the line of $\lambda/2$. This is not surprising, for we already know that impedances (and admittances) repeat every $\lambda/2$ on the line. We can now conclude that all possible values of normalized admittances also lie inside, or on, the unit circle in the U, V plane. In fact, since z_{in} and y_{in} are $\lambda/4$ apart, they must be *diametrically opposite* in the U, V plane.

Several important conclusions may now be reached concerning the Smith chart (the interior of the unit circle in the U, V plane).

1. Counterclockwise increasing angle measured from the U axis is $\phi_l - 2kz$. Thus, distance toward the load (z decreasing) is scaled linearly on the outside of the unit circle and measured in the counterclockwise direction. The actual location of $z = 0$ on the chart is arbitrary. On more expensive charts, the position of $z = 0$ can be set at will. As shown in Figure 5.22, the angle ϕ_l is scaled like $-z$, but its total excursion is 2π rad. It is easy to show that $\phi_l = 0$ if $X_l = 0$ and $R_l > R_0$. Values of z_{in} for these cases lie on the $+U$ axis and thus $\phi_l = 0$ also corresponds to the $+U$ axis.
2. Distance toward the generator is obviously measured in the clockwise direction (z increasing).

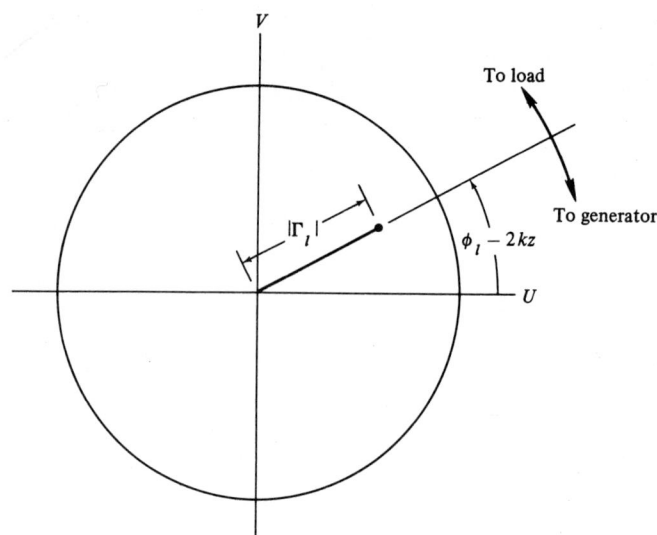

Figure 5.22. $|\Gamma_l| \underline{/\phi_l - 2kz}$ in the U, V plane.

3. Since there is a one-to-one correspondence between Γ_l and z_{in} (or y_{in}), including $z_l = Z_l/R_0$ itself, rotation at constant radius $|\Gamma_l|$ in the U, V plane consists of moving along a circle that uniquely specifies all possible values of z_{in} (or y_{in}) for a given $|\Gamma_l|$. (The effect of loss in the line might be pondered at this point in the development.)

4. Constant radius $|\Gamma_l|$ in the U, V plane also uniquely determines such things as the VSWR, reflected power, and so forth. A large number of radially scaled parameters may be used with the chart if desired. A normalized impedance or admittance Smith chart is shown in Figure 5.24, which is used for Examples 5.8 and 5.9.

Figure 5.23(a) shows the loci of constant r (or constant g) circles on the normalized Smith chart. The largest circle represents $r = 0$, while the smallest circle (a point) represents $r = \infty$. The circles are labeled as they are for an actual chart (e.g., Figure 5.24). The loci of constant x (or constant b) circles are shown in Figure 5.23(b). They are incomplete circles. The largest circle represents $x = 0$ and is actually a straight line (a circle with an infinite radius), while the smallest circle (a point) represents $x = \pm\infty$. Note carefully that positive reactances or susceptances lie *above* the horizontal axis, while negative reactances or susceptances lie *below* the horizontal axis. All numbers on the chart are labeled as positive, however. This can be confusing. Figure 5.23(c) demonstrates the loci of two points corresponding to $z = 1 + j3$ (or $y = 1 + j3$) and $z = 1 - j1$ (or $y = 1 - j1$). These points lie at the intersection of an r_a or an x_a circle. Some typical examples will help clarify any remaining questions about the use of the Smith chart.

5.11 THE SMITH CHART

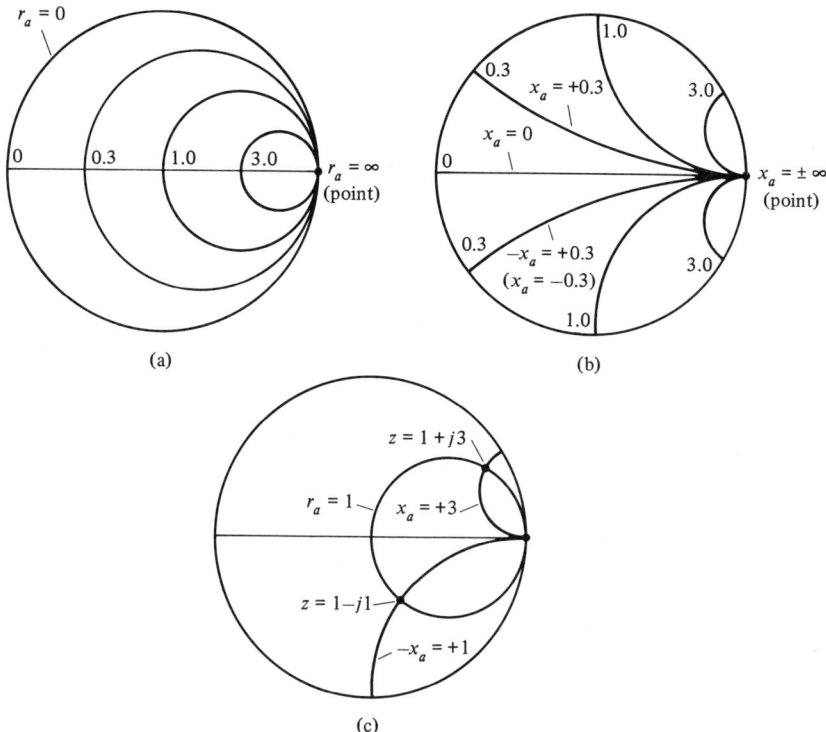

Figure 5.23. Loci of constant r (or g) and constant x (or b) circles on the normalized Smith chart. (a) Constant r (or g) = 0, 0.3, 1.0, 3.0, ∞ circles. (b) Constant x (or b) = 0, 0.3, 1.0, 3.0, ∞ circles (incomplete). Positive x (or b) values lie above the horizontal axis, while negative values lie below. (c) The locus of z (or y) = $1 + j3$ and z (or y) = $1 - j1$.

Example 5.8

If $Z_l = 100 + j50$ Ω and $R_0 = 50$ Ω, find Γ_l and the VSWR. The normalized load impedance is $z_l = (100 + j50)/50 = 2 + j1$. This point is located on the chart as the intersection of the circles $r_a = 2$ and $x_a = +1$ as shown in Figure 5.24. From the previous discussion, it is obvious that the distance from the center of the chart to z_l is $|\Gamma_l|$. In Figure 5.24, this is indicated on the upper right-hand scale below the chart labeled "refl. coef. vol." Thus, $|\Gamma_l| = 0.45$ as shown. The angle ϕ_l is measured from the $+U$ axis and is $+26.5°$. Thus, we have

$$\Gamma_l = 0.45 \underline{|26.5°}.$$

$|\Gamma_l|$ also uniquely determines the VSWR which is indicated on the lower left-hand scale labeled "standing wave vol. ratio." Then, we have

$$\text{VSWR} = 2.6.$$

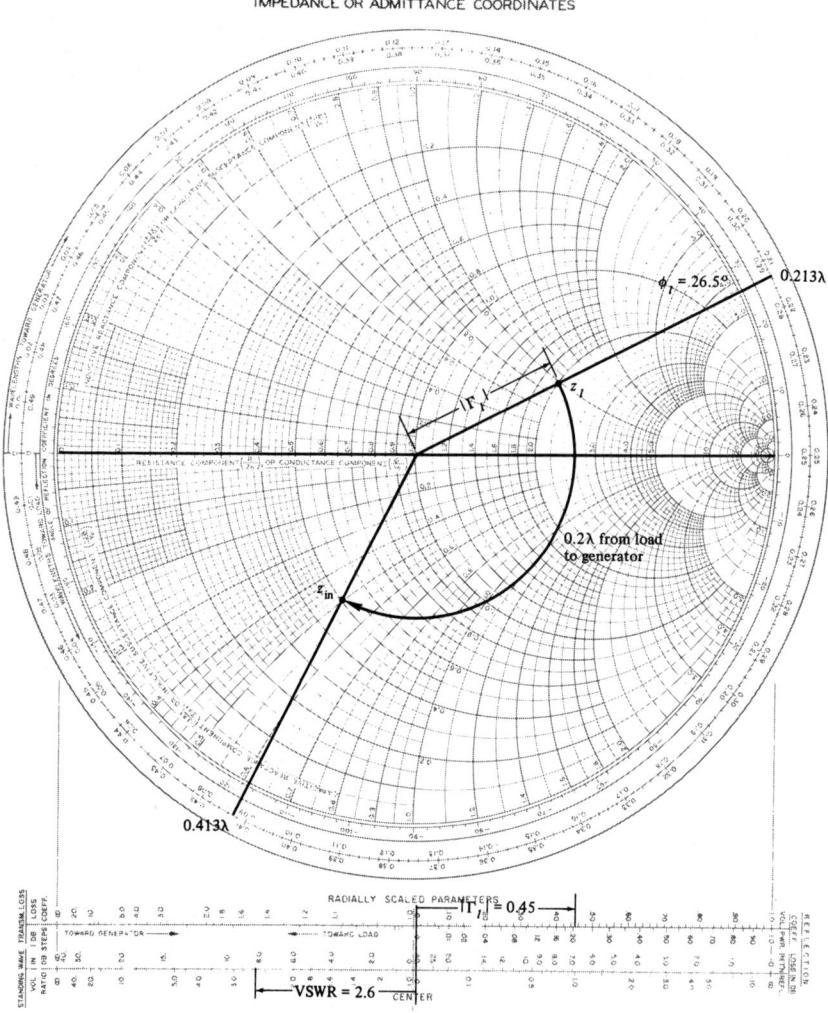

Figure 5.24. Smith chart data for Examples 5.8 and 5.9. *Source:* Copyright © 1949 by Kay Electric Company, Pine Brook, NJ. Renewal copyright © 1976 by P. H. Smith, New Providence, NJ.

Equations (5.42) and (5.60) give

$$\Gamma_l = \frac{Z_l - R_0}{Z_l + R_0} = \frac{50 + j50}{150 + j50} = 0.447 \underline{/26.57°},$$

$$\text{VSWR} = \frac{1 + |\Gamma_l|}{1 - |\Gamma_l|} = \frac{1.447}{0.553} = 2.618.$$

∎

Example 5.9

Suppose we desire the input impedance 0.2λ from the load in Example 5.8. Note that the radial line through z_l intersects the "wavelength toward generator scale" at 0.213λ. This is only a reference reading and corresponds to $z = 0$. Consequently, the desired point where the input impedance is to be found is at $0.213\lambda + 0.200\lambda = 0.413\lambda$ on the same scale. At the same $|\Gamma_l|$, we find $z_{in} = 0.5 - j0.5$ or $Z_{in} = R_0 z_{in}$, so we have

$$Z_{in} = 25 - j25 \quad (\Omega).$$

Equation (5.67) gives

$$Z_{in} = R_0 \frac{Z_l \cos kz + jR_0 \sin kz}{R_0 \cos kz + jZ_l \sin kz} = 24.81 - j24.62 \quad (\Omega).$$

Equation (5.84) (with the results of Example 5.8) gives the same result. ∎

Note that both wavelength scales end at 0.5λ, and when rotating past this point in either direction, care must be taken to avoid losing the references. This difficulty is not encountered with those charts having a rotatable wavelengths scale.

Example 5.10

The following data are taken with an impedance bridge:

$$Z_{in}^{sc} = +j106 \quad (\Omega),$$
$$Z_{in}^{oc} = -j23.6 \quad (\Omega),$$

and

$$Z_{in} = 25 - j70 \quad (\Omega) \quad \text{(actual load in place)}.$$

The actual load impedance Z_l is desired. From Equation (5.70), we get

$$R_0 = \sqrt{Z_{in}^{oc} Z_{in}^{sc}} = 50 \quad (\Omega).$$

The distance from the measuring point at the bridge to the actual location of the load can be found from Z_{in}^{sc} (or Z_{in}^{oc}) as follows:

$$z_{in}^{sc} = \frac{Z_{in}^{sc}}{R_0} = +j2.12$$

is located on the chart. This point is located on the *unit* circle since the line is lossless. The short circuit $Z_l = z_l = 0$, in place when Z_{in}^{sc} was measured, is located

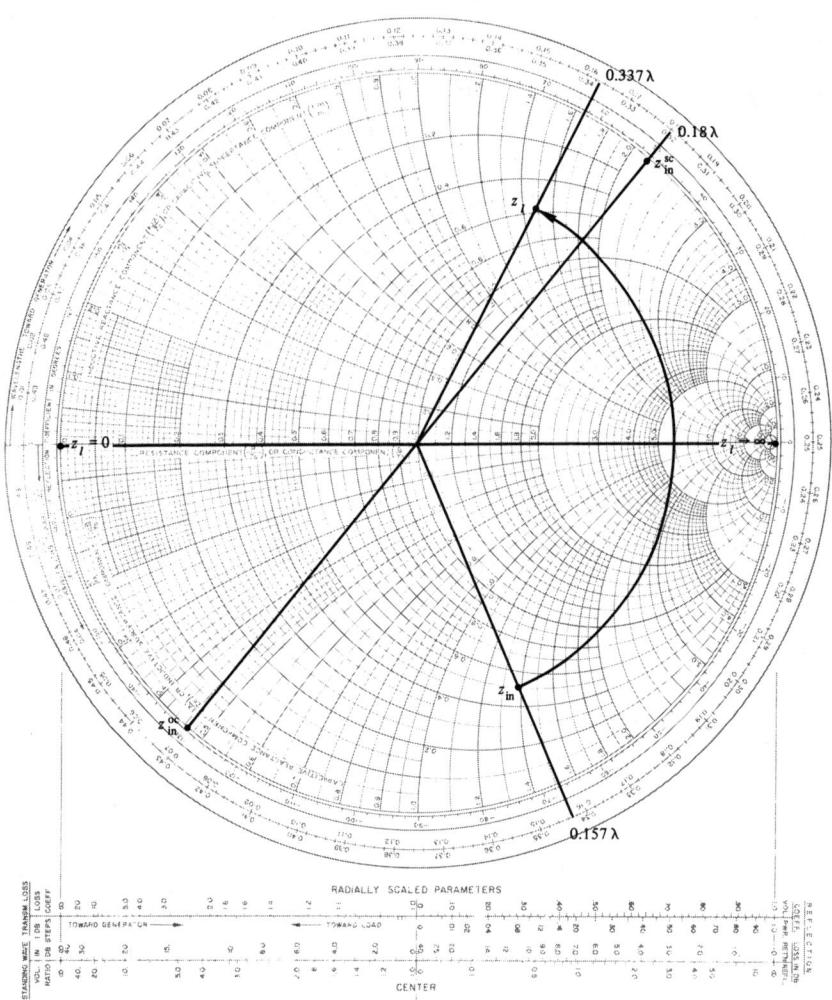

Figure 5.25. Smith chart data for Example 5.10. *Source:* Copyright © 1949 by Kay Electric Company, Pine Brook, NJ. Renewal copyright © 1976 by P. H. Smith, New Providence, NJ.

at $U = -1$, $V = 0$ as shown in Figure 5.25. Then the distance desired is found as that distance from $z_l = 0$ toward z_{in}^{sc}, rotating in the "toward-the-generator" (clockwise) direction. Then $z = 0.18\lambda$ as shown. Note that this information is also available from z_{in}^{oc} and the location of $z_l \to \infty$ ($U = +1, V = 0$).

The point $z_{in} = Z_{in}/R_0 = 0.5 - j1.4$ is next located on the chart as shown. A radial line through z_{in} gives a wavelength reference on the "toward-the-load" scale of 0.157λ. The load then is located at $0.18\lambda + 0.157\lambda = 0.337\lambda$ as shown. Thus $z_l = 0.57 + j1.5$ or $Z_l = R_0 z_l$, or

$$Z_l = 28.5 + j75 \quad (\Omega).$$

5.11 THE SMITH CHART

Equation (5.71) gives

$$Z_l = Z_{in}^{oc} \frac{Z_{in}^{sc} - Z_{in}}{Z_{in} - Z_{in}^{oc}} = 27.53 + j74.69 \quad (\Omega).$$

∎

■ Example 5.11

The following data are taken with a 50-Ω slotted line and calibrated detector with an unknown load impedance Z_l in place:

$$|V|_{max} = 0 \quad (dB),$$
$$|V|_{min} = -6 \quad (dB),$$

and $|V|_{min}$ occurs at relative slotted line positions of 0.10 m, 0.35 m, 0.60 m, and 0.85 m *measured toward the load*. When the unknown load is replaced with a short circuit, the positions of $|V|_{min}$ are 0 m, 0.25 m, 0.50 m, and 0.75 m, *measured toward the load*. Determine Z_l and the frequency.

Frequency is normally determined by measuring $\lambda/2$ as the distance between adjacent voltage minima when the short circuit is in place. The VSWR is infinite in this case, and thus the "nulls" will be sharpest giving a very accurate position of $|V|_{min}$. Then, $\lambda/2 = 0.25$ m, or $\lambda = 0.5$ m. Since the dielectric is air, we get

$$f = \frac{3 \times 10^8}{0.5} = 600 \quad \text{MHz}.$$

The Smith chart may be used to determine Z_l in the following way. The VSWR is 2 (or 6 dB), and so z_l is somewhere on the Smith chart circle corresponding to VSWR $= 2$. Note in Figure 5.26 that the VSWR scale corresponds to the r_a scale on the $+U$ axis and a 1/(VSWR) scale corresponds to the r_a scale on the $-U$ axis. This occurs because, according to Equations (5.43) and (5.61), we have

$$V_{max} = V^- e^{jkz}(1 + |\Gamma_l|),$$

while at the same location on the line (z_{max}), we have

$$I_{min} = \frac{V^-}{R_0} e^{jkz}(1 - |\Gamma_l|)$$

using Equations (5.44) and (5.62). Using Equation (5.60)

$$Z = Z_{max} = \frac{V_{max}}{I_{min}} = R_0(\text{VSWR})$$

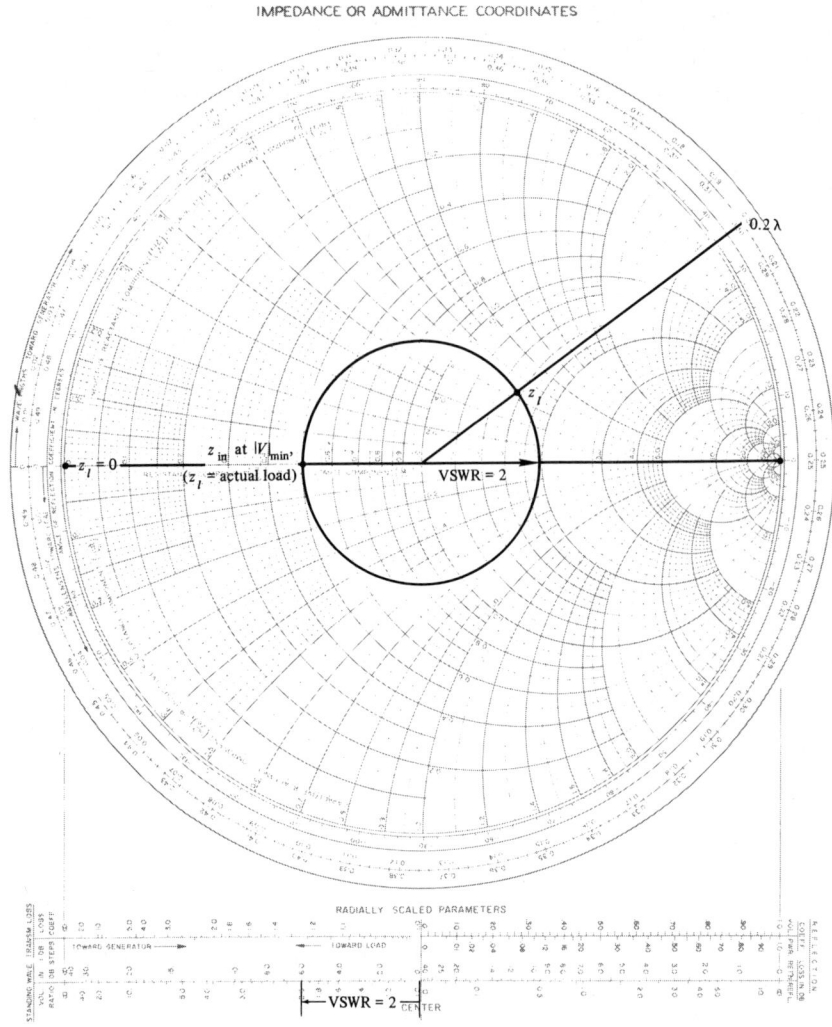

Figure 5.26. Smith chart data for Example 5.11. *Source:* Copyright © 1949 by Kay Electric Company, Pine Brook, NJ. Renewal copyright © 1976 by P. H. Smith, New Providence, NJ.

Thus, the maximum *normalized* impedance equals the VSWR, and points of maximum impedance, corresponding to maximum voltage, are located on the $+U$ axis. In the same way, we have

$$Z_{\min} = \frac{V_{\min}}{I_{\max}} = \frac{R_0}{\text{VSWR}}.$$

Thus, the minimum *normalized* impedance equals the reciprocal of the VSWR, and points of minimum impedance, corresponding to minimum voltage, are located on the $-U$ axis. With $Z_l = 0$ the voltage minimum or null occurs at 0λ "toward gen-

5.11 THE SMITH CHART

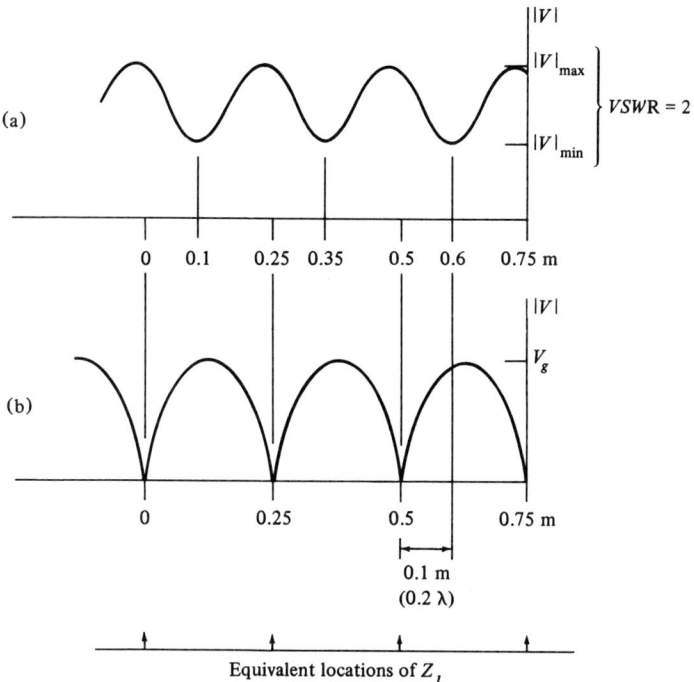

Figure 5.27. (a) $|V|$ versus z when Z_l = actual load. VSWR = 2. (b) $|V|$ versus z when $Z_l = 0$. (Example 5.11) $Z_g = R_0$.

erator" or "toward load" since z_l itself is located at $U = -1$, $V = 0$. This is shown in Figure 5.26. The slotted line data indicate that $|V|_{\min}$ shifts $(0.10/0.50)\lambda = 0.2\lambda$ toward the load when the short circuit is replaced by the actual load. This is shown in Figure 5.27. Then z_l on the chart is 0.2λ toward the generator from the location of $|V|_{\min}$ with the actual load in place. This is shown in Figure 5.26 where it can be seen that $z_l = 1.56 + j0.69$ or $Z_l = 77.5 + j34.5$ Ω. The situation is clarified by Figure 5.27. Note that only the "artificial" end of the line, or load location, is used. Equations (5.65) and (5.66) give

$$\Gamma_l = 0.333 \underline{|36°}, \quad Z_l = 77.73 + j34.27 \ (\Omega).$$

∎

Before proceeding to a new topic, some general statements concerning Smith charts and other transmission line calculators need to be made.

1. A *normalized* Smith chart applies to a line of *any* characteristic resistance and serves just as well for normalized admittances as for normalized impedances.
2. If 50-Ω (or 20-m℧) lines are worked with almost exclusively, then a 50-Ω (or 20-m℧) Smith chart should be used.

3. Some impedance bridges read out in polar form. It is not difficult to show that a polar coordinate Smith chart contains circles of constant $|Z|$ and circles of constant $\underline{/Z}$, and is available.
4. Quite often difficulty is encountered in accurately reading a normal Smith chart, especially in the neighborhood of $U = +1$, $V = 0$. This difficulty can be overcome by using an inverted circle (or Blanchard) chart.

5.12 TRANSMISSION LINE MATCHING

Transmission lines are primarily used to guide energy or information from a source to a load. It was previously shown that maximum power transfer occurs for $Z_l = R_0$ (when $Z_g = R_0$). Quite often the load on the line is such that its impedance cannot be easily altered. It is still desirable to dissipate maximum power *in* the load (or more correctly, in the *real* part of the load impedance). If we can design a lossless matching system, whose input impedance is R_0, and place this system on the generator side of the load, then we will have achieved maximum power transfer. The matching system and the line are lossless. There is no reflection back to the generator, and *all* of the power available in the *incident* wave *must* be ultimately dissipated in the *real* part of the load impedance. This matching scheme is indicated in Figure 5.28 by successive equivalent circuits (neglecting time delays).

Some examples of lossless elements, or at least the equivalents of lossless elements, suitable for use in a matching scheme, are the short- and open-circuited loss-

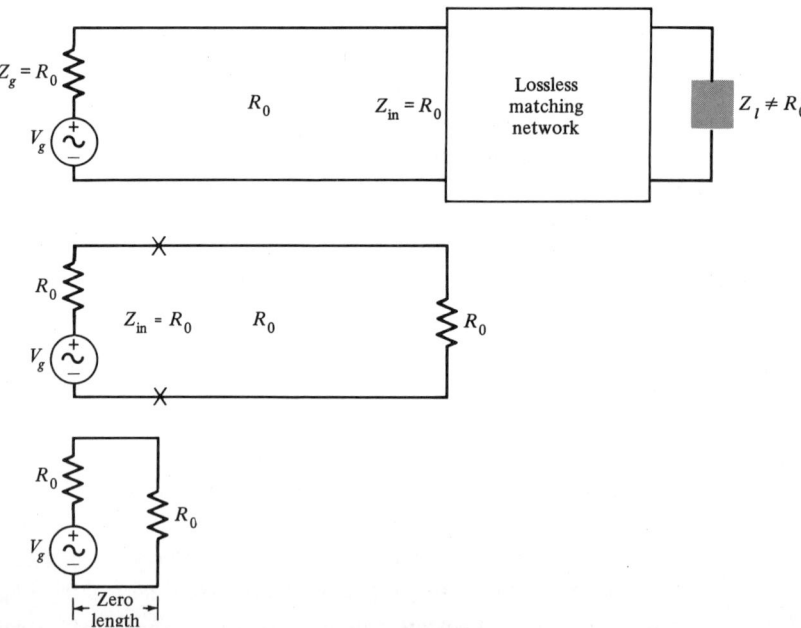

Figure 5.28. Successive equivalent circuits (neglecting time delays) for a lossless matching system.

5.12 TRANSMISSION LINE MATCHING

less lines. From Equations (5.68) and (5.69) *normalized* values of input impedance and admittance for these elements are

$$z_{in}^{sc} = y_{in}^{oc} = +j \tan kz \qquad (5.92)$$

and

$$z_{in}^{oc} = y_{in}^{sc} = -j \cot kz. \qquad (5.93)$$

Equations (5.92) and (5.93) reveal that any reactance $-\infty \leq x_{in} \leq +\infty$ is available from these line sections, called "stubs." Two cases where these stubs are utilized will be examined separately.

The basic philosophy in any matching scheme is the same: we attempt to make $z_{in} + 1$ ($Z_{in} = R_0$) at some point on the line, or equivalently, we attempt to make $y_{in} = +1$ at some point on the line. Since the stubs will normally be *shunted across the line*, and admittances *add* in parallel, it is much easier to work with normalized admittances.

Since we are attempting to make $y_{in} = +1$ at some point on the line, the locus, $y = 1 + jb$ should be located on the Smith chart first. This locus is shown in Figure 5.29.[3] Inspection of Figure 5.29 reveals that for any y_l (with the exception of the special cases $y_l = +1$ and $y_l = +jb$), the $1 + jb$ circle is intersected twice (points P and Q) in rotating from y_l toward the generator. Thus, in any practical case, two values of z ($< \lambda/2$) can be found on the line where $y_{in} = 1 + jb$. Then, a shunt stub introduced at either of these points on the line will give $y_{in} = 1 + j0$ if the stub admittance is $y_s = -jb$. That is, $y_{in} = 1 + jb + y_s = 1 + jb - jb = 1 + j0$. If $y_{in} = 1 + j0$, then $z_{in} = 1 + j0$ and $Z_{in} = R_0$. This reproduces exactly the situation shown in the second circuit of Figure 5.28. An example is in order.

[3]Refer to Figure 5.23(b) for the loci of constant b (or x).

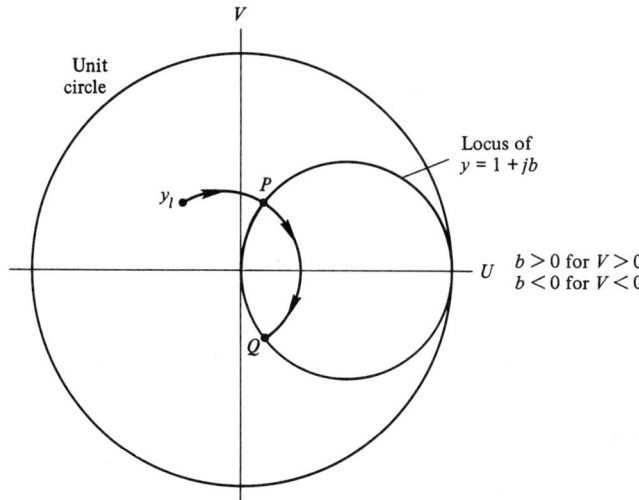

Figure 5.29. The $y = 1 + jb$ locus on the Smith chart.

Example 5.12

Suppose $R_0 = 50\ \Omega$ and $Z_l = 25 + j75\ \Omega$. Then $\Gamma_l = 0.74\underline{|64°}$ and the VSWR = 6.7. The standing wave is excessively large, which will cause additional difficulties if the line has loss. Furthermore, the ratio of the actual load power to that which could be obtained under matched conditions is $1 - |\Gamma_l|^2 = 0.45$ [from Equations (5.57) and (5.59)]. It is desired to match the line to the load with a single shorted ($R_0 = 50\text{-}\Omega$) stub. The steps in the procedure are numbered:

1.
$$z_l = (25 + j75)/50 = 0.5 + j1.5,$$
$$y_l = (0.5 + j1.5)^{-1} = 0.2 - j0.6.$$

Note here that z_l and y_l are diametrically opposite on the Smith chart (Why?).

2. y_l is located on the chart and a wavelength reference of 0.412λ on the "toward generator" scale is obtained as shown in Figure 5.30).

3. A constant radius circle ($|\Gamma_l| = 0.74$) is drawn through y_l and intersects the $1 + jb$ locus at two points:

$$\begin{bmatrix} y''_{in} = 1 + j2.2 \text{ at} \\ 0.192\lambda \text{ reference} \end{bmatrix} \text{ or } \begin{bmatrix} y''_{in} = 1 - j2.2 \text{ at} \\ 0.308\lambda \text{ reference} \end{bmatrix}.$$

4. The point of stub connection is the distance rotated or

$$\begin{bmatrix} l'_1 = 0.5\lambda - 0.412\lambda + 0.192\lambda \\ l'_1 = 0.28\lambda \end{bmatrix}$$

or

$$\begin{bmatrix} l''_1 = 0.5\lambda - 0.412\lambda + 0.308\lambda \\ l''_1 = 0.396\lambda. \end{bmatrix}$$

5. The required stub susceptance is

$$y'_s = -j2.2 \quad \text{or} \quad y''_s = +j2.2.$$

6. The stub lengths are found by simply treating the stubs separately. The stub load is $y_l \to \infty$ (short circuit) which is located at $U = +1, V = 0$. Rotating from this point to the required susceptance gives the stub length. Thus, we have

$$\begin{bmatrix} s'_1 = 0.318\lambda - 0.25\lambda \\ s'_1 = 0.068\lambda \end{bmatrix} \text{ or } \begin{bmatrix} s''_1 = 0.25\lambda + 0.182\lambda \\ s''_1 = 0.432\lambda \end{bmatrix}.$$

This gives the matched system shown in Figure 5.31.

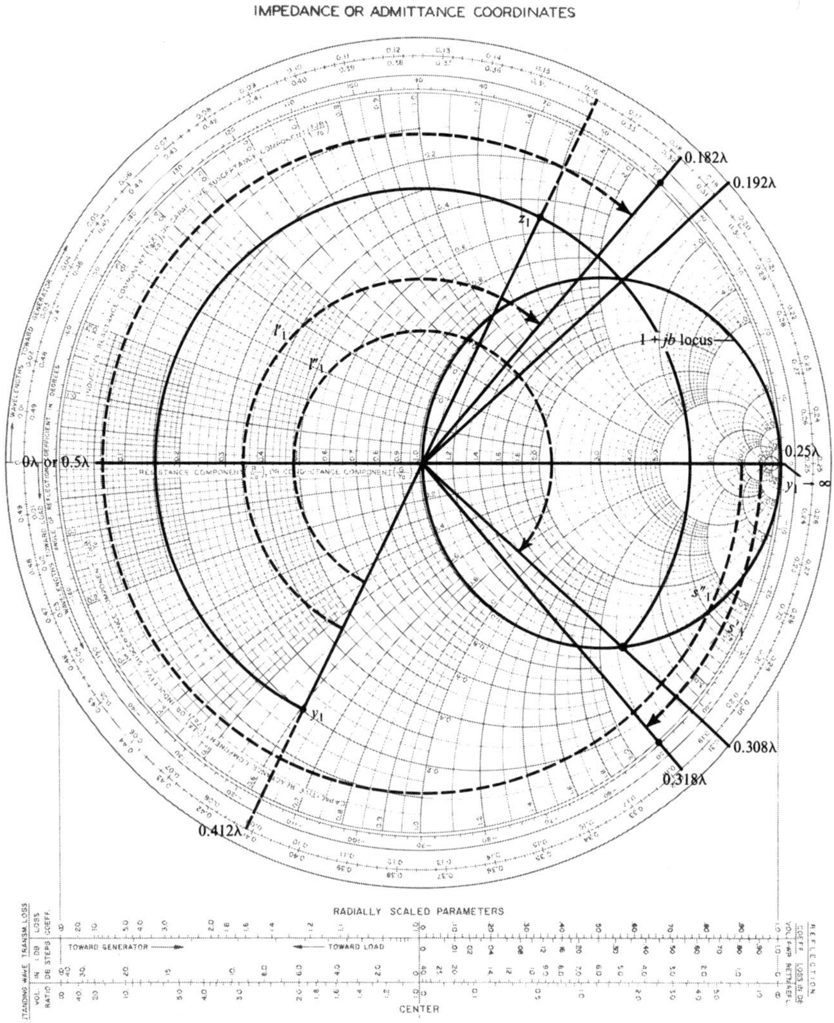

5.12 TRANSMISSION LINE MATCHING

Figure 5.30. Single stub matching problem, Example 5.12. *Source:* Copyright © 1949 by Kay Electric Company, Pine Brook, NJ. Renewal copyright © 1976 by P. H. Smith, New Providence, NJ.

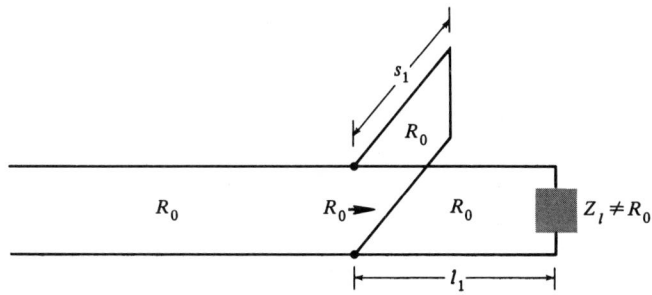

Figure 5.31. Single stub matching system.

The choice between the two possible solutions is not important unless the frequency is low enough so that the physical length of the line is an economic factor, or unless one solution has a better frequency bandwidth characteristic. A shorted stub is usually chosen over the open stub, because (in coaxial systems at least) it has less tendency to radiate energy. In practice, l_1 and s_1 are adjusted to minimize reflected power utilizing a directional coupler.

The stub length s_1 and point of connection l_1 can also be calculated analytically. From Equation (5.85), we get

$$y_{\text{in}} = \frac{1 - |\Gamma_l|e^{j(\phi_l - 2kz)}}{1 + |\Gamma_l|e^{j(\phi_l - 2kz)}} = g_{\text{in}} + jb_{\text{in}}; \tag{5.94}$$

or, rationalizing, we get

$$g_{\text{in}} = \frac{1 - |\Gamma_l|^2}{1 + |\Gamma_l|^2 + 2|\Gamma_l|\cos(\phi_l - 2kz)} \tag{5.95}$$

and

$$b_{\text{in}} = \frac{-2|\Gamma_l|\sin(\phi_l - 2kz)}{1 + |\Gamma_l|^2 + 2|\Gamma_l|\cos(\phi_l - 2kz)}. \tag{5.96}$$

For a match, $g_{\text{in}} = 1$ at $z = l_1$; therefore, from Equation (5.95), we get

$$\cos(\phi_l - 2kl_1) = -|\Gamma_l|.$$

That is,

$$\boxed{l'_1 = \frac{\lambda}{4\pi}(\phi_l + \pi - \cos^{-1}|\Gamma_l|) \quad \text{or} \quad l''_1 = \frac{\lambda}{4\pi}(\phi_l + \pi + \cos^{-1}|\Gamma_l|)}.$$

$$\tag{5.97}$$

The susceptance b_{in} at l_1 is obtained by substituting the last result into Equation (5.96):

$$b'_{\text{in}} = +\frac{2|\Gamma_l|}{\sqrt{1 - |\Gamma_l|^2}} \quad \text{or} \quad b''_{\text{in}} = -\frac{2|\Gamma_l|}{\sqrt{1 - |\Gamma_l|^2}}.$$

For a shorted stub having the same R_0 as the main line $b_s = -\cot ks_1 = -b_{\text{in}}$ or $\tan ks_1 = 1/b_{\text{in}}$:

$$\boxed{s'_1 = \frac{\lambda}{2\pi}\tan^{-1}\frac{\sqrt{1 - |\Gamma_l|^2}}{2|\Gamma_l|} \quad \text{or} \quad s''_1 = \frac{\lambda}{2\pi}\left(\pi - \tan^{-1}\frac{\sqrt{1 - |\Gamma_l|^2}}{2|\Gamma_l|}\right)}. \tag{5.98}$$

5.12 TRANSMISSION LINE MATCHING

Using numbers from the preceding example gives[4]

$$l'_1 = 0.280\lambda \quad \text{or} \quad l''_1 = 0.396\lambda,$$
$$s'_1 = 0.067\lambda \quad \text{or} \quad s''_1 = 0.433\lambda.$$

The values taken from the Smith chart agree well with the calculated values.

Before leaving the single-stub design of Example 5.12, it is well worth the effort to calculate the current (or the voltage) at the load. Let us assume that $Z_g = R_0 = 50\ \Omega$, $V_g = 10\ \underline{|0°}$ V peak, and the line length from the generator to the point where the stub is connected is 3.2λ. The Norton equivalent circuit with respect to the junction (where the stub is connected) and looking to the left (Figure 5.31) is given by Equations (5.51) and (5.52) with $Z_g = R_0$, $\Gamma_g = 0$, and $l = 3.2\lambda$:

$$I_N = (V_g/R_0)e^{-j3.2(k)\lambda}$$
$$Z_T = Z_N = R_0, \qquad Y_N = 1/R_0, \qquad y_N = 1.$$

The normalized stub admittance is $y_s = +jb_s$ and is in parallel with $y_N = 1$. Therefore, the equivalent current source at the junction is I_N, in parallel with the impedance $Z'_T = R_0/(1 + jb_s)$, or $V'_T = I_N Z'_T$ in series with Z'_T, as shown in Figure 5.32, and the load voltage can be found by using Thévenin's or Norton's theorem once more, or by simply using Equation (5.43) as applied to Figure 5.31. Choosing the latter course, and using l'_1, and s'_1 from Example 5.12, we get

$$b_s = -\cot ks_1 = -2.196, \quad Z'_T = R_0/(1 - j2.196) = 8.59 + j18.86\ (\Omega),$$

$$\Gamma'_g = \frac{Z'_T - R_0}{Z'_T + R_0} = 0.739\ \underline{|137.68°}, \qquad \Gamma_l = \frac{Z_l - R_0}{Z_l + R_0} = 0.745\ \underline{|63.43°},$$

$$V_l = \frac{V'_T R_0}{R_0 + Z'_T}\ \frac{e^{-j0.28(k)}(1 + \Gamma_l)}{1 - \Gamma'_g \Gamma_l e^{-j0.56(k)}} = 11.18\ \underline{|-99.3°}\ (V),$$

$$I_l = V_l/Z_l = 0.141\ \underline{|-170.90°}\ (A).$$

---

[4]The calculator will give principal values for $\tan^{-1} x$. These are angles in the first and *fourth* quadrant. We want angles in the first and *second* quadrants. Thus, whenever $\tan^{-1} x < 0$, we merely add π radians to the angle to place it in the second quadrant.

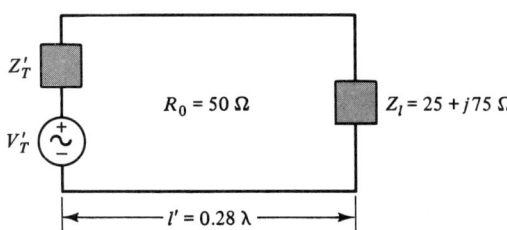

Figure 5.32. Equivalent circuit for the single-stub matched system (Example 5.12 and Figure 5.31).

The load power is maximized and is given by

$$<\mathcal{P}_l> = \tfrac{1}{2}|I_l|^2 R_l = 250 \quad \text{mW}$$

or

$$<\mathcal{P}_l> = V_g^2/(8R_0) = 250 \quad \text{mW}$$

as expected.

Equations (5.97) and (5.98) are the equations to be used when calculating l_1 and s_1 with a calculator. If the necessary input data are taken from slotted line measurements (VSWR and $z_{\min}$), then Equation (5.65) will also be needed to convert VSWR and $z_{\min}$ to Γ_l. On the other hand, if the data are taken from impedance bridge measurements, then it is perhaps simplest to use Equation (5.71) to calculate Z_l and then use Equation (5.42) to calculate Γ_l. The procedures are summarized below. They can be used as a guide for writing simple computer programs for calculating l_1 and s_1 from given or measured data.

Given Z_l	Impedance Bridge	Slotted Line
Data → Z_l, R_0, λ	Z_{in}, $Z_{\text{in}}^{\text{sc}}$, $Z_{\text{in}}^{\text{oc}}$, R_0, λ	VSWR, $z_{\min}$, R_0, λ
(5.42)	(5.71)	(5.65)
↓	↓	↓
Γ_l	Z_l	Γ_l
↙ ↘	↓	↙ ↘
(5.97) (5.98)	(5.42)	(5.97) (5.98)
↓ ↓	↓	↓ ↓
l_1 s_1	Γ_l	l_1 s_1
	↙ ↘	
	(5.97) (5.98)	
	↓ ↓	
	l_1 s_1	

A double stub matching system is shown in Figure 5.33. The spacings l_1 and l_2 are normallly fixed, while the stub lengths s_1 and s_2 are normally adjustable. The normalized admittance at $x - x$ with s_1 in place, but s_2 not in place *must* be $1 + jb$. Then s_2 is attached and adjusted so that its susceptance is $-jb$ (as in the single stub system), giving a matched system. Thus, the admittance at $y - y$, with s_1 in place, but s_2 not in place, must lie on the Smith chart locus of all points l_2 wavelengths *toward the load from the* $1 + jb$ locus (L_1). The new locus L_2 is just a circle of the same size as the $1 + jb$ locus. It follows then that the stub s_1 has the function of moving the admittance at $y - y$ *along a contour of constant g* to the locus L_2. An example will clarify the procedure.

5.12 TRANSMISSION LINE MATCHING

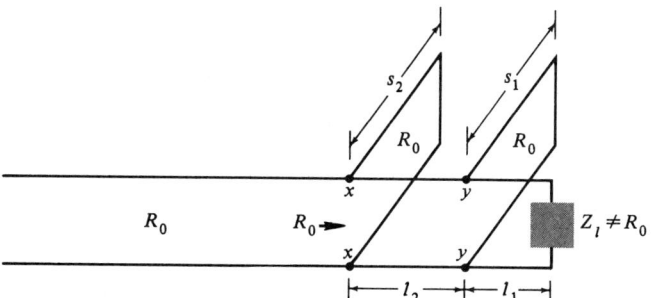

Figure 5.33. Double stub matching system.

Example 5.13

Suppose $R_0 = 50\,\Omega$ and $Z_l = 25 + j75\,\Omega$ (as in the single stub example), so that $\Gamma_l = 0.74\,\underline{/64°}$ and the VSWR = 6.7. Let $l_1 = 0.2\lambda$ and $l_2 = 0.375\lambda$. The steps in the design of this double stub system are as follows below for Figure 5.34.

1. The first step in a double stub design is that of locating the locus L_2. This locus is just the $L_1(1 + jb)$ locus rotated 0.375λ toward the load as shown in Figure 5.34. The locus L_2 is drawn on the chart.
2. The normalized load admittance is $y_l = 0.2 - j0.6$ and is located on the chart.
3. y_l is rotated 0.2λ toward the generator. This gives $y_{in} = 0.25 + j0.82$ (at $y - y$) at a reference of 0.112λ.
4. The stub s_1 at $y - y$ can move the admittance at $y - y$ to *either* of the two points on the constructed locus L_2. These points are A, $y = 0.25 - j0.34$, and B, $y = 0.25 - j1.65$.

[It is worthwhile at this point to note that if y_{in} at $y - y$ without s_1 has a real part g_{in} greater than 2 (for example, $y_{in} = 3 + j1 = y_c$ in Figure 5.34), a solution does not exist! Thus, sometimes a solution is not possible for a double stub system, unless l_2 is altered.]

5. For point A, the stub s_1 must change the susceptance b from $+0.82$ to -0.34, while for point B, the stub s_1 must change the susceptance from $+0.82$ to -1.65. Therefore,

$$(b'_{s_1} = -1.16) \quad \text{or} \quad (b''_{s_1} = -2.47).$$

6. The required stub lengths are found as the single stub system and are

$$(s'_1 = 0.114\lambda) \quad \text{or} \quad (s''_1 = 0.061\lambda).$$

7. Point A is rotated 0.375λ toward the generator to A' where $y_{in} = 1 - j1.65$ at $x - x$. Point B is rotated 0.375λ toward the generator to B', where $y_{in} = 1 + j3.6$.

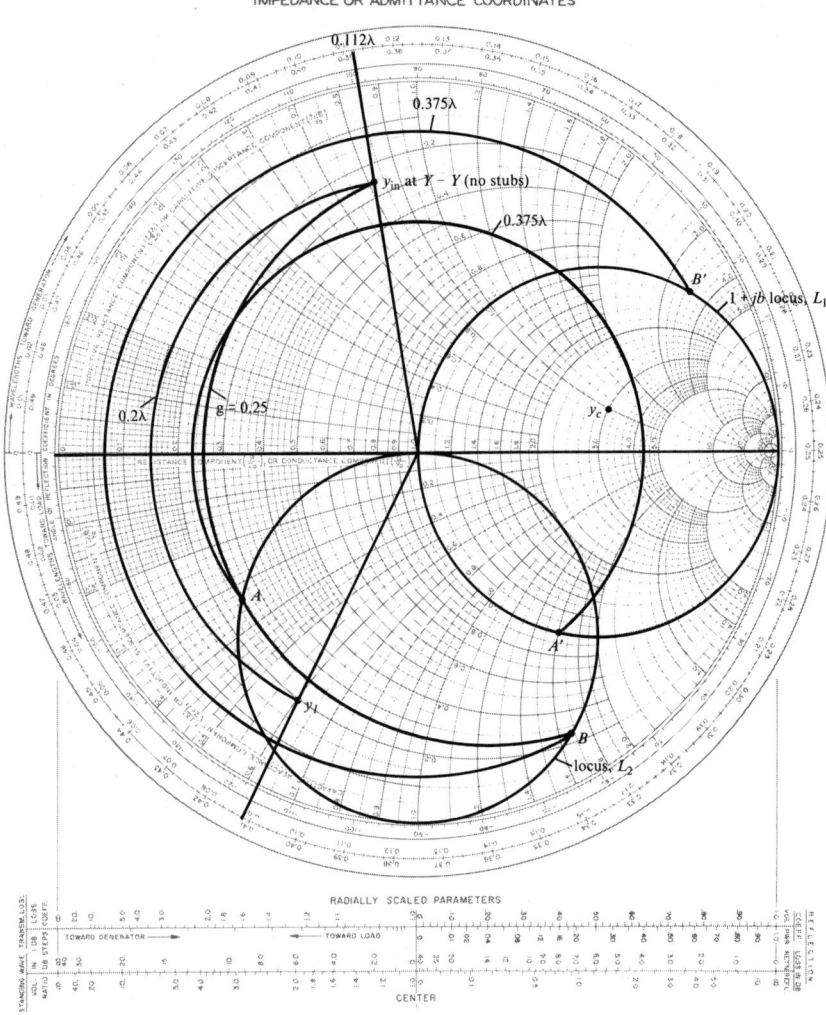

Figure 5.34. Double stub matching problem, Example 5.13. *Source:* Copyright © 1949 by Kay Electric Company, Pine Brook, NJ. Renewal copyright © 1976 by P. H. Smith, New Providence, NJ.

8. The stub s_2 must cancel the susceptance at $x - x$, thus,

$$(y'_{s_2} = +j1.65) \quad \text{or} \quad (y''_{s_2} = -j3.6).$$

9. The stub lengths are found as before:

$$(s'_2 = 0.414\lambda) \quad \text{or} \quad (s''_2 = 0.044\lambda).$$

s'_1 and s'_2 must be used together or s''_1 and s''_2 must be used together.

The same remarks, concerning a choice of one of the two possible solutions for the single stub system, apply here. ∎

5.12 TRANSMISSION LINE MATCHING

The double-stub tuner can also be designed without the aid of the Smith chart. It is assumed that l_1 and l_2 are fixed and are given quantities, and we want to determine s_1 and s_2. The first step in the analysis is that of determining the normalized admittance at $y - y$ without s_1. This is accomplished by inverting Equation (5.67) and normalizing:

$$y_{\text{in}} = \frac{y_l \cos kz + j \sin kz}{\cos kz + j y_l \sin kz}, \qquad (5.99)$$

$$y_{\text{in}} = y'_{yy} = g'_{yy} + j b'_{yy}. \qquad (5.100)$$

The admittance at $y - y$ with s_1 is $y_{yy} = y'_{yy} - j \cot k s_1$ or

$$y_{yy} = g'_{yy} + j(b'_{yy} - \cot k s_1). \qquad (5.101)$$

Now, Equation (5.99) is general and can be used backward. That is, it can be used to find the normalized admittance at any point on the line in terms of the normalized admittance at any other point. Here we know that the normalized admittance at $x - x$ without s_2 must be $1 + jb$. Therefore, the normalized admittance at a distance l_2 *toward the load* from $x - x$ must be y_{yy} again. Setting $y_{\text{in}} = 1 + jb$ and $y = y_{yy}$ in Equation (5.99), and then solving for y_{yy} gives

$$y_{yy} = \frac{(1 + jb) \cos kl_2 - j \sin kl_2}{\cos kl_2 - j(1 + jb) \sin kl_2} \qquad (5.102)$$

It is a straightforward (but lengthy) matter to equate the right-hand sides of Equations (5.101) and (5.102), and then equate the real and imaginary parts. This leaves two equations in two unknowns: b and s_1. Solving for s_1 gives the result

$$\boxed{s_1 = \frac{\lambda}{2\pi} \tan^{-1} \left\{ \left[b'_{yy} - \cot kl_2 \mp g'_{yy} \left(\frac{1}{g'_{yy} \sin^2 kl_2} - 1 \right)^{1/2} \right]^{-1} \right\}}, \qquad (5.103)$$

and

$$b = -\cot kl_2 \mp \left(\frac{1}{g'_{yy} \sin^2 kl_2} - 1 \right)^{1/2}. \qquad (5.104)$$

The second stub must have the normalized admittance $y_{s_2} = -jb = -j \cot k s_2$, or $s_2 = (\lambda/2\pi) \tan^{-1}(1/b)$, or

$$\boxed{s_2 = \frac{\lambda}{2\pi} \tan^{-1} \left\{ \left[-\cot kl_2 \mp \left(\frac{1}{g'_{yy} \sin^2 kl_2} - 1 \right)^{1/2} \right]^{-1} \right\}}, \qquad (5.105)$$

where the $-$ or $+$ signs in Equations (5.103) and (5.105) go together.

Returning to Example 5.13 we will now calculate s_1 and s_2. Equation (5.99), with $z = l_1 = 0.2\lambda$ and $y_l = 0.2 - j0.6$ gives[4]

$$y'_{yy} = g'_{yy} + jb'_{yy} = 0.247 + j0.817.$$

Using Equations (5.103) and (5.104) with $l_2 = 0.375\lambda$ gives

$$(s'_1 = 0.113\lambda) \quad \text{or} \quad (s''_1 = 0.061\lambda)$$

or

$$(s'_2 = 0.414\lambda) \quad \text{or} \quad (s''_2 = 0.042\lambda).$$

These results agree with those taken from the Smith chart.

General procedures for the double stub design are outlined below and can be used as a guide for writing simple computer programs for calculating s_1 and s_2 from given or measured data.

Given Z_l	Impedance Bridge	Slotted Line
Data $\rightarrow Z_l, R_0, \lambda$	$Z_{\text{in}}, Z_{\text{in}}^{\text{sc}}, Z_{\text{in}}^{\text{oc}}, R_0, \lambda$	VSWR, $z_{\min}, R_0, \lambda$
$\downarrow$	(5.71)	(5.65)
$y_l = R_0/Z_l$	$\downarrow$	$\downarrow$
$\downarrow$	Z_l	Γ_l
(5.99)	$\downarrow$	(5.66)
$\downarrow$	$y_l = R_0/Z_l$	$\downarrow$
y'_{yy}	$\downarrow$	Z_l
↙ ↘	(5.99)	$\downarrow$
(5.103) (5.105)	$\downarrow$	$y_l = R_0/Z_l$
$\downarrow$ $\downarrow$	y'_{yy}	$\downarrow$
s_1 s_2	↙ ↘	(5.99)
	(5.103) (5.105)	$\downarrow$
	$\downarrow$ $\downarrow$	y'_{yy}
	s_1 s_2	↙ ↘
		(5.103) (5.105)
		$\downarrow$ $\downarrow$
		s_1 s_2

5.13 PULSES ON THE LOSSLESS LINE

The time delay encountered by a sinusoidal signal traveling a distance l on a lossless line is $\tau_D = l/u_e = l/u_p = l/u_g$. The group velocity is $u_g = 1/(d\beta/d\omega) = 1/(dk/d\omega) = 1/\sqrt{\mu\varepsilon}$, so all the velocities are the same and independent of frequency. Thus, we have

$$\boxed{\tau_D = l\sqrt{\mu\varepsilon} = l\sqrt{LC}} \qquad (5.106)$$

5.13 PULSES ON THE LOSSLESS LINE

and, since all frequency components that make up a pulse (for example) travel at the same velocity, the *pulse shape* is preserved as it travels, and the line is said to be *distortionless*.

Another way to reach the same conclusion from a somewhat different approach is to note that the phase constant for a lossless line is the *intrinsic* phase constant $k = \omega\sqrt{\mu\varepsilon} = \omega\sqrt{LC}$. The phase shift over a distance l is $-kl = -\omega\sqrt{LC}\,l$ since it represents a phase *lag* in the direction of propagation. The time delay may now be defined in a general way as

$$\tau_D = -\frac{d}{d\omega}(\text{phase shift}) = \text{group time delay}. \tag{5.107}$$

For the present case, we have

$$\tau_D = -\frac{d}{d\omega}(-\omega\sqrt{LC}\,l) = l\sqrt{LC}$$

as before. We may now conclude that if the phase constant is linear function of frequency, or equivalently, if the phase velocity is independent of frequency, then the time delay is independent of frequency and the line is distortionless.

The term $(1 - \Gamma_l\Gamma_g e^{j2kl})^{-1}$ that appears in Equation (5.47) can be put in series form by simple long division so that when Equation (5.47) is substituted into Equations (5.43) and (5.44), we obtain

$$V(z, \omega) = \frac{V_g R_0}{Z_g + R_0}\Big(e^{-jk(l-z)} + \Gamma_l e^{-jk(l+z)} + \Gamma_l\Gamma_g e^{-jk(3l-z)} + \Gamma_l^2\Gamma_g e^{-jk(3l+z)}$$
$$+ \Gamma_l^2\Gamma_g^2 e^{-jk(5l-z)} + \Gamma_l^3\Gamma_g^2 e^{-jk(5l+z)} + \ldots\Big), \tag{5.108}$$

$$I(z, \omega) = \frac{V_g}{Z_g + R_0}\Big(e^{-jk(l-z)} - \Gamma_l e^{-jk(l+z)} + \Gamma_l\Gamma_g e^{-jk(3l-z)} - \Gamma_l^2\Gamma_g e^{-jk(3l+z)}$$
$$+ \Gamma_l^2\Gamma_g^2 e^{-jk(5l-z)} - \Gamma_l^3\Gamma_g^2 e^{-jk(5l+z)} + \ldots\Big). \tag{5.109}$$

The various traveling waves (and their associated phase shifts or time delays) that are due to reflections from the ends of the line can be identified in these equations.

■ Example 5.14

Suppose that $v_g(t)$ is a 3-V peak pulse that is 1 ns wide and begins at $t = 0$. The lossless (air) line has $Z_g = R_0 = 50\,\Omega$ ($\Gamma_g = 0$), $Z_l = 25\,\Omega$ ($\Gamma_l = -\frac{1}{3}$), and is 600 m in length (said to be 2 μs long since $u = 300$ m/μs). We want to plot the voltage at the middle of the line $v(300, t)$ as a function of time. The line is shown in Figure 5.35. Since the pulse width is so small, we will show the pulses as vertical lines. Taking an *informal* approach, we find by voltage division that the first pulse that reaches the middle of the line at 1 μs is given by $[(3)(50)/(50 + 50)] = 1.5$ V peak. This pulse travels to the end of the line where it arrives after a total time of 2 μs. There, it is reflected as pulse that has a peak value of $(-1/3)(1.5) =$

−1/2 V. This reflected pulse then arrives back at the middle of the line after a total time of 3 μs. After a total time of 4 μs this pulse arrives at the generator where it is completely absorbed ($\Gamma_g = 0$). The voltage $v(300, t)$ is shown in Figure 5.35.

Equation (5.108) gives the *phasor* result:

$$V(300, \omega) = \frac{V_g(\omega)}{2}\left(e^{-j10^{-6}\omega} - \tfrac{1}{3}e^{-j(3\times10^{-6})\omega}\right).$$

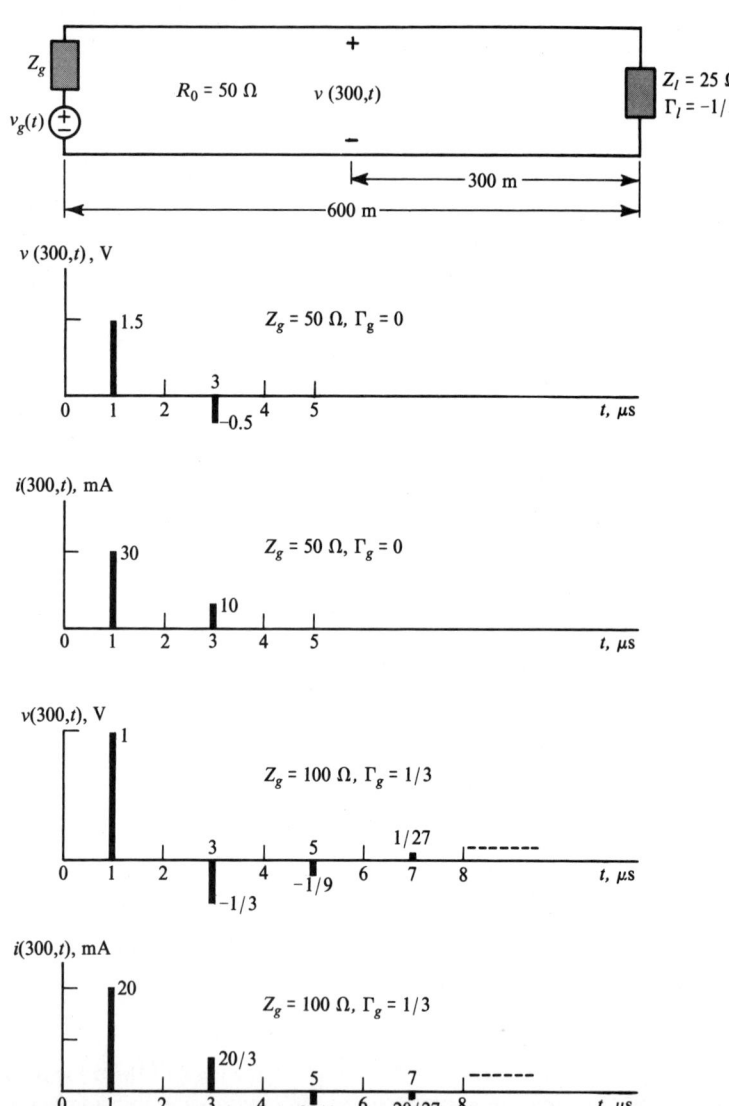

Figure 5.35. Voltage and current pulses at the middle of a 600-m long line.

This *formal* approach has left us with a problem. What is $V_g(\omega)$? Now $v_g(t)$ is the 3-V peak pulse described at the beginning of this example, and this is *not* $V_g(\omega)$. Actually, $v_g(t)$ is the *inverse Fourier transform* of $V_g(\omega)$, and, if we all were all familiar with Fourier methods, it would be a very simple matter to obtain $v(300,t)$ from the preceding equation. Here, we will be satisfied to obtain $v(300,t)$ from the discussion in the preceding paragraph:

$$V(300,t) = \tfrac{1}{2}[v_g(t - 10^{-6}) - \tfrac{1}{3}v_g(t - 3 \times 10^{-6})].$$

The current at the middle of the line $i(300,t)$ consists of a 30-mA (3/100) pulse at $t = 1$ μs followed by a 10-mA [3/100][1/3] pulse at $t = 3$ μs.

If Z_g is changed to 100 Ω ($\Gamma_g = \tfrac{1}{3}$), then $v(300,t)$ consists of a 1-V [3(50)/(50 + 100)] pulse at $t = 1$ μs, a $-\tfrac{1}{3}$-V [3(50)/(50 + 100)][$-\tfrac{1}{3}$] pulse at $t = 3$ μs, a $-\tfrac{1}{9}$-V [3(50)/150][+1/3][1/3] at $t = 5$ μs, a $+\tfrac{1}{27}$-V pulse at $t = 7$ μs, and so on. $i(300,t)$ consists of a 20-mA [3/(50 + 100)] pulse at 1 μs, a $\tfrac{20}{3}$-mA pulse at $t = 3$ μs, a $-\tfrac{20}{9}$-mA pulse at $t = 5$ μs, a $-\tfrac{20}{27}$-mA pulse at $t = 7$ μs, and so on. ∎

$v(z,t)$ and $i(z,t)$ can be found for any $v_g(t)$ by using Equations (5.108) and (5.109) with Fourier methods. It should be mentioned, however, that if $Z_l(\omega)$ or $Z_g(\omega)$ are complicated functions of ω, numerical integration will probably be required. The *fast Fourier transform* (FFT) is particularly useful in this regard.

5.14 LOSSES

The losses in a given material are extremely difficult to categorize in the general case, and, here, we will be satisfied with a very simple treatment. We will define a frequency-dependent complex permittivity $\varepsilon(\omega) = \varepsilon_1(\omega) - j\varepsilon_2(\omega) = |\varepsilon(\omega)|e^{-j\delta_d}$. $\varepsilon_1(\omega)$ is the ac *capacitivity*, $\varepsilon_2(\omega) > 0$ is the *dielectric loss factor*, and δ_d is the *dielectric loss angle*. Thus, Maxwell's curl **H** equation may be written for a linear, isotropic, and homogeneous medium

$$\nabla \times \mathbf{H} = [(\sigma + \omega\varepsilon_2) + j\omega\varepsilon_1]\mathbf{E}.$$

If one is attempting to measure σ, ε_2, or ε_1, then it may be very difficult to distinguish them. In the same way, $\mu(\omega) = \mu_1(\omega) - j\mu_2(\omega) = |\mu(\omega)|e^{j\delta_m}$, where $\mu_1(\omega)$ is the ac *inductivity*, $\mu_2(\omega)$ is the *magnetic loss factor*, and δ_m is the *magnetic loss angle*. Thus, under the same conditions as above, Maxwell's curl **E** equation may be written

$$\nabla \times \mathbf{E} = [-j\omega(\mu_1 - j\mu_2)]\mathbf{H}.$$

For our purposes, and in the interest of simplicity, we will neglect the magnetic loss ($\mu_2 = 0$, $\mu = \mu_1$), lump the two terms that contribute to loss together, calling the sum σ ($\sigma + \omega\varepsilon_2 \to \sigma$), and, finally call ε_1 by the name ε ($\varepsilon_1 \to \varepsilon$). In this case Maxwell's equations reduce to the form that was used in Chapter 4:

$$\nabla \times \mathbf{E} = -j\omega\mu\mathbf{H}, \tag{5.110}$$

$$\nabla \times \mathbf{H} = j\omega\varepsilon'\mathbf{E}, \tag{5.111}$$

where $\varepsilon' \equiv \varepsilon(1 + \sigma/j\omega\varepsilon)$ and $\sigma/(\omega\varepsilon)$ is the *loss tangent*.

A damped uniform plane wave propagating in the $+z$ direction and satisfying Equations (5.110) and (5.111) is given by the set

$$E_x(z,\omega) = E^+ e^{-\alpha z} e^{-j\beta z}, \qquad H_y(z,\omega) = (E^+/\eta) e^{-\alpha z} e^{-j\beta z}.$$

This is *still* a TEM *to z mode*, and, furthermore, it can be supported by a pair of *perfectly* conducting parallel planes as in Section 5.1. Thus, we conclude that a uniform two-conductor transmission line can support the TEM mode in a lossy dielectric if the conductors are perfectly conducting.

If we next recognize that real conductors are not perfectly conducting, then a TEM mode is no longer possible because a component of $\mathbf{E}$ must exist *inside* the conductor in the direction of the current density ($\mathbf{J} = \sigma\mathbf{E}$) which is the same as the direction of propagation. Continuity of tangential $\mathbf{E}$ at the conductor surface then requires a component of $\mathbf{E}$ *outside* the conductor in the direction of propagation. Thus, if propagation is in the $+z$ direction, then $E_z \neq 0$, and this is not a TEM to z mode.

Faced with this situation, we can either attempt an exact solution, which will not be easy because of the complicated boundary conditions, or we can make suitable approximations. The engineer is quite often willing to choose the latter course. In pursuing that course here, we will use a *perturbation* method. If the loss in the conductors is small, then the field external to the conductors is not altered, or perturbed, much from the field in the lossless case. An alternate statement is that if the losses are small, the power transmitted along the *system* is much greater than the power lost in the conductors of the *system*. (The use of the word *system* is meant to indicate that the technique being developed here applies to any electromagnetic guiding system.) If the magnetic field at the conductor surface is that of the perfect conductor case (as it almost is), then we can calculate the *power lost* as was done in Chapter 4. The *power flow* along the system may also be calculated using the lossless values of $\mathbf{E}$ and $\mathbf{H}$. With this information, an attenuation constant α may be obtained in the following *general* way.

The time-average power flow along the system is proportional to $e^{+2\alpha z}$, for a case where the power flow is in the $-z$ *direction*. Thus, we have

$$<\mathcal{P}_f> = P_0 e^{+2\alpha z} \quad \text{(W)}. \tag{5.112}$$

The time-average power dissipated per unit system length $<\mathcal{P}_d>$, is the rate of decrease of $<\mathcal{P}_f>$ with $-z$, or

$$<\mathcal{P}_d> = -\frac{\partial <\mathcal{P}_f>}{\partial(-z)} = 2\alpha P_0 e^{+2\alpha z}$$

or

5.14 LOSSES

$$<\mathcal{P}_d> = 2\alpha <\mathcal{P}_f> \quad \text{(W/m)}. \tag{5.113}$$

Therefore, we have

$$\boxed{\alpha = \frac{1}{2}\frac{<\mathcal{P}_d>}{<\mathcal{P}_f>}} \quad \text{(Np/m)}. \tag{5.114}$$

Equation (5.114) is exact. The results become approximate when we use approximate values for $<\mathcal{P}_d>$ and $<\mathcal{P}_f>$.

Besides the obvious effect of adding series resistance to the equivalent circuit of the line, another effect of finite conductor conductivity is the addition of an *internal inductance*. This occurs because the *internal* magnetic flux cannot link all of the current. Insofar as the transmission line equivalent circuit and equations are concerned, the internal and external inductances are merely added. The equivalent circuit of the general line is as shown in Figure 5.36.

Following the same procedure as was used for the lossless line, we get

$$V(z,\omega) = V^-(e^{+(\alpha+j\beta)z} + \Gamma_l e^{-(\alpha+j\beta)z}) \tag{5.115}$$

and

$$I(z,\omega) = \frac{V^-}{Z_0}(e^{+(\alpha+j\beta)z} - \Gamma_l e^{-(\alpha+j\beta)z}), \tag{5.116}$$

where

$$\boxed{\Gamma_l = \frac{Z_l - Z_0}{Z_l + Z_0} = |\Gamma_l|e^{j\phi_l}}, \tag{5.117}$$

$$\boxed{Z_0 = \sqrt{\frac{Z}{Y}} = \sqrt{\frac{R+j\omega L}{G+j\omega C}}}, \tag{5.118}$$

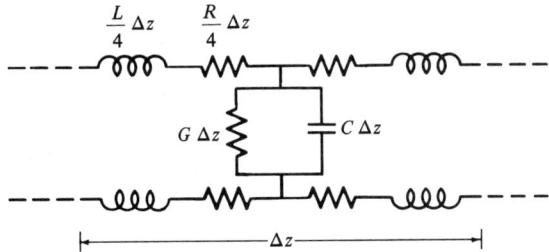

Figure 5.36. Equivalent circuit of distributed parameters for a general transmission line.

$$\gamma = \sqrt{ZY} = \sqrt{(R + j\omega L)(G + j\omega C)} = \alpha + j\beta, \qquad (5.119)$$

$$\alpha = \operatorname{Re}(\gamma) = \{[RG - \omega^2 LC + \sqrt{(RG - \omega^2 LC)^2 + \omega^2 (LG + RC)^2}]/2\}^{1/2}, \qquad (5.120)$$

and

$$\beta = \operatorname{Im}(\gamma) = \{[\omega^2 LC - RG + \sqrt{(RG - \omega^2 LC)^2 + \omega^2 (LG + RC)^2}]/2\}^{1/2}. \qquad (5.121)$$

Taking the time-domain value of the $-z$ traveling wave from the phasor form [Equation (5.115)]

$$v(z,t) = \operatorname{Re}(V^{-} e^{+j\omega t} e^{+\alpha z} e^{+j\beta z})$$

or, when V^{-} is real,

$$v(z,t) = V^{-} e^{+\alpha z} \cos(\omega t + \beta z). \qquad (5.122)$$

The phase velocity is obtained as in the lossless case and is

$$u_p = \frac{\omega}{\beta} = \frac{\omega}{\operatorname{Im}(\gamma)} = \frac{\omega}{\operatorname{Im}\left\{\sqrt{(R + j\omega L)(G + j\omega C)}\right\}}. \qquad (5.123)$$

Note that, in general, u_p *is a function of* frequency, and thus, the line is not (in general) distortionless! This is another effect of the loss.

Taking the generator characteristics into account will eliminate the constant V^{-} from Equations (5.115) and (5.116). This is done just like it was for the lossless case (Section 5.4) but rather than following that procedure, it is easier to simply replace R_0 by Z_0 and jk by $\alpha + j\beta$ in Equation (5.47). This results in

$$V^{-}(\omega) = \frac{V_g Z_0}{Z_g + Z_0} \frac{e^{-(\alpha + j\beta)l}}{1 - \Gamma_l \Gamma_g e^{-j2(\alpha + j\beta)l}}, \qquad (5.124)$$

where

$$\Gamma_g = \frac{Z_g - Z_0}{Z_g + Z_0}. \qquad (5.125)$$

5.14 LOSSES

Also, the input impedance is the ratio of V to I and is

$$Z_{in} = Z_0 \frac{e^{+(\alpha+j\beta)z} + \Gamma_l e^{-(\alpha+j\beta)z}}{e^{+(\alpha+j\beta)z} - \Gamma_l e^{-(\alpha+j\beta)z}}. \tag{5.126}$$

The equivalent circuit, as far as the generator is concerned, is identical to Figure 5.11. Note carefully, however, that the input impedance in the present case is given by Equation (5.126) with $z = l$, or in a more convenient computational form

$$Z_{in} = Z_0 \frac{1 + \Gamma_l e^{-2(\alpha+j\beta)l}}{1 - \Gamma_l e^{-2(\alpha+j\beta)l}}. \tag{5.127}$$

As far as the load is concerned, the Thévenin equivalent circuit is found as before, and is identical to Figure 5.10, but in the present case, we have

$$V_T = V_g \frac{2Z_0}{Z_0 + Z_g} \frac{e^{-2(\alpha+j\beta)l}}{1 - \Gamma_g e^{-2(\alpha+j\beta)l}} \tag{5.128}$$

and

$$Z_T = Z_0 \frac{1 + \Gamma_g e^{-2(\alpha+j\beta)l}}{1 - \Gamma_g e^{-2(\alpha+j\beta)l}}. \tag{5.129}$$

■ Example 5.15

We return to Example 5.1, where $V_g = 10$ V, $Z_g = 100 + j50$ Ω, $Z_l = 75 + j100$ Ω, and $l = 5\lambda$. We now assume that $Z_0 = 50\underline{|0°}$ Ω, $\alpha = 0.01$ Np/m and $\beta = 2\pi/\lambda$ (rad/m), with $\lambda = 1$ m. We shall show a little later that this is a *distortionless line* because Z_0 is real. The average load power is desired, and we first calculate Γ_g to find the Thévenin equivalent circuit parameters

$$\Gamma_g = \frac{Z_g - Z_0}{Z_g + Z_0} = \frac{50 + j50}{150 + j50} = 0.447 e^{+j0.464}.$$

Equations (5.128) and (5.129) give

$$V_T = 10 \frac{100}{150 + j50} \frac{e^{-0.1 - j20\pi}}{1 - 0.447 e^{-0.1 - j(20\pi - 0.464)}},$$

$$V_T = 8.629 \underline{|-2.6°} \quad (V)$$

and

$$Z_T = 50\frac{1 + 0.447e^{-0.1-j(20\pi-0.464)}}{1 - 0.447e^{-0.1-j(20\pi-0.464)}},$$

$$Z_T = 103.58\underline{|23.40°}\quad(\Omega),$$

$$I_l = \frac{V_T}{Z_T + Z_l} = 39.04\underline{|-42.29°}\quad\text{mA},$$

$$<\mathcal{P}_l> = \tfrac{1}{2}|I_l|^2 R_l = 57.17\quad\text{mW}.$$

This is the average power dissipated in the load. The average power dissipated in the *line and load* can be found from the generator end:

$$\Gamma_l = \frac{Z_l - Z_0}{Z_l + Z_0} = \frac{25 + j100}{125 + j100} = 0.644e^{-j0.651},$$

$$Z_{in} = 50\,\frac{1 + 0.644e^{-0.1-j(20\pi-0.651)}}{1 - 0.644e^{-0.1-j(20\pi-0.651)}} = 117.19\underline{|46.91°}\quad\Omega,$$

$$I_{in} = \frac{V_g}{Z_g + Z_{in}} = 44.37\underline{|-36.98°}\quad\text{mA},$$

$$<\mathcal{P}_d> = \tfrac{1}{2}|I_{in}|^2 \text{Re}\{Z_{in}\} = 78.79\quad\text{mW}.$$

Thus, the power lost in the line is $78.79 - 57.17 = 21.62$ mW. ∎

If $Z_g = Z_0$, then $\Gamma_g = 0$ and the generator is matched to the line. In this case, Equations (5.115) and (5.116) with Equation (5.124) become

$$V = \frac{V_g}{2}e^{-(\alpha+j\beta)l}(e^{+(\alpha+j\beta)z} + \Gamma_l e^{-(\alpha+j\beta)z}) \tag{5.130}$$

and

$$I = \frac{V_g}{2Z_0}e^{-(\alpha+j\beta)l}(e^{+(\alpha+j\beta)z} - \Gamma_l e^{-(\alpha+j\beta)z}). \tag{5.131}$$

Some special cases with $Z_g = Z_0$ may now be investigated and compared to their lossless counterparts. V_g is real. A plot of $|V|$ versus z ($Z_g = Z_0$, $Z_l \to \infty$) is shown in Figure 5.37 as an example of the general behavior. Note that the VSWR

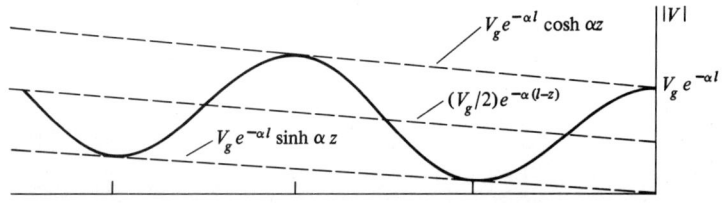

Figure 5.37. $|V|$ versus z, $Z_g = Z_0$, $Z_l \to \infty$, small attenuation.

is rather nebulous. When the loss is small then the lossless value of the VSWR [Equation (5.60)] is usually quoted. Also note that maximum power transfer to the load occurs when $Z_l = Z_0$ because there is no reflection from the load.

Open Circuit	Short Circuit	Matched Case
$Z_l \to \infty$	$Z_l = 0$	$Z_l = Z_0$
$\Gamma_l = +1$	$\Gamma_l = -1$	$\Gamma_l = 0$
$Z_{in}^{oc} = Z_0 \coth \gamma z$	$Z_{in}^{sc} = Z_0 \tanh \gamma z$	$Z_{in} = Z_0$
$\|V\| = V_g e^{-\alpha l} \|\cosh \gamma z\|$	$\|V\| = V_g e^{-\alpha l} \|\sinh \gamma z\|$	$\|V\| = \dfrac{V_g}{2} e^{-\alpha(l-z)}$
$\|I\| = \dfrac{V_g}{\|Z_0\|} e^{-\alpha l} \|\sinh \gamma z\|$	$\|I\| = \dfrac{V_g}{\|Z_0\|} e^{-\alpha l} \|\cosh \gamma z\|$	$\|V\| = \dfrac{V_g}{2\|Z_0\|} e^{-\alpha(l-z)}$

5.15 SPECIAL TWO-WIRE LINES

Some special values of R, L, G, and C, the distributed line parameters, give special results and are summarized here.

(1) $R = G = 0$. This is the lossless case.

(2) A distortionless line can be approximated in spite of the loss. Distortionless conditions are met if the time delay is frequency independent, or equivalently [see Equation (5.106)] if β is directly proportional to frequency and if α is frequency independent. If it is possible to make

$$LG = RC; \qquad (5.132)$$

then Equations (5.121) and (5.120) reduce to

$$\beta = \omega \sqrt{LC} \qquad (5.133)$$

and

$$\alpha = \sqrt{RG} \qquad (5.134)$$

and a distortionless line results. At frequencies where it is practical to "load" a line to satisfy Equations (5.132), it is usually true that $R \gg G$. Then L is usually increased by adding lumped inductance to the line at periodic intervals. Reducing R has the same effect, but is more expensive. Why?

(3) If the line loss is small, that is, $R \ll \omega L$ and $G \ll \omega C$, then a binomial expansion may be advantageously employed. From Equation (5.118), we get

$$Z_0 = \sqrt{\dfrac{L}{C}} \sqrt{\dfrac{1 + R/j\omega L}{1 + G/j\omega C}}$$

or

$$Z_0 = R_0\left(1 + \frac{R}{j2\omega L} + \cdots\right)\left(1 - \frac{G}{j2\omega C} - \cdots\right).$$

Retaining only the first-order terms, we get

$$Z_0 \approx R_0\left[1 + j\left(\frac{G}{2\omega C} - \frac{R}{2\omega L}\right)\right]. \quad (5.135)$$

In the same way, we have

$$\gamma \approx \frac{R}{2R_0} + \frac{GR_0}{2} + j\omega\sqrt{LC}. \quad (5.136)$$

The results of these cases are summarized in Table 5.1.

According to Equation (5.115), the voltage at some point z on the line (measured from the load) for $Z_l = Z_0$ ($\Gamma_l = 0$) is

$$V = \frac{V_g Z_0}{Z_0 + Z_g} e^{-(\alpha + j\beta)(l-z)}$$

or

$$\frac{V(z,\omega)}{V_g(\omega)} = H(\omega) = \frac{Z_0(\omega)}{Z_0(\omega) + Z_g(\omega)} e^{-[\alpha(\omega) + j\beta(\omega)](l-z)}. \quad (5.137)$$

$H(\omega)$ is the *transfer function* for the line when $v(z,t)$ is the response or "output" and $v_g(t)$ is the excitation or "input." It is simply the ratio of the *phasor* output to the phasor input.

TABLE 5.1
Special Cases—Transmission Line Losses

Case	Z_0	$\gamma = \alpha + j\beta$
Lossless	$\sqrt{\dfrac{L}{C}} \equiv R_0$	$j\omega\sqrt{LC} = j\beta,\ \alpha = 0$
Distortionless $\dfrac{L}{C} = \dfrac{R}{G}$	$\sqrt{\dfrac{L}{C}} = \sqrt{\dfrac{R}{G}} = R_0$	$\sqrt{RG} + j\omega\sqrt{LC}$
Small loss $R \ll \omega L,\ G \ll \omega C$	$R_0\left[1 + j\left(\dfrac{G}{2\omega C} - \dfrac{R}{2\omega L}\right)\right]$	$\dfrac{R}{2R_0} + \dfrac{GR_0}{2} + j\omega\sqrt{LC}$
General	$\sqrt{\dfrac{R + j\omega L}{G + j\omega C}}$	$\sqrt{(R + j\omega L)(G + j\omega C)}$

5.15 SPECIAL TWO-WIRE LINES

The preceding equation shows that *amplitude* distortion as well as time delay distortion may cause difficulties.[5] Amplitude distortion arises because of the presence of Z_0 in this equation. In Table 5.1, it can be seen that the lossless line has no distortion and no attenuation. The distortionless line has no distortion but does have attenuation (independent of frequency, however). The "small loss" case has no time delay distortion, but does have amplitude distortion as well as attenuation. The general case has both types of distortion as well as attenuation. A further complication which does not appear explicitly in Table 5.1 is that due to R, L, G, and C being *frequency*-dependent parameters rather than constants.

As an example, suppose that we have a pulse generator connected to a lossy line which is matched at the load end. Let v_g be the repetitive pulse shown in Figure 5.38 (an even function). The Fourier series representation of v_g is

$$v_g(t) = \frac{a_0}{2} + \sum_{n=1}^{\infty} a_n \cos n\omega_T t, \qquad (5.138)$$

where

$$a_n = \frac{V_0 \omega_T d}{\pi} \frac{\sin n\omega_T d/2}{n\omega_T d/2}, \qquad \omega_T = 2\pi/T. \qquad (5.139)$$

Since each term in Equation (5.138) represents a sinusoidal excitation, the phasor concept applies to each term separately, and superposition can then be employed. That is, the input $a_n \cos n\omega_T t$ has the phasor form $a_n \lfloor 0° $ and for this input the phasor output is $[a_n \lfloor 0°]|H(n\omega_T)| \lfloor H(n\omega_T) = a_n|H(n\omega_T)| \lfloor H(n\omega_T)$ whose time-domain form is simply $a_n|H(n\omega_T)|\cos[n\omega_T t + \lfloor H(n\omega_t)]$. Thus, by superposition,

$$v(z,t) = \frac{a_0}{2} H(0) + \sum_{n=1}^{\infty} a_n |H(n\omega_T)| \cos[n\omega_T t + \lfloor H(n\omega_T)] \qquad (5.140)$$

If R, L, C, and G are constant, then

---

[5]It has already been pointed out in Chapter 4 that the lossy line is anomalously dispersive.

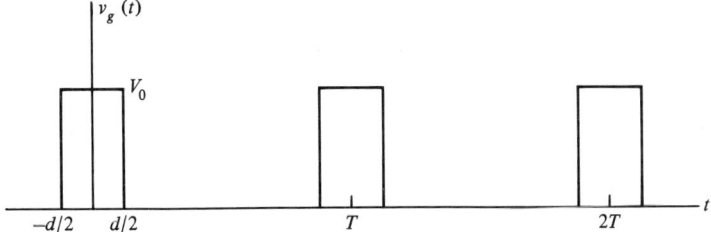

Figure 5.38. Periodic rectangular pulse representing $v_g(t)$.

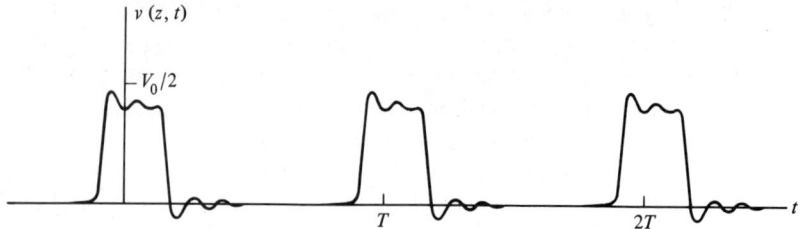

Figure 5.39. $v(z,t)$ versus t at a fixed value of z when loss is present and $v_g(t)$ is the rectangular pulse of Figure 5.38.

$$H(n\omega_T) = \frac{Z_0(n\omega_T)}{Z_0(n\omega_T) + Z_g(n\omega_T)} e^{-[\alpha(n\omega_T) + j\beta(n\omega_T)](l-z)}, \quad (5.141)$$

where

$$Z_0(n\omega_T) = \sqrt{\frac{R + jn\omega_T L}{G + jn\omega_T C}}, \quad (5.142)$$

$$\gamma(n\omega_T) = \alpha(n\omega_T) + j\beta(n\omega_T) = \sqrt{(R + jn\omega_T L)(G + jn\omega_T C)}. \quad (5.143)$$

Results for a typical case are shown in Figure 5.39. The degradation of the input pulse because of the attenuation and distortion is obvious.

5.16 QUALITY FACTOR

A general definition for the quality factor Q of a propagating system is

$$\boxed{Q \equiv \omega \frac{<\mathcal{W}>}{<\mathcal{P}_d>}}, \quad (5.144)$$

where $<\mathcal{W}>$ is the time-average total energy stored per unit length of the system (joules per meter) and $<\mathcal{P}_d>$ is the time-average power dissipated per unit length of the system (watts per meter). Equation (5.113) may now be used in Equation (5.144), giving

$$Q = \frac{1}{2\alpha} \frac{\omega <\mathcal{W}>}{<\mathcal{P}_f>}, \quad (5.145)$$

where $<\mathcal{P}_f>$ is the time-average power flow in the system (watts). The velocity of energy flow as defined in Equation (4.33) is $u_e = <\mathcal{P}_f> / <\mathcal{W}>$, so that $<\mathcal{P}_f> = u_e <\mathcal{W}>$ and thus

5.16 QUALITY FACTOR

$$Q = \frac{1}{2\alpha}\frac{\omega}{u_e}. \tag{5.146}$$

Equation (5.146) is exact[6] and applies to *any* propagating system.

We now assume that the losses are small and the mode of propagation on the line is *essentially* TEM. *In this case*, $u_e \approx \omega/\beta$, and Equation (5.146) becomes

$$Q = \frac{\beta}{2\alpha}, \tag{5.147}$$

which is *independent of length*. Since we have already assumed that α is small, the Q may be large, perhaps in the neighborhood of 10^3.

If we use the circuit theory result,

$$Q = \frac{\text{resonant frequency}}{\text{half-power bandwidth}} \tag{5.148}$$

instead of Equation (5.144), and apply this to a resonant line length, the same result as that given by Equation (5.147) is obtained.

The Q of a transmission line, along with Z_0, may be determined by measurements similar to those used to obtain R_0. (The line will always have loss, and whether it is neglected or not, is a matter of judgment.) These measurements consist of determining the input impedance with short-circuit and open-circuit load impedances. If these impedances have real parts then the line is bossy. The data listed below Equation (5.131) indicate that

$$Z_{\text{in}}^{\text{oc}} = Z_0 \coth \gamma z \tag{5.149}$$

and

$$Z_{\text{in}}^{\text{sc}} = Z_0 \tanh \gamma z, \tag{5.150}$$

so that

$$Z_0 = \sqrt{Z_{\text{in}}^{\text{sc}} Z_{\text{in}}^{\text{oc}}}. \tag{5.151}$$

Substituting for Γ_l in Equation (5.126) using Equation (5.117):

$$Z_{\text{in}} = Z_0 \frac{Z_l + Z_0 \tanh \gamma z}{Z_0 + Z_l \tanh \gamma z}. \tag{5.152}$$

---

[6]Note that both $<\mathcal{P}_f>$ and $<\mathcal{W}>$ depend on z, so u_e is somewhat nebulous.

Solving for Z_l in terms of the measured quantities Z_{in}, Z_{in}^{sc}, Z_{in}^{oc} gives

$$Z_l = Z_{in}^{oc} \frac{Z_{in} - Z_{in}^{sc}}{Z_{in}^{oc} - Z_{in}}, \tag{5.153}$$

which is identical in form to Equation (5.71). The quantity $\tanh \gamma z$ can be found as Z_{in}^{sc}/Z_0 or as the square root of the ratio of Equations (5.150) to (5.149):

$$\tanh(\alpha z + j\beta z) = \sqrt{Z_{in}^{sc}/Z_{in}^{oc}} = Z_{in}^{sc}/Z_0 = A + j B \tag{5.154}$$

or

$$\alpha z + j\beta z = \tanh^{-1}\sqrt{Z_{in}^{sc}/Z_{in}^{oc}}. \tag{5.155}$$

It can be shown that[7]

$$\alpha z = \frac{1}{2}\tanh^{-1}\left(\frac{2A}{1 + A^2 + B^2}\right) \tag{5.156}$$

and

$$\beta z = \frac{1}{2}\tan^{-1}\left(\frac{2B}{1 - A^2 - B^2}\right) + m\pi/2, \qquad m = 0, 1, 2, \ldots. \tag{5.157}$$

Equation (5.157) has many solutions, one of which is correct. That is, it may be difficult to tell the difference between a short, but very lossy line, and a long line with little loss. An observation of the physical length will usually determine the correct solution. If this cannot be done, a low-frequency measurement of the total shunt capacitance of the line section will determine the correct solution. Using Equations (5.147), (5.156), and (5.157), we have

$$Q = \frac{1}{2}\frac{\tan^{-1}[2B/(1 - A^2 - B^2)] + m\pi}{\tanh^{-1}[2A/(1 + A^2 + B^2)]} \tag{5.158}$$

Suppose we want to determine the input characteristics for a resonant shorted line with "small losses." We assume that shorted means $Z_l = 0$, even though a short piece of wire or a short plate used in practice to provide the "short circuit" does not actually represent zero impedance:

$$Z_{in}^{sc} = Z_0 \tanh \gamma z = (R_0 + jX_0)\frac{\tanh(\alpha z) + j\tan(\beta z)}{1 + j\tanh(\alpha z)\tan \beta z} = R_{in}^{sc} + jX_{in}^{sc}. \tag{5.159}$$

---

[7] See Everitt, 1937 in the references at the end of this chapter.

5.16 QUALITY FACTOR

Rationalizing and then equating the imaginary part to zero gives

$$\frac{\tan \beta z}{\cosh^2 \alpha z} = -\frac{X_0 \tanh \alpha z}{R_0 \cos^2 \beta z} \tag{5.160}$$

or

$$\boxed{\sin 2\beta z = -\frac{X_0}{R_0} \sinh 2\alpha z}, \tag{5.161}$$

an *exact* equation that must be solved for $z = z_0$ to find the exact resonant length. Equation (5.159) can then be used to find $Z_{\text{in}}^{\text{sc}}$.

We next use the fact that the loss is small. Now $\sin 2\beta z = \sin(\pi - 2\beta z) \approx \pi - 2\beta z$, since $\pi - 2\beta z = \pi - 4\pi z/\lambda$ is very small ($z = z_0 \approx \lambda/4$). Also $\sinh 2\alpha z \approx 2\alpha z$, since $2\alpha z$ is also small. Equation (5.161) then reduces to the approximate result

$$\pi - 2\beta z \approx -\frac{X_0}{R_0}(2\alpha z) \tag{5.162}$$

or

$$\boxed{z_0 \approx \frac{\lambda/4}{1 - (X_0/R_0)(\alpha/\beta)}}. \tag{5.163}$$

Using Equation (5.159), we get

$$R_{\text{in}}^{\text{sc}} = \left(R_0 \frac{\tanh \alpha z}{\cos^2 \beta z} - X_0 \frac{\tan \beta z}{\cosh^2 \alpha z} \right) / (1 + \tanh^2 \alpha z \tan^2 \beta z).$$

Using Equation (5.160), and the fact that αz is small, gives

$$R_{\text{in}}^{\text{sc}} \approx -X_0 \frac{(1 + R_0^2/X_0^2) \tan \beta z}{1 + (\alpha z)^2 \tan^2 \beta z} \tag{5.164}$$

At the resonant length, Equation (5.162) gives $\beta z_0 \approx \pi/2 + \alpha z_0 X_0/R_0$, and $\tan \beta z_0 = -[\tan(\alpha z_0 X_0/R_0)]^{-1} \approx -R_0/(X_0 \alpha z_0)$. In this case, Equation (5.164) reduces to

$$\boxed{R_{\text{in}}^{\text{sc}} \approx 4R_0/(\alpha\lambda)}. \tag{5.165}$$

The resonant length [Equation (5.163)] and input resistance can be written in terms of the distributed parameters by using Table 5.1:

$$z_0 = \frac{\lambda/4}{1 + \frac{1}{4}[(G/C)^2 - (R/L)^2]}, \qquad (5.166)$$

$$R_{in}^{sc} = \frac{8}{\lambda}\left(\frac{1}{C}\right)\frac{1}{R/L + G/C}. \qquad (5.167)$$

Note that for the distortionless line, Equation (5.166) gives $z_0 = \lambda/4$.
The resonant Q is determined by $\beta/2\alpha$, and using Table 5.1:

$$Q_0 = \frac{\beta}{2\alpha} = \frac{\omega_0 \sqrt{LC}}{R/R_0 + GR_0} = \frac{\omega_0}{R/L + G/C}$$

or

$$\boxed{\frac{1}{Q_0} = \frac{R}{\omega_0 L} + \frac{G}{\omega_0 C} = \frac{1}{Q_c} + \frac{1}{Q_d}}. \qquad (5.168)$$

That is, $Q_c = \omega_0 L/R$ is the Q due to *conductor* losses, while $Q_d = \omega_0 C/G = \omega_0 \varepsilon/\sigma_d$ is the Q due to dielectric losses, and is, in fact, just the Q of the dielectric.

■ Example 5.16

Measurements on a transmission line show that

$$Z_0 = 50.5 \underline{|1.5°}\ \Omega,$$

$$Q = \beta/2\alpha = 95 \quad \text{and} \quad \lambda = 1 \text{ m} \quad \text{or} \quad \beta = 2\pi \text{ (rad/m)}.$$

These data give

$$R_0 = \text{Re}(Z_0) = 50.48\ \Omega, \quad X_0 = \text{Im}(Z_0) = 1.322\ \Omega$$

and

$$\alpha = \beta/2Q = \pi/Q = 33.07 \times 10^{-3} \quad \text{(Np/m)}.$$

The exact resonant length using Equation (5.161) (with trial and error or *root-solve* with the programmable calculator) is $z_0 = 0.250$ m, while Equation (5.163) gives the same result (actually, to more decimal places than needed).

The exact input resistance at resonance [using Equation (5.159)] is $R_{in}^{sc} = 6.106$ kΩ, and the approximate value from Equation (5.164) is the same. Thus, the approximate formulas are more than sufficient for this example.

If the original data remain unchanged, except that $Q = 10$, then $\alpha = \pi/Q = 0.1(\pi)$. The exact and approximate resonant lengths still agree ($z_0 = 0.250$ m), whereas the exact input resistance is $R_{in}^{sc} = 644.1\ \Omega$, but the approximate value is 642.8 Ω. We conclude that if $Q \geq 10$, the approximate formulas are sufficient for engineering purposes. ■

5.16 QUALITY FACTOR

■ Example 5.17

The following measurements are made on a transmission line whose length is known to be less than $\lambda/4$:

$$Z_{in}^{oc} = 28.8 \underline{|-75°} \ \Omega,$$
$$Z_{in}^{sc} = 80 \underline{|+85°} \ \Omega,$$

and

$$Z_{in} = 96 \underline{|50°} \ \Omega \quad \text{(actual load in place)}.$$

We want to find Z_l and Q. Substituting the measured values into Equation (5.153) gives

$$Z_l = \frac{28.8 \underline{|-75°} \ (96 \underline{|50°} - 80 \underline{|85°})}{28.8 \underline{|-75°} - 96 \underline{|50°}}$$

or

$$Z_l = 13.8 \underline{|36.74°} \ \Omega.$$

Using Equation (5.155) directly, or, alternatively, using Equations (5.156), (5.157), and (5.154), we find that $\alpha z = 77.2 \times 10^{-3}$ and $\beta z = 1.034$. This is the correct value of βz, since we know that $z < \lambda/4$, or $\beta z < \pi/2$. Therefore, from Equation (5.147), we get

$$Q = \frac{\beta}{2\alpha} = \frac{\beta z}{2\alpha z} = 6.695.$$

Suppose that we knew that the line was between $\lambda/2$ and $3\lambda/4$ in length, or $\pi < \beta z < 3\pi/2$. In this case the correct value of βz is $1.034 + \pi$ or $\beta z = 4.176$ according to Equation (5.157). Thus,

$$Q = \frac{\beta}{2\alpha} = \frac{\beta z}{2\alpha z} = 27.04.$$

Most quantities can be determined rapidly using the Smith chart. The question now arises as to what modifications must be made in the use of the chart to account for the losses. The chart was originally derived using Equation (5.85), or

$$z_{in} = \frac{1 + |\Gamma_l| e^{j(\phi_l - 2kz)}}{1 - |\Gamma_l| e^{j(\phi_l - 2kz)}} = r_a + j x_a \tag{5.169}$$

for the lossless line. For the *general* line Equation (5.126) gives

$$z_{in} = \frac{Z_{in}}{Z_0} = \frac{1 + \Gamma_l e^{-2(\alpha + j\beta)z}}{1 - \Gamma_l e^{-2(\alpha + j\beta)z}}. \tag{5.170}$$

Equation (5.170) may be written

$$z_{in} = \frac{1 + (|\Gamma_l|e^{-2\alpha z})e^{j(\phi_l - 2\beta z)}}{1 - (|\Gamma_l|e^{-2\alpha z})e^{j(\phi_l - 2\beta z)}}. \tag{5.171}$$

A comparison of Equations (5.169) and (5.171) reveals that k has been replaced by β and $|\Gamma_l|$ has been replaced by $|\Gamma_l|e^{-2\alpha z}$. The effects of these changes, insofar as the Smith chart is concerned, are rather obvious. Motion on the chart, *toward the generator*, must now be accompanied by an *inward* spiral ($|\Gamma_l|e^{-2\alpha z}$), rather than constant radius motion ($|\Gamma_l|$), as in the lossless case. The effect of the damping term $e^{-2\alpha z}$ can be radially scaled, but usually transmission loss $e^{-\alpha z}$ is what is actually scaled. ∎

The same quantities as in Example 5.17 can be found using the Smith chart. From Equation (5.151), we get

$$Z_0 = 48 \,\underline{|{+5°}}.$$

The normalized measurements are

$$z_{in}^{oc} = \frac{Z_{in}^{oc}}{Z_0} = 0.6\,\underline{|{-80°}} = 0.1042 - j0.591,$$

$$z_{in}^{sc} = \frac{Z_{in}^{sc}}{Z_0} = 1.667\,\underline{|{+80°}} = 0.29 + j1.641,$$

and

$$z_{in} = \frac{Z_{in}}{Z_0} = 2\,\underline{|{45°}} = 1.414 + j1.414.$$

These points are located on the Smith chart in Figure 5.40. Both z_{in}^{oc} and z_{in}^{sc} show that $z = 0.164\lambda$ ($< \lambda/4$). Motion from the short or open circuit has followed the exponential spiral on the chart. Using the reflection coefficient scale shows that

$$|\Gamma_l|e^{-2\alpha z} = 0.857.$$

We know that $|\Gamma_l| = 1$ for either the open or short circuit case, so

$$e^{-2\alpha z} = 0.857$$

or

$$\alpha z = 0.0772 \quad (\text{Np})$$

5.16 QUALITY FACTOR

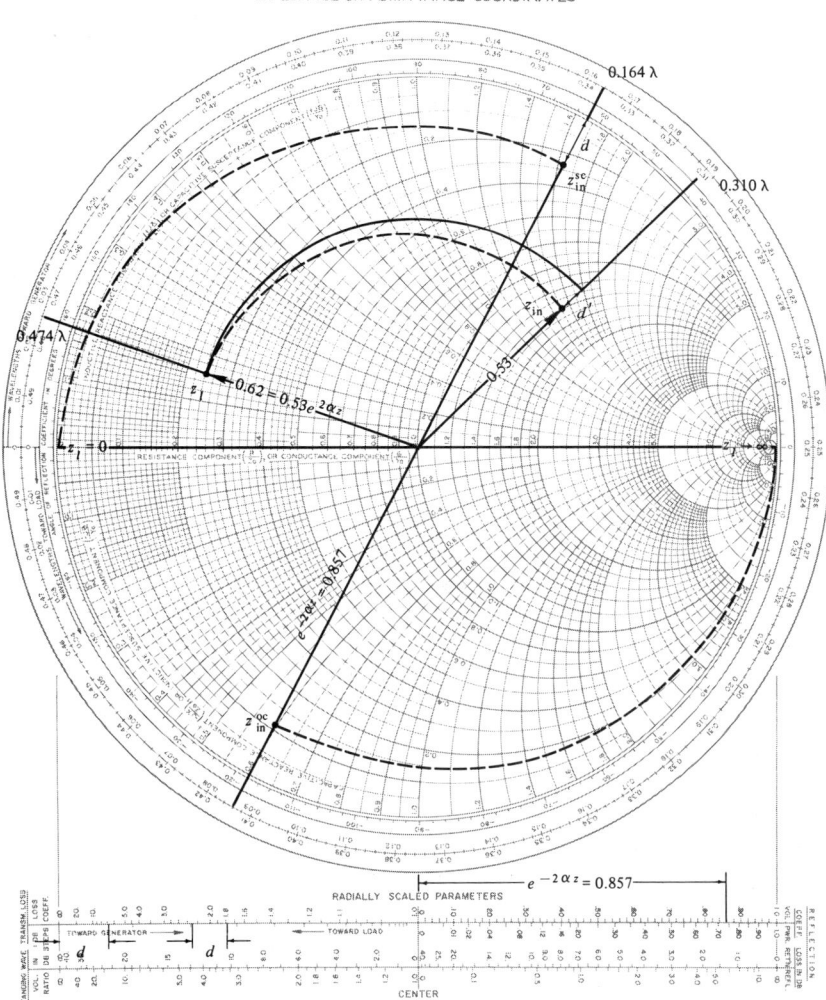

Figure 5.40. Smith chart data for Example 5.17. *Source:* Copyright © 1949 by Kay Electric Company, Pine Brook, NJ. Renewal copyright © 1976 by P. H. Smith, New Providence, NJ.

Also, we have

$$\beta z = \frac{2\pi z}{\lambda} = \frac{2\pi}{\lambda}(0.164\lambda) = 1.03 \quad \text{rad.}$$

Thus, we get

$$Q = \frac{\beta}{2\alpha} = \frac{\beta z}{2\alpha z} = 6.68.$$

z_l is determined by rotating toward the load an amount 0.164λ from z_{in} along an *outward* spiraling exponential curve. The coefficient of reflection magnitude corresponding to z_{in} is 0.53, so we must spiral out to a radius $0.53e^{+2\alpha z} = 0.53/0.857 = 0.62$. Here we find

$$z_l = 0.24 + j0.155 = 0.286 \underline{|32.86°},$$
$$Z_l = Z_0 z_l = 13.7 \underline{|37.9°} \ \Omega.$$

These answers compare favorably to those obtained by direct calculation. The *transmission loss radial scale* may be used instead of the $|\Gamma_l|$ scale. This scale is calibrated in terms of $e^{-\alpha z}$ (not $e^{-2\alpha z}$) in 1-dB steps. The distance d in Figure 5.40 gives a transmission loss of about 0.67 dB, so

$$e^{-\alpha z} = 0.924 \quad (-0.67 \quad \text{dB})$$

or

$$\alpha z = 0.077 \quad \text{Np}.$$

This checks the previous result. Note that the distances d and d' are both 0.67 dB, but are *not* the same physical length! Also, α and β have not been determined, but αz and total phase shift βz have been determined.

In Example 5.17, the value of Q that was determined was rather low. In fact, it is so low that its accuracy may be questioned, since the formula, $Q = \beta/2\alpha$, was based on the assumption that the losses are small, or the Q is high. In this example, the short-circuit and open-circuit impedances were purposely chosen to make the Q rather low so that the Smith chart could be easily used. Normally, Q will be the order of 10^3.

It should be emphasized that the Q we have calculated in Example 5.17 is that for a *propagating system*. Quite often, as has already been mentioned, sections of lines are used as equivalent lumped elements in certain applications.

■ Example 5.18

Suppose that the same line that we just considered is to be used (with a short circuit at its load end) to represent an inductive reactance. What is the "inductor" Q in this application? We have

$$Z_{in}^{sc} = 80 \underline{|85°} = R_{in}^{sc} + jX_{in}^{sc}$$
$$= 6.97 + j79.7 \ \Omega.$$

The "inductor" Q is defined here as being

$$Q_L = \frac{\omega L_{in}^{sc}}{R_{in}^{sc}} = \frac{X_{in}^{sc}}{R_{in}^{sc}} = \tan \underline{|85°} = 11.43.$$

■

5.17 DISTRIBUTED PARAMETERS

The Q we have just calculated is a *nonresonant* value for an equivalent lumped inductor ($z < \lambda/4$) which the line represents at its input terminals. $Z_{\text{in}}^{\text{sc}}$ can be found from the Smith chart by taking values of $z_{\text{in}}^{\text{sc}}$ from the short-circuit spiral and multiplying by Z_0. From $Z_{\text{in}}^{\text{sc}}$ the Q_l for any length may be found. Note particularly that this Q will depend on the line length!

We have also assumed that the short circuit truly represents zero impedance in the preceding examples. The loss in the actual "short circuit" can be calculated with previously used techniques, but it is obvious that this loss will be small compared to the other losses unless the line length is very small. An example that considers all losses will be presented in Section 5.18.

5.17 DISTRIBUTED PARAMETERS

Our work in earlier chapters showed that it is relatively easy to calculate C and, hence G, whereas internal inductance per unit length and R are not so simple to calculate because of the existence of internal flux linkage. We will consider three cases: (i) the lossless case ($R = 0$, $G = 0$), (ii) the low-frequency case where the conduction current density is essentially uniform throughout the conductor, and (iii) the high-frequency case where the current density is assumed to be uniform in a layer of thickness δ, the skin depth. In this latter case there are four points to keep in mind from previous work:

(a) The external inductance per unit length can be easily calculated from C:

$$L_{\text{ext}} = \mu\varepsilon/C \quad (\text{H/m}). \tag{5.172}$$

(b) The shunt conductance per unit length is also easily obtainable from C:

$$G = \sigma_d C/\varepsilon \quad (\mho/\text{m}), \tag{5.173}$$

(c) The skin effect (Section 4.5) is pronounced so that the resistance per unit length is calculated as if the current density were uniform in a cross-sectional area that is δ (m) deep and w (m) wide at the surface of the conductors. The length of the conductors is obviously 1 m. We are dealing with two conductor systems, so this formulation must be used *twice*, once for each conductor. The total resistance per unit length is

$$R = \frac{1}{\sigma_{c1}\delta_1 w_1} + \frac{1}{\sigma_{c2}\delta_2 w_2} \quad (\Omega/\text{m}). \tag{5.174}$$

(d) The "conductor impedance" is the surface impedance (Section 4.5) where the resistance per unit length and the reactance per unit length are equal. Thus, the *internal* inductance per unit length is related to R [Equation (5.174)] by

$$L_{\text{int}} = R/\omega \quad (\text{H/m}). \tag{5.175}$$

The distributed parameters are listed in Table 5.2 for three common geometries that are shown in Figure 5.41. These are the planar (parallel-plate), coaxial, and two-wire line.

TABLE 5.2
Common Transmission Line Parameters

Case		Parallel Plane System	Coaxial Cable System	Two-Wire Line System
Lossless		$C = \dfrac{\varepsilon w}{d}$	$C = \dfrac{2\pi\varepsilon}{\ln(b/a)}$	$C = \dfrac{\pi\varepsilon}{\cosh^{-1} D/d} \approx \dfrac{\pi\varepsilon}{\ln(2D/d)}, \; D \gg d$
		$L = \dfrac{\mu d}{w}$	$L = \dfrac{\mu}{2\pi} \ln \dfrac{b}{a}$	$L = \dfrac{\mu}{\pi} \cosh^{-1} \dfrac{D}{d} \approx \dfrac{\mu}{\pi} \ln \dfrac{2D}{d}, \; D \gg d$
		$C = \dfrac{\varepsilon w}{d}$	$C = \dfrac{2\pi\varepsilon}{\ln(b/a)}$	$C = \dfrac{\pi\varepsilon}{\cosh^{-1} D/d}$
		$G = \dfrac{\sigma_d w}{d}$	$G = \dfrac{2\pi\sigma_d}{\ln(b/a)}$	$G = \dfrac{\pi\sigma_d}{\cosh^{-1} D/d}$
High-frequency		$R = \dfrac{2}{\sigma_c \delta w}$	$R = \dfrac{1}{\sigma_c 2\pi\delta}\left(\dfrac{1}{a} + \dfrac{1}{b}\right)$	$R = \dfrac{2}{\pi\sigma_c \delta(d-\delta)} \approx \dfrac{2}{\pi\sigma_c \delta d}$
		$L = \dfrac{\mu d}{w} + \dfrac{2}{\omega\sigma_c \delta w} = \dfrac{\mu}{w}(d+\delta)$	$L = \dfrac{\mu}{2\pi}\left[\ln \dfrac{b}{a} + \dfrac{\delta}{2}\left(\dfrac{1}{a} + \dfrac{1}{b}\right)\right]$	$L = \dfrac{\mu}{\pi}\left(\cosh^{-1} \dfrac{D}{d} + \dfrac{\delta}{d}\right)$ $\approx \dfrac{\mu}{\pi}\left(\ln \dfrac{2D}{d} + \dfrac{\delta}{d}\right), \; D \gg d$
		$C = \dfrac{\varepsilon w}{d}$	$C = \dfrac{2\pi\varepsilon}{\ln(b/a)}$	$C = \dfrac{\pi\varepsilon}{\cosh^{-1} D/d}$
		$G = \dfrac{\sigma_d w}{d}$	$G = \dfrac{2\pi\sigma_d}{\ln(b/a)}$	$G = \dfrac{\pi\sigma_d}{\cosh^{-1} D/d}$
Low-frequency		$R = \dfrac{2}{\sigma_c t w}$	$R = \dfrac{1}{\pi\sigma_c}\left(\dfrac{1}{a^2} + \dfrac{1}{c^2 - b^2}\right)$	$R = \dfrac{8}{\pi d^2 \sigma_c}$
		$L = \dfrac{\mu}{w}\left(d + \dfrac{t}{6}\right)$	$L = \dfrac{\mu}{2\pi}\left[\ln \dfrac{b}{a} + \dfrac{1}{4} + \dfrac{1}{4(c^2-b^2)}\right.$ $\left. \times \left(b^2 - 3c^2 + \dfrac{4c^4}{c^2-b^2}\ln \dfrac{c}{b}\right)\right]$	$L = \dfrac{\mu}{\pi}\left(\ln \dfrac{2D}{d} + \dfrac{1}{4}\right)$

5.17 DISTRIBUTED PARAMETERS

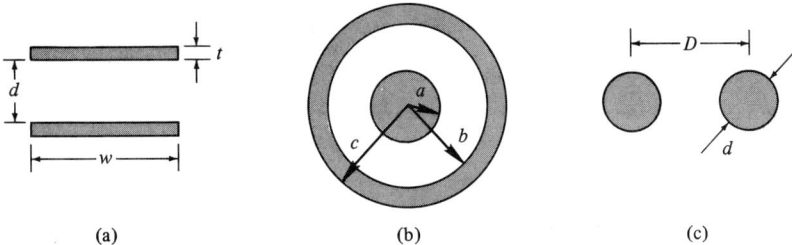

Figure 5.41. Common transmission lines. (a) Planar. (b) Coaxial. (c) Two-wire.

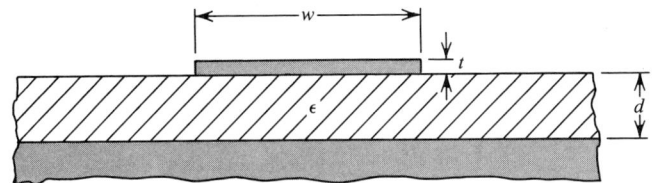

Figure 5.42. Stripline.

Striplines[8] can be fabricated by automated printed circuit techniques, and are widely used to interconnect high-speed (hence, high-frequency) logic circuits in digital computers. Generally speaking, there are four types: The *parallel plane* (or plate, already mentioned) *stripline*; the *coplanar stripline* which consists of coplanar conducting strips mounted on a dielectric substrate; the *shielded stripline* which is a single conducting strip in a dielectric substrate mounted between large conducting planes; and the *microstrip line* which is a single conducting strip (width w, thickness t) mounted on a large dielectric substrate (thickness d) with a large conducting plane beneath the dielectric.

A microstrip line is shown in Figure 5.42. Approximate values for the distributed parameters can be obtained in a simple manner for the high-frequency case if we ignore fringing and employ the skin-effect concepts. These require that $w \gg h$ and $w \gg t \gg \delta$, respectively. If these conditions are satisfied, the analysis of the microstrip line is identical to that of the parallel-plate line (Figure 5.41, Table 5.2).

■ Example 5.19

The dimensions of a microstrip line are $w = 5$ mm, $d = 1$ mm, and $t = 0.2$ mm. The dielectric substrate is characterized by $\varepsilon_R = 5$ and $\sigma_d = 10 \times 10^{-6}$ (℧/m) at 100 MHz, and the conductors are copper, $\sigma_c = 5.8 \times 10^7$ (℧/m). The capacitance per unit length is

$$C = \frac{\varepsilon_R \varepsilon_0 (w)(1)}{d} = \frac{[10^{-9}/(36\pi)](5)(5 \times 10^{-3})}{10^{-3}} = 221 \quad \text{pF/m}.$$

---

[8] See Liao in the references at the end of this chapter

The external inductance per unit length [Equation (5.172)] is

$$L_{\text{ext}} = \frac{\mu_0 \varepsilon_0 \varepsilon_R}{C} = \frac{\mu_0 d}{w} = \frac{(4\pi \times 10^{-7})(10^{-3})}{5 \times 10^{-3}} = 251.3 \quad \text{nH/m}.$$

The shunt conductance per unit length [Equation (5.173)] is

$$G = \sigma_d C/(\varepsilon_0 \varepsilon_R) = \frac{\sigma_d w}{d} = \frac{(10^{-5})(5 \times 10^{-3})}{10^{-3}} = 50 \quad (\mu\mho/\text{m}).$$

The resistance per unit length is obtained from Equation (5.174) with $\sigma_{c1} = \sigma_{c2}$, $\delta_1 = \delta_2$, and $w_1 = w_2$. Note that $w_1 = w_2$ because we have ignored fringing. Thus, using Equations (4.75) and (5.174), we get

$$\delta = \frac{1}{\sqrt{\pi f \mu_0 \sigma_c}}, \qquad R = \frac{2}{\sigma_c \delta w},$$

$$R = \frac{2}{w}(\pi f \mu_0 / \sigma_c)^{1/2} = \frac{2}{5 \times 10^{-3}} \left[\frac{(\pi \times 100 \times 10^6 \times 4\pi \times 10^{-7})}{(5.8 \times 10^7)} \right]^{1/2},$$

$$R = 1.044 \quad \Omega/\text{m}.$$

The internal inductance per unit length [Equation (5.175)] is

$$L_{\text{int}} = \frac{R}{\omega} = \frac{1}{\pi w}\sqrt{\frac{\pi \mu_0}{\sigma_c f}} = 1.661 \quad \text{nH/m} \quad (\ll L_{\text{ext}}).$$

The characteristic impedance [Equation (5.118)] is

$$Z_0 = \sqrt{\frac{R + j\omega L}{G + j\omega C}} = 33.83 \underline{|-0.18°} \quad \Omega.$$

The complex propagation constant [Equation (5.119)] is

$$\gamma = \sqrt{(R + j\omega L)(G + j\omega C)} = 16.27 \times 10^{-3} + j4.7 \quad \text{m}^{-1}.$$

∎

5.18 THE COAXIAL CAVITY[9]

A very well-shielded resonant cavity results from a shorted (nominal) quarter-wave section of rigid coaxial line. It may be used to measure σ_d or ε, or it may be used to represent a parallel resonant circuit of lumped parameters. It is shown in Figure 5.43. Our first task will be that of finding Q_d. The Q was defined in

---

[9] See Hayt, 1973 in the references at the end of this chapter.

5.18 THE COAXIAL CAVITY

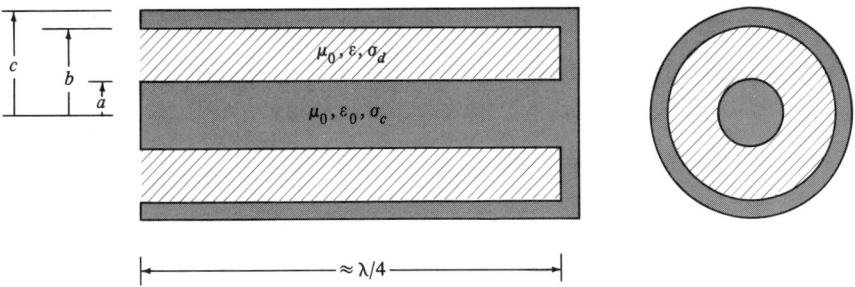

Figure 5.43. Resonant ($\lambda/4$) coaxial cavity.

Equation (5.144) for a propagating system: $Q = \omega <\mathcal{W}> / <\mathcal{P}_d>$. In the present situation, $<\mathcal{W}>$ is the total of the time-average energy stored in the cavity and $<\mathcal{P}_d>$ is the total of the average power dissipated in the cavity. We will employ the *perturbation method* here whereby the *lossless* fields are used to calculate both $<\mathcal{W}>$ and $<\mathcal{P}_d>$. This will be accurate so long as the loss is small. We already know that **E** and **H** are distributed (for the lossless case) in transverse planes precisely the same way as they are statically. We also know that they are quadrature standing waves longitudinally. Thus, as can easily be verified, we get

$$E_\rho(\rho, z, \omega) = \frac{V_{\text{in}} \sin kz}{\rho \ln b/a}, \tag{5.176}$$

$$H_\phi(\rho, z, \omega) = j\frac{V_{\text{in}} \cos kz}{\eta\rho \ln b/a}, \tag{5.177}$$

where V_{in} is the voltage at the input to the line. Using Equations (4.72) and (5.144) the dielectric Q is[10]

$$Q_d = \omega \frac{\dfrac{\varepsilon}{2}\int_a^b\int_0^{2\pi}\int_0^{\lambda/4}\dfrac{1}{2}|E_\rho|^2\rho\,d\rho\,d\phi\,dz + \dfrac{\mu}{2}\int_a^b\int_0^{2\pi}\int_0^{\lambda/4}\dfrac{1}{2}|H_\phi|^2\rho\,d\rho\,d\phi\,dz}{\dfrac{1}{2}\sigma_d\int_a^b\int_0^{2\pi}\int_0^{\lambda/4}|E_\rho|^2\rho\,d\rho\,d\phi\,dz}, \tag{5.178}$$

where the lossless resonant $\lambda/4$ length has been used. We already know from Section 5.16, however, that the dielectric Q is just the Q of the dielectric. Eliminating H_ϕ with Equation (5.177) gives

$$Q_d = \omega\frac{\dfrac{\varepsilon}{2}\iiint\cdots + \dfrac{\mu}{2\eta^2}\iiint\cdots}{\sigma_d\iiint\cdots} = \omega\frac{\varepsilon\iiint\cdots}{\sigma_d\iiint\cdots}$$

---

[10] Do not forget to use "$\frac{1}{2}$ real part of"; for example, $<\mathcal{W}_E> = \frac{1}{2}\text{Re}(\frac{1}{2}\iiint \mathbf{D}\cdot\mathbf{E}^*\,dv)$.

or

$$Q_d = \frac{\omega\varepsilon}{\sigma_d}. \qquad (5.179)$$

We next find the Q due to conductor losses Q_c. Using our skin effect concepts, we have a "uniform" current density

$$J_{za} = \left.\frac{H_\phi}{\delta}\right|_{\rho=a} = j\frac{V_0}{\eta a\delta}\cos kz \qquad [V_0 = V_{in}/\ln(b/a)]$$

flowing in the inner conductor, while in the outer conductor, the current density is

$$J_{zb} = -\left.\frac{H_\phi}{\delta}\right|_{\rho=b} = -j\frac{V_0}{\eta b\delta}\cos kz.$$

The shorting plug at $z = 0$ has a current density

$$J_{\phi 0} = \frac{H_\phi|_{z=0}}{\delta} = j\frac{V_0}{\eta\rho\delta}.$$

Then, the power loss for these three conductors will be, respectively [see Equation (4.72)],

$$<\mathcal{P}_d>_{\rho=a} = \frac{1}{2}\left(\frac{1}{\sigma_c}\int_{a-\delta}^{a}\int_0^{2\pi}\int_0^{\lambda/4}|J_{za}|^2\rho\,d\rho\,d\phi\,dz\right),$$

$$<\mathcal{P}_d>_{\rho=b} = \frac{1}{2}\left(\frac{1}{\sigma_c}\int_b^{b+\delta}\int_0^{2\pi}\int_0^{\lambda/4}|J_{zb}|^2\rho\,d\rho\,d\phi\,dz\right),$$

$$<\mathcal{P}_d>_{z=0} = \frac{1}{2}\left(\frac{1}{\sigma_c}\int_a^b\int_0^{2\pi}\int_{-\delta}^{0}|J_{\phi 0}|^2\rho\,d\rho\,d\phi\,dz\right).$$

Using the total average power dissipated and the same total average energy stored as in finding Q_d we have

$$Q_c = \frac{(2/\delta)\ln b/a}{(1/a)(1 - \delta/2a) + (1/b)(1 + \delta/2b) + (8/\lambda)\ln(b/a)}. \qquad (5.180)$$

The overall Q $(=Q_0)$ is found as in Equation (5.168).

5.18 THE COAXIAL CAVITY

This resonant line must have an equivalent in terms of parallel lumped parameters that we can find. The maximum energy stored in the capacitor C_e (when the energy stored in the inductor L_e is zero) is

$$\tfrac{1}{2} C_e V_{in}^2 = \frac{\pi \varepsilon V_{in}^2}{\ln b/a}\left(\frac{\lambda}{8}\right).$$

The right-hand side of the preceding equation is the numerator of Equation (5.178) (not including ω). Solving for C_e, we get

$$C_e = \frac{\pi \varepsilon \lambda}{4 \ln (b/a)}. \tag{5.181}$$

Also, we have

$$Q_0 = \omega_0 C_e R_e$$

or

$$R_e = \frac{Q_0}{\omega_0 C_e}, \tag{5.182}$$

and finally, we have

$$\omega_0^2 = \frac{1}{L_e C_e}$$

or

$$L_e = \frac{1}{\omega_0^2 C_e}. \tag{5.183}$$

A numerical example is in order.

Example 5.20

Let us determine the important quantities for a resonant quarter-wave coaxial resonator where $f_0 = 300$ MHz, $\mu = \mu_0$, $a = 1$ cm, $b = 2.3$ cm. In the first part we assume that the dielectric is air so that $\sigma_d \approx 0$, $\varepsilon = \varepsilon_0$. The walls are silver ($\sigma_c = 6.17 \times 10^7$) plated. Then, we have

$$\lambda = \frac{u}{f} = \frac{300 \times 10^6}{300 \times 10^6} = 1 \text{ m}, \quad l = \frac{\lambda}{4} = \frac{1}{4} = 0.25 \text{ m},$$

$$\delta = (\pi f \mu \sigma_c)^{-1/2} = 3.7 \times 10^{-6} \text{ m},$$

$$Q_0 = Q_c = 3000 \quad (Q_d \to \infty),$$

$$C_e = 8.34 \text{ pF},$$
$$R_e = 191 \text{ k}\Omega,$$
$$L_e = 33.7 \text{ nH}.$$

Circuit theory tells us that the bandwidth f_0/Q_c is 0.1 MHz.

Now suppose that the interior of this cavity is filled with a dielectric that we are testing. The new resonant frequency is found to be 190 MHz and the new bandwidth is measured as 0.240 MHz. Then, from Equation (5.180) ($\delta = 4.65 \times 10^{-6}$ m), we get

$$Q_c = 2387,$$

while the overall Q is found from the resonant frequency and the bandwidth ($= B$):

$$Q_0 = f_0/B = 792.$$

Therefore, we have

$$\frac{1}{Q_d} = \frac{1}{Q_0} - \frac{1}{Q_c}, \qquad Q_d = 1184.$$

The relative permittivity is found from the velocity

$$f_0 \lambda = (190 \times 10^6)(1) = (3 \times 10^8)/\sqrt{\varepsilon_R}, \qquad \varepsilon_R = 2.49,$$

and so, from Equation (5.179), we get

$$\sigma_d = \omega \varepsilon / Q_d = 22.22 \quad \mu \mho/\text{m}.$$

Thus, the dielectric is characterized by $\varepsilon_r = 2.49$ and $\sigma_d = 22.22$ $\mu \mho$/m at 190 MHz.

The equivalent lumped parameters now become

$$C_e = 20.76 \text{ pF},$$
$$R_e = 31.96 \text{ k}\Omega,$$
$$L_e = 33.8 \text{ nH}.$$

Before leaving this example it is worthwhile to consider the distributed parameters and make the same calculations. Using Table 5.2, we know

$$R = 79.6 \text{ m}\Omega/\text{m},$$
$$L = 166.6 \text{ nH/m},$$
$$C = 166.3 \text{ pF/m},$$
$$G = 167.6 \text{ }\mu\Omega/\text{m}.$$

Therefore, we have

$$Q_c = \omega L/R = 2499, \qquad Q_d = \omega C/G = 1184, \qquad Q_0 = 804.$$

Note that Q_c (here) is greater than the value calculated from the field equations because the loss in the short circuit has been ignored. On the other hand, Q_d is the same as before. Proceeding, we get

$$C_e = 20.76 \quad \text{pF},$$
$$R_e = 32.44 \quad \text{k}\Omega,$$
$$L_e = 33.8 \quad \text{nH}.$$

■

5.19 CONCLUDING REMARKS

The general behavior of transmission lines operating in the TEM mode is similar to the behavior of uniform plane waves propagating in a homogeneous medium although we are usually concerned with voltage and current rather than the field quantities when working with the transmission line. Such things as coefficient of reflection, standing wave ratio, power (density) flow, and impedance are common to the behavior of the transmission line and the uniform plane wave. The analysis of the transmission line can be expedited by means of the Smith chart, or, if preferred, the computer or programmable calculator with complex mode can be used.

Normally, the losses for a transmission line are small and can be ignored in most cases. If the losses are to be included in the analysis, then this can easily be done at a single frequency or over a narrow band of frequencies. On the other hand, since the distributed parameters are actually frequency dependent, a general time-domain analysis may be very difficult to perform.

REFERENCES

Brown, R. G., Sharpe, R. A., and Hughes, W. L. *Lines, Waves, and Antennas*, 2nd ed. New York: Ronald Press, 1973. A good reference for transmission lines and waveguides.

Chipman, R. A. *Transmission Lines*. Schaum Outline Series. New York: McGraw-Hill, 1968. Many solved problems are included.

Everitt, E. E. *Communication Engineering*, 2nd ed. New York: McGraw-Hill, 1937. See page 169.

Hayt, Jr., W. H. *Engineering Electromagnetics*, 2nd ed. New York: McGraw-Hill, 1973. The $\lambda/4$ resonant line is discussed in Chapter 13.

International Telephone and Telegraph Co. Inc. *Reference Data for Radio Engineers*, 5th ed. Indianapolis: Howard W. Sams and Co., 1968. A handbook containing much data on transmission lines of various shapes and dielectric materials.

Liao, S. Y. *Engineering Applications of Electromagnetic Theory*. St. Paul, MN: West Publishing, 1988. A good discussion of striplines.

Moore, R. K. *Traveling-Wave Engineering*. New York: McGraw-Hill, 1960. A simultaneous development of plane waves and transmission lines is given.

Ryder, J. D. *Networks, Lines and Fields*. Englewood Cliffs, NJ: Prentice Hall, 1949. A still valuable book for transmission line and waveguide studies.

Skilling, H. H. *Electric Transmission Lines*. New York: McGraw-Hill, 1951.

PROBLEMS

1. The distributed capacitance per unit length for a lossless line is measured as 60 pF/m in air. Find $L = L_{ext}$ and R_0.

2. A lossless line has a measured length of 25 m. A low-frequency bridge measurement shows that the capacitance between the conductors is 1500 pF. A *time-domain reflectometer* (TDR) measurement shows that a pulse travels from the sending end to the receiving end and back in 250 ns. Find the characteristic resistance of the line.

3. Find the phasor voltage and current in the middle of a lossless 50-Ω line that is one wavelength long if $V_g = 1\underline{|0°}$ V, $Z_g = 100\,\Omega$, and $Z_l = 0$.

4. (a) Find the phasor voltage and current in the middle of a lossless 50-Ω air line that is 300 m long at 1 MHz if $V_g = 1\underline{|0°}$ V, $Z_g = 100\,\Omega$, and $Z_l = 0$.
   (b) Find the time-domain voltage in the middle of the line if $v_g(t) = 10\sin(2\omega t)$.

5. If $Z_{in} = 30 + j70\,\Omega$ at $z = 0.2\lambda$ for a lossless 50-Ω line, find Z_l.

6. A lossless 300-Ω line has $u_p = 0.85/\sqrt{\mu_0\varepsilon_0}$. At 30 MHz the input impedance is a pure resistance equal to $0.5\,R_0$ at a point 1.3 m from the load. Find (a) Z_l, (b) Γ_l, (c) VSWR.

7. A lossless 50-Ω line is terminated in $Z_l = R_l + jX_l$, so that the VSWR = 2.5 when $R_l = 100\,\Omega$. (a) Find X_l. (b) Find the location of $|V|_{min}$ nearest the load.

*8. Two co-planar loops are parallel and very closely spaced. Their diameters are $3\lambda/8$. What is the impedance looking in between two points, one on one loop and one at the same point on the other loop?

*9. An AM broadcast transmitter operates at a carrier frequency of 1 MHz. It is connected through a lossless 50-Ω line that is 0.165λ long to a monopole antenna whose impedance is $36.5 + j21.2\,\Omega$. The output impedance of the transmitter is $100 + j25\,\Omega$ and cannot be easily altered. An (ideal) matching transformer and an adjustable series (pure) reactance are available at the load; $V_g = 800\underline{|0°}$ V.
   (a) What is the maximum (average) power that can be radiated by the antenna under ideal conditions?

PROBLEMS

(b) Specify the primary-to-secondary turns ratio of the transformer and reactance in series with the antenna to achieve maximum power radiated by the antenna.

(c) What is the inductance (or capacitance) represented by the series reactance?

10. Derive a formula for $<\mathcal{P}_l>/<\mathcal{P}_{inc}>$ as a function of VSWR and plot this function.

*11. A push–pull amplifier is designed to operate at 300 MHz. It has an output capacitance of 5 pF.
   (a) Find the length of 300-Ω air line (shorted at the far end) required to provide parallel resonance.
   (b) The short circuit is removed and replaced with a tuning capacitor. The line length is adjusted to be 10 cm. What value of C gives parallel resonance?
   (c) Where is the magnetic field intensity greatest [part (b)]? **Hint:** Make a plot of $|I(z)|$ versus z.
   (d) Where should a pickup loop be placed to couple energy from the line?

12. Prove that $|\Gamma_l| = 1$ for any purely reactive load.

*13. (a) Prove that the input impedance at the $x - x$ terminals in Figure 5.18(b) is infinite for any $d\ (> 0)$.
   (b) Specify d to provide a filter that is band pass at 100 MHz and band reject at 120 MHz; see Figure 5.44.
   (c) Use the results of parts (a) and (b) to design a system that allows two receiving antennas (one at 100 MHz, the other at 120 MHz) to be connected to a common transmission line that leads to the receiver. R_0 is the same everywhere. Does the system also operate as a transmitting system?
   (d) It is arbitrarily decided that bandwidth is to be measured between frequencies where $<\mathcal{P}_l>/<\mathcal{P}_{inc}> \geq 0.5$. What is the bandwidth of the system in Figure 5.44? Plot the bandwidth $B = (f_2/f_0) - (f_1/f_0)$, versus d/l, where f_1 and f_2 are the lower and upper half-power frequencies, respectively, and $f_0 = (4l\sqrt{\mu_0\varepsilon_0})^{-1}$.

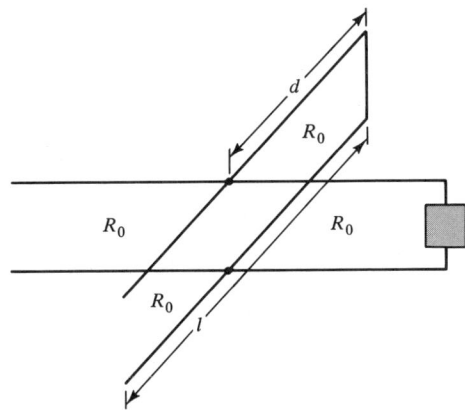

Figure 5.44. Transmission line with a line section across it [Figure 5.18(b)] (Problem 13).

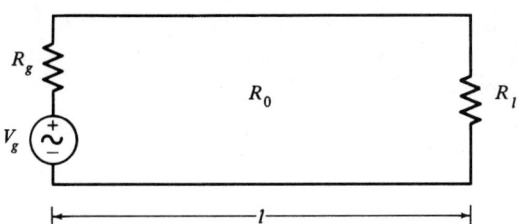

Figure 5.45. Lossless line terminated in R_l (Problem 14).

14. For Figure 5.45,
    (a) Plot $<\mathcal{P}_l>$ versus R_g/R_0 if $R_l = R_0$.
    (b) Plot $<\mathcal{P}_l>$ versus R_l/R_0 if $R_g = R_0$.

15. (a) Show that

$$\mathcal{E}_\rho(\rho,z,t) = \frac{v(z,t)}{\rho \ln(b/a)}$$

for a lossless coaxial cable. $v(z,t)$ is the line voltage.
    (b) Find $\mathcal{H}$.
    (c) The maximum value of E_ρ occurs for $\rho = a$. Show that a coaxial cable designed for maximum voltage breakdown rating (minimum E_ρ) has $R_0 = 60\ \Omega$ ($\varepsilon = \varepsilon_0$).

*16. (a) Show that

$$\Phi(x,y,z,t) = \frac{v(z,t)}{4\cosh^{-1}(D/d)} \ln \frac{(x+a)^2 + y^2}{(x-a)^2 + y^2}$$

for a two wire line. See Example 2.25. Note that $a = [(D/2)^2 - (d/2)^2]^{1/2}$.
    (b) Find $\mathcal{E}$.
    (c) Make a normalized plot of $<\mathcal{S}>$ versus y/a for $x = 0$.
    (d) Make a normalized plot of $<\mathcal{S}>$ versus x/a for $y = 0$.

*17. (a) Find the VSWR on the line in Figure 5.46 for $0 \le z \le \lambda/4$.
    (b) Find the power dissipated in the 25-Ω resistor.

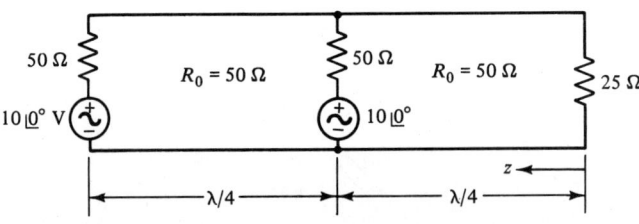

Figure 5.46. Lossless line with several lumped elements (Problem 17).

PROBLEMS

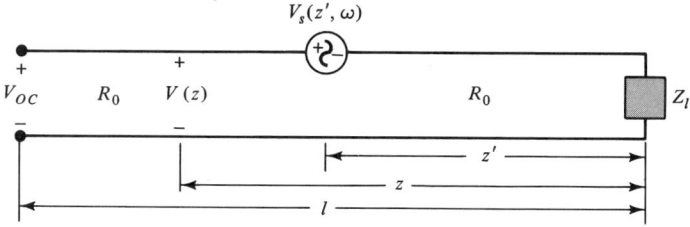

Figure 5.47. Lossless line with the generator located in the middle and in series with one side (Problem 18).

*18. Find the Thévenin equivalent circuit at the open-circuit terminals in Figure 5.47.

19. A standing wave indicator in the line from a radio transmitter to the antenna shows that the VSWR is 1.2 and the incident power is 900 W. What is the reflected power?

20. The equivalent circuit for a television receiving antenna, transmission line, and receiver is shown in Figure 5.48. Suppose that neither the antenna impedance (Z_g) nor the receiver input impedance (Z_l) is matched to the line, so that there will be reflections from both ends. As a result of this, multiple "ghosts" will appear in the television picture. Assuming that the transmission line is 100 ft long, the velocity of propagation is 2.5×10^8 m/s, the time for a horizontal line on the screen is 53.3 μs, and the width of the raster (picture) is 12 in., how far apart are the ghosts?

21. Suppose that the television signal (Problem 20) reaches the antenna by a direct path from the transmitter and also by a *longer* path by way of reflection from a large conducting body. This latter path is 450λ longer at 83 MHz. If the receiving antenna is matched to the line, how far apart on the screen are the primary image and the single ghost?

*22. A *tapered* transmission line can be used as a transition to match transmission lines of different characteristic impedance if the taper is gradual.
   (a) Let $L = L(z)$ and $C = C(z)$, and show that Equations (5.34) and (5.35) lead to

$$\frac{d^2V}{dz^2} - \frac{1}{L}\frac{dL}{dz}\frac{dV}{dz} + \omega^2 LCV = 0$$

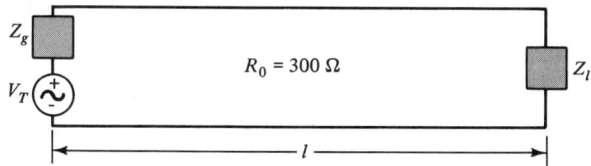

Figure 5.48. Equivalent circuit of a television receiving system. Z_l represents the receiver input impedance (Problem 20).

and

$$\frac{d^2 I}{dz^2} - \frac{1}{C}\frac{dC}{dz}\frac{dI}{dz} + \omega^2 L C I = 0.$$

(b) Let $L(z) = L(0)e^{-\alpha_1 z}$ and $C(z) = C(0)e^{+\alpha_1 z}$, giving the *exponential taper*, and show that for the $-z$ traveling wave:

$$V(z) = V(0)e^{-\alpha_1 z/2 + j\sqrt{\beta^2 - (\alpha_1/2)^2}\, z}$$

and

$$I(z) = I(0)e^{+\alpha_1 z/2 + j\sqrt{\beta^2 - (\alpha_1/2)^2}\, z},$$

where

$$\beta^2 = \omega^2 LC = \omega^2 L(0)C(0).$$

(c) What is the lowest frequency that propagates?

23. Amplitude modulation is often used with the slotted line to expedite voltage measurements. Let $v_g(t) = V_0(1 + m_a \cos \omega_m t) \cos \omega_c t$, where ω_m is the modulation frequency, ω_c is the carrier frequency, and m_a is the percent modulation. Assuming a lossless line, $Z_g = R_0$, and a frequency-dependent load (e.g., $Z_l = R_l + jX_l$), can the slotted line give exact results? Use phasor concepts and superposition to find a time-domain expression for the reflected voltage.

24. If $Z_l = 200 - j50\,\Omega$ on a lossless 50-Ω line, find Γ_l, VSWR, and Z_{in} at 0.3λ from the load with the Smith chart and without the Smith chart.

25. The following data are measured with an impedance bridge: $Z_{in}^{sc} = +j100\,\Omega$, $Z_{in}^{oc} = -j25\,\Omega$, and $Z_{in} = 100\lfloor 30°\,\Omega$ (actual load in place). Find Z_l with and without the Smith chart.

26. The following data are measured with a calibrated slotted line: $|V|_{max} = 0$ dB. $|V|_{min} = -10$ dB, $z_{min} = 0, 0.375, 0.75, \ldots$ m (with short circuit as load), and $z_{min} = 0.10, 0.475, 0.85, \ldots$ m (with actual load in place). Find the frequency. Find Z_l with and without the Smith chart.

27. (a) Design a single (shunt, shorted) stub system to match a load $Z_l = 10 - j50\,\Omega$ to a 50-Ω lossless line. Does maximum power transfer from the source occur?

    (b) Repeat using double (shunt, shorted) stubs that are spaced $\lambda/4$ with one stub placed directly across the load.

28. It is found that an open-circuited stub of length 0.2λ that is connected 0.3λ from the load provides a match to the line. What is Z_l?

PROBLEMS

*29. A $\lambda/4$ matching section is designed to match a 50-Ω line to $Z_l = 200\lfloor 0°\ \Omega$ at 500 MHz. The bandwidth is arbitrarily measured between the frequencies for which VSWR $\leq$ 1.5. Use the Smith chart to complete (a)–(d).
   (a) Find this bandwidth.
   (b) What is the maximum VSWR?
   (c) What is the minimum value of $<\mathscr{P}_l>/<\mathscr{P}_{inc}>$?
   (d) What is the bandwidth if it is measured between frequencies for which $<\mathscr{P}_l>/<\mathscr{P}_{inc}> \geq 0.5$?

30. In a certain two-element linear antenna array it is found that the antenna impedances are $Z_1 = 100 + j83\ \Omega$ and $Z_2 = 46 + j2\ \Omega$. The elements are spaced by $\lambda/4$, and the currents that are required to produce the desired radiation pattern are $I_1 = 1\lfloor 0°$ A and $I_2 = 1\lfloor 90°$ A (peak). Use Figure 5.49 to complete (a)–(d).
   (a) Specify R_0' and X.
   (b) What power is radiated?
   (c) What is the VSWR on the main feed line?
   (d) Should it be matched?

31. It is desired to drive three UHF voltage amplifiers from a single voltage source even though the amplifiers may have different input impedances. Devise a scheme to accomplish the desired result if one of the amplifiers is to be driven out of phase with the other two.

32. Three impedances ($Z_1 = 100\lfloor 0°$, $Z_2 = 75\lfloor 50°$, and $Z_3 = 150\lfloor -40°$) are driven by lossless lines ($R_{01} = 50$, $R_{02} = 75$, and $R_{03} = 100$) that are $\lambda/2$ in length and connected in parallel to the main line ($R_0 = 50$). The voltage across each of the impedances is $1\lfloor 0°$ (peak). What is the VSWR on the main line?

33. A television cable company runs a cable to a private home that is equipped with a distribution amplifier whose input impedance matches the cable impedance (ideally, under all conditions). Insofar as the power that is delivered by the cable is concerned, does it matter how many receivers are connected to the distribution amplifier?

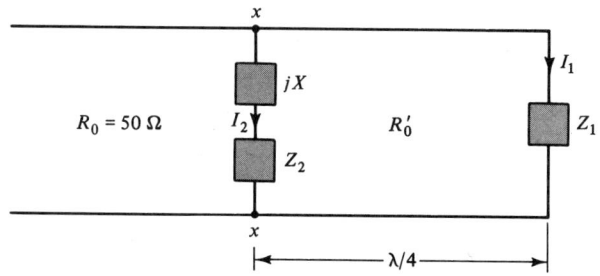

Figure 5.49. Transmission line feed system for Problem 30.

34. A load $Z_l = 30 + j\omega L$ (Ω) on a 50-Ω line produces a VSWR of 3 at 300 MHz. Find L with, and without, the Smith chart.

*35. A load $Z_l = 100 + j50$ Ω is connected to a line of length d and characteristic resistance R_0'. This line is then connected to the generator through a 50-Ω line. Use the Smith chart to find R_0' and d for a matched system.

36. An air-filled coaxial cable has $R_0 = 50$ Ω. A lossless dielectric slug ($\varepsilon_R = 4$) that is 0.2 m long is inserted with one end 0.1 m from the load. It fills the air space between the inner and outer conductors. If $f = 300$ MHz and $Z_l = R_0$, what is the VSWR in the region between the dielectric slug and the generator?

37. Using the smaller of the two possible values of l_1 and s_1, design a single (shorted) stub matching system for $Z_l = 200 \underline{|0°}$ Ω and $R_0 = 50$ Ω. Plot VSWR versus normalized frequency.

38. If $Z_l = 100 + j200$ Ω, $L = 0.1$ μH, $C = 20$ pF, $R_0 = 50$ Ω, and $f = 300$ MHz, find the VSWR to the left of the capacitor in Figure 5.50. What power is dissipated in the load?

39. It is possible for the reflected voltage to be larger than the incident voltage on a line if the equivalent impedance at the load has a negative real part. The tunnel diode, for example, can represent a negative resistance. If the reflected voltage (or part of it) is removed from the line with a directional coupler, then a voltage amplifier results. If $Z_l = -60$ Ω, $R_0 = 50$ Ω, and the directional coupler has a 3-dB loss in the backward direction, what is the voltage gain?

40. (a) A telephone line has $R = 6.3$ mΩ/m, $L = 2.27$ μH/m, $G = 180$ p℧/m, and $C = 5.2$ pF/m at 1 kHz. Find α, β, Z_0, u_p, and λ.

(b) How much inductance per mile should be added to this line to make it distortionless? What is the new value of α?

41. A lossy transmission line is 10 m long and has $Z_0 = 51 \underline{|2°}$ Ω, $\alpha = 10^{-3}$ Np/m, and $\beta = 2\pi$ rad/m. Find the average load power if $Z_g = Z_0$, $V_g = 10$ V, rms and

(a) $Z_l = Z_0$,
(b) $Z_l = 100$ Ω,
(c) $Z_l = 25$ Ω,
(d) What is the input impedance at $z = \lambda/2$ when $Z_l = 100$ Ω?
(e) If $Z_l = 0$, plot $|V|$ versus z for $0 \le z \le \lambda$. Find $|V|_{max}$ and $|V|_{min}$.

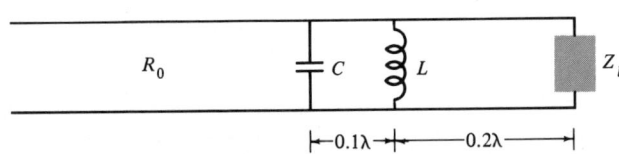

Figure 5.50. Lossless line with a lumped L and C inserted across the line (Problem 38).

PROBLEMS

42. Measurements on a lossy line show that $Z_0 = 50 + j0\ \Omega$, $\alpha = 10^{-3}$ Np/m, and $\beta = 2.5$ rad/m at 100 MHz. Find R, L, G, and C.

*43. Suppose that $\alpha = 5 \times 10^{-3}$ Np/m for the tapped resonant $\lambda/4$ line of Figure 5.18. If $d = 0.125$ m, find the bandwidth between the points where the impedance magnitude is 0.707 times its value at the resonant frequency. The line is 0.25 m long. Does the result agree with $B = f_0/Q_0$?

44. (a) Assume copper conductors and a polyethylene dielectric: $\varepsilon_R = 2.26$, $\sigma_d = 5 \times 10^{-6}$ ℧/m, and find the distributed parameters for a two-wire line at 100 MHz: $D = 10^{-2}$ m, $d = 10^{-3}$ m.

(b) Repeat for a coaxial cable: $b = 10^{-2}$ m, $a = 3 \times 10^{-3}$ m.

(c) Repeat for a microstrip line that is fabricated out of a thin conducting strip of width w located at a height h above a large conducting ground plane.

45. Why is the "velocity factor" for commercial two-wire line with polyethylene dielectric about 0.85 rather than $1/\sqrt{\varepsilon_R} = 0.67$?

*46. (a) Show that

$$\frac{|V|}{|V|_{max}} = \frac{1}{s}\sqrt{1 + (s^2 - 1) \sin^2 [k(z - z_{min})]}, \quad s = \text{VSWR}$$

for a lossless line.

(b) Show that

$$s \approx \lambda/\pi \Delta z$$

for a line with small loss, $s > 10$, and $|V|/|V|_{max} = \sqrt{2}/s$ (that is, 3 dB above $|V|_{min}$). The distance between two points 3 dB above $|V|_{min}$ is Δz. This is called the *width of minimum method of determining the VSWR*.

47. The incident field in the vicinity of a receiving antenna is such that the antenna can deliver 10^{-12} W into a matched load. If the transmission line is characterized by $\alpha = 3$ dB/100 ft and is 50 ft long, what is the maximum power that can be delivered to the receiver?

*48. Suppose that $Z_0 = 50\ \underline{|1°}\ \Omega$, $\alpha = 5 \times 10^{-3}$ Np/m, $\beta = 2\pi$ (rad/m), and $\lambda = 1$ m.

(a) Suppose that it is assumed that this line is lossless in order to expedite the design of a single (shorted) stub matching system $Z_l = 300\ \Omega$. Proceed with the design on this basis, using $R_0 = \text{Re}(Z_0)$ and the smaller of the two possible values of l_1 and s_1. Finally, calculate the input impedance at the stub junction.

(b) Comment on the results of (a).

49. A "dummy" resistance can be obtained for dissipating energy by using the input impedance to a very lossy line. Suppose such a two-wire line is constructed in air with $f = 30$ MHz, $D = 0.3$ m, $d = 10^{-2}$ m, $\mu_r = 400$ (iron), and

$\sigma_c = 10^7$ ℧/m (iron). If such a line is open at the far end, and is 100 m long, what is the input impedance?

50. A rigid coaxial cable with silver-plated conductors is used between an FM transmitter and the transmitting antenna. The line length is 100 m, $a = 2 \times 10^{-2}$ m, and $b = 5 \times 10^{-2}$ m. If the power available from the transmitter is 10 kW, and the antenna is matched to the line, what is the power delivered to the antenna? The carrier frequency is 97.5 MHz. The dielectric is air.

51. Follow the same procedure as that used in Section 5.8, and show that the *transmission parameters* for a lossy line are given in the set of equations:

$$V_1 = \cosh[\gamma(z_1 - z_2)]V_2 + Z_0 \sinh[\gamma(z_1 - z_2)]I_2,$$
$$I_1 = (1/Z_0) \sinh[\gamma(z_1 - z_2)]V_2 + \cosh[\gamma(z_1 - z_2)]I_2.$$

6
WAVEGUIDES AND CAVITIES

In the microwave region of the spectrum, beginning roughly at 1 GHz, ordinary two-conductor systems, commonly called transmission lines, become inefficient, so that another guiding system is sought. A hollow conducting pipe or cylinder of some arbitrary, but uniform, cross section with fields *inside* might be expected to have less loss than the ordinary line since only one conductor is involved. This pipe is commonly called a *waveguide*, but all guiding systems are really waveguides. A dielectric rod of uniform cross section can also support or guide waves. We will investigate the propagation characteristics of these waveguides in this chapter.

There are many methods for solving boundary value problems of the type posed here, but they are all essentially the same in that what they accomplish is a solution to Maxwell's equations fitting the boundary conditions for the particular problem. We will begin this chapter with a simplified treatment of the hollow rectangular waveguide operating in the dominant mode. This will demonstrate the important features without the detailed mathematics of the general case. The general case is presented in Section 6.5, and can be omitted if desired.

6.1 THE RECTANGULAR WAVEGUIDE, TE_{10} MODE

The hollow one-conductor rectangular waveguide is shown in Figure 6.1, and our problem is that of finding the propagating electromagnetic fields that it can support. In particular, we want to find the *simplest* field that it can support, and this, as it turns out, means finding that mode with lowest *cutoff frequency*, for the hollow waveguide acts like a high-pass filter. Since we already know that a lossless *two-conductor* system can propagate a TEM to z mode even as $\omega \to 0$ (where $\pm z$ is the direction of propagation, and is the axial direction in Figure 6.1), we begin by examining this mode.

6 WAVEGUIDES AND CAVITIES

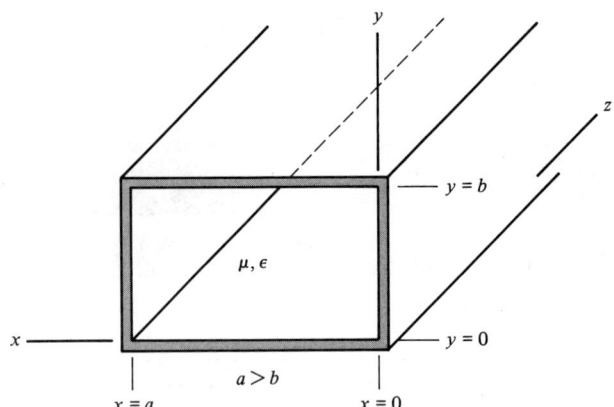

Figure 6.1. Hollow rectangular waveguide ($a > b$).

It is not difficult to show that a hollow conducting waveguide of *any* uniform cross section (not necessarily rectangular as in Figure 6.1) will *not* support a TEM to z mode inside. If it did, then $E_z = H_z \equiv 0$ by the definition of a TEM to z mode. Then, the magnetic field lines, which must form closed loops ($\nabla \cdot \mathbf{H} = 0$ for any simple material in the hollow region) in $z =$ constant planes, must encircle current. This current cannot be due to conduction current density since there are no conductors *inside* the waveguide. It cannot be due to displacement current density ($j\omega\varepsilon E_z$) since $E_z \equiv 0$ inside the guide. Hence, the TEM to z mode cannot exist, and we can proceed to examine other possibilities.

It was pointed out in Section 4.7 (Figure 4.14) that a two-conductor parallel-plane system could support a mode that was not TEM by simply allowing plane waves to bounce back and forth between the planes. The last paragraph of Section 4.7 is of particular importance in this regard. Let us see if this same principle can be made to apply to the present geometry, which is certainly different, but not very different. We begin by assuming that the walls of Figure 6.1 are perfectly conducting and that the hollow region contains at most a perfect dielectric. These assumptions mean that there are no losses anywhere.

The boundary conditions that must be satisfied at the perfectly conducting surfaces are

$$E_x = 0 \begin{cases} y = 0, \\ y = b, \end{cases} \quad E_y = 0 \begin{cases} x = 0, \\ x = a, \end{cases} \quad E_z = 0 \begin{cases} x = 0, x = a, \\ y = 0, y = b. \end{cases} \quad (6.1)$$

In accordance with our desire to obtain a simple solution like that in Section 4.7, we simply choose $E_x = E_z = 0$ and assume that there are no variations with y ($\partial/\partial y \equiv 0$). Note that in Section 4.7 and Figure 4.14 we had $E_y = E_z = 0$ and $\partial/\partial x \equiv 0$. Expanding Maxwell's equations, using the fact that $\mathbf{E} = \mathbf{a}_y E_y$ and $\partial/\partial y \equiv 0$ gives

6.1 THE RECTANGULAR WAVEGUIDE, TE_{10} MODE

$$E_y = \frac{1}{j\omega\varepsilon}\left(\frac{\partial H_x}{\partial z} - \frac{\partial H_z}{\partial x}\right), \quad (6.2)$$

$$H_x = \frac{1}{j\omega\mu}\frac{\partial E_y}{\partial z}, \quad (6.3)$$

$$H_z = -\frac{1}{j\omega\mu}\frac{\partial E_y}{\partial x}, \quad (6.4)$$

with $E_x = E_z = H_y \equiv 0$. Note that the boundary conditions on E_x and E_z [in Equations (6.1)] are automatically satisfied. Substituting Equations (6.3) and (6.4) into Equation (6.2) gives

$$E_y = -\frac{1}{k^2}\left(\frac{\partial^2 E_y}{\partial x^2} + \frac{\partial^2 E_y}{\partial z^2}\right), \quad (6.5)$$

where $k^2 = \omega^2\mu\varepsilon$.

Returning once more to Figure 4.14, we see that because of the fact that the plane waves are bouncing back and forth between the conducting plane at $z = 0$ and the conducting plane imagined to be at $z = 1/(2f\sqrt{\mu_1\varepsilon_1}\cos\theta_i)$ ($p = 1$), the phase constant in Equation (4.132) is $k_1\sin\theta_i$, and not simply k_1. In the same way, the phase constant in our present problem cannot be simply k, but must be k multiplied by the sine or cosine of some angle. Since we are looking for propagation in the $+z$ direction (the $-z$ direction could be chosen as well), we choose to call this phase constant k_z, and therefore propagation must occur in the form $e^{-jk_z z}$. Thus, we are assuming that

$$E_y(x, z, \omega) = f(x)e^{-jk_z z} \quad (6.6)$$

and Equation (6.5) becomes

$$f(x)e^{-jk_z z} = \frac{1}{k^2}e^{-jk_z z}\left(k_z^2 f(x) - \frac{\partial^2 f(x)}{\partial x^2}\right)$$

or, simplifying,

$$\frac{d^2 f(x)}{dx^2} + (k^2 - k_z^2)f(x) = 0. \quad (6.7)$$

This is an ordinary second-order linear differential equation, and, as such, possesses two independent solutions: $\sin(\sqrt{k^2 - k_z^2}\,x)$ and $\cos(\sqrt{k^2 - k_z^2}\,x)$ for $k^2 - k_z^2 \neq 0$. The cosine must be discarded because it cannot satisfy the second set of boundary conditions in Equations (6.1). Thus, we are left with $f(x) = E_0\sin(\sqrt{k^2 - k_z^2}\,x)$ and the boundary condition $E_y = 0$ for $x = a$ which is satisfied if

$$f(a) = E_0 \sin(\sqrt{k^2 - k_z^2}\, a) = 0, \tag{6.8}$$

where E_0 is the peak value of E_y. The roots of $\sin(\sqrt{k^2 - k_z^2}\, a)$ are $\sqrt{k^2 - k_z^2}\, a = m\pi$, $m = 1, 2, 3, \ldots$. The case $m = 0$ gives a null field. Thus,

$$\sqrt{k^2 - k_z^2} = m\pi/a, \quad m = 1, 2, 3, \ldots \tag{6.9}$$

Each integer m designates a distinct mode with a different variation with x. Since there are no variations with y in the present case, the designation for variations with y is $n = 0$, but, in general, variations with y can also exist (as we shall see later). Thus, we designate the modes that we have found as TE to z ($E_z \equiv 0$), or, more particularly, TE$_{mo}$ modes. The simplest of these occurs for $m = 1$, giving the TE$_{10}$ mode, and Equation (6.9) leads to the determination of k_z:

$$k_z = \sqrt{k^2 - (\pi/a)^2} \quad (m = 1) \tag{6.10}$$

and $f(x)$:

$$f(x) = E_0 \sin(\pi x/a). \tag{6.11}$$

The complete field for the TE$_{10}$ mode is found by substituting $f(x)$ [Equation (6.11)] into Equation (6.6), and then substituting the result into Equations (6.3) and (6.4). The result is

$$\boxed{\begin{aligned} E_y &= E_0 \sin(\pi x/a) e^{-jk_z z}, \\ H_x &= -\frac{k_z E_0}{\omega \mu} \sin\left(\frac{\pi x}{a}\right) e^{-jk_z z}, \\ H_z &= j\frac{E_0 \pi}{\omega \mu a} \cos\left(\frac{\pi x}{a}\right) e^{-jk_z z}, \\ k_z &= \sqrt{k^2 - (\pi/a)^2}, \end{aligned}} \quad \text{(TE}_{10}\text{ mode)}. \tag{6.12}$$

The electric field varies as $\sin(\pi x/a)$ with x, but does not vary at all with y. This is shown in Figure 6.2. Note how neatly the boundary conditions on the tangential electric field are satisfied. The electric field is maximum in magnitude at $x = a/2$ ($|E_y| = E_0$), and if arcing occurs, it will occur there.

6.2 PROPERTIES OF THE TE$_{10}$ MODE

The phase velocity is the velocity of a point of constant phase. Taking E_y out of the set of Equations (6.12) and findings its time-domain form from the phasor form:

6.2 PROPERTIES OF THE TE$_{10}$ MODE

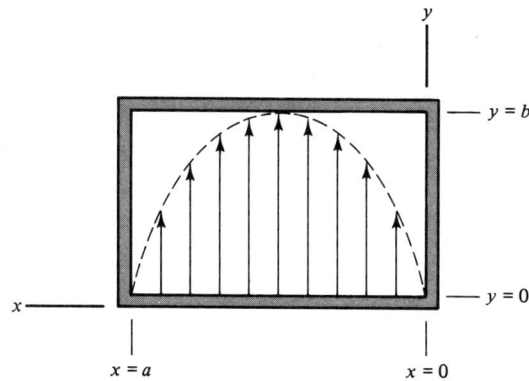

Figure 6.2. Electric field intensity (E_y) distribution (at fixed z) for the TE$_{10}$ rectangular waveguide mode.

$$\mathcal{E}_y(x,z,t) = E_0 \sin(\pi x/a) \cos(\omega t - k_z z),$$

and, for a point of constant phase, we have

$$\omega t - k_z z = \text{constant}.$$

Differentiating with time:

$$\omega - k_z \frac{dz}{dt} = 0,$$

$$\frac{dz}{dt} = u_p = \frac{\omega}{k_z} \tag{6.13}$$

The phase constant k_z is given by Equation (6.10):

$$k_z = \begin{cases} \sqrt{k^2 - (\pi/a)^2} = \beta, & k > \pi/a, \\ -j\sqrt{(\pi/a)^2 - k^2} = -j\alpha, & k < \pi/a. \end{cases} \tag{6.14}$$

Note the choice of signs for k_z. For $k > \pi/a$ we choose $k_z = +\beta$ to give a $+z$ traveling wave $e^{-j\beta z}$. For $k < \pi/a$ we choose $k_z = -j\alpha$ to give $e^{-\alpha z}$, an exponentially increasing term being impossible since z is not bounded. The transition occurs where $k_z = 0$, or $k = \pi/a$. The particular frequency at this transition is called the *cutoff frequency*:

$$k_c = \pi/a = \omega_c \sqrt{\mu\varepsilon} \tag{6.15}$$

or

$$\boxed{\omega_c = \pi/(a\sqrt{\mu\varepsilon})} \quad (\textit{cutoff frequency}). \tag{6.16}$$

Substituting Equation (6.15) into (6.14):

$$k_z = \begin{cases} k\sqrt{1-(\omega_c/\omega)^2} = \beta, & \omega > \omega_c, \\ -jk_z\sqrt{1-(\omega/\omega_c)^2} = -j\alpha, & \omega < \omega_c. \end{cases} \quad (6.17)$$

The interpretation of the last result is easy. We have a propagating mode when the frequency is above the cutoff frequency, and we have a nonpropagating, or *evanescent mode*, below the cutoff frequency. Thus, the waveguide does act like a high-pass filter, as stated earlier.

Returning to Equation (6.13), we can express the phase velocity as

$$\boxed{u_p = \frac{1}{\sqrt{\mu\varepsilon}} \frac{1}{\sqrt{1-(\omega_c/\omega)^2}} \geq \frac{1}{\sqrt{\mu\varepsilon}}} \quad \textit{(phase velocity)}. \quad (6.18)$$

Note that phase velocity has no meaning for $\omega < \omega_c$. The time-domain electric field in this case is evanescent:

$$\begin{aligned}\mathcal{E}_y(x,z,t) &= \text{Re}[E_0 \sin(\pi x/a)e^{j\omega t}e^{-\alpha z}], \\ \mathcal{E}_y(x,z,t) &= E_0 \sin(\pi x/a)e^{-\alpha z}\cos\omega t, \quad \omega < \omega_c.\end{aligned} \quad (6.19)$$

Note also that as the cutoff frequency is approached from above ($\omega > \omega_c$), the phase velocity approaches infinity. The phase velocity is always greater than the velocity of light for the medium inside the waveguide. This is not surprising for it occurred for oblique incidence in Chapter 4. We will examine this phenomenon again shortly.

■ Example 6.1

The rectangular waveguide of Figure 6.1 is air-filled and has dimensions $a = 2$ cm and $b = 1$ cm. The cutoff frequency for the dominant TE_{10} mode is $\omega_c = \pi/(a\sqrt{\mu\varepsilon}) = (3\pi \times 10^8)/(2 \times 10^{-2}) = 47.12 \times 10^9$ rad/s, or $f_c = 7.5$ GHz. If this waveguide is operated in the TE_{10} mode at a frequency of 10 GHz, then $k_z = k\sqrt{1-(\omega_c/\omega)^2} = \beta = (2\pi \times 10^{10}/3 \times 10^8)\sqrt{1-(7.5/10)^2} = 138.5$ rad/m. On the other hand, if the frequency is 5 GHz, the mode is nonpropagating, and $\alpha = k_c\sqrt{1-(\omega/\omega_c)^2} = (\pi/2 \times 10^{-2})\sqrt{1-(5/7.5)^2} = 117.1$ Np/m or 1.017×10^3 dB/m. This is a very large attenuation constant, and the (nonpropagating) fields would only be appreciable very close to the source. ■

The system or guide wavelength λ_g is the distance in the z direction for 2π rad of phase shift. Therefore,

$$k_z(z_1 - z_2) = 2\pi = k_z\lambda_g$$

or

6.2 PROPERTIES OF THE TE$_{10}$ MODE

$$\lambda_g = \frac{2\pi}{k_z} = \frac{1}{f\sqrt{\mu\varepsilon}\sqrt{1-(\omega_c/\omega)^2}} \geq \lambda. \qquad (6.20)$$

The velocity of energy flow is defined in Equation (4.33) as

$$u_e \equiv \frac{<\mathcal{P}_f>}{<\mathcal{W}>} \qquad (6.21)$$

or

$$u_e = \frac{\int_0^a \int_0^b \frac{1}{2}\mathrm{Re}(\mathbf{E}\times\mathbf{H}^*)\cdot \mathbf{a}_z\, dx\, dy}{\int_0^a \int_0^b \int_0^1 \frac{1}{2}\mathrm{Re}\left(\frac{\mathbf{D}}{2}\cdot\mathbf{E}^* + \frac{\mathbf{B}}{2}\cdot\mathbf{H}^*\right) dx\, dy\, dz}. \qquad (6.22)$$

After some labor this reduces to

$$u_e = \frac{2\omega}{k_z}\frac{(a/2)}{(k^2/k_z^2)(a)} = \frac{\omega k_z}{k^2}$$

or

$$u_e = \frac{\sqrt{1-(\omega_c/\omega)^2}}{\sqrt{\mu\varepsilon}} \leq \frac{1}{\sqrt{\mu\varepsilon}}. \qquad (6.23)$$

Equation (6.23) states that the velocity of energy flow is less than the plane wave velocity of light in the same medium. It follows from Equations (6.23) and (6.18) that

$$u_p u_e = \frac{1}{\mu\varepsilon}. \qquad (6.24)$$

The wave impedance equations follow from the definitions in Chapter 4:

$$\eta^+ = -\frac{E_y}{H_x} = \begin{cases} \eta/\sqrt{1-(\omega_c/\omega)^2}, & \omega > \omega_c, \\ j\eta(\omega/\omega_c)/\sqrt{1-(\omega/\omega_c)^2}, & \omega < \omega_c. \end{cases} \qquad (6.25)$$

It is not surprising that η^+ is imaginary for $\omega < \omega_c$.

It is interesting and informative to plot the important quantities versus normalized frequency. Figure 6.3 shows that for $\omega > \omega_c$ the wave impedance is real, indicating power flow, whereas for $\omega < \omega_c$ the impedance is imaginary, indicating the lack of

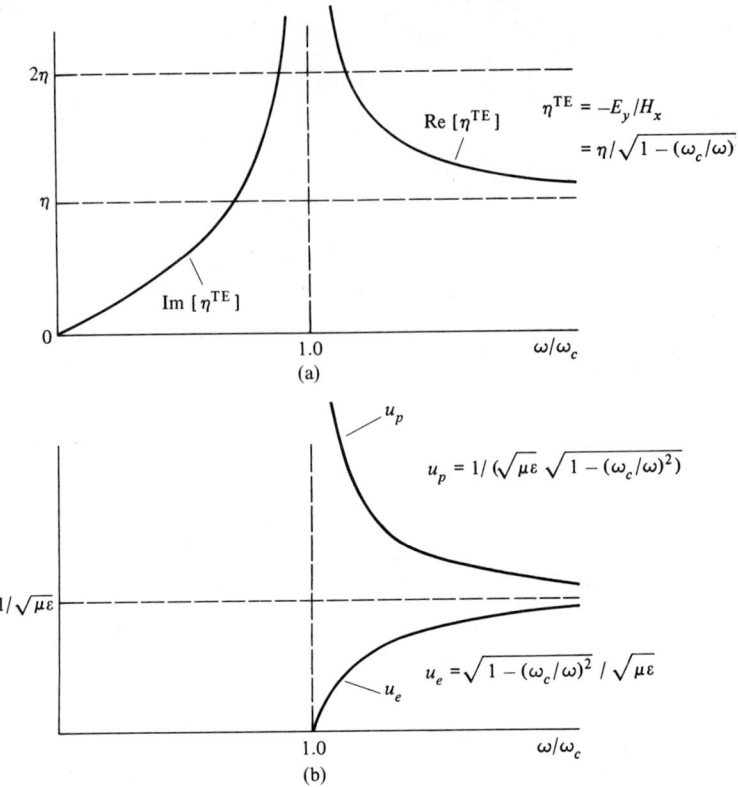

Figure 6.3. (a) TE$_{10}$ wave impedance ($-E_y/H_x$) rectangular waveguide. (b) TE$_{10}$ phase velocity and velocity of energy flow, rectangular waveguide.

power flow. For $\omega \gg \omega_c$ the wave impedance approaches the intrinsic impedance, $\sqrt{\mu/\varepsilon}$. The phase velocity is infinite at the cutoff frequency, but approaches the (plane wave) velocity of light for the medium for $\omega \gg \omega_c$. The velocity of energy flow is zero at the cutoff frequency, but approaches the velocity of light for $\omega \gg \omega_c$. The three parts of Figure 6.3 show the same thing: propagation without loss above cutoff and no propagation below cutoff.

The phasor electric field is given in the set of Equations (6.12):

$$E_y = E_0 \sin \frac{\pi x}{a} e^{-jk_z z}, \tag{6.26}$$

which can be written

$$E_y = \frac{E_0}{2j}(e^{j(\pi x/a)} - e^{-j(\pi x/a)})e^{-j\sqrt{k^2-(\pi/a)^2}\,z}. \tag{6.27}$$

This may be rearranged to

6.2 PROPERTIES OF THE TE$_{10}$ MODE

$$E_y = \frac{E_0}{2j} e^{jk(\pi x/ka)} e^{-jk\sqrt{1-(\pi/ka)^2} z} - \frac{E_0}{2j} e^{-jk(\pi x/ka)} e^{-jk\sqrt{1-(\pi/ka)^2} z} \quad (6.28)$$

or

$$E_y = \frac{E_0}{2j} e^{-jks'} - \frac{E_0}{2j} e^{-jks''}, \quad (6.29)$$

where s' and s'' refer to directions shown in Figure 6.4. The first term is a *uniform plane wave* traveling in the s' direction, while the second is a *uniform plane wave* traveling in the s'' direction. In fact, one plane wave is just a reflection of the other plane wave. Note that the angle ξ is

$$\xi = \cos^{-1} \sqrt{1 - \left(\frac{\pi}{ka}\right)^2} \quad (6.30)$$

or

$$\xi = \cos^{-1} \sqrt{1 - \left(\frac{\omega_c}{\omega}\right)^2}. \quad (6.31)$$

Another simple interpretation of what happens at the cutoff frequency is now apparent. When $\omega = \omega_c$, then $\xi = \pi/2$, and there is no *net* plane wave propagation in the $+z$ direction (and thus no power flow). When $\omega > \omega_c$, the component of *plane wave* velocity in the $+z$ direction is $(1/\sqrt{\mu\varepsilon}) \cos \xi$ since the plane wave velocities in the s' and s'' directions are $1/\sqrt{\mu\varepsilon}$. This component velocity can be written

$$\frac{1}{\sqrt{\mu\varepsilon}} \cos \xi = \frac{1}{\sqrt{\mu\varepsilon}} \sqrt{1 - \left(\frac{\omega_c}{\omega}\right)^2} = u_e. \quad (6.32)$$

It should not be surprising to find that this is the same as the velocity of energy flow.

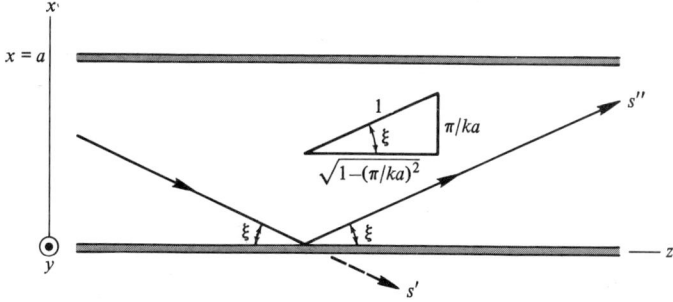

Figure 6.4. Plane wave directions of propagation (s' and s'') for the TE$_{10}$ mode.

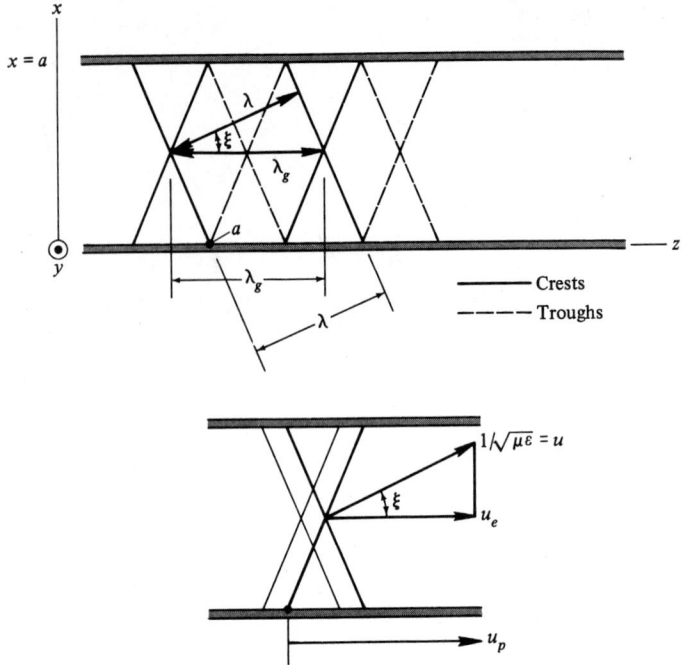

Figure 6.5. Plane waves (TE$_{10}$ mode) showing u_p, u_e, u, λ, and λ_g for the rectangular waveguide.

What about the phase velocity? A better picture of what this phase velocity actually is may be obtained by adding lines of constant phase onto the plane waves of Figure 6.4. These are shown in Figure 6.5. The phase velocity is the velocity of a point of constant phase for the composite wave. Point a (the point of intersection of the plane wave equiphase trough and the perfectly conducting plane at $x = 0$) is such a point. The plane wave travels a distance λ while the intersection point a travels a distance λ_g. Geometrically, $\lambda = \lambda_g \cos \xi$. Therefore, the velocities are related by

$$\frac{1/\sqrt{\mu\varepsilon}}{u_p} = \frac{\lambda}{\lambda_g} = \frac{\lambda \cos \xi}{\lambda}. \tag{6.33}$$

Again, recall the similar discussion for oblique incidence in Section 4.7. Equations (6.31) and (6.33) can be combined to give

$$u_p = \frac{1}{\sqrt{\mu\varepsilon}\sqrt{1 - (\omega_c/\omega)^2}} > \frac{1}{\sqrt{\mu\varepsilon}}. \tag{6.34}$$

We have also shown that

$$\lambda_g = \frac{\lambda}{\sqrt{1 - (\omega_c/\omega)^2}} > \lambda. \tag{6.35}$$

It is helpful in viewing Figure 6.5 to consider the analogous picture of an ocean wave striking a straight shore line at an angle slightly off the normal. Visualize the TE_{10} mode between perfectly conducting planes at $x = 0$ and $x = a$ as being nothing more than plane waves being reflected back and forth, while having a component velocity u_e in the direction of propagation. Note that the time delay encountered by a sinusoidal signal traveling a distance l in the waveguide, because of the zigzag paths, will be given by

$$\tau_D = \frac{l}{u_e}. \qquad (6.36)$$

Thus, a signal is delayed more by the zigzag path of the TE_{10} to z mode than by a TEM mode in the same medium. More importantly, the TE_{10} mode *cannot be distortionless* even if the medium is lossless, because u_e and u_p are *frequency dependent*.

It is interesting that the waveguide is *normally dispersive*[1] even in the absence of losses. The *group velocity*, given by Equation (4.57), is $d\omega/d\beta$, and this becomes

$$u_g = \frac{\sqrt{1 - (\omega_c/\omega)^2}}{\sqrt{\mu\varepsilon}} = u_e,$$

so that the group time delay is

$$\tau_D = \frac{l}{u_g} = \frac{l\sqrt{\mu\varepsilon}}{\sqrt{1 - (\omega_c/\omega)^2}}.$$

Additional light is shed on the concept of time delay and on the concept of dispersion in the problems at the end of the chapter.

Example 6.2

Returning to Example 6.1, it was found that $k_z = \beta = 138.5$ rad/m at 10 GHz. The time delay at this frequency is $\tau_D = l/u_e$ or

$$\tau_D = \frac{l\sqrt{\mu\varepsilon}}{\sqrt{1 - (f_c/f)^2}} = \frac{l}{3 \times 10^8 \sqrt{1 - (7.5/10)^2}} = 5.039526l \quad (ns).$$

For a length $l = 10$ m this corresponds to a phase shift of 3.166427×10^3 rad. At 10.001 GHz the phase shift is 3.166337×10^3 rad. The difference in phase shift for these two frequencies is 90.4×10^{-3} rad or $5.18°$. For a length $l = 100$ m this phase difference becomes $51.8°$. The point is that the signal to be propagated down the waveguide is usually not a single frequency, but, rather, is a modulated signal. This means that it will not consist of a single sinusoid (at the carrier frequency), but may contain parts (sidebands) at nearby frequencies. That is, a modulated signal

---

[1] $(u_g < 1/\sqrt{\mu\varepsilon})$, see Section 4.9 and Figure 4.17.

requires the use of a finite region of the spectrum, and, over a given distance l, the various frequencies suffer different time delays and phase shifts, and this must be considered. ■

The time-average power density is

$$\langle \mathscr{S} \rangle = \tfrac{1}{2}\mathrm{Re}\,(\mathbf{E} \times \mathbf{H}^*),$$

which reduces to

$$\langle \mathscr{S}_z \rangle = \begin{cases} \dfrac{E_0^2}{2\eta^+} \sin^2\left(\dfrac{\pi x}{a}\right), & \omega > \omega_c, \\ 0, & \omega < \omega_c, \end{cases} \qquad (6.37)$$

where η^+ is given by Equation (6.25). So, once again, we see real power flow above cutoff and none below cutoff.

■ **Example 6.3**

An air-filled x-band waveguide is operated in the TE_{10} mode at a frequency of 10 GHz. The nominal dimensions of this guide are $a = 0.9$ inch and $b = 0.4$ inch. The important quantities are

$$\omega_c = \pi/(a\sqrt{\mu\varepsilon}) = 41.23 \times 10^9 \quad \mathrm{rad/s} \qquad (6.15)$$

or

$$f_c = 6.56 \quad \mathrm{GHz},$$

$$\eta^+ = \frac{\eta}{\sqrt{1-(\omega_c/\omega)^2}} = 499.6 \ \Omega, \qquad (6.25)$$

$$u_p = \frac{1}{\sqrt{\mu\varepsilon}\sqrt{1-(\omega_c/\omega)^2}} = 3.976 \times 10^8 \quad \mathrm{m/s}, \qquad (6.18)$$

$$u_e = \frac{1}{\sqrt{\mu\varepsilon}}\sqrt{1-(\omega_c/\omega)^2} = 2.264 \times 10^8 \quad \mathrm{m/s}, \qquad (6.23)$$

$$\lambda_g = \frac{\lambda}{\sqrt{1-(\omega_c/\omega)^2}} = 3.976 \quad \mathrm{cm}, \qquad (6.20)$$

and, finally,

$$\langle \mathscr{S}_z \rangle = \frac{E_0^2}{2\eta^+} \sin^2(\pi x/a), \qquad (6.37)$$

$$\langle \mathscr{S}_z \rangle = 1.001\, E_0^2 \sin^2(\pi x/a) \quad (\mathrm{mW/m^2}).$$

■

6.3 WAVEGUIDE CURRENT, TE_{10} MODE

We have said nothing to this point about how the TE_{10} mode (or any other mode) is excited in the waveguide. The answer, of course, is that electric current must be introduced into the waveguide. Ordinarily, this is accomplished by inserting a short monopole antenna (probe) at $x = a/2$ in the walls at $y = 0$ or $y = b$ (Figure 6.6). This produces the required y component of electric field for the TE_{10} mode. Many other modes are unavoidably introduced near the antenna, but, as we shall see a little further on, these modes are normally well below cutoff and decay very rapidly with distance, and are of no concern, other than affecting the monopole impedance. The induced TE_{10} field, in turn, induces currents on the waveguide walls. How are these currents distributed?

It was found in the case of the coaxial line that an axial slot could be cut in the outer conductor for probing the line. Can an axial slot be cut in the rectangular waveguide to examine the TE_{10} field? To answer this question, we must examine the current flow on the inside of the waveguide walls. Since the waveguide has been assumed lossless (the walls are perfect conductors), this current will exist in the form of a surface current density according to

$$\mathbf{J}_s = \mathbf{a}_n \times \mathbf{H} \Big|_{\text{on walls}}, \tag{6.38}$$

where $\mathbf{a}_n$ is a unit normal vector as shown in Figure 6.7. From the set of Equations (6.12):

$$J_{sz} = +H_x \Big|_{y=b} = -\frac{k_z E_0}{\omega \mu} \sin \frac{\pi x}{a} e^{-jk_z z} \tag{6.39}$$

and

$$J_{sx} = -H_z \Big|_{y=b} = -j \frac{E_0 \pi}{\omega \mu a} \cos \frac{\pi x}{a} e^{-jk_z z}. \tag{6.40}$$

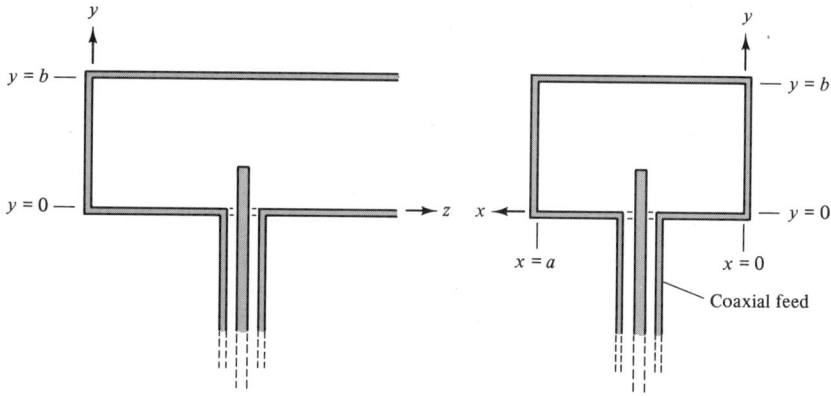

Figure 6.6. Monopole coupling into a waveguide from a coaxial cable.

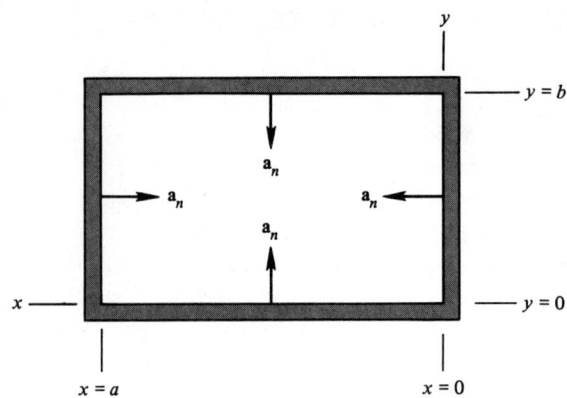

Figure 6.7. Unit normal vectors $\mathbf{a}_n$ inside the rectangular waveguide.

In particular, at the top of the waveguide in the center ($x = a/2$, $y = b$)

$$J_{sz} = -\frac{k_z E_0}{\omega \mu} e^{-jk_z z} \tag{6.41}$$

and

$$J_{sx} = 0. \tag{6.42}$$

Therefore, a narrow axial slot may be cut in the top (or the bottom) of the waveguide at $x = a/2$ without appreciably disturbing the flow of current (or the field).

A slotted waveguide (TE$_{10}$ mode) may be probed in much the same way as the slotted coaxial line. If a standing wave is set up in the slotted section, the distance between minima *is not* $\lambda/2$, but is $\lambda_g/2$ (see Figure 6.5). Measurement of λ_g and the dimension a are sufficient to determine the operating frequency by simply using Equations (6.15) and (6.20). An example at this point is helpful.

■ Example 6.4

Suppose that an air-filled x-band waveguide ($a = 0.9$ inch, $b = 0.4$ inch) is probed with a slotted section, and it is found that the distance between minima is 2 cm. The minimum position does not shift when the actual load is replaced by a shorting plane (short circuit). The standing wave ratio is 2. What is the load impedance and what is the operating frequency?

$$\frac{\lambda_g}{2} = 2 \times 10^{-2} \quad \text{or} \quad \lambda_g = 4 \quad \text{cm},$$

$$a = 0.9(2.54 \times 10^{-2}) \quad \text{or} \quad 2.286 \quad \text{cm}.$$

From Equation (6.15), we have

$$\omega_c = \frac{3\pi \times 10^8}{2.286 \times 10^{-2}} = 41.23 \times 10^9 \quad \text{rad/s}$$

6.3 WAVEGUIDE CURRENT, TE$_{10}$ MODE

or

$$f_c = 6.56 \quad \text{GHz}.$$

From Equation (6.35), we get

$$\lambda_g = \frac{\lambda}{\sqrt{1-(\omega_c/\omega)^2}} = \frac{1}{f\sqrt{\mu\varepsilon}\sqrt{1-(\omega_c/\omega)^2}},$$

or solving for f, we get

$$f = \sqrt{f_c^2 + \frac{1}{\mu\varepsilon(\lambda_g)^2}} = \sqrt{(6.56)^2 + (7.5)^2} \times 10^9$$

or

$$f = 9.97 \quad \text{GHz}.$$

The *normalized* load impedance is on the Smith chart circle corresponding to VSWR = 2 as shown in Figure 6.8. Since the minimum position *did not shift* when the short was replaced by the actual load, we have the *equivalent* of a voltage minimum at the load. This means the impedance at the load is minimum, or $z_l = \frac{1}{2}$. This is a normalized impedance and must be related to a waveguide characteristic resistance. The measurements in the slotted section were made at $x = a/2$ and, in this case,

$$<\mathcal{S}_z> = \frac{E_0^2}{2\eta^+}$$

from Equation (6.37). A logical decision then is to call η^+ the waveguide impedance. That is,

$$(Z_0)_{\text{TE}_{10}} = \eta^+ = \frac{\eta}{\sqrt{1-(\omega_c/\omega)^2}}$$

or

$$(Z_0)_{\text{TE}_{10}} = \frac{377}{\sqrt{1-(6.56/9.97)^2}} = \frac{377}{0.752} = 501 \quad \Omega$$

Therefore, the equivalent load impedance is

$$Z_l = z_l(Z_0)_{\text{TE}_{10}} = \tfrac{1}{2}(501) = 250.5 \quad \Omega.$$

∎

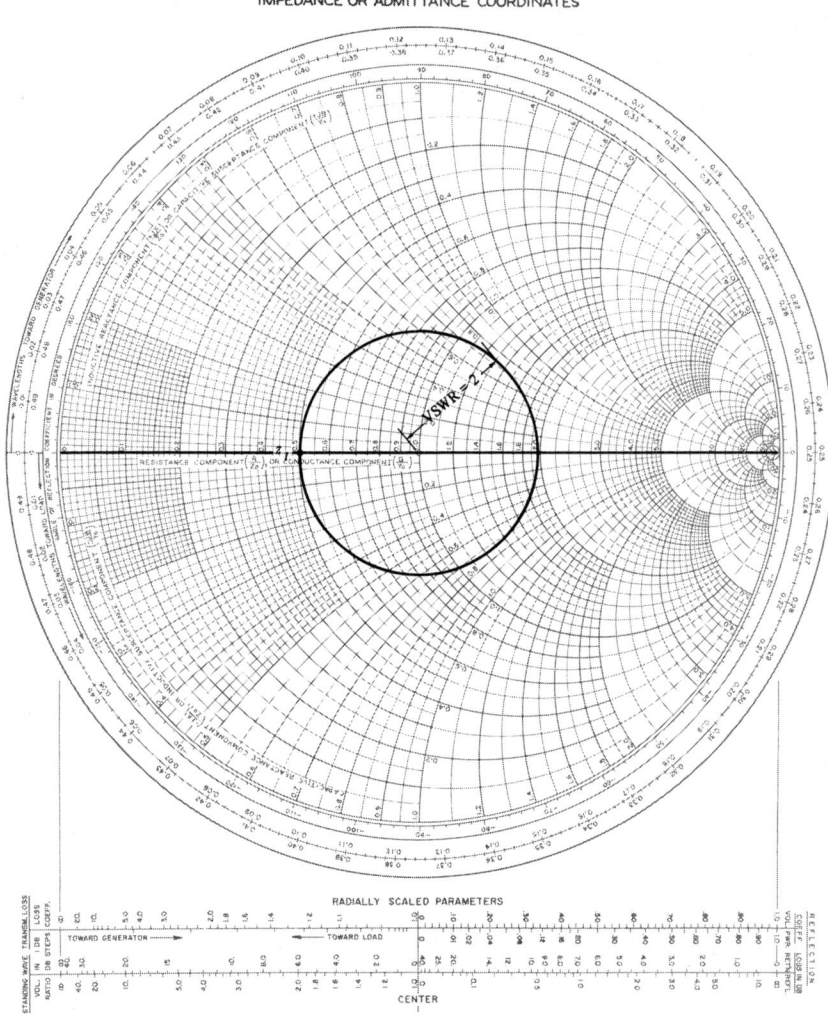

Figure 6.8. Waveguide impedance data, Example 6.4. *Source:* Copyright ©1949 by Kay Electric Company, Pine Brook, NJ. Renewal copyright ©1976 by P. H. Smith, New Providence, NJ.

Equations (6.39) and (6.40) give the z and x components of the surface current density on the top of the waveguide ($y = b$ in Figure 6.7). On the sidewall at $x = a$, we have

$$J_{sy} = H_z \bigg|_{x=a} = -j\frac{E_0 \pi}{\omega \mu a} e^{-jk_z z}.$$

The surface current density is plotted in Figure 6.9. What appear to be charge sinks and sources, $\lambda_g/2$ apart, are really not regions of current discontinuity. The current

6.3 WAVEGUIDE CURRENT, TE₁₀ MODE

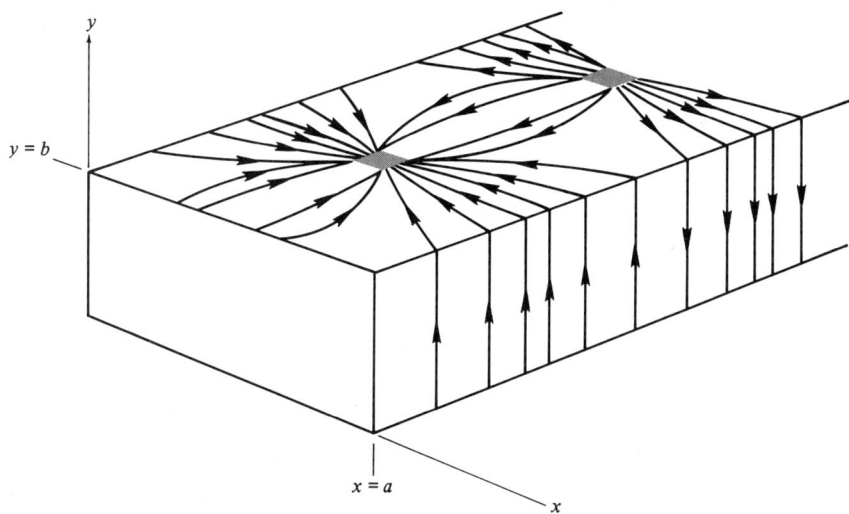

Figure 6.9. Surface current density (inside), TE$_{10}$ mode (rectangular waveguide).

is continuous by means of the displacement current $j\omega\varepsilon E_y$ inside the waveguide according to the continuity equation

$$\nabla \cdot \mathbf{J} = -j\omega\rho_v, \tag{6.43}$$

which for surface current densities becomes

$$\nabla \cdot \mathbf{J}_s = -j\omega\rho_s. \tag{6.44}$$

The surface charge density is ρ_s. The sidewall current obviously flows without divergence. It is relatively simple to verify that Equation (6.44) holds for the top (or bottom) of the waveguide. The left-hand side of Equation (6.44) may be written (for the top)

$$\begin{aligned}\nabla \cdot \mathbf{J}_s &= \nabla \cdot (\mathbf{a}_n \times \mathbf{H}) = \nabla \cdot [-\mathbf{a}_y \times (\mathbf{a}_x H_x + \mathbf{a}_z H_z)] \\ &= \nabla \cdot [\mathbf{a}_z H_x - \mathbf{a}_x H_z] \\ &= -\frac{\partial H_z}{\partial x} + \frac{\partial H_x}{\partial z},\end{aligned}$$

while the right-hand side may be written (for the top)

$$-j\omega\rho_s = -j\omega D_n = j\omega D_y.$$

Thus, we have

$$\left(\frac{\partial H_x}{\partial z} - \frac{\partial H_z}{\partial x}\right)_{y=b} = j\omega D_y \bigg|_{y=b}$$

or

$$|(\nabla \times \mathbf{H})_y|_{y=b} = j\omega\varepsilon E_y\bigg|_{y=b}. \qquad (6.45)$$

Equation (6.45) is just the y component of one of Maxwell's equations (evaluated at $y = b$), and this was satisfied initially!

The field plots for the TE_{10} mode (and other modes) are shown later in Figure 6.11. We next turn to the more general case where higher-order modes will appear, but first we must reexamine the vector potential.

6.4 WAVEGUIDE VECTOR POTENTIALS

In a region free of isolated electric charges and conduction current densities, such as the interior of a hollow one conductor system, Maxwell's equations are

$$\nabla \times \mathbf{E} = -j\omega\mu\mathbf{H} \qquad (6.46)$$

and

$$\nabla \times \mathbf{H} = j\omega\varepsilon\mathbf{E}. \qquad (6.47)$$

Taking the divergence of both sides of Equations (6.46) and (6.47) gives

$$\nabla \cdot \nabla \times \mathbf{E} = -j\omega\mu(\nabla \cdot \mathbf{H}) \equiv 0 \qquad (6.48)$$

and

$$\nabla \cdot \nabla \times \mathbf{H} = j\omega\varepsilon(\nabla \cdot \mathbf{E}) \equiv 0 \qquad (6.49)$$

by vector identity. We have assumed that μ and ε are scalar quantities, independent of position and direction. Then,

$$\nabla \cdot \mathbf{H} = 0 = \nabla \cdot \mathbf{B} \qquad (6.50)$$

and

$$\nabla \cdot \mathbf{E} = 0 = \nabla \cdot \mathbf{D} \qquad (6.51)$$

or, *both* $\mathbf{D}$ and $\mathbf{B}$ are *solenoidal* fields! In this case, it is possible to define

$$\mathbf{D} \equiv -\nabla \times \mathbf{F} \qquad (6.52)$$

and

$$\mathbf{B} \equiv \nabla \times \mathbf{A} \qquad (6.53)$$

6.4 WAVEGUIDE VECTOR POTENTIALS

and Equations (6.50) and (6.51) will always be satisfied. **F** is the *electric vector potential* (C/m) and **A** is, of course, the *magnetic vector potential* (webers per meter). Note that it would be possible to define $\mathbf{E} = \pm \nabla \times \mathbf{F}$ and $\mathbf{H} = \pm \nabla \times \mathbf{A}$, and these potentials would only differ from those of Equations (6.52) and (6.53) by constants. We will use $\mathbf{B} = \nabla \times \mathbf{A}$ since we have already done so in Chapter 2. It is also important to note that *isolated magnetic charges* do not exist in nature, so for the μ and ε used here, **B** and **H** are always *solenoidal* and **B** or $\mathbf{H} = \nabla \times \mathbf{A}$ is always possible. On the other hand, Equation (6.52) was possible *only* because we assumed a region free of isolated *electric charges*.

Equations (6.52) and (6.53) do not completely define **F** and **A**, for *both* the *divergence and the curl* of a vector are required to completely specify it. We can specify the divergence of **F** and **A** at a more advantageous point in the development. It is convenient here to remember that the equations are still *linear*, so part of the field may be determined from **A**, with $\mathbf{F} = 0$; then the rest of the field is determined from **F** with $\mathbf{A} = 0$. By superposition, a total field may be found as the sum of the two parts *if necessary*. Quite often, it is not necessary to add the two partial fields. For this reason, it is desirable to consider the two cases separately.

(1) $\mathbf{F} = 0$. This case was treated in considerable detail for the time domain (t shown explicitly) with sources (ρ_v and **J**) present. All we need for present purposes is to use phasor notation and set all the sources equal to zero. From Equation (3.78), the Lorentz condition is

$$\nabla \cdot \mathbf{A} = -j\omega\mu\varepsilon\Phi_a, \tag{6.54}$$

where Φ_a is the ordinary *electric scalar* potential, and the added subscript merely reminds us to associate it with **A**. Using Equations (3.80), (3.92), and (3.91), we have

$$\boxed{\begin{aligned} &\mathbf{F} = 0, \\ &\nabla^2 \mathbf{A} + k^2 \mathbf{A} = 0 \quad \text{(wave equation)}, \\ &\mathbf{E} = -j\omega\mathbf{A} + \frac{1}{j\omega\mu\varepsilon}\nabla(\nabla \cdot \mathbf{A}), \\ &\mathbf{B} = \nabla \times \mathbf{A}. \end{aligned}} \tag{6.55}$$

(2) $\mathbf{A} \equiv 0$. From Equations (6.52) and (6.47), we have

$$\nabla \times \mathbf{H} = -j\omega(\nabla \times \mathbf{F})$$

or

$$\nabla \times (\mathbf{H} + j\omega\mathbf{F}) = 0 \tag{6.56}$$

and $\mathbf{H} + j\omega\mathbf{F}$ is a conservative field. As in case (1), we define

$$\mathbf{H} + j\omega\mathbf{F} \equiv -\nabla\Phi_f \tag{6.57}$$

where Φ_f is *magnetic scalar potential* (amperes). Now substitute Equations (6.52) and (6.57) into Equation (6.46), so that

$$\nabla \times \nabla \times \mathbf{F} = j\omega\mu\varepsilon(-j\omega\mathbf{F} - \nabla\Phi_f)$$

or

$$\nabla \times \nabla \times \mathbf{F} - k^2\mathbf{F} + j\omega\mu\varepsilon\nabla\Phi_f = 0.$$

Expanding by vector identity gives

$$\nabla(\nabla \cdot \mathbf{F}) - \nabla^2\mathbf{F} - k^2\mathbf{F} + j\omega\mu\varepsilon\nabla\Phi_f = 0$$

or

$$\nabla(\nabla \cdot \mathbf{F} + j\omega\mu\varepsilon\Phi_f) - \nabla^2\mathbf{F} - k^2\mathbf{F} = 0. \tag{6.58}$$

An obvious choice for $\nabla \cdot \mathbf{F}$ is (again) the Lorentz condition,

$$\nabla \cdot \mathbf{F} \equiv = -j\omega\mu\varepsilon\Phi_f, \tag{6.59}$$

so that

$$\nabla^2\mathbf{F} + k^2\mathbf{F} = 0 \quad \text{(wave equation)}. \tag{6.50}$$

Substituting Equation (6.59) into Equation (6.57) gives

$$\mathbf{H} = -j\omega\mathbf{F} + \frac{1}{j\omega\mu\varepsilon}\nabla(\nabla \cdot \mathbf{F}). \tag{6.61}$$

Case (2) may be summarized

$$\boxed{\begin{aligned} \mathbf{A} &= 0, \\ \nabla^2\mathbf{F} + k^2\mathbf{F} &= 0 \quad \text{(wave equation)}, \\ \mathbf{H} &= -j\omega\mathbf{F} + \frac{1}{j\omega\mu\varepsilon}\nabla(\nabla \cdot \mathbf{F}), \\ \mathbf{D} &= -\nabla \times \mathbf{F}. \end{aligned}} \tag{6.62}$$

Note that these results could have been written directly from the set of Equations (6.55) after comparing Equations (6.46), (6.47), (6.52), and (6.53) (principle of duality).

As previously mentioned, the sum of the fields given by Equations (6.55) and (6.62) is a general solution. We will find that Equations (6.55) and (6.62) will *separately* accommodate all the waveguide problems of interest here.

6.5 THE RECTANGULAR WAVEGUIDE, HIGHER-ORDER MODES

The rectangular waveguide is shown in Figure 6.1. The boundary conditions are given by the set of Equations (6.1). The choice $\mathbf{F} \equiv 0$, $\mathbf{A} = \mathbf{a}_z A_z$, $A_x = A_y \equiv 0$ will give $H_z = 0$, and is case (1) of Section 6.4. Substituting this value of $\mathbf{A}$ into Equations (6.55), and then expanding in Cartesian coordinates gives

$$\frac{\partial^2 A_z}{\partial x^2} + \frac{\partial^2 A_z}{\partial y^2} + \frac{\partial^2 A_z}{\partial z^2} + k^2 A_z = 0 \tag{6.63}$$

and

$$\begin{aligned}
E_x &= \frac{1}{j\omega\mu\varepsilon} \frac{\partial^2 A_z}{\partial x \partial z}, & H_x &= \frac{1}{\mu} \frac{\partial A_z}{\partial y}, \\
E_y &= \frac{1}{j\omega\mu\varepsilon} \frac{\partial^2 A_z}{\partial y \partial z}, & H_y &= -\frac{1}{\mu} \frac{\partial A_z}{\partial x}, \quad \text{(TM to } z\text{)}. \\
E_z &= \frac{1}{j\omega\mu\varepsilon} \left(k^2 + \frac{\partial^2}{\partial z^2}\right) A_z, & H_z &\equiv 0,
\end{aligned} \tag{6.64}$$

This is always a TM *to z mode* since the magnetic field is entirely transverse ($H_z = 0$) to the z direction.

The boundary condition on the magnetic field will be automatically accounted for. A general solution to Equation (6.63), as derived in Appendix C by separation of variables is

$$A_z(x, y, z) = h(k_x x) h(k_y y) h(k_z z), \tag{6.65}$$

where the *separation equation* is

$$k_z^2 = k^2 - k_x^2 - k_y^2. \tag{6.66}$$

k_x, k_y, and k_z are called *eigenvalues* (characteristic values) and h is a *harmonic* function, or

$$h(k_x x) = A \cos k_x x + B \sin k_x x, \tag{6.67}$$

$$h(k_y y) = C \cos k_y y + D \sin k_y y, \tag{6.68}$$

and

$$h(k_z z) = E \cos k_z z + F \sin k_z z. \tag{6.69}$$

We expect propagation in the $+z$ direction, so we choose $F = -jE$ to make

$$h(k_z z) = E e^{-jk_z z}. \tag{6.70}$$

Boundary conditions on E_z will be satisfied if they are satisfied for A_z itself. This requires that $A = C = 0$,

$$k_x = \frac{m\pi}{a}, \quad m = 1, 2, 3, \ldots \tag{6.71}$$

and

$$k_y = \frac{n\pi}{b}, \quad n = 1, 2, 3, \ldots \tag{6.72}$$

The use of *either* m or $n = 0$ gives a trivial solution. Thus,

$$A_z = BDE \sin k_x x \sin k_y y \, e^{-jk_z z} \tag{6.73}$$

and boundary conditions on E_x and E_y are also satisfied. It is convenient to let $BDE = 1$, so that

$$A_z = \sin k_x x \sin k_y y \, e^{-jk_z z}. \tag{6.74}$$

Substituting Equations (6.71) and (6.72) into Equation (6.66) gives

$$k_z^2 = k^2 - \left(\frac{m\pi}{a}\right)^2 - \left(\frac{n\pi}{b}\right)^2. \tag{6.75}$$

The field is given by substitution of Equation (6.74) into Equations (6.64) and will be presented in tabular form.

Following the procedure used in Section 6.1, we have

$$k_z = \begin{cases} \sqrt{k^2 - \left(\frac{m\pi}{a}\right)^2 - \left(\frac{n\pi}{b}\right)^2} = \beta, & k^2 > \left(\frac{m\pi}{a}\right)^2 + \left(\frac{n\pi}{b}\right)^2, \\ -j\sqrt{\left(\frac{m\pi}{a}\right)^2 + \left(\frac{n\pi}{b}\right)^2 - k^2} = -j\alpha, & k^2 < \left(\frac{m\pi}{a}\right)^2 + \left(\frac{n\pi}{b}\right)^2. \end{cases} \tag{6.76}$$

The transition or cutoff frequency occurs for $k_z = 0$, or

$$\boxed{\omega_c = \frac{\pi}{\sqrt{\mu\varepsilon}} \sqrt{\left(\frac{m}{a}\right)^2 + \left(\frac{n}{b}\right)^2}.} \tag{6.77}$$

Substituting Equation (6.77) into Equation (6.76) gives [see Equation (6.17)]

6.5 THE RECTANGULAR WAVEGUIDE, HIGHER-ORDER MODES

$$k_z = \begin{cases} k\sqrt{1 - \left(\dfrac{\omega_c}{\omega}\right)^2} = \beta, & \omega > \omega_c, \\ -jk_c\sqrt{1 - \left(\dfrac{\omega}{\omega_c}\right)^2} = -j\alpha, & \omega < \omega_c. \end{cases} \quad (6.78)$$

The choice $\mathbf{A} \equiv 0$, $\mathbf{F} = \mathbf{a}_z F_z$, $F_x = F_y \equiv 0$ will give $E_z = 0$ and is case (2) of Section 6.4. Substituting this value of $\mathbf{F}$ into Equations (6.62) gives

$$\frac{\partial^2 F_z}{\partial x^2} + \frac{\partial^2 F_z}{\partial y^2} + \frac{\partial^2 F_z}{\partial z^2} + k^2 F_z = 0 \quad (6.79)$$

and

$$\boxed{\begin{aligned} E_x &= -\frac{1}{\varepsilon}\frac{\partial F_z}{\partial y}, & H_x &= \frac{1}{j\omega\mu\varepsilon}\frac{\partial^2 F_z}{\partial x\,\partial z}, \\ E_y &= \frac{1}{\varepsilon}\frac{\partial F_z}{\partial x}, & H_y &= \frac{1}{j\omega\mu\varepsilon}\frac{\partial^2 F_z}{\partial y\,\partial z}, \\ E_z &\equiv 0, & H_z &= \frac{1}{j\omega\mu\varepsilon}\left(k^2 + \frac{\partial^2}{\partial z^2}\right)F_z, \end{aligned}} \quad \text{(TE to } z.\text{)} \quad (6.80)$$

This is always a TE to z mode since the electric field is transverse to z ($E_z = 0$). Note the duality between the TM mode [Equations (6.64)] field and the TE mode [Equations (6.80)] field.

For reasons already given, the solution to F_z must be of the form

$$F_z = h(k_x x)h(k_y y)e^{-jk_z z}. \quad (6.81)$$

The boundary conditions are, of course, the same as those given by the set of Equations (6.1). Boundary conditions on E_z are automatically satisfied since $E_z = 0$ for any TE to z mode. Inspection of Equations (6.80) reveals that boundary conditions on E_x and E_y are satisfied if $h(k_x x)$ and $h(k_y y)$ are such that

$$F_z = \cos\frac{m\pi x}{a}\cos\frac{n\pi y}{b}e^{-jk_z z}. \quad (6.82)$$

In this case, $m = 0, 1, 2, 3, \ldots$ and $n = 0, 1, 2, 3, \ldots$, but *both* m and n cannot be zero, or a trivial solution will result.

The separation equation is the same as Equation (6.66) since the eigenvalues (k_x, k_y, and k_z) are the same. For this reason the cutoff frequencies are again given by Equation (6.77). The field is obtained by substituting Equation (6.82) into Equations (6.80). The general results for both the TE to z and TM to z mode are shown in Table 6.1. The peak values of E_x in the TM case and E_y in the TE case have been normalized to E_0 for convenience.

TABLE 6.1
Rectangular Waveguide Field Equations with Propagation in the $+z$ Direction

$$k_z = \sqrt{k^2 - (m\pi/a)^2 - (n\pi/b)^2}, \quad k = \omega\sqrt{\mu\varepsilon}$$

| TM to z | $m = 1, 2, \ldots$ $m = 1, 2, \ldots$ | TE to z | $m = 0, 1, 2, \ldots$ $n = 0, 1, 2, \ldots$ | m, n not both zero |

$$E_x = E_0 \cos\frac{m\pi x}{a} \sin\frac{n\pi y}{b} e^{-jk_z z}$$

$$E_x = -\frac{na}{mb} E_0 \cos\frac{m\pi x}{a} \sin\frac{n\pi y}{b} e^{-jk_z z}$$

$$E_y = \frac{na}{mb} E_0 \sin\frac{m\pi x}{a} \cos\frac{n\pi y}{b} e^{-jk_z z}$$

$$E_y = E_0 \sin\frac{m\pi x}{a} \cos\frac{n\pi y}{b} e^{-jk_z z}$$

$$E_z = ja\frac{k^2 - k_z^2}{m\pi k_z} E_0 \sin\frac{m\pi x}{a} \sin\frac{n\pi y}{b} e^{-jk_z z}$$

$$E_z \equiv 0$$

$$H_x = -\frac{na}{mb} E_0 \sin\frac{m\pi x}{a} \cos\frac{n\pi y}{b} e^{-jk_z z}$$

$$H_x = -\frac{k_z}{\omega\mu} E_0 \sin\frac{m\pi x}{a} \cos\frac{n\pi y}{b} e^{-jk_z z}$$

$$H_y = \frac{\omega\varepsilon}{k_z} E_0 \cos\frac{m\pi x}{a} \sin\frac{n\pi y}{b} e^{-jk_z z}$$

$$H_y = -\frac{na}{mb}\frac{k_z}{\omega\mu} E_0 \cos\frac{m\pi x}{a} \sin\frac{n\pi y}{b} e^{-jk_z z}$$

$$H_z \equiv 0$$

$$H_z = ja\frac{k^2 - k_z^2}{m\pi\omega\mu} E_0 \cos\frac{m\pi x}{a} \cos\frac{n\pi y}{b} e^{-jk_z z}$$

Example 6.5

Suppose that $a = b = 10$ cm and the air-filled waveguide is excited in the TM_{11} mode (Figure 6.11). Then, we have

$$f_c = \frac{1}{2\sqrt{\mu\varepsilon}}\sqrt{\left(\frac{1}{a}\right)^2 + \left(\frac{1}{b}\right)^2} = (1.5 \times 10^8)\sqrt{200} = 2.121 \quad \text{GHz}.$$

If we choose an operating frequency of 3 GHz, then $k = (2\pi \times 3 \times 10^9)/(3 \times 10^8) = 20\pi$ (rad/m), and $k_z = 20\pi\sqrt{1 - (2.121/3)^2} = 44.43$ rad/m. Using Table 6.1, we get

$$E_x = E_0 \cos(10\pi x)\sin(10\pi y)e^{-j44.43z},$$
$$E_y = E_0 \sin(10\pi x)\cos(10\pi y)e^{-j44.43z},$$
$$E_z = 1.414 E_0 \sin(10\pi x)\sin(10\pi y)e^{-j44.43z},$$
$$H_x = -3.751 \times 10^{-3} E_0 \sin(10\pi x)\cos(10\pi y)e^{-j44.43z},$$
$$H_y = 3.751 \times 10^{-3} E_0 \cos(10\pi x)\sin(10\pi y)e^{-j44.43z},$$
$$H_z \equiv 0.$$

Note that all boundary conditions are satisfied, and $\nabla \cdot \mathbf{E} = \nabla \cdot \mathbf{H} \equiv 0$. ∎

6.5 THE RECTANGULAR WAVEGUIDE, HIGHER-ORDER MODES 341

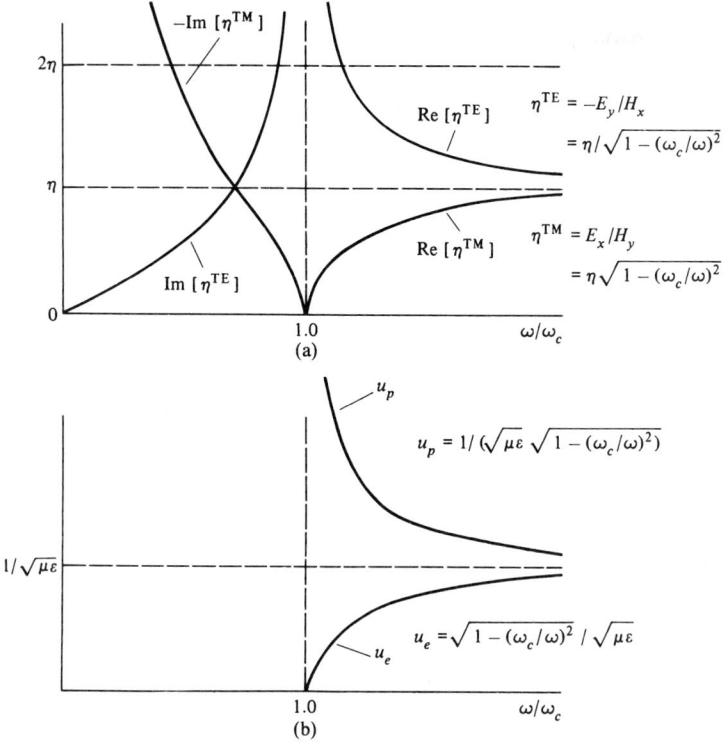

Figure 6.10. (a) TM (E_x/H_y) and TE ($-E_y/H_x$) wave impedances. (b) Phase velocity and velocity of energy flow (rectangular waveguide).

The equations for the wave impedance, phase velocity, energy velocity, group velocity, and guide wavelength are identical in form to those for the TE_{10} mode. $<\mathcal{S}_z>$ is now more complicated, but may always be determined from

$$<\mathcal{S}_z> = \tfrac{1}{2}\text{Re}(\mathbf{E} \times \mathbf{H}^*)_z. \tag{6.83}$$

Plots of η^+, u_p, and u_e are shown in Figure 6.10.

Modes are designated TE_{mn} or TM_{mn}. The dominant mode (lowest ω_c) for the rectangular waveguide is the TE_{10} for $a > b$, or the TE_{01} for $b > a$. This is determined by inspection of Equation (6.77). Field plots for the TE_{10}, TE_{20}, and TM_{11} modes are shown in Figure 6.11 using the field equations in Table 6.1. These field distributions are obtained from the time domain forms at $t = 0$, and are, therefore, like "snapshots." Higher-order field plots may be obtained in the same manner, if desired.

■ Example 6.6

The x-band waveguide of Examples 6.3 and 6.4 ($a = 0.9$ inch, $b = 0.4$ inch) is normally operated at frequencies near 10 GHz. Can any other modes propagate at this frequency? Equation (6.77) gives the cutoff frequency for any TE to z or TM

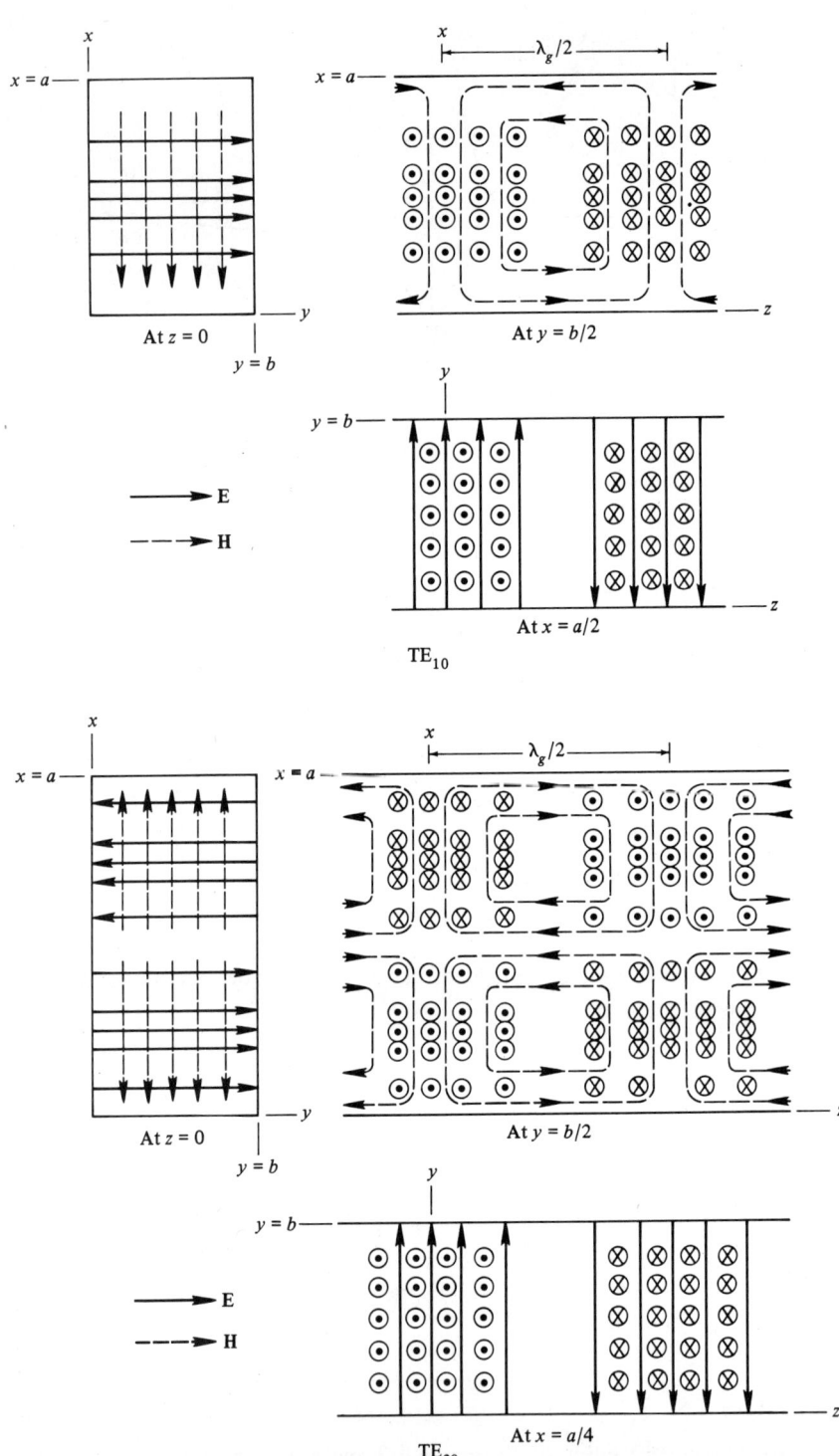

Figure 6.11. Field distributions for the TE$_{10}$, TE$_{20}$, and TM$_{11}$ rectangular waveguide modes.

6.5 THE RECTANGULAR WAVEGUIDE, HIGHER-ORDER MODES

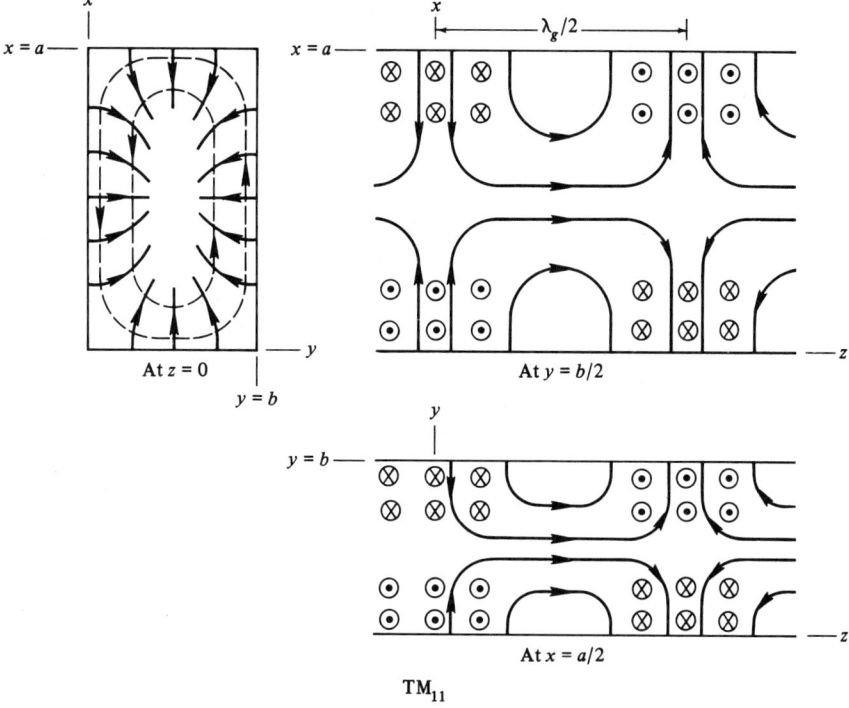

Figure 6.11. (*Continued*)

to z mode:

$$f_c = \frac{1}{2\sqrt{\mu\varepsilon}}\sqrt{(m/a)^2 + (n/b)^2}.$$

For the TE_{01} mode ($m = 0$, $n = 1$), we have

$$f_c = \frac{1.5 \times 10^8}{(0.4)(2.54 \times 10^{-2})} = 14.76 \quad \text{GHz.}$$

For the TE_{20} mode ($m = 2$, $n = 0$), we have

$$f_c = \frac{3 \times 10^8}{(0.9)(2.54 \times 10^{-2})} = 13.12 \quad \text{GHz.}$$

For the TE_{11} or TM_{11} mode ($m = n = 1$), we have

$$f_c = 1.5 \times 10^8 \sqrt{(1.914 \times 10^3) + (9.688 \times 10^3)} = 16.16 \quad \text{GHz.}$$

Other higher-order modes will have higher cutoff frequencies. Thus, we see that the x-band waveguide operated at 10 GHz can propagate only the dominant TE_{10} mode, all other modes having cutoff frequencies above this frequency. ∎

Following exactly the same procedure that was used in this section to derive the field equations for the rectangular waveguide, the field equations for the circular waveguide (and cavity) are derived in Appendix C for the interested reader. Here, we are only interested in comparing the dominant mode from that geometry to the TE_{10} rectangular waveguide mode (dominant for $a > b$). The geometry is shown in Figure C.2. The TM mode fields are given by Equations (C.18), where Equation (C.25) gives A_z. The TE mode fields are given by Equations (C.31), where Equation (C.36) gives F_z. In particular, the dominant mode is the TE_{11} mode ($n = 1, l = 1$), and is given in normalized form by

$$E_\rho = E_0 J_1(x'_{11}\rho/a) \sin\phi e^{-jk_z z},$$

$$E_\phi = (x'_{11}\rho/a) E_0 J'_1(x'_{11}\rho/a) \cos\phi e^{-jk_z z},$$

$$E_z \equiv 0,$$

$$H_\rho = -\frac{(x'_{11}\rho/a) k_z}{\omega\mu} E_0 J'_1(x'_{11}\rho/a) \cos\phi e^{-jk_z z}, \qquad (6.84)$$

$$H_\phi = \frac{k_z}{\omega\mu} E_0 J_1(x'_{11}\rho/a) \sin\phi e^{-jk_z z},$$

$$H_z = \frac{(x'_{11}\rho/a)^2}{j\omega\mu\rho} E_0 J_1(x'_{11}\rho/a) \cos\phi e^{-jk_z z},$$

where $J_1(x)$ (see Figure C.1) is a Bessel function of the first kind, order one, and argument x, and $J'_1(x)$ is dJ_1/dx. In Equations (6.84), $x'_{11} = 1.841$ is the *first* root of $J'_1(x)$, so that when $\rho = a$, E_ϕ and H_ρ in Equations (6.84) will vanish as required by the perfectly conducting cylindrical surface. The cutoff frequency, propagation constant, and waveguide impedance are given by

$$f_c = \frac{x'_{11}}{2\pi a \sqrt{\mu\varepsilon}}, \qquad (6.85)$$

$$k_z = \begin{cases} k\sqrt{1 - (\omega_c/\omega)^2} = \beta, & \omega > \omega_c, \\ -jk_c\sqrt{1 - (\omega/\omega_c)^2} = -j\alpha, & \omega < \omega_c, \end{cases} \qquad (6.86)$$

$$(Z_0)^{TE} = \eta^+ = E_\rho/H_\phi = -E_\phi/H_\rho = \frac{\omega\mu}{k_z}. \qquad (6.87)$$

Note that the last two equations are identical in form to those for rectangular waveguides.

The field is plotted in Figure 6.12. Comparing this field to that for the TE_{10} rectangular waveguide mode (Figure 6.11) shows that they are very similar. In fact, they are so similar that we might expect that a gradual transition from the rectangular to the circular geometry (or vice versa) could be accomplished with little difficulty, and the resultant fields would be very close to what our theory predicts. Suppose that the rectangular waveguide is actually square and its perimeter

6.6 DIELECTRIC WAVEGUIDES

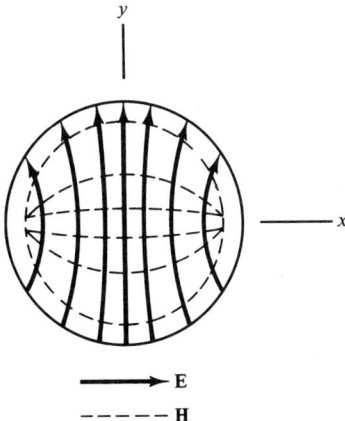

Figure 6.12. Field distribution for the TE_{11} cylindrical waveguide mode.

is identical to the circumference of the circular waveguide. It is easy to show that the cutoff frequency of the circular waveguide is 0.92 times the cutoff frequency of the rectangular waveguide. That is, they are almost the same. The implication of this discussion is that bends, twists, and other distortions of a rigid waveguide can be tolerated to a reasonable extent in practice.

Circular cylindrical waveguides have a number of applications besides the normal use as a structure for conveying microwave power from one point to another. Any time mechanical rotation is required (such as for feeding a rotating antenna), a circular geometry is called for. Another application of the circular waveguide that takes advantage of the circular symmetry is found in the "waveguide beyond cutoff" type attenuator. When a waveguide is operated beyond cutoff, the attenuation is exponentially distributed (an evanescent field) along the axial direction from the source toward the receiving end. A pickup probe can be mounted in a barrel which is threaded onto the waveguide. Thus, the probe can be moved along the waveguide by rotating the barrel. The attenuation *in decibels* will be linear with axial distance, and therefore linear with barrel turns.

6.6 DIELECTRIC WAVEGUIDES

It was suggested in Example 4.8 that a dielectric slab could serve as a waveguide. This is possible in principle because a plane wave in a dielectric that strikes a plane interface with air (for example) at an incident angle θ_i that is greater than the critical angle θ_{ic} will be totally reflected, leaving only damped fields in the air space. If the angle θ_i in Figure 6.13 is greater than the critical angle, then the dielectric slab can guide plane waves along a zigzag path in a manner similar to that for the hollow rectangular waveguide or the parallel plane system.

The circular cylindrical dielectric rod would certainly be the more practical geometry to consider, but its analysis would lead us to Bessel functions, and we choose to avoid that course here. Instead, we consider the much simpler two-

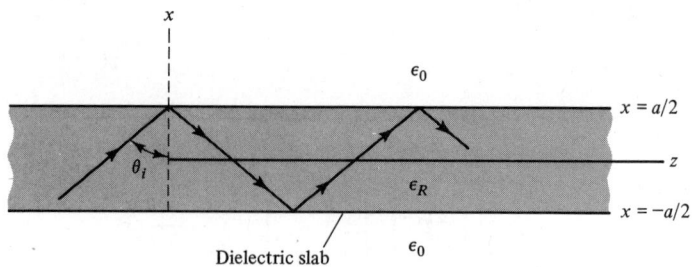

Figure 6.13. Dielectric slab waveguide.

dimensional dielectric slab of Figure 6.13 ($\partial/\partial y \equiv 0$). We also assume that $\mu = \mu_0$ and $\sigma = 0$ everywhere (no losses). In order to utilize the mathematics that we spent considerable time and effort on in Section 6.4, we can consider TM to z modes ($\mathbf{A} = \mathbf{a}_z A_z$, $\mathbf{F} \equiv 0$) or TE to z modes ($\mathbf{F} = \mathbf{a}_z F_z$, $\mathbf{A} \equiv 0$). For each of these we can consider two subcases: A_z (or F_z) is an odd function of x, or A_z (or F_z) is an even function of x. Of these four possibilities we will only explicitly consider the TM to z case where A_z is an odd function of x.

Equation (6.63), with $\partial/\partial y = 0$, becomes

$$\frac{\partial^2 A_z}{\partial x^2} + \frac{\partial^2 A_z}{\partial z^2} + k^2 A_z = 0, \tag{6.88}$$

whose solution is (once again) the product of harmonic functions:

$$A_z(x,z) = h(k_x x) h(k_z z), \tag{6.89}$$

where the separation equation is

$$k_z^2 = k^2 - k_x^2 \quad (k_y = 0). \tag{6.90}$$

We choose

$$h(k_z z) = E e^{-j k_z z} \quad (-\infty < x < \infty) \tag{6.91}$$

for propagation in the $+z$ direction (as before). The field is found from the set of Equations (6.64) with $\partial/\partial y = 0$:

$$\begin{aligned} E_x &= -\frac{k_z}{\omega\mu\varepsilon} \frac{\partial A_z}{\partial x}, \\ E_z &= \frac{k_x^2}{j\omega\mu\varepsilon} A_z, \\ H_y &= -\frac{1}{\mu} \frac{\partial A_z}{\partial x}. \end{aligned} \tag{6.92}$$

6.6 DIELECTRIC WAVEGUIDES

In the dielectric slab we have

$$\left.\begin{array}{l} k = k_d = \omega\sqrt{\mu_0\varepsilon_0\varepsilon_R} = k_0\sqrt{\varepsilon_R}, \\ k_x = k_{xd}, \\ k_z^2 = k_d^2 - k_{xd}^2, \end{array}\right\} \quad |x| < a/2, \quad k_0 = \omega\sqrt{\mu_0\varepsilon_0}, \quad (6.93)$$

while in the surrounding medium, we have

$$\left.\begin{array}{l} k = k_0 = \omega\sqrt{\mu_0\varepsilon_0}, \\ k_x = k_{x0}, \\ k_z^2 = k_0^2 - k_{x0}^2, \end{array}\right\} \quad |x| > a/2. \quad (6.94)$$

In order to make A_z an odd function of x, as stipulated, we choose

$$h(k_x x) = C \sin(k_{xd} x), \quad |x| < a/2$$

and, with $CE \equiv A$, Equation (6.89) becomes

$$A_{zd}(x,z) = A \sin(k_{xd} x) e^{-jk_z z}, \quad |x| < a/2. \quad (6.95)$$

The field then becomes

$$\begin{aligned} E_x &= -\frac{k_z k_{xd}}{\omega\mu_0\varepsilon_0\varepsilon_R} A \cos(k_{xd} x) e^{-jk_z z}, \\ E_z &= \frac{k_{xd}^2}{j\omega\mu_0\varepsilon_0\varepsilon_R} A \sin(k_{xd} x) e^{-jk_z z}, \\ H_y &= -\frac{k_{xd}}{\mu_0} A \cos(k_{xd} x) e^{-jk_z z}. \end{aligned} \quad (6.96)$$

It is expected that above the cutoff frequency the guide propagates unattenuated modes (k_z is real), while below cutoff there is attenuated propagation ($k_z = \beta - j\alpha$). This behavior is different from that of the rectangular waveguide. The loss here can only be accounted for by *radiation* since there is no heat loss anywhere. Thus, the dielectric slab can be thought of as an antenna when operated below cutoff.

We are then led to choose

$$A_{z0}(x,z) = \begin{cases} B e^{+jk_{x0} x} e^{-jk_z z} & (x > a/2), \quad (6.97) \\ B e^{-jk_{x0} x} e^{-jk_z z} & (x < -a/2), \quad (6.98) \end{cases}$$

as the odd function of x for the air space outside the dielectric slab. Note that for unattenuated wave propagation (above cutoff) k_{x0} will be imaginary, and the fields for $|x| > a/2$ will be exponentially damped. Below cutoff k_{x0} will be real, giving the radiated fields mentioned in the preceding paragraph. Equations (6.97) and (6.98) give the fields for $|x| > a/2$:

$$\left. \begin{aligned} E_x &= -j \frac{k_z k_{x0}}{\omega \mu_0 \varepsilon_0} B e^{+jk_{x0}x} e^{-jk_z z}, \\ E_z &= -j \frac{k_{x0}^2}{\omega \mu_0 \varepsilon_0} B e^{+jk_{x0}x} e^{-jk_z z}, \\ H_y &= -j \frac{k_{x0}}{\mu_0} B e^{+jk_{x0}x} e^{-jk_z z}, \end{aligned} \right\} \quad x > a/2; \qquad (6.99)$$

$$\left. \begin{aligned} E_x &= -j \frac{k_z k_{x0}}{\omega \mu_0 \varepsilon_0} B e^{-jk_{x0}x} e^{-jk_z z}, \\ E_z &= +j \frac{k_{x0}^2}{\omega \mu_0 \varepsilon_0} B e^{-jk_{x0}x} e^{-jk_z z}, \\ H_y &= -j \frac{k_{x0}}{\mu_0} B e^{-jk_{x0}x} e^{-jk_z z}, \end{aligned} \right\} \quad x < -a/2. \qquad (6.100)$$

Boundary conditions are satisfied when E_z and H_y are continuous at $x = \pm a/2$. Using Equations (6.96), (6.99), and (6.100) gives

$$\left. \begin{aligned} \frac{k_{xd}^2}{\varepsilon_R} A \sin(k_{xd} a/2) &= k_{x0}^2 B e^{+jk_{x0} a/2}, \\ k_{xd} A \cos(k_{xd} a/2) &= j k_{x0} B e^{+jk_{x0} a/2}, \end{aligned} \right\} \quad x = \pm a/2. \qquad (6.101)$$

The ratio of the first to the second equation in the set gives

$$\frac{j k_{xd}}{\varepsilon_R} \tan(k_{xd} a/2) = k_{x0}. \qquad (6.102)$$

This last equation along with the last equation in the sets of Equations (6.93) and (6.94),

$$k_z^2 = k_d^2 - k_{xd}^2, \qquad (6.103)$$
$$k_z^2 = k_0^2 - k_{x0}^2, \qquad (6.104)$$

determine the characteristic equation of the system.
The last two equations can be combined:

$$k_d^2 - k_{xd}^2 = k_0^2 - k_{x0}^2 \qquad (6.105)$$

6.6 DIELECTRIC WAVEGUIDES

or

$$k_{xd}^2 - k_{x0}^2 = k_d^2 - k_0^2 = k_0^2(\varepsilon_R - 1) > 0. \qquad (6.106)$$

Equation (6.102) shows that either k_{x0} will be imaginary and k_{xd} real (above cutoff) or k_{x0} will be real and k_{xd} imaginary (below cutoff). Equations (6.103) and (6.104) give

$$k_{xd}^2 = \sqrt{k_d^2 - k_z^2}$$

and

$$k_{x0}^2 = \sqrt{k_0^2 - k_z^2}.$$

Therefore, above cutoff $k_d > k_z$ and $k_0 < k_z$, or

$$k_0 < k_z < k_d \qquad (6.107)$$

and cutoff occurs when $k_z = k_0$ and $k_{x0} = 0$. It is worth mentioning that the inequalities in (6.107) hold for all "cylindrical" dielectric waveguides when in a propagating mode. We finally have found that cutoff occurs [using Equation (6.102)] for

$$\frac{k_{xd}}{\varepsilon_R} \tan(k_{xd}a/2) = 0$$

or

$$k_{xd}a/2 = n\pi, \quad n = 0, 1, 2, \ldots .$$

When $k_{x0} = 0$, $k_z = k_0$ and $k_{xd} = \sqrt{k_d^2 - k_0^2} = k_0\sqrt{\varepsilon_R - 1}$. So $k_0\sqrt{\varepsilon_R - 1}\, a/2 = n\pi$, or $\omega_c\sqrt{\mu_0\varepsilon_0}\sqrt{\varepsilon_R - 1}\, a/2 = n\pi$, or $\omega_c\sqrt{\mu_0\varepsilon_0}\sqrt{\varepsilon_R - 1}\, a/2 = n\pi$, or

$$\boxed{\omega_c = \frac{2n\pi}{a\sqrt{\mu_0\varepsilon_0}\sqrt{\varepsilon_R - 1}}}. \qquad (6.108)$$

Modes can be designated TM_n^0 (0 for *odd*). For $n = 0$ the cutoff frequency is zero and the lowest-order TM_n mode propagates without attenuation, regardless of how small a becomes. This is another property of all cylindrical dielectric waveguides. Note, however, that for small a, k_{x0} is very small and imaginary so the field extends to relatively large distances from the slab. In other words, it does not "stick" closely to the slab.

The propagation constant k_z that we usually need to know, is determined by solving Equation (6.102) using Equation (6.106):

$$j\frac{k_{xd}}{\varepsilon_R} \tan(k_{xd}a/2) = k_{x0} = \sqrt{k_{xd}^2 - k_0^2(\varepsilon_R - 1)} = j\sqrt{k_0^2(\varepsilon_R - 1) - k_{xd}^2}$$

or

$$\frac{k_{xd}a/2}{\varepsilon_R} \tan(k_{xd}a/2) = \sqrt{(k_0a/2)^2(\varepsilon_R - 1) - (k_{xd}a/2)^2} = -jk_{x0}. \quad (6.109)$$

■ Example 6.7

Suppose that a dielectric slab 1 mm thick has $\varepsilon_R = 4$, and the operating frequency is $\omega = 300 \times 10^9$ rad/s. We want to determine what TM_n^0 modes can propagate, and how the field configuration looks. We have

$$\frac{k_0 a}{2}\sqrt{\varepsilon_R - 1} = (\omega\sqrt{\mu_0\varepsilon_0}\,a/2)\sqrt{\varepsilon_R - 1} = \sqrt{3}/2,$$

and with Equation (6.109) ($u \equiv k_{xd}a/2$):

$$\frac{u}{4}\tan u = \sqrt{3/4 - u^2}.$$

The solution (using root-solve, trial and error, or graphical techniques) is

$$u = k_{xd}a/2 = 0.8348 \quad \text{or} \quad k_{xd} = 1.670 \times 10^3 \quad \text{rad/m}.$$

Next, we have

$$k_z = \sqrt{k_0^2\varepsilon_R - k_{xd}^2} = \sqrt{4 \times 10^6 - 2.788 \times 10^6} = 1.101 \times 10^3 \quad \text{rad/m},$$

$$k_{x0} = j\sqrt{k_0^2(\varepsilon_R - 1) - k_{xd}^2}$$

$$= j\sqrt{3 \times 10^6 - 2.788 \times 10^6} = j0.461 \times 10^3 \quad (\text{rad/m}),$$

$$k_0 = 10^3 \quad \text{rad/m},$$

$$k_d = k_0\sqrt{\varepsilon_R} = 2 \times 10^3 \quad \text{rad/m}.$$

Note that $k_0 < k_z < k_d$ as required. The cutoff frequencies are given by Equation (6.108):

$$\omega_c = \frac{2n\pi}{a\sqrt{\mu_0\varepsilon_0}\sqrt{\varepsilon_R - 1}} = 1.088 \times 10^{12}n \quad (\text{rad/s}),$$

so the only propagating TM_n^0 mode is the TM_0^0 mode ($n = 0$).

6.6 DIELECTRIC WAVEGUIDES

Returning to the second equation of the set of Equations (6.101),

$$B = \frac{k_{xd} \cos(k_{xd}a/2)}{jk_{x0}e^{+jk_{x0}a/2}} A = -3.062A.$$

Using this value of B, substituting the previously calculated values for the various phase constants, and normalizing the field quantities in Equations (6.96), (6.99), and (6.100) to a peak value of E_0 for $E_x(|x| < a/2)$ gives

$$\left.\begin{aligned} E_x &= E_0 \cos(k_{xd}x)e^{-jk_z z}, \\ E_z &= +j1.517 E_0 \sin(k_{xd}x)e^{-jk_z z}, \\ H_y &= 9.637 \times 10^{-3} E_0 \cos(k_{xd}x)e^{-jk_z z}, \end{aligned}\right\} \quad |x| < a/2;$$

$$\left.\begin{aligned} E_x &= 3.381 E_0 e^{-461x} e^{-jk_z z}, \\ E_z &= +j1.416 E_0 e^{-461x} e^{-jk_z z}, \\ H_y &= 8.146 \times 10^{-3} E_0 e^{-461x} e^{-jk_z z}, \end{aligned}\right\} \quad x > a/2;$$

$$\left.\begin{aligned} E_x &= 3.381 E_0 e^{+461x} e^{-jk_z z}, \\ E_z &= -j1.416 E_0 e^{+461x} e^{-jk_z z}, \\ H_y &= 8.146 \times 10^{-3} E_0 e^{+461x} e^{-jk_z z}, \end{aligned}\right\} \quad x < -a/2.$$

The TM_0^0 mode is shown in Figure 6.14. Note that it is relatively easy to verify that boundary conditions are satisfied at $x = \pm a/2$. ■

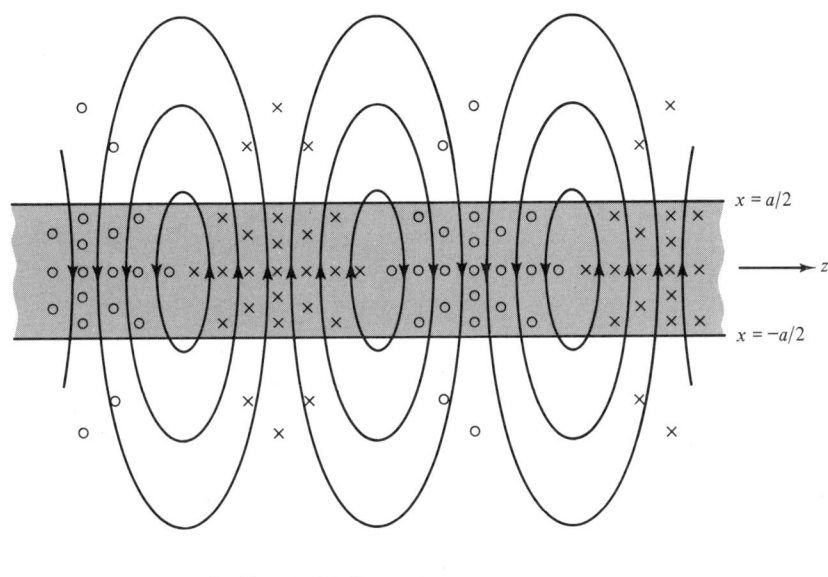

Figure 6.14. TM_0^0 dielectric slab field distribution.

The field in the dielectric slab can be reduced to that of plane waves in a manner similar to that used for the rectangular waveguide. Using the field of Example 6.7 for $|x| < a/2$:

$$2E_x = E_0 e^{+jk_{xd}x} e^{-jk_z z} + E_0 e^{-jk_{xd}x} e^{-jk_z z},$$

$$2E_z = 1.517 E_0 e^{+jk_{xd}x} e^{-jk_z z} - 1.517 E_0 e^{-jk_{xd}x} e^{-jk_z z},$$

$$2H_y = 9.637 \times 10^{-3} E_0 e^{+jk_{xd}x} e^{-jk_z z} + 9.637 \times 10^{-3} E_0 e^{-jk_{xd}x} e^{-jk_z z}.$$

As can be seen in Figure 6.15, $k_{xd} = k_d \sqrt{1 - (k_z/k_d)^2} = k_d \cos\theta$, and $k_z = k_d(k_z/k_d) = k_d \sin\theta$, so

$$2E_x = E_0 e^{+jk_d(x\cos\theta - z\sin\theta)} + E_0 e^{-jk_d(x\cos\theta + z\sin\theta)},$$

$$2E_z = 1.517 E_0 e^{+jk_d(x\cos\theta - z\sin\theta)} - 1.517 E_0 e^{-jk_d(x\cos\theta + z\sin\theta)},$$

$$2H_y = 9.637 \times 10^{-3} E_0 e^{+jk_d(x\cos\theta - z\sin\theta)}$$
$$+ 9.637 \times 10^{-3} E_0 e^{-jk_d(x\cos\theta + z\sin\theta)},$$

where $\theta = \sin^{-1}(k_z/k_d) = \cos^{-1}(k_{xd}/k_d) = 33.4°$. Equation (4.129) gives the critical angle:

$$\theta_{ic} = \sin^{-1}\sqrt{1/\varepsilon_R} = \sin^{-1}(0.5) = 30°.$$

Thus $\theta = \theta_i > \theta_{ic}$, as it must be for total reflection. On the other hand, when $\theta = \theta_{ic}$, $k_{x0} = \sqrt{k_0^2 - k_z^2} = \sqrt{k_0^2 - k_d^2 \cos^2\theta_{ic}} = k_0^2 - k_0^2 = 0$, which is exactly the expected result. Finally, note that

$$|\mathbf{E}/\mathbf{H}| = 188.5 = \eta_0/2 = \eta_d,$$

as expected.

It is possible to produce guidance of electromagnetic waves along "coated" surfaces or even along natural layers in a variety of ways. A simple demonstration

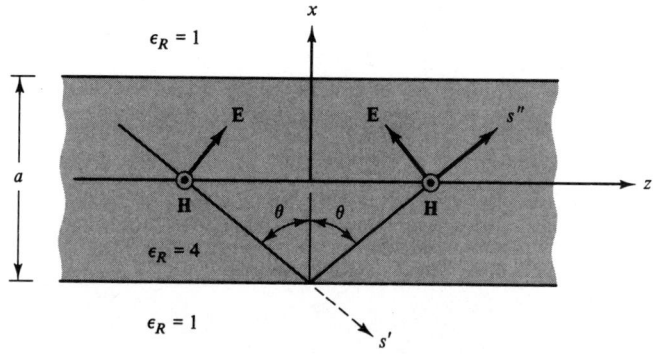

Figure 6.15. Plane waves in the dielectric slab waveguide.

6.6 DIELECTRIC WAVEGUIDES

using results that have already been derived will serve here. Consider the TM_n^0 modes of this section. The z component of the electric field ($|x| < a/2$) was

$$E_z = \frac{k_{xd}^2}{j\omega\mu_0\varepsilon_0\varepsilon_R} A \sin(k_{xd}x) e^{-jk_z z}$$

and is identically zero for $x = 0$. A perfectly conducting plane can therefore be placed at $x = 0$, and the field for $x > 0$ (for example) can still be supported (image theory). This is shown in Figure 6.16. The field is given by the equation sets (6.93) and (6.99) with no field for $x < 0$. These *surface guided waves* can also be supported in the TE_n^e mode by the geometry of Figure 6.16.

The analysis of the case where A_z is an even function of z proceeds in a similar manner. We choose

$$A_{zd} = A \cos k_{xd} e^{-jk_z z}, \quad |x| < a/2, \quad (6.110)$$

$$A_{z0} = B e^{+jk_{x0}|x|}, \quad |x| > a/2. \quad (6.111)$$

The field is still given by Equations (6.92), and the separation equations are still given by Equations (6.103) and (6.104). Satisfying the boundary conditions at the interfaces gives

$$-\frac{k_{xd}a/2}{\varepsilon_R} \cot(k_{xd}a/2) = \sqrt{(k_0 a/2)^2(\varepsilon_R - 1) - (k_{xd}a/2)^2}. \quad (6.112)$$

The cutoff frequencies are given by $[k_{x0} = 0, k_{xd}a/2 = (2n + 1)\pi/2, n = 0, 1, 2, \ldots]$

$$\omega_c = \frac{(2n + 1)\pi}{a\sqrt{\mu_0\varepsilon_0}\sqrt{\varepsilon_R - 1}}. \quad (6.113)$$

The TE case is dual to the TM case, and the results for the dielectric slab waveguide are summarized in Table 6.2.

The two-dimensional dielectric slab waveguide is not so important from a practical point of view, but its analysis is important because the principles that are involved carry over to the very important cylindrical dielectric rod waveguide.

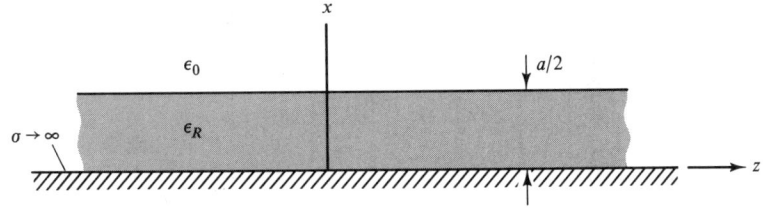

Figure 6.16. Dielectric slab above a perfectly conducting ground plane.

TABLE 6.2
Dielectric Slab Waveguide Properties

	TM (A_z odd)	TM (A_z even)	TE (F_z odd)	TE (F_z even)
Vector potential (times $e^{jk_z z}$)	$A\sin(k_{xd}x)$, $\|x\| < a/2$ $B\exp(+jk_{x0}x)$, $x > a/2$ $B\exp(-jk_{x0}x)$, $x < -a/2$	$A\cos(k_{xd}x)$, $\|x\| < a/2$ $B\exp(+jk_{x0}\|x\|)$, $\|x\| > a/2$	$A\sin(k_{xd}x)$, $\|x\| < a/2$ $B\exp(+jk_{x0}x)$, $x > a/2$ $B\exp(-jk_{x0}x)$, $x < -a/2$	$A\cos(k_{xd}x)$, $\|x\| < a/2$ $B\exp(+jk_{x0}\|x\|)$, $\|x\| > a/2$
Field $(\partial/\partial y) \equiv 0$	Eq. (6.64)	Eq. (6.64)	Eq. (6.80)	Eq. (6.80)
Propagation constant (k_z)	$u \tan u = \varepsilon_R v$	$-u \cot u = \varepsilon_R v$	$u \tan u = v$	$-u \cot u = v$
Cutoff frequency	$\omega_c = \dfrac{2n\pi}{a\sqrt{\mu_0\varepsilon_0(\varepsilon_R - 1)}}$	$\omega_c = \dfrac{(2n+1)\pi}{a\sqrt{\mu_0\varepsilon_0(\varepsilon_R - 1)}}$	$\omega_c = \dfrac{2n\pi}{a\sqrt{\mu_0\varepsilon_0(\varepsilon_R - 1)}}$	$\omega_c = \dfrac{(2n+1)\pi}{a\sqrt{\mu_0\varepsilon_0(\varepsilon_R - 1)}}$

$n = 0, 1, 2, \ldots, k_0 = \omega\sqrt{\mu_0\varepsilon_0}$, $k_d = \omega\sqrt{\mu_0\varepsilon_0\varepsilon_R}$, $k_z^2 = k_d^2 - k_{xd}^2 = k_0^2 - k_{x0}^2$, $u = k_{xd}a/2$, $v = \sqrt{(k_0 a/2)^2(\varepsilon_R - 1) - u^2}$.

6.7 LOSSLESS CAVITY RESONATORS

The latter can be used at optical frequencies for communications systems. With dielectrics such as quartz fibers these waveguides offer low loss, wide bandwidth, and the ability to take sharp bends without loss in performance.

6.7 LOSSLESS CAVITY RESONATORS

Energy storage elements are required for engineering purposes in the microwave range of frequencies. It was found in Chapter 5 that certain resonant lengths of transmission line behave as resonant circuits at certain resonant frequencies. We spent a good deal of time and effort analyzing the resonant quarter-wave coaxial cavity. Another example might be a parallel plane system shorted at both ends. We can visualize the two shorting ends as capacitor plates and the planes as inductances. In extending the resonant frequency upwards, we would naturally attempt to reduce the inductance by paralleling more conductors between the ends (like a cage). Ultimately, this leads to a closed box or cavity. It is this geometry, called a rectangular cavity, that we will investigate in this section.

The geometry of the rectangular cavity is shown in Figure 6.17. Note that this cavity is merely a section of rectangular waveguide shorted at both ends. Instead of a traveling wave in the axial (z) direction, we expect two traveling waves in *opposite* directions, or a *standing wave*.

(1) For a TM to z mode, Equations (6.65) and (6.66) are still valid. They are

$$A_z(x,y,z) = h(k_x x)h(k_y y)h(k_z z) \tag{6.114}$$

and

$$k_z^2 = k^2 - k_x^2 - k_y^2. \tag{6.115}$$

The field is given by Equations (6.64). As before, the boundary conditions at $x = 0, a$ and $y = 0, b$ are satisfied by the choice

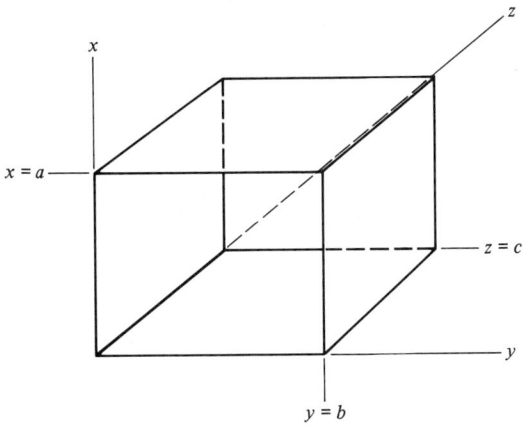

Figure 6.17. Geometry for the retangular cavity.

$$h(k_x x) = \sin \frac{m\pi x}{a}, \quad m = 1, 2, 3, \ldots \tag{6.116}$$

and

$$h(k_y y) = \sin \frac{n\pi y}{b}, \quad n = 1, 2, 3, \ldots \tag{6.117}$$

We how have the additional boundary condition

$$E_x = E_y = 0 \begin{cases} z = 0, \\ z = c. \end{cases} \tag{6.118}$$

Inspection of Equations (6.64) reveals that Equation (6.118) is satisfied if

$$h(k_z z) = \cos \frac{p\pi z}{c}, \quad p = 0, 1, 2, \ldots \tag{6.119}$$

Therefore, Equation (6.114) becomes

$$\boxed{A_z = \sin \frac{m\pi x}{a} \sin \frac{n\pi y}{b} \cos \frac{p\pi z}{c}}. \tag{6.120}$$

The separation equation becomes

$$\left(\frac{p\pi}{c}\right)^2 = k^2 - \left(\frac{m\pi}{a}\right)^2 - \left(\frac{n\pi}{b}\right)^2, \tag{6.121}$$

which can *only* be satisfied now at *discrete* frequencies called the *resonant* frequencies. This is a typical characteristic of lossless infinite Q systems. Solving Equation (6.121) for the resonant frequencies ($k = \omega\sqrt{\mu\varepsilon}$), we get

$$\boxed{f_r = \frac{1}{2\sqrt{\mu\varepsilon}} \sqrt{\left(\frac{m}{a}\right)^2 + \left(\frac{n}{b}\right)^2 + \left(\frac{p}{c}\right)^2}}. \tag{6.122}$$

The dominant TM_{mnp} mode is the TM_{110}.

(2) For a TE to z mode, we follow the same procedure and find that

$$\boxed{F_z = \cos \frac{m\pi x}{a} \cos \frac{n\pi y}{b} \sin \frac{p\pi z}{c}} \tag{6.123}$$

6.7 LOSSLESS CAVITY RESONATORS

is the potential required to satisfy both the wave equation and boundary conditions. In this case, $m = 0, 1, 2, \ldots, n = 0, 1, 2, \ldots,$ and $p = 1, 2, 3, \ldots,$ but *both* m and n cannot be zero. The field is given by Equations (6.80), and the separation equation is identical to Equation (6.115). Then, the resonant frequencies are the same and are given by Equation (6.122). If $a < b < c$, the dominant TE_{mnp} mode is the TE_{011} mode. It is easy to show that more than one mode may have the same resonant frequency. If this occurs, one mode will usually dominate the other because of the manner in which the cavity is excited. For example, an electric loop may excite one mode while being oriented in such a position as to make it impossible to couple to the other mode.

■ Example 6.8

When $a > b > c$, the dominant TE mode is the TE_{101} mode. Equation (6.123) gives ($m = 1, n = 0, p = 1$) $F_z = \cos(\pi x/a) \sin(\pi z/c)$, and when F_z is substituted into Equations (6.80), the field is

$$E_y = -\frac{\pi}{\varepsilon a} \sin(\pi x/a) \sin(\pi z/c),$$

$$H_x = -\frac{\pi^2}{j\omega\mu(ac)\varepsilon} \sin(\pi x/a) \cos(\pi z/c),$$

$$H_z = \frac{\pi^2}{j\omega\mu a^2 \varepsilon} \cos(\pi x/a) \sin(\pi z/c),$$

or, when normalized

$$E_y = E_0 \sin(\pi x/a) \sin(\pi z/c),$$

$$H_x = -j\frac{E_0}{2f_r \mu c} \sin(\pi x/a) \cos(\pi z/c),$$

$$H_z = +j\frac{E_0}{2f_r \mu a} \cos(\pi x/a) \sin(\pi z/c),$$

where

$$f_r = \frac{1}{2\sqrt{\mu\varepsilon}} \sqrt{(1/a)^2 + (1/c)^2}.$$

Note that the electric and magnetic fields are 90° out of phase. The field distribution for the TE_{101} mode for the case $a = c > b$ is shown in Figure 6.18. ■

The circular cavity is obtained by simply adding conducting planes at $z = 0$ and $z = d$ in Figure C.2. It is analyzed in Appendix C. The TM field is given by Equations (C.18), (C.40), and (C.41), while the TE field is given by Equations

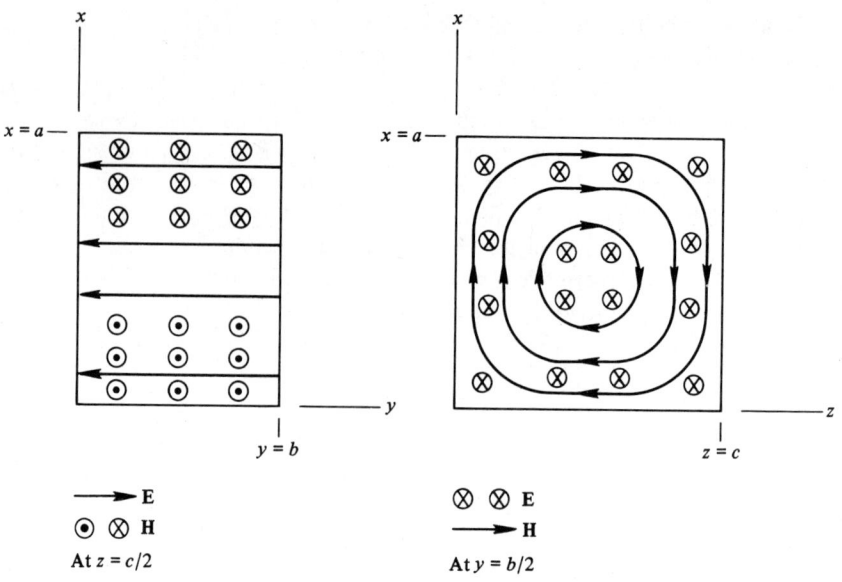

Figure 6.18. TE_{101} mode fields ($a = c > b$) inside a rectangular cavity.

(C.31), (C.42), and (C.43). When $d \leq 2a$, the dominant mode is the TM_{010} mode, and its field has a very simple form (Problem 20):

$$E_z = E_0 J_0(2.405\rho/a),$$
$$H_\phi = j(E_0/\eta) J_1(2.405\rho/a)$$

and its resonant frequency is

$$f_r = \frac{2.405}{2\pi a \sqrt{\mu\varepsilon}}. \tag{6.125}$$

The field is plotted in Figure 6.19. Note the similarity between this field and that for TE_{101} rectangular cavity mode (Figure 6.18). Note also that $J_0(2.405) = 0$ so that $E_z = 0$ when $\rho = a$, as required. Finally, note that E_z and H_ϕ are in phase quadrature.

6.8 WAVEGUIDE AND CAVITY LOSSES

If waveguide and cavity losses are small, as will generally be the case, the perturbation method introduced in Chapter 5 will give good engineering results. This method assumes that the field quantities are essentially unchanged from the lossless case when a small loss is present. The loss is then calculated from the lossless field equations and the known parameters of the surrounding media. In

6.8 WAVEGUIDE AND CAVITY LOSSES

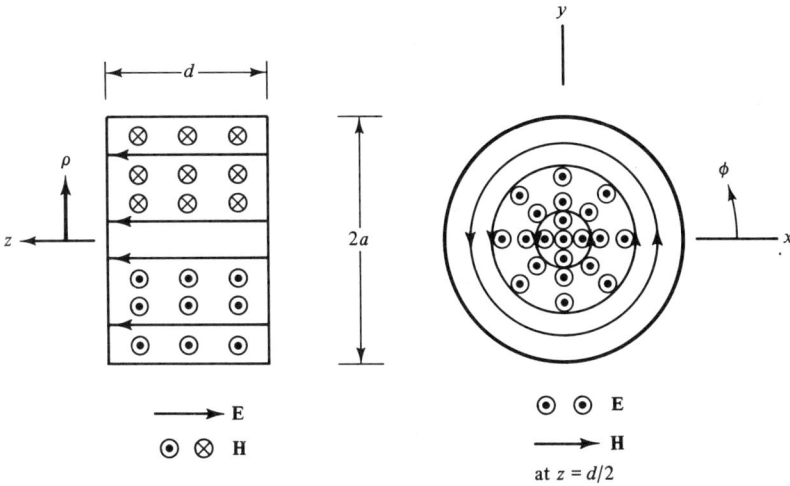

Figure 6.19. TM_{010} cylindrical cavity mode field distribution.

this chapter we will consider the most important cases and derive the attenuation constants and quality factors for them.

An exact equation (5.114), derived in Chapter 5 for the attenuation constant for waves propagating with damping or loss, is

$$\alpha = \frac{1}{2}\frac{<\mathcal{P}_d>}{<\mathcal{P}_f>} \quad \text{(Np/m)}, \tag{5.114}$$

where $<\mathcal{P}_d>$ is the time-average power dissipated per unit length and $<\mathcal{P}_f>$ is the time-average power flow in watts. The dissipation arises from several sources. The most important of these for hollow conducting waveguides and cavities is the finite conductivity of the walls because it is unavoidable.

The interior of the hollow waveguide or cavity may be a lossy dielectric. The interior medium may also have magnetic loss. Thus, it might be difficult to identify the source of the loss. Another factor that complicates the situation is that σ, μ, and ε all depend on frequency in general. It is important to recognize that there are no conductors used with the dielectric rod waveguide, and therefore the losses should be smaller than those for the conducting waveguide with a comparable dielectric inside.

An equation suitable for finding the Q of *any propagating system* of uniform cross section is Equation (5.146),

$$Q = \frac{\omega}{2\alpha u_e}. \tag{5.146}$$

In Equation (5.146), α is the attenuation constant due to losses in the propagating system *during propagation* and is not to be confused with the attenuation constant for a (lossless) evanescent mode below cutoff (therefore nonpropagating). An approximate formula for the Q may be found by noting that for any of the lossless systems we have studied *that required conductors*

$$u_e = \frac{\sqrt{1 - (\omega_c/\omega)^2}}{\sqrt{\mu\varepsilon}}, \quad f > f_c, \quad (6.126)$$

so that Equation (5.146) may be written

$$\boxed{Q \approx \frac{\omega\sqrt{\mu\varepsilon}}{2\alpha\sqrt{1 - (\omega_c/\omega)^2}}}. \quad (6.127)$$

Equation (6.127) is approximate since Equation (6.126) applies only to lossless systems. Note that Equation (6.127) reduces to Equation (5.147) for transmission lines, $Q \approx \beta/2\alpha$, because in the "small loss" TEM mode case $\beta \approx \omega\sqrt{\mu\varepsilon} = k$ and $\omega_c = 0$! In order to use Equation (6.127), we must calculate ω_c using lossless formulas. It should be pointed out here that if there is any loss, then there is no true cutoff frequency, since γ can never be zero. The procedure we will follow will be that of calculating the attenuation constant due to the waveguide walls separately from the attenuation constant due to the interior medium. When finished, we may use

$$\alpha = \alpha_c + \alpha_d \quad (6.128)$$

and

$$\frac{1}{Q_0} = \frac{1}{Q_c} + \frac{1}{Q_d}. \quad (6.129)$$

We must use the general form, Equation (5.114), when calculating the attenuation constant due to the finite conductivity of the waveguide walls α_c. The result will depend greatly on the waveguide geometry and the particular mode being employed. A calculation of α_c, then, will be made shortly for a specific example.

Suppose now the interior region of the waveguide has loss. In Chapter 5 we found for the resonant quarter-wave line that

$$Q_d = \frac{\omega\varepsilon}{\sigma_d}.$$

This is a general result, being independent of mode and geometry, so we will not repeat its calculation for the rectangular waveguide. Does this result hold for the dielectric rod waveguide?

1. TE₁₀ Mode, Rectangular Waveguide

The field equations for the TE_{10} to z dominant mode propagating in a lossless waveguide are given by the equation set 6.12, where k_z is given by Equation (6.17) and ω_c is given by Equation (6.16). If losses in the internal medium are neglected for the time being, we need consider only the losses in the conductor. Then, we use Equation (5.114),

$$\alpha_c = \frac{1}{2}\frac{<\mathscr{P}_d>}{<\mathscr{P}_f>}.$$

Now, from Equation (3.57), we get

$$<\mathscr{P}_f> = -\tfrac{1}{2}\text{Re}\left(\int_0^a\int_0^b E_y H_x^* \, dx \, dy\right)$$

$$= \tfrac{1}{2}\text{Re}\left(\int_0^a\int_0^b \frac{E_0^2}{\omega\mu} k_z \sin^2\frac{\pi x}{a} \, dx \, dy\right),$$

$$<\mathscr{P}_f> = \frac{ab}{4\eta^+}E_0^2 \quad (\text{W}), \tag{6.130}$$

a *lossless* value. Also, from Equation (4.71), we get

$$<\mathscr{P}_d> = \tfrac{1}{2}R_s\oint_{\text{walls}} |H_{\tan}|^2 \, ds, \tag{6.131}$$

where R_s is the surface resistance [see Equation (4.77)] and $|H_{\tan}|$ is the lossless value of the magnetic field tangent to the conducting surface. Thus, we have

$$<\mathscr{P}_d> = \tfrac{1}{2}R_s\left\{\int_0^1\int_0^b |H_z|^2_{x=0} \, dy \, dz + \int_0^1\int_0^b |H_z|^2_{x=a} \, dy \, dz\right.$$

$$+ \int_0^1\int_0^a (|H_x|^2 + |H_z|^2)_{y=0} \, dx \, dz$$

$$\left.+ \int_0^1\int_0^a (|H_x|^2 + |H_z|^2)_{y=b} \, dx \, dz\right\}$$

$$= \frac{R_s}{2}\left[2\left(\frac{\pi E_0}{\omega\mu a}\right)^2 b + a\left(\frac{\beta E_0}{\omega\mu}\right)^2 + a\left(\frac{\pi E_0}{\omega\mu a}\right)^2\right]$$

or

$$<\mathcal{P}_d> = \frac{R_s E_0^2}{2}\left[\left(\frac{\pi}{\omega\mu a}\right)^2 (a+2b) + \left(\frac{\beta}{\omega\mu}\right)^2 a\right] \quad \text{(W/m)}. \quad (6.132)$$

Substituting Equations (6.130) and (6.132) into Equation (5.114) and rearranging,

$$\alpha_c = \frac{R_s \eta^+}{ab}\left[\left(\frac{f_c}{\eta f}\right)^2 (a+2b) + \frac{a}{(\eta^+)^2}\right] \quad \text{(Np/m)}. \quad (6.133)$$

Remember that R_s and η^+ are both frequency dependent.

■ Example 6.9

As an example, suppose a silver plated waveguide is operated in the TE_{10} mode at 10 GHz with $a = 1.25$ cm and $b = 0.625$ cm. The waveguide is "loaded" with polystyrene to lower the cutoff frequency. From Equation (4.77), we get

$$R_s = \sqrt{\frac{\pi f \mu}{\sigma_c}} = 25.3 \quad \text{m}\Omega$$

for silver, a "good conductor" at 10 GHz. Also, the (lossless) intrinsic impedance is ($\varepsilon_R = 2.56$, Appendix B)

$$\sqrt{\frac{\mu}{\varepsilon}} = 235.6 \quad \Omega.$$

The cutoff frequency is

$$f_c = \frac{1}{2a\sqrt{\mu\varepsilon}} = 7.5 \quad \text{GHz}$$

and

$$\eta^+ = \frac{\eta}{\sqrt{1-(f_c/f)^2}} = 356.2 \quad \Omega.$$

Substituting these values into Equation (6.133) gives

$$\alpha_c \approx 4.06 \times 10^{-2} \quad \text{Np/m}.$$

Using Equation (6.127), gives

$$Q_c \approx 6241.$$

Using Equation (5.179), with $\sigma_d/\omega\varepsilon \approx 4.33 \times 10^{-4}$ (Appendix B), we get

$$Q_d = 2309.$$

6.8 WAVEGUIDE AND CAVITY LOSSES

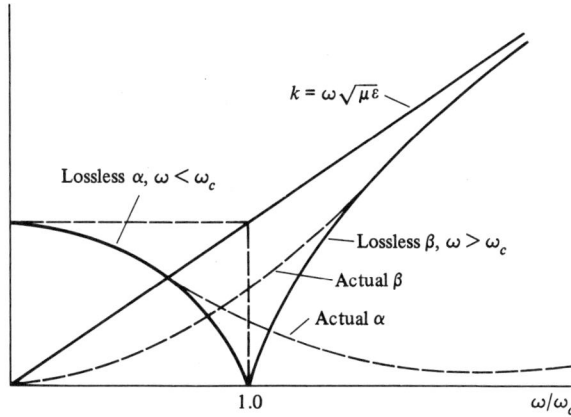

Figure 6.20. α, β, k versus ω/ω_c for the rectangular waveguide.

Then, from Equation (6.129), we have

$$Q_0 = \frac{Q_c Q_d}{Q_c + Q_d} = 1686.$$

∎

A plot of α and β against frequency for a general case is shown in Figure 6.20.

2. TE$_{101}$ Mode, Rectangular Cavity

The general expression for cavity Q is

$$\boxed{Q = \omega \frac{<\mathcal{W}_T>}{<\mathcal{P}_{dT}>}}, \qquad (6.134)$$

where $<\mathcal{W}_T>$ is the time-average *total* energy stored in the cavity (joules) and $<\mathcal{P}_{dT}>$ is the time-average total power dissipated in the cavity (watts). The procedure followed here is that of examining one of the more practical cases, rather than attempting to solve the general cases.

When $a > b > c$, the lowest resonant frequency for TE to z modes in the rectangular cavity occurs when $m = 1$, $n = 0$, and $p = 1$, giving the TE$_{101}$ mode. The (lossless) field for this mode was obtained in Example 6.8 and its distribution ($a = c > b$) is shown in Figure 6.18. The electric and magnetic fields are 90° out of phase, indicating no loss. When the magnetic field is zero (in time), the electric field is maximum, and all of the stored energy is in the electric field. When the electric field is zero, the reverse is true. The total energy is conserved, or constant, and independent of time. The time-average energies in the electric field and magnetic field are equal with sinusoidal excitation. Thus,

$$\mathcal{W}_T = <\mathcal{W}_T> = 2<\mathcal{W}_e> = 2<\mathcal{W}_m> \qquad (6.135)$$

and

$$<\mathcal{W}_e> = \tfrac{1}{2}\text{Re}\iiint_{\text{vol}} \frac{\mathbf{D}}{2}\cdot\mathbf{E}^*\,dv. \qquad (6.136)$$

For the cavity, the *lossless* equations give

$$<\mathcal{W}_e> = \frac{\varepsilon}{4}\int_0^a\int_0^b\int_0^c |E_y|^2\,dx\,dy\,dz$$

or, with the results of Example 6.8,

$$<\mathcal{W}_e> = <\mathcal{W}_m> = \frac{\varepsilon}{16}E_0^2 abc \qquad (6.137)$$

and

$$<\mathcal{W}_T> = \frac{\varepsilon}{8}E_0^2 abc \quad (\text{J}). \qquad (6.138)$$

For the time being, we neglect any loss in the internal medium and consider only conductor loss in the cavity walls. We now need to calculate the *total* time-average power dissipated in the walls. This is done in a manner similar to that for determining the time-average power dissipated per unit length $<\mathcal{P}_d>$ for the waveguide. That is,

$$<\mathcal{P}_{dT}> = \tfrac{1}{2}R_s\oiint_{\text{walls}} |H_{\text{tan}}|^2\,ds$$

$$= \tfrac{1}{2}R_s\int_0^c\int_0^b |H_z|^2_{x=0}\,dy\,dz + \int_0^c\int_0^b |H_z|^2_{x=a}\,dy\,dz$$

$$+ \int_0^c\int_0^a (|H_x|^2 + |H_z|^2)_{y=0}\,dx\,dz$$

$$+ \int_0^c\int_0^a (|H_x|^2 + |H_z|^2)_{y=b}\,dx\,dz$$

$$+ \int_0^b\int_0^a |H_x|^2_{z=0}\,dx\,dy + \int_0^b\int_0^a |H_x|^2_{z=c}\,dx\,dy$$

$$= \tfrac{1}{2}R_s\left(\frac{E_0}{2f_r\mu}\right)^2\left(\frac{bc}{a^2} + \frac{c}{2a} + \frac{a}{2c} + \frac{ab}{c^2}\right),$$

$$<\mathcal{P}_{dT}> = \tfrac{1}{2}R_s\frac{E_0^2}{\eta^2}\frac{a^2c^2}{a^2+c^2}\left(\frac{bc}{a^2} + \frac{c}{2a} + \frac{a}{2c} + \frac{ab}{c^2}\right) \quad (\text{W}). \qquad (6.139)$$

6.8 WAVEGUIDE AND CAVITY LOSSES

From Equation (6.134), we get

$$Q_c = 2\pi f_r \frac{<\mathcal{W}_T>}{<\mathcal{P}_{dT}>}$$

$$= \frac{\pi \varepsilon f_r \eta^2 b(a^2 + c^2)}{2R_s \left(\frac{bc^2}{a} + \frac{c^2}{2} + \frac{a^2}{2} + \frac{a^2 b}{c} \right)},$$

$$Q_c = \frac{\pi \eta}{4R_s} \frac{b(a^2 + c^2)^{3/2}}{\left(bc^3 + \frac{ac^3}{2} + \frac{ca^3}{2} + ba^3 \right)}. \quad (6.140)$$

If $a = c$, then

$$Q_c = \frac{1.11\eta}{R_s \left(1 + \frac{a}{2b}\right)}. \quad (6.141)$$

Example 6.10

Suppose a copper rectangular cavity has $a = c = 2$ cm and $b = 1$ cm, and is operated in the TE_{101} mode. If the interior is a vacuum, find Q_c.

The resonant frequency is

$$f_r = \frac{1}{2\sqrt{\mu\varepsilon}} \sqrt{\frac{1}{a^2} + \frac{1}{c^2}} = 10.61 \quad \text{GHz},$$

so

$$R_s = \sqrt{\frac{\pi f_r \mu}{\sigma_c}} = 26.87 \quad \text{m}\Omega.$$

Therefore,

$$Q_c = 7787.$$

∎

Quite often it is desirable, or even necessary, to resort to *dielectric loading* of a cavity to reduce its physical size for a given resonant frequency. If the frequency is high enough so that the dielectric loss must be considered, then, as we have already seen,

$$Q_d = \frac{\omega\varepsilon}{\sigma_d}.$$

■ **Example 6.11**

Repeat Example 6.10 if the cavity is loaded with polystyrene ($\varepsilon_R = 2.56$, $\sigma_d/\omega\varepsilon \approx 4.33 \times 10^{-4}$).

$$f_r = \frac{1}{2\sqrt{\mu\varepsilon}}\sqrt{\frac{1}{a^2} + \frac{1}{c^2}} = 6.629 \quad \text{GHz},$$

$$R_s = \sqrt{\frac{\pi f_r \mu}{\sigma_c}} = 21.24 \quad \text{m}\Omega.$$

Therefore, we have

$$Q_c = \frac{1.11\eta}{R_s(1 + a/2b)} = 6156, \quad Q_d = \frac{\omega\varepsilon}{\sigma_d} = 2309, \quad \text{and} \quad Q_0 = 1679.$$

The bandwidth is $f_r/Q_0 = 3.95$ MHz. The Q_0 is an optimistic value because no consideration was given to the fact that the cavity cannot be constructed exactly as the theory requires, and because the loss in the networks required to couple energy into and out of the cavity was not considered. The actual bandwidth will be larger than that calculated. If modulation frequencies up to 5 MHz (video, for example) are present, then the wider bandwidth is desirable. ■

Other cavity resonators are treated similarly, but we can predict here that for similar field distributions (they cannot be identical) the circular cavity will have a higher Q than the rectangular cavity, and the spherical cavity will have a higher Q than the cylindrical cavity. This follows from the smaller surface areas (same volume) and reduced joule heating for the spherical compared to the circular cavity, and the circular compared to the rectangular cavity. An example of this is considered in the Problems for the ambitious student.

6.9 CONCLUDING REMARKS

We have examined the propagation characteristics of the lossless hollow rectangular waveguide in this chapter. We began by treating the dominant TE_{10} mode in a relatively simple manner, and for many purposes this is sufficient. The more important characteristics of these waveguides such as the cutoff frequency, phase velocity, velocity of energy flow, and so on, were considered. A vector potential technique that is much more general was introduced because it is applicable to many more geometries.

Most of the characteristics of the hollow waveguide are similar to those of two-wire systems, with the notable exception that a TEM mode can never propagate inside a hollow one-conductor system. Many modes can be generated unintentionally in a waveguide around the source (a probe, for example), or around some other discontinuity. In a uniform section of the waveguide, only those modes "above

6.9 CONCLUDING REMARKS

cutoff" can propagate energy. It is worth noting that many systems in nature can support "normal modes." The familiar vibrating string satisfies a wave equation, and can certainly vibrate in many normal modes.

The same vector potential technique that was used for the hollow conducting waveguide was used to analyze the dielectric slab waveguide. The results are important in fiber optic techniques. The behavior of this type of waveguide was found to be very similar to that of the rectangular waveguide in most respects. An exception was found to be that when operated below cutoff, the dielectric waveguide "leaks" energy, and can be considered to be an antenna or radiator in that case.

Practical cavities are only sections of shorted waveguide (the spherical cavity is a notable exception), and are therefore usually easy to treat mathematically. We can expect the Q of resonant cavities to be rather high. The actual mode that is excited in a cavity or waveguide not only depends on the *geometry* of the particular cavity or waveguide being used, but also on the *geometry* of the exciting source (loop, monopole, etc.). The quality factors Q_c and Q_d as calculated in this chapter are intrinsic values, depending in general on mode, material, and geometry. We can expect in practice to obtain smaller values of Q and larger values of α than predicted by the theory. For example, coupling energy into, or out of, a waveguide requires some kind of network. If the effects of the coupling network are included and lumped into an external Q, then

$$\frac{1}{Q_{\text{eff}}} = \frac{1}{Q_c} + \frac{1}{Q_d} + \frac{1}{Q_{\text{ext}}},$$

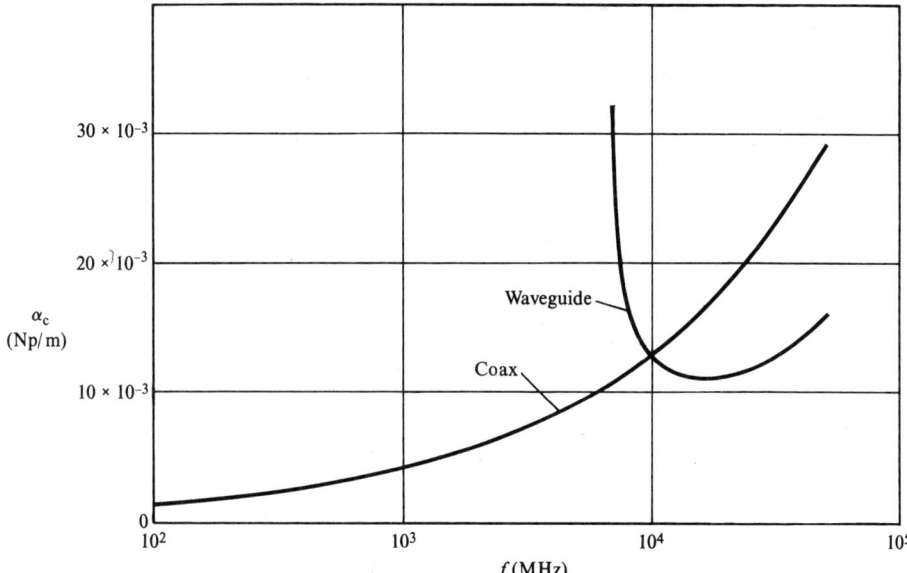

Figure 6.21. Comparison of rectangular waveguide (TE_{10} mode) losses with coaxial cable (TEM) mode losses.

where Q_{eff} is the overall or effective Q. Q_{eff} will obviously be smaller than any of the Q's which contribute to it. It should also be remembered that the perturbation technique we have employed here is approximate. Further considerations are beyond the scope of this material.

We conclude this chapter by comparing an air-filled x-band waveguide ($a = 0.9$ inch, $b = 0.4$ inch, copper conductor) operating in the dominant TE_{10} mode with an air-filled rigid coax ($a = 4.75$ mm, $b = 10.51$ mm, copper conductors) operating in the dominant TEM mode. The comparison is made in terms of the attenuation constant $\alpha = \alpha_c$, and is presented in Figure 6.21. The waveguide is more efficient than the coax above 10 GHz and the reverse is true below 10 GHz. The waveguide cutoff frequency is 6.56 GHz.

REFERENCES

Jordan (see references for Chapter 4).

Marcuvitz, N. *Waveguide Handbook*. MIT Radiation Laboratory Series, Vol. 10. New York: McGraw-Hill, 1951. A standard reference for guiding systems.

Ramo (see references for Chapter 2).

Skilling (see references for Chapter 1).

Thomassen, K. I. *Introduction to Microwave Fields and Circuits*. Englewood Cliffs, NJ: Prentice-Hall, 1971.

PROBLEMS

1. Show that the dominant mode for the rectangular waveguide is the TE_{10} mode for $a > b$, but is the TE_{01} mode for $b > a$.

2. Define a voltage as the line integral of **E** across the center of the rectangular waveguide (TE_{10} mode), and define a current as the total current in the wall at $y = 0$. Show that

$$V = bE_0 e^{-j\beta z}$$

and

$$I = (2aE_0)e^{-j\beta z}/(\pi Z_0), \quad Z_0 = \eta^+$$

3. (a) A lossless rectangular waveguide has $a = 2$ cm, $b = 1$ cm, and is operating in the TE_{10} mode at 10 GHz. What effective load impedance should it have to avoid reflections?

   (b) If the load impedance is 250 Ω, what is the SWR?

   (c) Design a $\lambda/4$ matching section using a rectangular dielectric slug for part (b).

4. What is the difference in time delay per unit length for two signals, one at 10 GHz and one at 10.006 GHz? The waveguide is x band, and the mode is TE_{10} (Example 6.1). The width of a television channel is 6 MHz.

*5. What is the maximum (average) power that the x-band waveguide of Example 6.3 can handle? The operating frequency is 10 GHz and the dielectric is air.

6. A hollow conducting waveguide is square with $a = b = 2$ cm. If $\mu = \mu_0$ and $\varepsilon = \varepsilon_0$, find the first five lowest cutoff frequencies.

7. Show that

$$(1/\lambda)^2 = (1/\lambda_c)^2 + (1/\lambda_g)^2$$

for the rectangular waveguide.

8. Design a rectangular waveguide to have a center frequency of 3 GHz (s-band) with a 2:1 frequency range of single-mode operation and maximum power handling capacity ($\mu = \mu_0$, $\varepsilon = \varepsilon_0$).

*9. A signal is *suddenly* applied to a lossless two-conductor guiding system A. The same signal is simultaneously applied to a lossless one-conductor guiding system B. Both systems have uniform cross sections and identical lengths (l). Observers with appropriate equipment are located at the load end of each system. Which observer first detects a signal, and what is the total time delay?

10. Show that $u_g = u_e$ for a lossless rectangular waveguide.

11. Find the time-domain form for the entire electromagnetic field for the TM dielectric-slab waveguide of Example 6.7.

*12. Show that electromagnetic waves propagate in a "cylindrical" dielectric waveguide with a velocity between that for plane waves in the dielectric of the waveguide and that for plane waves in the dielectric outside the waveguide.

13. A dielectric-slab waveguide has $\mu = \mu_0$ everywhere, $\varepsilon = \varepsilon_0$ outside the guide, and $\varepsilon_R = 3.0$ inside. Find the minimum thickness that allows a TM or TE even mode at 25 GHz to propagate.

*14. Derive an expression for k_{x0} for the dominant TM mode of the waveguide of Problem 13 when $k_{xd}a \ll 1$ (very thin).

15. (a) Find the resonant frequency for the TE$_{101}$ rectangular cavity mode if $a = c = 2b = 4$ cm and the dielectric is air.

(b) Calculate new dimensions for the cavity if the resonant frequency remains unchanged, but the cavity is filled with Teflon.

*16. (a) If the TE$_{101}$ mode is to be excited by a small monopole antenna that is an extension of the center conductor of a coaxial cable, where should it be placed? See Figure 6.18 (rectangular cavity).

(b) Suppose that the voltage on the cable has the form

$$v(t) = \sum_{q=0}^{\infty} a_q \cos(q\omega_r t),$$

where ω_r is the resonant frequency for the TE$_{101}$ mode. How will the field distribution compare to that when $v(t) = a_1 \cos(\omega_r t)$? Consider only TE$_{mnp}$ modes.

(c) Calculate $<\mathcal{S}>$ for the TE_{101} mode.

(d) Find the time-domain forms for the TE_{101} mode.

17. Find the first five lowest-order modes (and their resonant frequencies) for an air-filled cavity with $a = 5$ cm, $b = 4$ cm, and $c = 3$ cm.

18. Where should a small loop be placed to couple to the rectangular cavity TE_{101} mode? The loop is formed by extending the inner conductor of a coaxial cable feed, bending it into a loop, and connecting to the inside of the cavity.

19. (a) Find the electromagnetic field of the TM_{110} rectangular cavity mode. Let the peak value of E_z be E_0.

    (b) Specify a method of coupling energy into the cavity.

    (c) Calculate $<\mathcal{S}>$.

*20. (a) Refer to Appendix C and show that the electromagnetic field for the TM_{010} cylindrical cavity mode is

$$E_z = E_0 J_0(2.405 \rho/a),$$

$$H_\phi = j(E_0/\eta) J_1(2.405 \rho/a).$$

(b) What radius must the cavity have if its resonant frequency (TM_{010} mode) is to be the same as that for the rectangular cavity (TE_{101} mode) of Problem 15.

(c) Calculate d for the cylindrical cavity so that it has the same volume as the cavity of Problem 15. Compare the surface areas.

(d) Which cavity will have the higher Q_c?

(e) Which cavity can handle the greater electric field?

*21. Find the value of ω/ω_c for which α_c is minimum for the TE_{10} rectangular waveguide mode (see Figure 6.20).

22. Calculate Q_c for an air-filled x-band waveguide with copper walls. The operating frequency is 10 GHz and the mode is TE_{10}. Repeat for 7 GHz.

23. (a) Calculate Q_c for the cavity of Problem 15(a) if the walls are silver plated. Repeat for brass walls.

    (b) Repeat (a) for the cavity of Problem 15(b) (silver plated walls).

24. Show that for a cubical cavity ($a = b = c$) in the TE_{101} mode

$$Q_c = \frac{2v}{\delta s},$$

where $v = a^3$ is the volume, $s = 6a^2$ is the surface area, and δ is the skin depth.

25. Plot α_c versus ω/ω_c for the rectangular waveguide TE_{10} mode with copper walls and an air dielectric if $a = 2b = 5.08$ cm.

PROBLEMS

*26. Calculate Q_c for the cylindrical cavity of Problem 20(a) (TM$_{010}$ mode) if the walls are silver plated, the dielectric is air, and $a = 2.16$ cm. Does this calculation agree with the result of Problem 20(d)?

27. Calculate the bandwidth of the cavities of problems 23(a) and 23(b).

*28. (a) Plot E_z/E_0 versus ρ for the cylindrical cavity of Problem 26.
  (b) Plot $H_\phi/(jE_0/\eta)$ versus ρ.
  (c) Calculate the bandwidth.

*29. Show that $Q_c = v/(\delta s)$ for the cylindrical cavity TM$_{010}$ mode. See Problems 24 and 26.

*30. (a) Compare α_c for the waveguide of Problem 25 at 5 GHz with that for a two-wire line with $D = 1$ cm, $d = 1$ mm, copper conductors, and air dielectric. Comment on the results.
  (b) Repeat for a coaxial cable with $a = 0.3$ cm, $b = 0.5$ cm, copper conductors, and air dielectric.

*31. Following the method of Section 5.18, find an equivalent lumped circuit for the cavities of Problems 23 and 26.

APPENDIX A
PHYSICAL CONSTANTS

Permittivity of free space (vacuum), $\varepsilon_0 = 8.854 \times 10^{-12}$ F/m $\approx \dfrac{10^{-9}}{36\pi}$ F/m

Permeability of free space (vacuum), $\mu_0 \equiv 4\pi \times 10^{-7}$ H/m

Electron charge magnitude, $e = 1.602 \times 10^{-19}$ C

Electron rest mass, $m_e = 9.109 \times 10^{-31}$ kg

Proton rest mass, $m_p = 1.673 \times 10^{-27}$ kg

Speed of light in free space, $\dfrac{1}{\sqrt{\mu_0 \varepsilon_0}} = 2.998 \times 10^8$ m/s

APPENDIX B
MATERIAL PARAMETERS

Representative values of conductivity, relative permittivity, loss tangent, and dielectric strength for various materials are listed below. Most of the data have been taken from the references listed below.[1-3]

**TABLE B.1
Representative Values of Conductivity, σ_c, at Room Temperature and $f = 0$**

Material	σ_c (℧/m)
Silver	6.17×10^7
Copper	5.80×10^7
Gold	4.10×10^7
Aluminum	3.82×10^7
Tungsten	1.82×10^7
Brass	1.5×10^7
Iron	1.0×10^7
Solder	0.7×10^7
Nichrome	0.1×10^7
Seawater	4×10^0
Ferrites	$\approx 10^{-2}$
Fresh water	$\approx 10^{-3}$
Porcelain	$\approx 10^{-10}$

**TABLE B.2
Representative Values of Relative Permittivity (Dielectric Constant), ε_R, at Room Temperature and $f = 100$ Hz**

Material	ε_R
Bakelite	8.2
Clay soil (dry)	4.73
Ice	4.2
Lucite	3.2
Mica (ruby)	5.4
Mylar	5.0
Nylon	3.88
Plexiglas	3.40
Polyethylene	2.25
Polystyrene	2.56
Sandy soil (dry)	3.42
Styrofoam	1.03
Teflon	2.1
Water	78.0

[1]*Dielectric Materials and Applications*. Part V, Cambridge, MA: Technology Press, MIT, 1954.

[2]D. G. Fink and J. M. Carroll. *Standard Handbook for Electrical Engineers*, 10th ed., New York: McGraw-Hill, 1968.

[3]International Telephone and Telegraph Company, Inc., *Reference Data for Radio Engineers*, 5th ed., Indianapolis, Ind.: Howard W. Sams & Co., 1968.

TABLE B.3
Representative Values of Loss Tangent $\sigma_d/\omega\varepsilon$ for Various Materials at Room Temperature. Frequency (Hz)

Material	10^2	10^4	10^6	10^8	10^{10}
Bakelite	0.134	0.0631	0.0593	0.0773	0.0369
Clay soil (dry)	0.121	0.119	0.0661		0.013
Loamy soil (dry)	0.0686	0.0353	0.0182		0.00139
Lucite	0.0625	0.0315	0.0144	0.00678	0.00319
Nylon	0.0144	0.0233	0.0258	0.0209	
Plexiglas	0.0603	0.0300	0.0139		0.00676
Polyethylene	4.89×10^{-4}	3.11×10^{-4}	4.00×10^{-4}		4.00×10^{-4}
Polystyrene	5.08×10^{-5}	5.08×10^{-5}	7.03×10^{-5}	1.18×10^{-4}	4.33×10^{-4}
Sandy soil (dry)	0.196	0.0342	0.0170		0.00364
Styrofoam	1.94×10^{-4}	1.03×10^{-4}	1.94×10^{-4}		1.46×10^{-4}
Teflon	5.24×10^{-4}	3.33×10^{-4}	1.90×10^{-4}	1.90×10^{-4}	3.85×10^{-4}
Water			0.0190	0.00701	1.03

TABLE B.4
Dielectric Strength (Uniform Electric Field)*

Material	Dielectric strength (MV/m)
Air (normal pressure)	3
Oil (mineral)	15
Paper (impregnated)	15
Polyethylene	20
Polystyrene	20
Rubber (hard)	21
Bakelite	25
Glass (plate)	30
Paraffin	30
Quartz (fused)	30
Mica (ruby)	200
Mylar	—

*If E is not uniform, a *corona* discharge may occur before arcing occurs. See Kraus in the references at the end of Chapter 2.

APPENDIX C
SPECIAL FUNCTIONS

RECTANGULAR COORDINATES

Separation of Variables

$$\frac{\partial^2 \alpha}{\partial x^2} + \frac{\partial^2 \alpha}{\partial y^2} + \frac{\partial^2 \alpha}{\partial z^2} + k^2 \alpha = 0 \quad \text{[scalar Helmholtz (wave) equation]}. \tag{C.1}$$

Assume $\alpha(x, y, z) = X(x)Y(y)Z(z)$ and substitute into Equation (C.1), then divide by α, giving

$$\frac{1}{X}\frac{d^2 X}{dx^2} + \frac{1}{Y}\frac{d^2 Y}{dx^2} + \frac{1}{Z}\frac{d^2 Z}{dz^2} + k^2 = 0. \tag{C.2}$$

Each term is separately a constant, or

$$\frac{1}{X}\frac{d^2 X}{dx^2} = -k_x^2, \quad \frac{1}{Y}\frac{d^2 Y}{dy^2} = -k_y^2, \quad \frac{1}{Z}\frac{d^2 Z}{dz^2} = -k_z^2, \tag{C.3}$$

so the separation equation is

$$k^2 = k_x^2 + k_y^2 + k_z^2. \tag{C.4}$$

The first of Equations (C.3) has as a solution the sum of any *two* harmonic functions $h(k_x x)$ where

$$h(k_x x) = \cos(k_x x), \quad \sin(k_x x), \quad e^{jk_x x} \quad \text{or} \quad e^{-jk_x x}. \tag{C.5}$$

The second and third of Equations (C.3) have similar solutions, so a general solution to C.1 is

$$\alpha(x, y, z) = h(k_x x) h(k_y y) h(k_z z). \tag{C.6}$$

CYLINDRICAL COORDINATES

Separation of Variables

$$\frac{1}{\rho}\frac{\partial}{\partial \rho}\left(\rho \frac{\partial \alpha}{\partial \rho}\right) + \frac{1}{\rho^2}\frac{\partial^2 \alpha}{\partial \phi^2} + \frac{\partial^2 \alpha}{\partial z^2} + k^2 \alpha = 0 \quad \text{[scalar Helmholtz (wave) equation]} \tag{C.7}$$

Assume $\alpha(\rho, \phi, z) = R(\rho)\Phi(\phi)Z(z)$ and substitute into Equation (C.7), then divide by α, giving

$$\frac{1}{\rho R}\frac{d}{d\rho}\left(\rho \frac{dR}{d\rho}\right) + \frac{1}{\rho^2 \Phi}\frac{d^2 \Phi}{d\phi^2} + \frac{1}{Z}\frac{d^2 Z}{dz^2} + k^2 = 0. \tag{C.8}$$

The third term of Equation (C.8) must be a constant, so

$$\frac{1}{Z}\frac{d^2 Z}{dz^2} = -k_z^2. \tag{C.9}$$

Equation (C.9) has a solution [see Equation (C.5)], which is the sum of any two harmonic functions, $h(k_z z)$. Then Equation (C.8) becomes

$$\frac{\rho}{R}\frac{d}{d\rho}\left(\rho \frac{dR}{d\rho}\right) + \frac{1}{\Phi}\frac{d^2 \Phi}{d\phi^2} + (k^2 - k_z^2)\rho^2 = 0. \tag{C.10}$$

The second term of Equation (C.10) must now be a constant, so

$$\frac{1}{\Phi}\frac{d^2 \Phi}{d\phi^2} = -n^2. \tag{C.11}$$

Equation (C.11) has a solution which is the sum of any two harmonic functions $h(n\phi)$. Then, Equation (C.10) becomes

$$\frac{\rho}{R}\frac{d}{d\rho}\left(\rho \frac{dR}{d\rho}\right) - n^2 + (k^2 - k_z^2)\rho^2 = 0. \tag{C.12}$$

With the separation equation

$$k^2 = k_\rho^2 + k_z^2, \tag{C.13}$$

APPENDIX C SPECIAL FUNCTIONS

Equation (C.12) can be written

$$\frac{\rho}{R}\frac{d}{d\rho}\left(\rho\frac{dR}{d\rho}\right) - n^2 + k_\rho^2\rho^2 = 0, \tag{C.14}$$

which is Bessel's equation of order n. This equation has as a solution the sum of *Bessel functions*[1] of the first kind $J_n(k_\rho\rho)$, and Bessel functions of the second kind, $N_n(k_\rho\rho)$ (Neumann functions). Particular linear combinations of J_n and N_n, are called Hankel functions. The first kind is $H_n^{(1)}(k_\rho\rho) = J_n(k_\rho\rho) + jN_n(k_\rho\rho)$, and the second kind is $H_n^{(2)}(k_\rho\rho) = J_n(k_\rho\rho) - jN_n(k_\rho\rho)$. The general solution to Equation (C.7) is

$$\alpha(\rho, \phi, z) = \begin{Bmatrix} J_n(k_\rho\rho) \\ N_n(k_\rho\rho) \end{Bmatrix} h(n\phi)h(k_z z). \tag{C.15}$$

The recurrence relations are frequently useful. They are ($f_n = J_n$ or N_n)

$$f_n'(x) = f_{n-1}(x) - \frac{n}{x}f_n(x),$$

$$f_n'(x) = -f_{n+1}(x) + \frac{n}{x}f_n(x). \tag{C.16}$$

$J_n(x)$ and $N_n(x)$ for $n = 0, 1,$ and 2 are shown in Figure C.1.

CYLINDRICAL WAVEGUIDE (FIGURE C.2)

TM to z Mode

In Equations (6.55) let $\mathbf{A} = \mathbf{a}_z A_z$, $A_\rho = A_\phi = 0$ and $\mathbf{F} = 0$. Expand Equations (6.55) in circular cylindrical coordinates. We then have

$$\frac{1}{\rho}\frac{\partial}{\partial\rho}\left(\rho\frac{\partial A_z}{\partial\rho}\right) + \frac{1}{\rho^2}\frac{\partial^2 A_z}{\partial\phi^2} + \frac{\partial^2 A_z}{\partial z^2} + k^2 A_z = 0 \tag{C.17}$$

and

$$E_\rho = \frac{1}{j\omega\mu\varepsilon}\frac{\partial^2 A_z}{\partial\rho\,\partial z}, \qquad H_\rho = \frac{1}{\mu\rho}\frac{\partial A_z}{\partial\phi},$$

$$E_\phi = \frac{1}{j\omega\mu\varepsilon\rho}\frac{\partial^2 A_z}{\partial\phi\,\partial z}, \qquad H_\phi = -\frac{1}{\mu}\frac{\partial A_z}{\partial\rho}, \tag{C.18}$$

$$E_z = \frac{1}{j\omega\mu\varepsilon}\left(k^2 + \frac{\partial^2}{\partial z^2}\right)A_z, \qquad H_z \equiv 0.$$

[1] See Spiegel in the references at the end of Chapter 1.

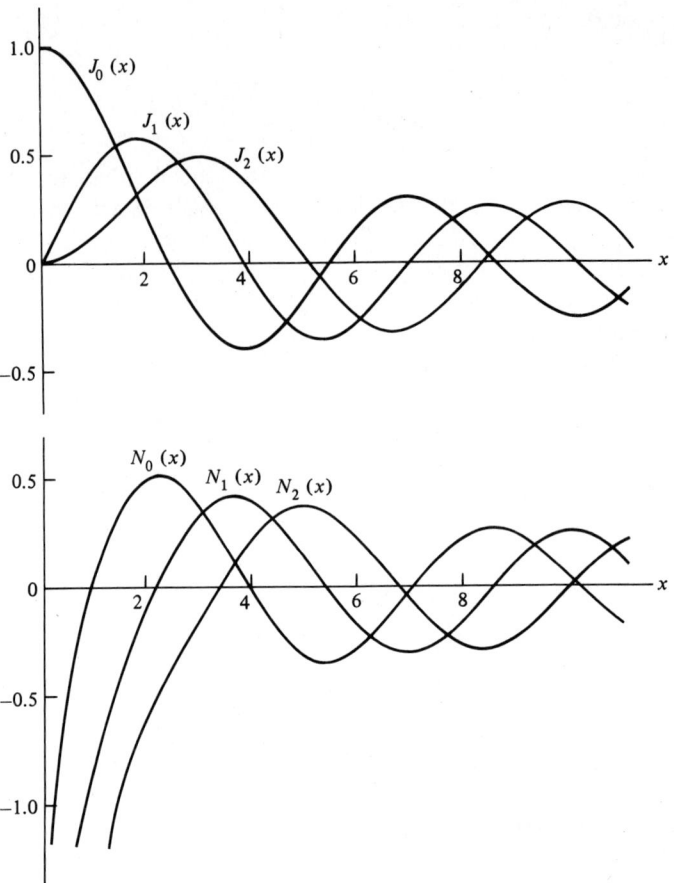

Figure C.1. $J_n(x)$ and $N_n(x)$ versus x, $n = 0, 1, 2$.

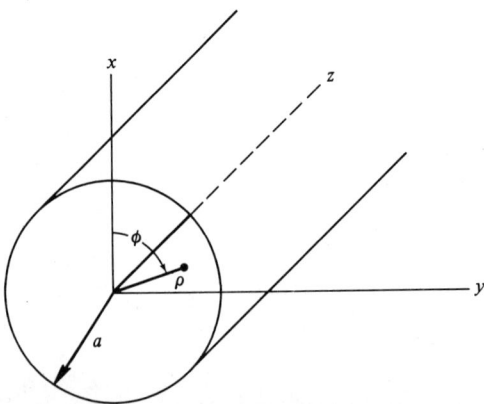

Figure C.2. Geometry of the circular cylindrical waveguide.

APPENDIX C SPECIAL FUNCTIONS

This, then, is always a TM *to z* mode. A general solution to Equation (C.17) as given by equation C.15 is

$$A_z(\rho, \phi, z) = [AJ_n(k_\rho\rho) + BN_n(k_\rho\rho)] \cdot (C \cos n\phi + D \sin n\phi)$$
$$\cdot (E \cos k_z z + F \sin k_z z), \quad (C.19)$$

where $n = 0, 1, 2, \ldots$ and the eigenvalues k_z and k_ρ are related by the separation Equation (C.13),

$$k_z^2 = k^2 - k_\rho^2.$$

No loss of generality occurs *in our problem* if we choose $D = 0$. We also choose $F = -jE$ (as in Chapter 6) to obtain the propagating form, $E e^{-jk_z z}$. $N_n(k_\rho\rho)$ is a Bessel function of the second kind. These functions all tend toward $(-)$ infinity as ρ approaches zero (see Figure C.1) We must include $\rho = 0$ in our solution, and must therefore exclude $N_n(k_\rho\rho)$. Thus, we set $B = 0$. With $ACE \equiv 1$, we have

$$A_z = J_n(k_\rho\rho) \cos n\phi \, e^{-jk_z z}. \quad (C.20)$$

The boundary conditions are (see Figure C.2)

$$E_\phi = E_z = 0, \quad \rho = a. \quad (C.21)$$

Equations (C.18) reveal that these conditions are satisfied if A_z is zero itself when $\rho = a$. Therefore,

$$J_n(k_\rho a) = 0. \quad (C.22)$$

Inspection of Figure C.1 reveals that $J_n(k_\rho a)$ has an infinite number of zeros for *each* value of n. We will denote these roots as x_{nl}. That is, x_{nl} is the lth root of $J_n(x) = 0$. The first few roots are shown in Table C.1. We now have

$$k_\rho = \frac{x_{nl}}{a} \begin{cases} n = 0, 1, 2, \ldots \\ l = 1, 2, 3, \ldots \end{cases} \quad (C.23)$$

TABLE C.1
x_{nl}, lth Root of $J_n(x) = 0$

l \ n	0	1	2	3
1	2.405	3.832	5.136	6.380
2	5.520	7.016	8.417	9.761
3	8.654	10.173	11.620	13.015

and

$$k_z^2 = k^2 - \left(\frac{x_{nl}}{a}\right)^2 \qquad (C.24)$$

The potential is now given by

$$A_z = J_n\left(x_{nl}\frac{\rho}{a}\right)\cos n\phi\, e^{-jk_z z}. \qquad (C.25)$$

The field is obtained by substituting Equation (C.25) into Equations (C.18). Cutoff occurs when $k_z = 0$, or $k_c = x_{nl}/a$. The cutoff frequency is

$$f_c = \frac{x_{nl}}{2\pi a \sqrt{\mu\varepsilon}}, \qquad (C.26)$$

so that

$$k_z = \begin{cases} k\sqrt{1 - \left(\dfrac{\omega_c}{\omega}\right)^2} = \beta, & \omega > \omega_c, \\ -jk_c\sqrt{1 - \left(\dfrac{\omega}{\omega_c}\right)^2} = -j\alpha, & \omega < \omega_c. \end{cases} \qquad (C.27)$$

The transverse wave impedance is

$$\eta_{TM}^+ = \frac{E_\rho}{H_\phi} = -\frac{E_\phi}{H_\rho} = \frac{k_z}{\omega\varepsilon}, \qquad (C.28)$$

which is the same form as for the rectangular waveguide.

TE to z Mode

In Equations (6.62), let $\mathbf{F} = \mathbf{a}_z F_z$, $F_\rho = F_\phi = 0$, and $\mathbf{A} = 0$. Expand Equation (6.62) in circular cylindrical coordinates. We then have

$$\frac{1}{\rho}\frac{\partial}{\partial\rho}\left(\rho\frac{\partial F_z}{\partial\rho}\right) + \frac{1}{\rho^2}\frac{\partial^2 F_z}{\partial\phi^2} + \frac{\partial^2 F_z}{\partial z^2} + k^2 F_z = 0, \qquad (C.29)$$

$$k_z^2 = k^2 - k_\rho^2, \qquad (C.30)$$

and

APPENDIX C SPECIAL FUNCTIONS

$$E_\rho = -\frac{1}{\varepsilon\rho}\frac{\partial F_z}{\partial \phi}, \qquad H_\rho = \frac{1}{j\omega\mu\varepsilon}\frac{\partial^2 F_z}{\partial \rho\,\partial z},$$

$$E_\phi = \frac{1}{\varepsilon}\frac{\partial F_z}{\partial \rho}, \qquad H_\phi = \frac{1}{j\omega\mu\varepsilon\rho}\frac{\partial^2 F_z}{\partial \phi\,\partial z}, \qquad (C.31)$$

$$E_z \equiv 0, \qquad H_z = \frac{1}{j\omega\mu\varepsilon}\left(k^2 + \frac{\partial^2}{\partial z^2}\right)F_z.$$

This is a TE to z mode. An appropriate solution to Equation (C.29) is

$$F_z = J_n(k_\rho\rho)\,\cos n\phi\, e^{-jk_z z}, \qquad (C.32)$$

which is the same as Equation (C.20). The boundary conditions are the same as those given by Equation (C.21). Equations (C.31) reveal that E_z is identically zero, and the boundary condition on E_ϕ is satisfied if

$$\left.\frac{\partial F_z}{\partial \rho}\right|_{\rho=a} = 0.$$

This condition can be satisfied if

$$\left.\frac{\partial J_n(k_\rho\rho)}{\partial \rho}\right|_{\rho=a} = 0. \qquad (C.33)$$

Inspection of Figure C.1 reveals that the derivative of any order (n) Bessel function has an infinite number of zeros. We denote the roots by x'_{nl}. That is, x'_{nl} is the lth root of $J'_n(x) = 0$. The first few values of x'_{nl} are shown in Table C.2. The eigenvalue k_ρ is

$$k_\rho = \frac{x'_{nl}}{a} \quad \begin{cases} n = 0, 1, 2, \ldots, \\ l = 1, 2, 3, \ldots. \end{cases} \qquad (C.34)$$

The separation equation is then

$$k_z^2 = k^2 - \left(\frac{x'_{nl}}{a}\right)^2 \qquad (C.35)$$

TABLE C.2
x'_{nl}, lth Root of $J'_n(x) = 0$

n \ l	0	1	2	3
1	3.832	1.841	3.054	4.201
2	7.016	5.331	6.706	8.015
3	10.173	8.536	9.969	11.346

and the potential is now given by

$$F_z = J_n\left(x'_{nl}\frac{\rho}{a}\right)\cos n\phi\, e^{-jk_z z} \tag{C.36}$$

The field is obtained by substituting Equation (C.36) into Equations (C.31).

Cutoff, as usual, occurs when $k_z = 0$, or $k_c = x'_{nl}/a$. The cutoff frequency is

$$f_c = \frac{x'_{nl}}{2\pi a\sqrt{\mu\varepsilon}}. \tag{C.37}$$

Note that Equation (C.37) is not the same as Equation (C.26). The propagation constant is

$$k_z = \begin{cases} k\sqrt{1-\left(\dfrac{\omega_c}{\omega}\right)^2} = \beta, & \omega > \omega_c, \\ -jk_c\sqrt{1-\left(\dfrac{\omega}{\omega_c}\right)^2} = -j\alpha, & \omega < \omega_c. \end{cases} \tag{C.38}$$

The wave impedance is

$$\eta_{\text{TE}}^+ = \frac{F_\rho}{H_\phi} = -\frac{E_\phi}{H_\rho} = \frac{\omega\mu}{k_z}, \tag{C.39}$$

which is the same form as the wave impedance for rectangular waveguides (TE mode).

CYLINDRICAL CAVITY (FIGURE C.3)

TM to z Mode

Based on results for the rectangular cavity,

$$A_z = J_n\left(x_{nl}\frac{\rho}{a}\right)\cos n\phi\,\cos\frac{q\pi z}{d}, \tag{C.40}$$

where $n = 0, 1, 2, 3, \ldots$, $l = 1, 2, \ldots$, $q = 0, 1, 2, 3, \ldots$, and the resonant frequency is

$$f_r = \frac{1}{2\pi a\sqrt{\mu\varepsilon}}\sqrt{(x_{nl})^2 + \left(\frac{q\pi}{d}\right)^2}. \tag{C.41}$$

The field is given by Equations (C.18).

APPENDIX C SPECIAL FUNCTIONS

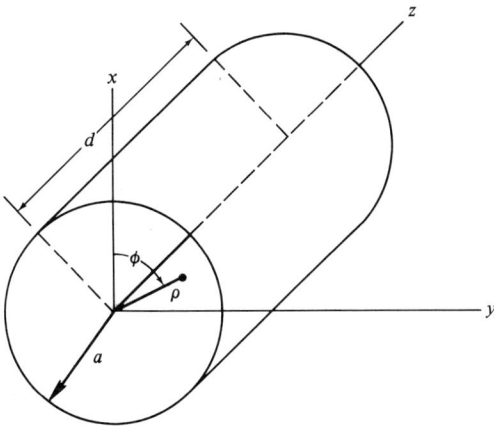

Figure C.3. Geometry of the circular cylindrical cavity.

TE to z Mode

$$F_z = J_n\left(x'_{nl}\frac{\rho}{a}\right)\cos n\phi \sin\frac{q\pi z}{d}, \quad \text{(C.42)}$$

where $n = 0, 1, 2, 3, \ldots, l = 1, 2, 3, \ldots, q = 1, 2, 3, \ldots,$ and

$$f_r = \frac{1}{2\pi a\sqrt{\mu\varepsilon}}\sqrt{(x'_{nl})^2 + \left(\frac{q\pi}{d}\right)^2}. \quad \text{(C.43)}$$

The field is given by Equations (C.31).

SPHERICAL COORDINATES

Separation of Variables

$$\frac{1}{r^2}\frac{\partial}{\partial r}\left(r^2\frac{\partial\alpha}{\partial r}\right) + \frac{1}{r^2\sin\theta}\frac{\partial}{\partial\theta}\left(\sin\theta\frac{\partial\alpha}{\partial\theta}\right) + \frac{1}{r^2\sin^2\theta}\frac{\partial^2\alpha}{\partial\phi^2}$$
$$+ k^2\alpha = 0 \quad \text{[scalar Helmholtz (wave) equation]}. \quad \text{(C.44)}$$

Assume $\alpha(r, \theta, \phi) = R(r)\Theta(\theta)\Phi(\phi)$ and substitute into Equation (C.44), then divide by α, giving

$$\frac{\sin^2\theta}{R}\frac{d}{dr}\left(r^2\frac{dR}{dr}\right) + \frac{\sin\theta}{\Theta}\frac{d}{d\theta}\left(\sin\theta\frac{d\Theta}{d\theta}\right) + \frac{1}{\Phi}\frac{d^2\Phi}{d\phi^2} + k^2 r^2\sin^2\theta = 0. \quad \text{(C.45)}$$

The third term of Equation (C.45) must be a constant, so

$$\frac{1}{\Phi}\frac{d^2\Phi}{d\phi^2} = -m^2. \qquad (C.46)$$

Equation (C.46) has as a solution the sum of any two harmonic functions $h(m\phi)$. Then Equation (C.45) becomes

$$\frac{1}{R}\frac{d}{dr}\left(r^2\frac{dR}{dr}\right) + \frac{1}{\Theta\sin\theta}\frac{d}{d\theta}\left(\sin\theta\frac{d\Theta}{d\theta}\right) - \frac{m^2}{\sin^2\theta} + k^2r^2 = 0. \qquad (C.47)$$

Now,

$$\frac{1}{\Theta\sin\theta}\frac{d}{d\theta}\left(\sin\theta\frac{d\Theta}{d\theta}\right) - \frac{m^2}{\sin^2\theta} = -n(n+1) \quad \text{(constant)}, \qquad (C.48)$$

which has as a solution the sum of any two *associated Legendre functions*,[2] $P_n^m(\cos\theta)$ (first kind), or $Q_n^m(\cos\theta)$ (second kind). Substitution of Equation (C.48) into Equation (C.47) leaves

$$\frac{d}{dr}\left(r^2\frac{dR}{dr}\right) + [(kr)^2 - n(n+1)]R = 0. \qquad (C.49)$$

Solutions to Equation (C.49) are *spherical Bessel functions*,[2] related to ordinary Bessel functions by

$$\begin{pmatrix} j_n(kr) \\ n_n(kr) \end{pmatrix} = \sqrt{\frac{\pi}{2kr}} \begin{pmatrix} J_{n+1/2}(kr) \\ N_{n+1/2}(kr) \end{pmatrix}. \qquad (C.50)$$

The general solution to Equation (C.44) then is

$$\alpha(r,\theta,\phi) = \begin{pmatrix} j_n(kr) \\ n_n(kr) \end{pmatrix} \cdot \begin{pmatrix} P_n^m(\cos\theta) \\ Q_n^m(\cos\theta) \end{pmatrix} \cdot h(m\phi). \qquad (C.51)$$

Spherical Cavity

Here we desire to find solutions, which are either TM or r ($H_r = 0$) or TE to r ($E_r = 0$), suitable for representing the fields inside a spherical cavity. For the rectangular and cylindrical cavity we were able to let $\alpha(x,y,z)$ or $\alpha(\rho,\phi,z)$ equal A_z (TM to z) or F_z (TE to z) since $\nabla^2 A_z$ (or $\nabla^2 F_z$) = $(\nabla^2 \mathbf{A})_z$ [or $(\nabla^2 \mathbf{F})_z$]. Now, $\nabla^2 A_r$ (or $\nabla^2 F_r$) $\neq (\nabla^2 \mathbf{A})_r$ [or $(\nabla^2 \mathbf{F})_r$], so a new formulation is necessary. The starting point for A_r is Equation (3.73) written in phasor form:

---

[2] See Spiegel in the references at the end of Chapter 1.

$$\nabla \times \nabla \times \mathbf{A} - k^2 \mathbf{A} = -j\omega\mu\varepsilon\nabla\Phi_a, \tag{C.52}$$

where $k^2 = \omega^2\mu\varepsilon$, $\mathbf{J} \equiv 0$ (inside the cavity), and

$$\mathbf{B} = \nabla \times \mathbf{A}. \tag{C.53}$$

Expanding Equation (C.52) in spherical coordinates gives

$$\frac{1}{r^2 \sin\theta} \frac{\partial}{\partial\theta}\left(\sin\theta \frac{\partial A_r}{\partial\theta}\right) + \frac{1}{r^2 \sin^2\theta} \frac{\partial^2 A_r}{\partial\phi^2} + k^2 A_r = j\omega\mu\varepsilon \frac{\partial\Phi_a}{\partial r}$$

$$-\frac{\partial^2 A_r}{\partial r\,\partial\theta} = j\omega\mu\varepsilon \frac{\partial\Phi_a}{\partial\theta} \tag{C.54}$$

$$-\frac{\partial^2 A_r}{\partial r\,\partial\phi} = j\omega\mu\varepsilon \frac{\partial\Phi_a}{\partial\phi},$$

if $A_\theta = A_\phi = 0$. The last two of Equations (C.54) are satisfied if we choose

$$\frac{\partial A_r}{\partial r} = -j\omega\mu\varepsilon\Phi_a, \tag{C.55}$$

which is *not* the Lorentz condition! Substituting Equation (C.55) into the first of Equations (C.54) gives

$$\frac{1}{r}\left[\frac{\partial^2 A_r}{\partial r^2} + \frac{1}{r^2 \sin\theta}\frac{\partial}{\partial\theta}\left(\sin\theta \frac{\partial A_r}{\partial\theta}\right) + \frac{1}{r^2 \sin^2\theta}\frac{\partial^2 A_r}{\partial\phi^2} + k^2 A_r\right] = 0. \tag{C.56}$$

It is a straightforward problem to show that the left-hand side of Equation (C.56) is identical with

$$\nabla^2\left(\frac{A_r}{r}\right) + k^2\left(\frac{A_r}{r}\right) = 0,$$

so A_r/r is a solution to the scalar Helmholtz equation, which, we have already seen, is α in Equation (C.51). Therefore,

$$A_r = r\alpha$$

or

$$A_r(r,\theta,\phi) = r \begin{pmatrix} j_n(kr) \\ n_n(kr) \end{pmatrix} \begin{pmatrix} P_n^m(\cos\theta) \\ Q_n^m(\cos\theta) \end{pmatrix} h(m\phi). \tag{C.57}$$

For TE to r modes we start with

$$\nabla \times \nabla \times \mathbf{F} - k^2 \mathbf{F} = -j\omega\mu\varepsilon\nabla\Phi_f \tag{C.58}$$

and

$$\mathbf{D} = -\nabla \times \mathbf{F} \qquad (C.59)$$

from the equations preceding Equations (6.58). If we choose

$$\frac{\partial F_r}{\partial r} = -j\omega\mu\varepsilon\Phi_f, \qquad (C.60)$$

we have

$$\nabla^2\left(\frac{F_r}{r}\right) + k^2\left(\frac{F_r}{r}\right) = 0$$

and F_r/r satisfies the scalar Helmholtz equation. Therefore,

$$F_r = r\alpha$$

or

$$F_r(r,\theta,\phi) = r \begin{pmatrix} j_n(kr) \\ n_n(kr) \end{pmatrix} \begin{pmatrix} P_n^m(\cos\theta) \\ Q_n^m(\cos\theta) \end{pmatrix} h(m\phi). \qquad (C.61)$$

The TM to r fields are given by Equation (C.53) and Maxwell's equation $\mathbf{E} = (1/j\omega\mu\varepsilon)\nabla \times \mathbf{B}$ as

$$\begin{aligned}
E_r &= \frac{1}{j\omega\mu\varepsilon}\left(k^2 + \frac{\partial^2}{\partial r^2}\right)A_r, & H_r &\equiv 0, \\
E_\theta &= \frac{1}{j\omega\mu\varepsilon r}\frac{\partial^2 A_r}{\partial r\,\partial\theta}, & H_\theta &= \frac{1}{\mu r \sin\theta}\frac{\partial A_r}{\partial\phi}, \\
E_\phi &= \frac{1}{j\omega\mu\varepsilon r \sin\theta}\frac{\partial^2 A_r}{\partial r\,\partial\phi}, & H_\phi &= -\frac{1}{\mu r}\frac{\partial A_r}{\partial\theta}.
\end{aligned} \qquad (C.62)$$

The TE to r fields are given by Equation (C.59) and Maxwell's equation $\mathbf{H} = -(1/j\omega\mu)\nabla \times \mathbf{E}$ as

$$\begin{aligned}
E_r &\equiv 0, & H_r &= \frac{1}{j\omega\mu\varepsilon}\left(k^2 + \frac{\partial^2}{\partial r^2}\right)F_r \\
E_\theta &= -\frac{1}{\varepsilon r \sin\theta}\frac{\partial F_r}{\partial\phi}, & H_\theta &= \frac{1}{j\omega\mu\varepsilon r}\frac{\partial^2 F_r}{\partial r\,\partial\theta} \\
E_\phi &= \frac{1}{\varepsilon r}\frac{\partial F_r}{\partial\theta}, & H_\phi &= \frac{1}{j\omega\mu\varepsilon r \sin\theta}\frac{\partial^2 F_r}{\partial r\,\partial\phi}
\end{aligned} \qquad (C.63)$$

APPENDIX C SPECIAL FUNCTIONS

It is convenient to use Schelkunoff's[3] Bessel functions,

$$\hat{J}_n(kr) = kr\, j_n(kr) = \sqrt{\frac{\pi kr}{2}} J_{n+1/2}(kr) \qquad (C.64)$$

and

$$\hat{N}_n(kr) = kr\, n_n(kr) = \sqrt{\frac{\pi kr}{2}} N_{n+1/2}(kr). \qquad (C.65)$$

Then the potentials may be written

$$A_r(r,\theta,\phi) \text{ or } F_r(r,\theta,\phi) = \begin{pmatrix} \hat{J}_n(kr) \\ \hat{N}_n(kr) \end{pmatrix} \begin{pmatrix} P_n^m(\cos\theta) \\ Q_n^m(\cos\theta) \end{pmatrix} h(m\phi). \qquad (C.66)$$

Since $Q_n^m(+1) \to \infty$, a finite field inside a sphere is given by

$$\begin{pmatrix} A_r(r,\theta,\phi) \\ F_r(r,\theta,\phi) \end{pmatrix} = \hat{J}_n(kr) P_n^m(\cos\theta) h(m\phi), \qquad (C.67)$$

with m and n integers.

Note that

$$j_0(kr) = \frac{\sin kr}{kr}, \qquad n_0(kr) = -\frac{\cos kr}{kr},$$

$$j_1(kr) = \frac{\sin kr}{(kr)^2} - \frac{\cos kr}{kr}, \qquad n_1(kr) = -\frac{\cos kr}{(kr)^2} - \frac{\sin kr}{kr}$$

$$P_0^0 = 1,$$

$$P_1^0 = u,$$

$$P_2^0(u) = \frac{3u^2 - 1}{2},$$

$$P_1^1(u) = -\sqrt{1 - u^2},$$

$$P_2^1(u) = -3\sqrt{1 - u^2}\, u.$$

The special functions of this section are tabulated in mathematical references.[4]

---

[3] S. A. Schelkunoff, *Electromagnetic Fields*. New York: Blaisdell Publishing Company, 1963.

[4] *Handbook of Mathematical Functions*. U.S. Department of Commerce, National Bureau of Standards, Applied Mathematics Series 55, 1964.

Appendix D
RADIATION

An antenna is the interface, or connecting link, between some guiding system and (usually) free space. As such, its function is either to *radiate* electromagnetic energy (in the transmitting case), or *receive* electromagnetic energy (in the receiving case). In the former case, a transmitting system feeds the guiding system to the antenna, and in the latter case, the guiding system feeds a receiving system from the antenna. Most antenna characteristics apply in a receiving case or a transmitting case, and can be determined, one from the other, by a generalized reciprocity relation.

Figure D.1 shows a source (a time-dependent current or charge) producing an electromagnetic field. We already know that this will occur. We also know that the phasor field will, in general, be given by the phasor forms

$$\mathbf{B} = \boldsymbol{\nabla} \times \mathbf{A} \tag{3.91}$$

and

$$\mathbf{E} = \frac{1}{j\omega\mu\varepsilon}\boldsymbol{\nabla}(\boldsymbol{\nabla} \cdot \mathbf{A}) - j\omega\mathbf{A} \tag{3.92}$$

as shown in Chapter 3. The magnetic vector potential $\mathbf{A}$ must satisfy the partial differential equation [phasor form of Equation (3.80)]

$$\nabla^2 \mathbf{A} + k^2 \mathbf{A} = \begin{cases} -\mu\mathbf{J}, & \mathbf{r} \text{ inside vol}', \\ 0, & \mathbf{r} \text{ outside vol}'. \end{cases}$$

A solution to the preceding equation is certainly the Helmholtz or superposition integral which we obtained in Chapter 3:

$$\mathbf{A}(\mathbf{r}, \omega) = \frac{\mu}{4\pi} \iiint_{\text{vol}'} \frac{\mathbf{J}(\mathbf{r}', \omega)e^{-jkR}}{R} dv', \tag{D.1}$$

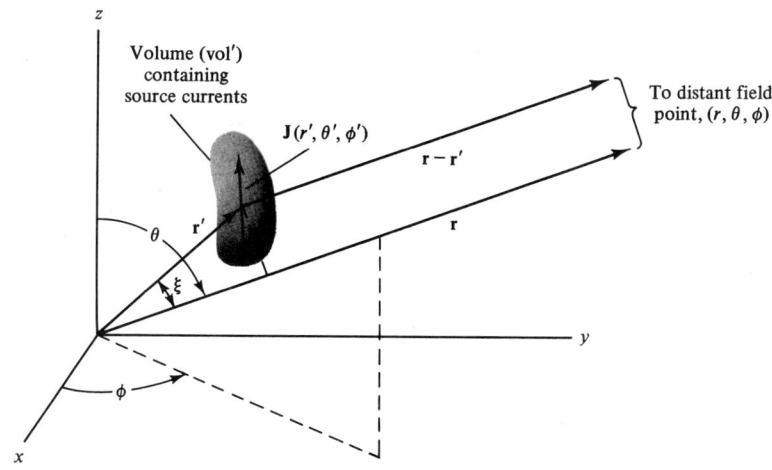

Figure D.1. Geometry for determining the radiation field from a source current density.

where $R = |\mathbf{r} - \mathbf{r}'|$ and we have assumed that μ and ε are scalar constants everywhere.

We will now point out explicitly those features that are special to the *radiation zone* or *far field*. The radiation zone is characterized by $|\mathbf{r}| \gg |\mathbf{r}'|$, in which case $\mathbf{r}$ and $\mathbf{r} - \mathbf{r}'$ are essentially parallel, or

$$|\mathbf{r} - \mathbf{r}'| \approx r - r'\cos\xi \qquad (D.2)$$

as Figure D.1 clearly shows. It is convenient at this point to list for future use the term $r'\cos\xi$ where the source coordinates are rectangular, cylindrical, and spherical, while the field coordinates are spherical in each case. In order to do this, we use

$$\frac{\mathbf{r}\cdot\mathbf{r}'}{r} = r'\cos\xi = \frac{xx' + yy' + zz'}{r}, \qquad (D.3)$$

with the coordinate transformations found on the inside front cover. The result is

$$\boxed{r'\cos\xi = (x'\cos\phi + y'\sin\phi)\sin\theta + z'\cos\theta} \qquad (D.4a)$$

or

$$\boxed{r'\cos\xi = \rho'\sin\theta\cos(\phi - \phi') + z'\cos\theta} \qquad (D.4b)$$

or

$$\boxed{r'\cos\xi = r'[\cos\theta\cos\theta' + \sin\theta\sin\theta'\cos(\phi - \phi')]}. \qquad (D.4c)$$

Insofar as the denominator of Equation (D.1) (a magnitude only) is concerned, we may go one step further and use

$$|\mathbf{r} - \mathbf{r}'| \approx r \quad \text{(denominator only)}. \tag{D.5}$$

The last approximation is not possible for the phase term in the numerator of Equation (D.1) because the difference between $\mathbf{r}$ and $\mathbf{r}'$ becomes extremely important if it amounts to an appreciable part of a wavelength. In other words, $(kr - kr')$ may represent a large and important phase angle. Since the denominator of Equation D.1 is a positive quantity, the difference between $|\mathbf{r} - \mathbf{r}'|$ and r is unimportant, so long as $r \gg r'$. Equation (D.1) is then represented accurately by

$$\mathbf{A}(\mathbf{r}, \omega) = \frac{\mu}{4\pi} \frac{e^{-jkr}}{r} \iiint_{\text{vol}'} \mathbf{J}(\mathbf{r}', \omega) e^{jkr' \cos \xi} dv' \tag{D.6}$$

for the radiation zone, or far-field.

Now, since $\mathbf{J}(\mathbf{r}', \omega)$ is independent of r, and, as Equations (D.4a), (D.4b), and (D.4c) show, $\cos \xi$ is independent of r, we are able to conclude that the triple integral in Equation (D.6) is independent of r!

Equation (D.6) will always produce the general form

$$\mathbf{A}(\mathbf{r}, \omega) = \frac{\mu}{4\pi} \frac{e^{-jkr}}{r} [f_r(\theta, \phi)\mathbf{a}_r + f_\theta(\theta, \phi)\mathbf{a}_\theta + f_\phi(\theta, \phi)\mathbf{a}_\phi]. \tag{D.7}$$

The special feature of Equation (D.6) is that r is *outside* the integral sign, and, therefore, many of the operations required to find $\mathbf{E}$ and $\mathbf{H}$ can be performed with no difficulty, regardless of whether $\mathbf{J}(\mathbf{r}')$ is known or not.

Equations (3.91) and (3.92) may be written in terms of the spherical components:

$$\begin{aligned} E_r &= \frac{1}{j\omega\mu\varepsilon}[\nabla(\nabla \cdot \mathbf{A})]_r - j\omega A_r, \\ E_\theta &= \frac{1}{j\omega\mu\varepsilon}[\nabla(\nabla \cdot \mathbf{A})]_\theta - j\omega A_\theta, \\ E_\phi &= \frac{1}{j\omega\mu\varepsilon}[\nabla(\nabla \cdot \mathbf{A})]_\phi - j\omega A_\phi, \\ H_r &= \frac{1}{\mu}(\nabla \times \mathbf{A})_r, \\ H_\theta &= \frac{1}{\mu}(\nabla \times \mathbf{A})_\theta, \\ H_\phi &= \frac{1}{\mu}(\nabla \times \mathbf{A})_\phi. \end{aligned} \tag{D.8}$$

As another specialization of the radiation zone, we retain only the terms that vary with $1/r$ from the preceding set of equations. The terms that vary as $1/r^2$ or $1/r^3$ are negligibly small for large r, and make up what is called the "near-field" and the "induction field." These fields, as opposed to the fields in the radiation zone, or

APPENDIX D RADIATION

far-field, will not contribute to the time-average power flow away from the source. When the indicated expansions are carried out (a rather lengthy proposition) using Equation (D.7), the result is

$$\boxed{\begin{aligned} E_r &= 0, & H_r &= 0 \\ E_\theta &= -j\omega A_\theta, & H_\theta &= \frac{-E_\phi}{\eta} \\ E_\phi &= -j\omega A_\phi, & H_\phi &= \frac{E_\theta}{\eta} \end{aligned}} \quad \text{(radiation field)} \tag{D.9}$$

for the radiation zone. Therefore, with Equations (D.9), the radiation field is obtained without differentiating the vector potential! That is, this operation has now been performed once and for all! Then, the radiation field consists of outward traveling spherical waves which decrease[1] as $1/r$. In order to avoid difficulties in integrating nonconstant unit vectors, it is always advisable to perform the integration using the Cartesian components (where the unit vectors are constant) of the current. In this case Equation (D.6) together with Equations (D.9) give the most general form of the radiation field.

$$\boxed{\begin{aligned} E_\theta(r,\theta,\phi) &= \frac{-j\omega\mu\, e^{-jkr}}{4\pi\, r}\Bigg(\cos\theta\cos\phi \iiint_{\text{vol}'} J_x(\mathbf{r}')e^{jkr'\cos\xi}\,dv' \\ &\quad + \cos\theta\sin\phi \iiint_{\text{vol}'} J_y(\mathbf{r}')e^{jkr'\cos\xi}\,dv' \\ &\quad - \sin\theta \iiint_{\text{vol}'} J_z(\mathbf{r}')e^{jkr'\cos\xi}\,dv'\Bigg), \\ E_\phi(r,\theta,\phi) &= \frac{-j\omega\mu\, e^{-jkr}}{4\pi\, r}\Bigg(-\sin\phi \iiint_{\text{vol}'} J_x(\mathbf{r}')e^{jkr'\cos\xi}\,dv' \\ &\quad + \cos\phi \iiint_{\text{vol}'} J_y(\mathbf{r}')e^{jkr'\cos\xi}\,dv'\Bigg), \\ H_\theta &= -\frac{E_\phi}{\eta}, \\ H_\phi &= \frac{E_\theta}{\eta}. \end{aligned}} \tag{D.10}$$

[1]This is a spreading effect.

Equations (D.10) indicate that if the current density **J** is known, then, in a formal sense at least, the solution is complete. This is true regardless of the type of antenna one is analyzing. For most antennas made of conductors in the form of wires, such as the dipole antenna, we can approximate the current density reasonably well. On the other hand, slot antennas and aperture type antennas are such that it is very difficult to estimate the current densities on the conducting surfaces. These antennas are usually analyzed in terms of the known (or approximated) fields over the aperture, for which equations other than those developed here will be needed.

In many cases of practical interest the current density has only a z component, and in this case Equations (D.10) simplify to

$$E_\theta = \frac{j\omega\mu}{4\pi} \frac{e^{-jkr}}{r} \sin\theta \iiint_{vol'} J_z(r') e^{jkr' \cos\xi} \, dv' \quad (\mathbf{J} = \mathbf{a}_z J_z), \tag{D.11}$$

$$H_\phi = E_\theta / \eta.$$

If the current is also filamentary, then it can only depend on z'. That is, $J_z(r') dv'$ becomes $I_z(z') dz'$, and Equation (D.4a) gives $r' \cos\xi = z' \cos\theta$, so Equations (D.11) reduce to

$$E_\theta = \frac{j\omega\mu}{4\pi} \frac{e^{-jkr}}{r} \sin\theta \int_{l'} I_z(z') e^{jkz' \cos\theta} \, dz', \tag{D.12}$$

$$H_\phi = E_\theta / \eta.$$

The time-average power density for the general case is

$$<\mathcal{S}> = \tfrac{1}{2} \text{Re}(\mathbf{E} \times \mathbf{H}^*)$$

and, when Equations (D.9) are substituted, we have the general result:

$$<\mathcal{S}> = \frac{\mathbf{a}_r}{2\eta}(|E_\theta|^2 + |E_\phi|^2) \quad (\text{W/m}^2). \tag{D.13}$$

That is, the time-average power density is apparently directed *radially outward*. The outward power flow, or *radiated power* is obtained by integrating the normal component of $<\mathcal{S}>$ over a sphere of (large) radius, r, or

$$<\mathcal{P}_r> = \int_0^{2\pi} \int_0^\pi <\mathcal{S}> \cdot \mathbf{a}_r r^2 \sin\theta \, d\theta \, d\phi \quad (\text{W}). \tag{D.14}$$

Substituting Equation (D.13) into Equation (D.14) gives

$$\boxed{<\mathcal{P}_r> = \frac{r^2}{2\eta} \int_0^{2\pi} \int_0^\pi (|E_\theta|^2 + |E_\phi|^2) \sin\theta \, d\theta \, d\phi} \quad (\textit{radiated power}). \tag{D.15}$$

APPENDIX D RADIATION

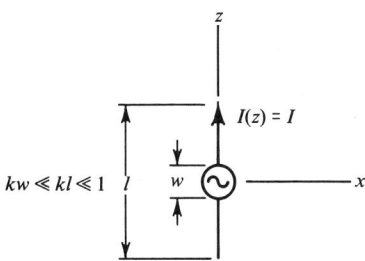

Figure D.2. Hertzian dipole centered on the z axis ($kw \ll kl \ll 1$).

If the source distribution (current density, surface current density, or current) is known explicitly, then the formal solution to the radiation problem is complete, being given by the preceding equations. Unfortunately, this is rarely the case, especially for those antenna geometries of practical importance. An *exact* solution for the source distribution requires the solution to a boundary value problem, and is almost never tractable. Quite often, we are willing to approximate this source distribution.

The Hertzian dipole is a filamentary current I that extends over a length l that is very small compared to a wavelength ($kl \ll 1$). Ideally it is a *point* source, and may be regarded as the elementary source for radiation studies. An ideal Hertzian dipole is shown in Figure D.2. Equation (D.12) gives

$$E_\theta = \frac{j\omega\mu}{4\pi}\frac{e^{-jkr}}{r}\sin\theta \int_{-l/2}^{l/2} I\, e^{jkz'\cos\theta}\, dz \approx \frac{j\omega\mu}{4\pi}\frac{e^{-jkr}}{r} I \sin\theta \int_{-l/2}^{l/2} dz,$$

since $kw \ll kl \ll 1$. The result is

$$E_\theta = j\frac{\eta(kl)I}{4\pi}\frac{e^{-jkr}}{r}\sin\theta, \quad H_\phi = E_\theta/\eta. \tag{D.16}$$

A plot of E_θ (normalized) versus θ is given in Figure D.3, and shows the directional characteristics of the Hertzian dipole. E_θ is independent of ϕ.

The radiated power is given by Equations (D.15) and (D.16) as

$$\langle \mathcal{P}_r \rangle = \frac{\eta(kl)^2|I|^2}{32\pi^2}\int_0^{2\pi}\int_0^\pi \sin^3\theta\, d\theta\, d\phi$$

and since the double integral is $8\pi/3$,

$$\langle \mathcal{P}_r \rangle = 10(kl)^2|I|^2 \quad \text{(W)}, \tag{D.17}$$

where $|I|$ is a peak value. If the same power is dissipated in a resistance R, then $\langle \mathcal{P}_r \rangle = |I|^2 R/2$. Call this resistance the *radiation resistance* R_{rad}:

$$R_{\text{rad}} = 20(kl)^2 \quad (\Omega) \quad (\textit{radiation resistance}). \tag{D.18}$$

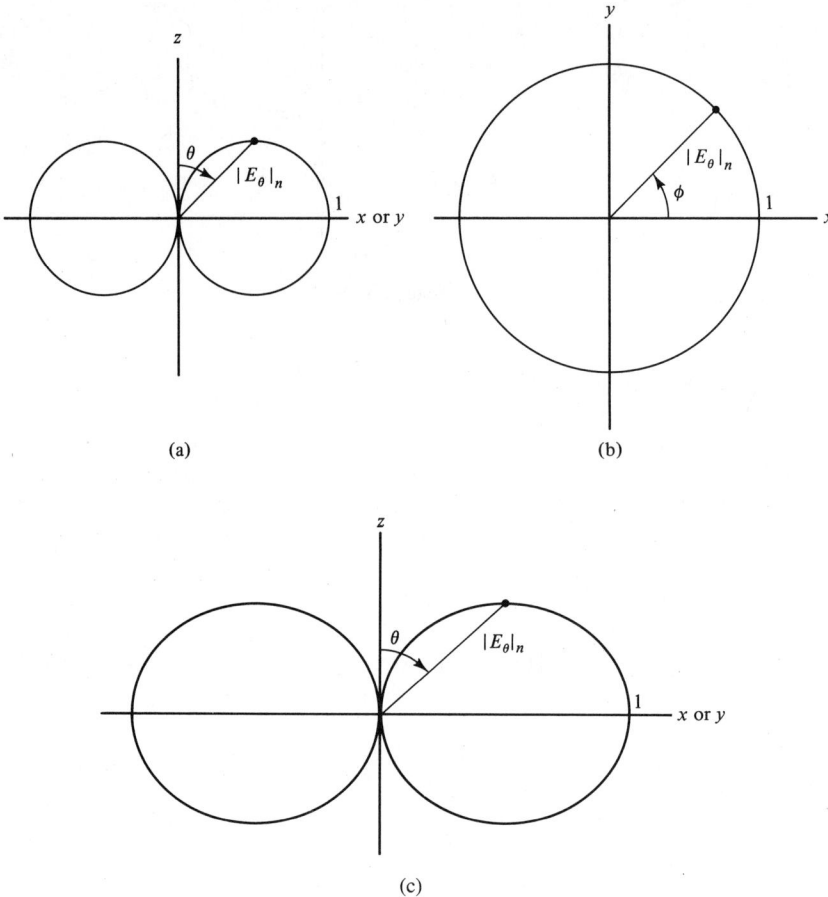

Figure D.3. (a) *E*-plane pattern of the Hertzian dipole of Figure D.2 (E_θ, normalized, versus θ, $\phi = 0, \pi$). (b) *H*-plane pattern of the Hertzian dipole (E_θ, normalized, versus ϕ, $\theta = \pi/2$). (c) *E*-plane pattern of the half-wave dipole wtih sinusoidal current.

Since $kl \ll 1$, R_{rad} is small and the Hertzian dipole is inefficient in practice. Typical values of R_{rad} do not match typical transmission line characteristic resistances.

The complete field (near and far) for the Hertzian dipole is obtained from the reduced form of Equation (D.6) for filamentary currents and the equation set (D.8).

$$E_r = \frac{Il}{j2\pi\omega\varepsilon} \cos\theta \frac{e^{-jkr}}{r^2}\left(jk + \frac{1}{r}\right)$$

$$E_\theta = \frac{Il}{j4\pi\omega\varepsilon} \sin\theta \frac{e^{-jkr}}{r}\left(-k^2 + \frac{jk}{r} + \frac{1}{r^2}\right) \qquad \text{(D.19)}$$

$$H_\phi = \frac{Il}{4\pi} \sin\theta \frac{e^{-jkr}}{r}\left(jk + \frac{1}{r}\right)$$

APPENDIX D RADIATION

When (only) the terms that vary with $1/r$ are retained, Equations (D.19) reduce to Equations (D.16).

The *half-wave* dipole is like the Hertzian dipole, except that its length is $\lambda/2$, and $\lambda/2 \gg a$, where a is the dipole radius. Thus, the current is essentially filamentary. Furthermore, it is assumed that the current distribution remains that of an open-circuited (open-wire) transmission line when its legs are bent 90° at a distance $\lambda/4$ from the end to form the half-wave dipole:

$$I_z(z) = I_t \cos(kz). \tag{D.20}$$

When this assumed current is substituted into Equation (D.12), the result is

$$E_\theta = \frac{j\omega\mu}{4\pi} \frac{e^{-jkr}}{r} \sin\theta \int_{-\lambda/4}^{\lambda/4} I_t \cos(kz') e^{jkz'\cos\theta} dz'.$$

The integration is not difficult and results in

$$E_\theta = j\frac{\eta}{2\pi} I_t \frac{e^{-jkr}}{r} \frac{\cos[(\pi/2)\cos\theta]}{\sin\theta}, \quad H_\phi = E_\theta/\eta. \tag{D.21}$$

The normalized plot of E_θ versus θ is shown in Figure D.3(c) and is remarkably similar to that of the Hertzian dipole. On the other hand, Equation (D.15) gives

$$<\mathcal{P}_r> = \frac{15|I_t|^2}{\pi} \int_0^{2\pi} \int_0^\pi \frac{\cos^2[(\pi/2)\cos\theta]}{\sin\theta} d\theta\, d\phi$$

$$= 30|I_t|^2 \int_0^\pi \frac{\cos^2[(\pi/2)\cos\theta]}{\sin\theta} d\theta.$$

The integral can be reduced to tabulated functions or simply evaluated numerically. The result is

$$<\mathcal{P}_r> = 36.56|I_t|^2 \quad (W) \tag{D.22}$$

and

$$R_{\text{rad}} = 73.13 \quad (\Omega). \tag{D.23}$$

This is much closer to transmission characteristic resistances.

SYMBOLS AND UNITS

The International System of Units is an expanded form of the rationalized meter-kilogram-second-ampere (MKSA) system of units, and is used throughout this textbook. It is based on six basic units: the meter (length), the kilogram (mass), the second (time), the ampere (electric current), the kelvin degree (temperature), and the candela (luminous intensity). All other units may be derived from these six. Definitions of these fundamental units may be found in the references. See Hayt in the references at the end of Chapter 2, for example.

The symbols used in this text are, more or less, conventional. That is, they are usually the same as those for the same quantities found in other textbooks. Occasionally, it is convenient, because of inference, to use the same symbol for two different quantities. In a situation such as this, these quantities will not appear together in the discussion, and certainly not in the same equations. There will never be any confusion about what is meant. The following table lists the quantities that are encountered in the text. It also includes the symbol used for the quantity, the unit of the quantity, and the abbreviation of the unit.

SYMBOLS AND UNITS

Names and Units of the Quantities Found in the Text in the International System

Quantity	Symbol	Unit	Abbreviation
ac capacitivity	$\varepsilon'(\omega)$	farads/meter	F/m
ac inductivity	$\mu'(\omega)$	henrys/meter	H/m
Acceleration	$\mathbf{a}$	meters/second^2	m/s^2
Admittance	Y	mhos	℧
Admittivity	$\hat{y}$	mhos/meter	℧/m
Angular frequency	ω	radians/second	rad/s
Angular velocity	$\boldsymbol{\omega}$	radians/second	rad/s
Associated Legendre function, first kind	$P_n^m(x)$	—	—
Associated Legendre function, second kind	$Q_n^m(x)$	—	—
Attenuation constant	α	nepers/meter	Np/m
Bessel function, first kind, order n	$J_n(x)$	—	—
Bessel function, second kind, order n	$N_n(x)$	—	—
Bound charge, density	ρ_v^b	coulombs/meter^3	C/m^3
Bound current density	$\mathbf{J}^b$	amperes/meter^2	A/m^2
Capacitance	C	farads	F
Characteristic impedance	Z_0	ohms	Ω
Characteristic resistance	R_0	ohms	Ω
Charge	$Q, q(t)$	coulombs	C
Coefficient of magnetic coupling	k_{mc}	—	—
Coefficient of reflection	Γ	—	—
Coefficient of transmission	T	—	—
Complex propagation constant	γ	meter^{-1}	m^{-1}
Conductance	G	mhos	℧
Conductivity	σ	mhos/meter	℧/m
Current	$I, i(t)$	amperes	A
Current density	$\mathbf{J}, \mathcal{J}$	amperes/meter^2	A/m^2
Cutoff angular frequency	ω_c	radians/second	rad/s
Cutoff frequency	f_c	hertz	Hz
Del operator	∇	—	—

Names and Units of the Quantities Found in the Text in the International System

Quantity	Symbol	Unit	Abbreviation
Dielectric loss angle	δ_d	radians	rad
Dielectric loss factor	$\varepsilon''(\omega)$	farads/meter	F/m
Drift velocity	$\mathbf{u}_d$	meters/second	m/s
Electric dipole moment	$\mathbf{p}$	coulomb-meter	C · m
Electric energy	W_E	joules	J
Electric energy density	w_E	joules/meter^3	J/m^3
Electric field intensity	$\mathbf{E}, \boldsymbol{\mathscr{E}}$	volts/meter	V/m
Electric flux	Ψ_E	coulombs	C
Electric flux density	$\mathbf{D}, \boldsymbol{\mathscr{D}}$	coulombs/meter^2	C/m^2
Electric potential difference	Φ_{ab}, Φ	volts	V
Electric scalar potential	$\Phi, \varPhi$	volts	V
Electric susceptibility	χ_E	—	—
Electric vector potential	$\mathbf{F}, \boldsymbol{\mathscr{F}}$	coulombs/meter	C/m
Electromotive force	emf	volts	V
Electron charge	$-e$	coulombs	C
Electron mobility	μ_e	meter^2/volt-second	m^2/V · s
Electron rest mass	m_e	kilograms	kg
Energy (work)	W	joules	J
Force	$\mathbf{F}, \boldsymbol{\mathscr{F}}$	newtons	N
Frequency	f	hertz	Hz
Group velocity	u_g	meters/second	m/s
Hankel function, first kind, order n	$H_n^{(1)}(x)$	—	—
Hankel function, second kind, order n	$H_n^{(2)}(x)$	—	—
Heat flux vector	$\mathbf{s}$	watts/meter^2	W/m^2
Impedance	Z	ohms	Ω
Inductance	L	henrys	H
Intrinsic impedance	η	ohms	Ω
Length	$l, d,$ etc.	meters	m
Line charge density	ρ_l	coulombs/meter	C/m
Mass	M, m	kilograms	kg
Magnetic dipole moment	$\mathbf{m}$	ampere-meter^2	A · m^2
Magnetic energy	W_H	joules	J
Magnetic energy density	w_h	joules/meter^3	J/m^3
Magnetic field intensity	$\mathbf{H}, \boldsymbol{\mathscr{H}}$	amperes/meter	A/m

SYMBOLS AND UNITS

Quantity	Symbol	Unit	Abbreviation
Magnetic flux	ψ_m	webers	Wb
Magnetic flux density	**B**, $\mathcal{B}$	webers/meter^2 (tesla)	Wb/m^2
Magnetic loss angle	δ_m	radians	rad
Magnetic loss factor	$\mu''(\omega)$	henrys/meter	H/m
Magnetic scalar potential	Φ_m, $\mathit{\Phi}_m$	amperes	A
Magnetic susceptibility	χ_m	—	—
Magnetic vector potential	**A**, $\mathcal{A}$	webers/meter	Wb/m
Magnetization	**M**	amperes/meter	A/m
Magnetomotive force	mmf	ampere-turns	A · t
Permeability	μ	henrys/meter	H/m
Permeability of vacua	μ_0	henrys/meter	H/m
Permittivity	ε	farads/meter	F/m
Permittivity of vacua	ε_0	farads/meter	F/m
Phase constant	β	radians/meter	rad/m
Phase velocity	u_p	meters/second	m/s
Polarization	**P**	coulombs/meter^2	C/m^2
Power	P, $\mathcal{P}$	watts	W
Power radiated	P_r	watts	W
Poynting vector (power density)	**S**, $\mathcal{S}$	watts/meter^2	W/m^2
Proton rest mass	m_p	kilograms	kg
Quality factor	Q	—	—
Radiation resistance	R_r	ohms	Ω
Relative permeability	μ_R	—	—
Relative permittivity	ε_R	—	—
Reluctance	$\mathcal{R}$	henrys^{-1}	H^{-1}
Resistance	R	ohms	Ω
Resonant frequency	f_r	hertz	Hz
Schelkunoff's Bessel function, first kind	$\hat{J}_n(x)$	—	—
Schelkunoff's Bessel function, second kind	$\hat{N}_n(x)$	—	—
Spherical Bessel function, first kind	$j_n(x)$	—	—
Spherical Bessel function, second kind	$n_n(x)$	—	—
Skin depth (depth of penetration)	δ	meters	m

Names and Units of the Quantities Found in the Text in the International System

Quantity	Symbol	Unit	Abbreviation
Standing wave ratio	SWR	—	—
Surface area	s	meter^2	m^2
Surface charge density	ρ_s	coulombs/meter^2	C/m^2
Surface current density	$\mathbf{J}_s$, $\boldsymbol{\mathcal{J}}_s$	amperes/meter	A/m
Surface impedance	η	ohms	Ω
Temperature	T	degrees kelvin	K
Time delay	t_D	seconds	s
Torque	$\mathbf{T}$	newton-meters	N·m
Transfer function	$H(\omega)$	varies	varies
Transverse wave impedance	$\eta^{\pm}$	ohms	Ω
Velocity	$\mathbf{u}$	meters/second	m/s
Velocity of energy flow	u_e	meters/second	m/s
Velocity of light	c	meters/second	m/s
Voltage	$v(t)$, V_0, V_{21}, etc.	volts	V
Voltage standing wave ratio	VSWR, s	—	—
Volume	v, vol	meter^3	m^3
Volume charge density	ρ_v	coulombs/meter^3	C/m^3
Waveguide wavelength	λ_g	meters	m
Wavelength	λ	meters	m
Wave number (intrinsic phase constant)	k	radians/meter	rad/m

ANSWERS TO SELECTED PROBLEMS

CHAPTER 1

1. $\mathbf{F} = -32.32\mathbf{a}_y$

3. $\theta_x = 64.62°$, $\theta_y = 149.0°$, $\theta_z = 73.4°$

4. (a) $2x + 3y + 6z = 35$
   (b) $\mathbf{B} \cdot (\mathbf{A}/A) = 5$ m

6. (a) $\mathbf{a}_\rho = \mathbf{a}_x \cos\phi + \mathbf{a}_y \sin\phi$, $\mathbf{a}_\phi = -\mathbf{a}_x \sin\phi + \mathbf{a}_y \cos\phi$,
   $\mathbf{a}_r = \mathbf{a}_x \sin\theta \cos\phi + \mathbf{a}_y \sin\theta \sin\phi + \mathbf{a}_z \cos\theta$,
   $\mathbf{a}_\theta = \mathbf{a}_x \cos\theta \cos\phi + \mathbf{a}_y \cos\theta \sin\phi - \mathbf{a}_z \sin\theta$

   (b) $\int_0^\pi \mathbf{a}_\rho \, d\phi = 2\mathbf{a}_y$, $\int_0^{\pi/2} \mathbf{a}_\phi d\phi = -\mathbf{a}_x + \mathbf{a}_y$

   $\int_0^1 \mathbf{a}_r \, dz = (x\mathbf{a}_x + y\mathbf{a}_y) \ln \dfrac{1 + (x^2 + y^2 + 1)^{1/2}}{(x^2 + y^2)^{1/2}}$
   $\qquad + \mathbf{a}_z[(x^2 + y^2 + 1)^{1/2} - (x^2 + y^2)^{1/2}]$

   $\int_0^{\pi/2} \mathbf{a}_\theta d\theta = \mathbf{a}_\rho - \mathbf{a}_z$

8. $V = 12$ m^3

9. (a) $V = \pi a^2 h/3$ (m^3)
   (b) $s = \pi a^2 + \pi a \sqrt{a^2 + h^2}$ (m^2)

11. (a) Flux $= 4\pi a$
    (b) Flux $= 4\pi a$
    (c) Flux $= 4\pi a$

401

13. (a) $\nabla \cdot \mathbf{C} = y^2 + 1$
    (b) $\nabla \cdot \mathbf{C} = 3$
    (c) $\nabla \cdot \mathbf{C} = 3\cos\phi + \cos\theta$
    (d) $\nabla \cdot \mathbf{C} = 6$

16. (a) conservative, (b) conservative, (c) conservative, (d) not conservative, (e) conservative, (f) not conservative, (g) conservative.

21. (a) Flux $= 4\pi a$
    (b) Flux $= 2\pi a$
    (c) Flux $= 2\pi a$
    (d) Flux $= 2\pi a$

23. (a) $5(\pi a)^2 = 5(\pi a)^2$
    (b) $10\pi ah = 10\pi ah$

25. (a) $2\pi a^2 = 2\pi a^2$
    (b) $2\pi a^2 = 2\pi a^2$

CHAPTER 2

2. $(0,0,0.601)$

4. (a) $E_z \approx 56.55$ V/m
   (b) $E_z \approx 1.8$ V/m
   (c) $E_z \approx 0.9 \times 10^{-6}$ V/m

5. $x_l = eE_0 l^2/(2mu_0^2)$, $x_a = eE_0 l(l_a - l/2)/(mu_0^2)$ (m)

7. $\Phi = -\rho_s |z|/(2\varepsilon_0)$ (V)

8. (a) $F = 19.78 \times 10^{19}$ N
   (b) $\mathbf{F}_m = -9.8m\mathbf{a}_r$ (N)
   (c) $\mathbf{g} = -9.8\mathbf{a}_r$ (m/s^2)
   (d) $\Phi_g = -3.985 \times 10^{14} m/r$ (N · m)
   (e) $u = 11.2$ km/s (25,000 mph)

10. $E_z = -1.058 V_0/d$ (V/m)

12. (a) $Q = 4\pi$ (C) at the origin
    (b) $\rho_l = 2\pi$ (C/m) on the z axis
    (c) $\rho_s = \varepsilon_0/(10a^2)$ (C/m^2) on the sphere $r = a$

13. (a) Proof
    (b) $\rho_{sa} = \varepsilon_0 V_0/[a \ln(b/a)]$ (C/m^2)

15. (a) $\mathbf{E} = \mathbf{a}_z 3E_0/(\varepsilon_R + 2)$ (V/m)
    (b) $\mathbf{E} = \mathbf{a}_z E_0 + \{E_0 a^3(\varepsilon_R - 1)/[r^3(\varepsilon_R + 2)]\}\{2\cos\theta\mathbf{a}_r + \sin\theta\mathbf{a}_\theta\}$ (V/m)
    (c) $\mathbf{E} = \mathbf{a}_z E_0$ (V/m)
    (d) Proof

ANSWERS TO SELECTED PROBLEMS

18. (a) $\Phi = V_0 \phi/\phi_0$ (V)
    (b) $\rho_s = -[\varepsilon V_0/(\rho\phi_0)]$ (C/m^2)
    (c) No. The capacitance per unit length is infinite.

20. (a) $\Phi = \dfrac{\rho_l}{4\pi\varepsilon_0} \ln \dfrac{\rho^2 - 2\rho(a^2/\rho_1)\cos\phi + (a^2/\rho_1)^2}{\rho^2 - 2\rho\rho_1 \cos\phi + \rho_1^2} + C$ (V)

   $C = -\dfrac{\rho_l}{4\pi\varepsilon_0} \ln(a/\rho_1)^2$ makes $\Phi(a,\phi) = 0$

   (b) $\Phi = \dfrac{\rho_l}{4\pi\varepsilon_0} \ln \dfrac{\rho^2 - 2\rho(a^2/\rho_1)\cos\phi + (a^2/\rho_1)^2}{\rho^2 - 2\rho\rho_1\cos\phi + \rho_1^2} \left(\dfrac{\rho_1}{a}\right)^2$ (V)

23. $\rho_s = -\dfrac{4\varepsilon V_0}{\pi a} \displaystyle\sum_{n=1,\text{odd}}^{\infty} (\rho/a)^{n-1} = -\dfrac{4\varepsilon V_0}{\pi a} \dfrac{1}{1-(\rho/a)^2}$ (C/m^2)

25. (a) $\nabla \cdot \mathbf{J} = 0$
    (b) $I = 2\pi a$ (A)
    (c) $\oiint \mathbf{J} \cdot d\mathbf{s} = 0$

27. (a) $\rho_v = -3.93 \times 10^{-12} V_0(zd^2)^{-2/3}$ (C/m^3)
    (b) $J_z = -2.33 \times 10^{-6} V_0^{3/2}/d^2$ (A/m^2)
    (c) $I = J_z s_a$ (A)
    (d) $t_t = 0.506$ ns
    (e) $I = 4.3$ mA

28. (a) 0
    (b) $2\pi a(\rho_a - a)J_{sz}$ (A)
    (c) 0

29. $x = (u_{z0}/\omega_c)(\cos\omega_c t - 1)$, $z = (u_{z0}/\omega_c)\sin\omega_c t$

31. (a) $\mathbf{T} = 7.54\mathbf{a}_z$ (N·m)
    (b) $P = 395$ W

33. $\mathbf{B}_z = 0.153\mathbf{a}_x + 0.153\mathbf{a}_y + 0.5\mathbf{a}_z$, $\mathbf{H}_z = \mathbf{B}_z/\mu_0$

34. (a) $\left.\dfrac{\partial \Phi_m}{\partial \phi}\right|_{\rho=a^-} = \left.\dfrac{\partial \Phi_m}{\partial \phi}\right|_{\rho=a^+}, \mu_R \left.\dfrac{\partial \Phi_m}{\partial \rho}\right|_{\rho=a^-} = \left.\dfrac{\partial \Phi_m}{\partial \rho}\right|_{\rho=a^+}$

    (b) $\Phi_m = \begin{cases} [-2/(\mu_R + 1)]H_0\rho\cos\phi, & \rho \leq a \\ [a^2/\rho][(\mu_R - 1)/(\mu_R + 1)] - H_0\rho\cos\phi, & \rho \geq a \end{cases}$

    (c) $\mathbf{H} = \mathbf{a}_x[2/(\mu_R + 1)]H_0$, $\rho \leq a$

36. $B_\text{left} = 0.406$, $B_\text{rt} = 0.094$, $B_\text{cent} = 0.313$ Wb/m^2

38. (a) $L = \dfrac{\mu s}{2l}(N_1 + N_2)^2$ (H)

    (b) $L = \dfrac{\mu_0 s}{l}N_1^2 + \dfrac{\mu_0 s}{l}N_2^2$ (H)

    (c) $L = \dfrac{\mu s}{2l}(N_1 - N_2)^2$ (H)

39. (a) $E_\rho(a) = V_0/(a \ln b/a)$ (V/m)
    (b) $E_\rho(a) = 2.718 V_0/b$ (V/m)
    (c) $C = 2\pi \varepsilon l$ (F)

41. (a) $R = \ln(b/a)/(2\pi \sigma t)$ (Ω)
    (b) $R = t/[\pi \sigma(b^2 - a^2)]$ (Ω)
    (c) $R = 2\pi/(\sigma t \ln b/a)$ (Ω)

42. $V_f = \varepsilon_R V_0$ (V)

43. $C = 11.26$ pF

45. $R_{\text{eff}} = 2.73$ mΩ, $\sigma_{\text{eff}} = 2.91 \times 10^7$ $\mho$/m

47. $\Delta f = 12.5$ kHz

48. $L_{12} = 61.4$ nH

50. $L_{\text{ext}} = 157$ nH/m

CHAPTER 3

1. $I = 7/4$ A

3. $i(t) = [\omega h l B_0/(2R)] \sin \omega t$ (clockwise)

6. $v_{\text{oc}}(t) = \pi \times 10^3 \sin(10^3 t)$ V

7. emf $= -2\pi t + 2\pi \times 10^{-2}$ V

9. (a) emf $= -\omega_0 B_0 a^2/2$ (V)
   (b) emf $= -(\omega_0 B_0 a^2/2) \sin \omega_b t - (\omega_b a b B_m) \cos \omega_b t$ (V)

13. (a) $\mathbf{H} = \sqrt{\varepsilon/\mu}\, E_\rho \mathbf{a}_\phi$ (A/m), $a \le \rho \le b$
    (b) $\mathbf{J}_s|_{\rho=a} = \mathbf{a}_z \dfrac{E_0}{\eta a} e^{-j\omega \sqrt{\mu\varepsilon} z}$, $\mathbf{J}_s|_{\rho=b} = -\mathbf{a}_z \dfrac{E_0}{\eta b} e^{-j\omega \sqrt{\mu\varepsilon} z}$ (A/m)

17. $d = 384 \times 10^6$ m

19. (a) Proof
    (b) $\mathscr{E} = \mathscr{H} = 0$

20. emf $= \oint_l \dfrac{\partial \mathbf{A}}{\partial t} \cdot d\mathbf{l}$ (V)

22. $\mathbf{E} = [j\beta \cos(\pi x)\mathbf{a}_x - \pi \sin(\pi x)\mathbf{a}_z] \dfrac{e^{-j\beta z}}{\sigma + j\omega\varepsilon}$ (V/m), $\sigma \ne 0$

24. (a) $\Phi = -\ln(\rho/b) E_0 e^{-j\omega\sqrt{\mu_0 \varepsilon_0} z}$ (V)
    (b) $\Phi_{ab} = \ln(b/a) E_0 e^{-j\omega\sqrt{\mu_0 \varepsilon_0} z}$ (V)
    (c) $V_{ab} = \Phi_{ab}$
    (d) Because $\mathbf{A} \cdot d\mathbf{l} = 0$

ANSWERS TO SELECTED PROBLEMS   405

26. $\mathcal{A}(0,0,z) = \dfrac{\mu_0 I_0 a}{4R} \cos \omega(t - \sqrt{\mu_0 \varepsilon_0}\, R)\mathbf{a}_y$ (Wb/m)

28. $<\mathcal{P}_f> = 36.54 I_t^2$ (W) (numerical integration)

CHAPTER 4

2. (a) $E_x = \begin{cases} 0, & z > 0 \\ jE_0 \sin(kz), & -\lambda/4 < z < 0 \\ E_0 e^{+jkz}, & z < -\lambda/4 \end{cases}$

   (b) $\mathbf{J}_s = \begin{cases} -\mathbf{a}_x(E_0/\eta)\ (\text{A/m}), & z = 0 \text{ plane} \\ -j\mathbf{a}_x(E_0/\eta)\ (\text{A/m}), & z = -\lambda/4 \text{ plane} \end{cases}$

4. (a) $\mathcal{H} = 26.5 \sin(\omega t - x)\mathbf{a}_z$ (mA/m)
   (b) $\mathcal{H} = 53.1 \sin(\omega t - x)\mathbf{a}_z$ (mA/m)

7. (a) $\alpha = 0.288$ (Np/m)
   (b) $\beta = \pi$ rad/m
   (c) $u_p = 200 \times 10^6$ m/s

11. (a) $\beta = 4.19$ rad/m
    (b) $\lambda = 1.5$ m
    (c) $\mathcal{H} = (-15.92\mathbf{a}_x + 21.22\mathbf{a}_y) \cos(2\pi \times 10^8 t - 4.19z)$ (mA/m)

13. $\mu_R = 1.099$, $\varepsilon_R = 4.068$, $\sigma = 10.24$ (m℧/m)

15. 86.47%

16. $\sigma = 13.33 \times 10^3$ ℧/m

18. (a) $\Gamma = 0.632\, \underline{|164°}$
    (b) $T = 0.429\, \underline{|23.8°}$
    (c) SWR $= 4.43$

20. (a) $\Gamma_0 = 0.268\ (z = 0, \lambda = 1/\sqrt{3})$, $\eta_{in} = 146.7 - j69.6\ \Omega$,
       $\Gamma_d = 0.455\, \underline{|-155.6°}, z = d$
    (b) SWR $= 2.67$
    (c) 79.3%

21. (a) $|\mathbf{E}| = 1.48$ V/m, $z = d$; $|\mathbf{E}| = 0$, $z = 0$
       $|\mathbf{H}| = 3.56$ mA/m, $z = d$; $|\mathbf{H}| = 7.13$ mA/m, $z = 0$
    (b) $d = n\lambda/2 = n/(2\sqrt{3})$ (m), $n = 0, 1, 2, \ldots$

22. (a) $t = 6.78$ mm
    (b) 99.92%

24. $\theta_t = 26.89°$ (red), $\theta_t = 26.77°$ (violet)

25. For total internal reflection and no losses, 100% of the incident power density is reflected.

27. $\mathbf{E} = (\mathbf{a}_x + \mathbf{a}_y - \mathbf{a}_z)\exp[(-jk_0)(x + 2y + 3z)/\sqrt{14}]$ (V/m)
    $\mathbf{H} = [(-5\mathbf{a}_x + 4\mathbf{a}_y - \mathbf{a}_z)/(\eta_0\sqrt{14}]\exp[(-jk_0)(x + 2y + 3z)/\sqrt{14}]$ (A/m)

28. (a) $<\mathcal{S}_z> = 50$ W/m^2
    (b) $<\mathcal{S}> = 10.9$ W/m^2

31. It appears closer than it actually is.

33. Brass must be 2.03 times as thick as silver.

36. (a) $\Gamma_H \rightarrow \left|\dfrac{\sqrt{\varepsilon_R - \sin^2\theta_i} - \varepsilon_R\cos\theta_i}{\sqrt{\varepsilon_R - \sin^2\theta_i} + \varepsilon_R\cos\theta_i}\right|\underline{|180°}$, $\theta_i < \theta_B$

    (b) $\Gamma_H \rightarrow 0 \underline{|90°}$, $\theta_i = \theta_B$

    (c) $\Gamma_H \rightarrow \left|\dfrac{\sqrt{\varepsilon_R - \sin^2\theta_i} - \varepsilon_R\cos\theta_i}{\sqrt{\varepsilon_R - \sin^2\theta_i} + \varepsilon_R\cos\theta_i}\right|\underline{|0°}$, $\theta_i > \theta_B$

38. (a) $\mathbf{H} = (-\mathbf{a}_x 39.8\underline{|30°} + \mathbf{a}_y 26.5)e^{-jkz}$ (mA/m)
    (b) $<\mathcal{S}> = 431\mathbf{a}_z$ (mW/m^2)

42. $z = \pi\delta$ (m)

43. (a) $<\mathcal{P}> = 1.48$ W/m^2
    (b) $<\mathcal{P}> = 8.22$ W/m^2

CHAPTER 5

1. $L_{ext} = 185.2$ nH/m, $R_0 = 55.6$ Ω

3. $V(\lambda/2) = 0$, $I(\lambda/2) = -10$ mA

4. (a) $V(\lambda/2) = 0$
   (b) $V(\lambda) = 0$

6. (a) $Z_l = 375.9 - j225$ Ω
   (b) $\Gamma_l = (1/3)\underline{|-52.94°}$
   (c) VSWR = 2

7. (a) $X_l = \pm 44.72$ Ω
   (b) $z_{min} = 0.285\lambda$ or 0.215λ

8. $Z_{in} = j0.313R_0$ (Ω)

9. (a) $<\mathcal{P}> = 800$ W
   (b) $N_p/N_s = 0.961$
   (c) $L = 1.46$ μH

11. (a) $l = 5.42$ cm
    (b) $C = 5.96$ pF

(c) $z = 4.58$ cm
(d) near $z = 4.58$ cm

17. (a) VSWR = 2
    (b) $<\mathcal{P}_l> = 160$ mW

18. $V_T = V_s(z',\omega)e^{-jkl}\dfrac{e^{jkz'} - \Gamma_l e^{-jkz'}}{1 - \Gamma_l e^{-j2kl}}$ (V)

    $Z_T = R_0 \dfrac{Z_l \cos kl + jR_0 \sin kl}{R_0 \cos kl + jZ_l \sin kl}$ (Ω)

20. 0.055 inch

21. 1.221 inch

24. $\Gamma_l = 0.62 \underline{|-7.13°}$, VSWR = 4.26, $Z_{in} = 22.88 \underline{|54.03°}$ Ω

25. $Z_l = 21.82 \underline{|19.11°}$ Ω

26. $Z_l = 54.48 \underline{|-54.75°}$ Ω

28. $Z_l = (10.37 + j3.616)R_0$ (Ω)

29. (a) $B = 175.5$ MHz
    (b) $\text{VSWR}_{max} = 4$, $f = 1$ GHz, 2 GHz, ...
    (c) $<\mathcal{P}_l>/<\mathcal{P}_{inc}> = 0.64$
    (d) B is infinite.

30. (a) $R_0' = 46$ Ω, $X = -2$ Ω
    (b) $<\mathcal{P}_{rad}> = 73$ W
    (c) VSWR = 4.64
    (d) Yes

32. VSWR = 1.376

34. $L = 21.22$ nH

35. $R_0' = 86.7$ (Ω), $d = 0.335\lambda$

38. VSWR = 7.9, $<\mathcal{P}_l> = 0.39 <\mathcal{P}_{inc}>$

39. $G_v = 17.8$ dB

40. (a) $\alpha = 4.72 \times 10^{-6}$ Np/m, $\beta = 22.07 \times 10^{-6}$ rad/m,
    $Z_0 = 691 \underline{|-11.8°}$ Ω, $u_p = 285 \times 10^6$ m/s, $\lambda = 284.7$ km
    (b) $L_{add} = 179.7$ μH/m, $\alpha = 1.065 \times 10^{-6}$ Np/m

42. $R = 50$ mΩ/m, $L = 198.9$ nH/m, $C = 79.58$ pF/m, $G = 20$ $\mu\mho$/m

43. $B = 1.592 \times 10^{-3} f_0$. This agrees with f_0/Q_0.

48. $Z_{in} = 49.36 - j0.323$ Ω. The results will be good.

49. $Z_{in}^{oc} = 553.1 - j17.44$ Ω

50. $<\mathcal{P}_l> = 9.49$ kW

CHAPTER 6

3. (a) $Z_{\text{eff}} = \eta^+ = 570\,\Omega$
   (b) SWR = 2.28
   (c) $\varepsilon_R = 1.56$, $\lambda g/4 = 0.75$ cm

4. $t_{d_1} - t_{d_2} = 1.139$ ps/m

5. $<\mathcal{P}_f> = 1.046$ MW

8. $a = 7.07$ cm, $b = a/2$

9. All systems of the type described have an impulse response containing the term $\delta(t - l\sqrt{\mu\varepsilon})$, so each observer detects a signal after a time delay of $l\sqrt{\mu\varepsilon}$ (s).

13. $a = 4.24$ mm, $n = 0$

14. $k_{x0} \approx j(k_{xd}a/2)^2 \varepsilon_R$

15. (a) $f_r = 5.303$ GHz
    (b) $a = c = 2.76$ cm $= 2b$

19. (a) $E_z = E_0 \sin(\pi x/a) \sin(\pi y/b)$ (V/m)
    $H_y = -jE_0(\pi/a)(1/\omega\mu) \cos(\pi x/a) \sin(\pi y/b)$ (A/m)
    $H_x = +jE_0(\pi/b)(1/\omega\mu) \sin(\pi x/a) \cos(\pi y/b)$ (A/m)
    (b) A monopole at $x = a/2$, $y = b/2$, $z = 0$ or c
    (c) $<\mathcal{S}_z> = 0$

20. (a) Proof
    (b) $a = 2.165$ cm, air
    (c) $d = 2.172$ cm, $s_{\text{cyl}} = 5.9 \times 10^{-3}$ m³, $s_{\text{rect}} = 6.4 \times 10^{-3}$ m²
    (d) Cylindrical
    (e) Cylindrical, $d > b$

21. $(f/f_c) = \{(3/2)(1 + 2b/a) + (1/2)[9(1 + 2b/a)^2 - 8b/a]^{1/2}\}^{1/2}$

23. (a) $Q_c = 11,340$, silver; $Q_c = 5580$, brass
    (b) $Q_c = 7825$

26. $Q_c = 12,340$, yes

27. $B = 0.468$ MHz, silver; $B = 0.951$ MHz, brass; $B = 0.678$ MHz, silver–Teflon

30. (a) $\alpha_c = 3.21 \times 10^{-3}$ Np/m, waveguide; $\alpha_c = 16.3 \times 10^{-3}$ Np/m, two-wire line. The waveguide is preferred.
    (b) $\alpha_c = 17.64 \times 10^{-3}$ Np/m, coax. The waveguide is preferred.

INDEX

Acceleration:
  charge, 71
  due to gravity, 114
Admittivity, 185
Air gap, 98, 148
Ampere's circuital law, 34, 80, 142
Ampere-turn, 97
Anisotropic material, 57, 60
Antenna, 347, 388
  array, 313
Associated Legendre functions, 384
Atom, 91, 97
Attenuation constant, 184, 283

Balmain, K. G., 221
Bandpass filter, 250, 309
Bandwidth, 250, 272
Bessel functions, 344, 377
Bicylindrical coordinates, 104
Biot-Savart law, 34, 72–75
Boast, W. B., 112
Boundary conditions:
  conductor-conductor, 72
  conductor-free space, 159
  dielectric-dielectric, 60–61
  dynamic, 159-161
  magnetic, 96
  perfect conductor, 161
Boundary value problem, 63–68
Bound charge, 57, 58
Bound current, 92, 94, 95
Bradshaw, M. D., 165
Brewster angle, 210
Brown, R. G., 307
Byatt, W. J., 165

Capacitance:
  between coaxial cylinders, 103, 147
  defined, 102, 147, 233
  between parallel planes, 103
  between parallel wires, 104
  of p-n junction, 107, 124
  between single wire and ground plane, 106
Capacitivity, 281
Capacitor:
  coaxial, 103, 115, 147, 168
  energy stored in, 147
  multiple dielectric, 123
  parallel plane, 103
  two-wire line, 104
Cartesian coordinate system, 5
  for expressing curl, 22
  for expressing divergence, 16
  for expressing gradient, 18
  transformation to other coordinate systems, 5, 6
Cauchy's equation, 224
Characteristic value, *see* Eigenvalue
Charge:
  configurations, 35–36
  differential, 35
  point, 35
Charge density:
  line, 35, 40, 42
  surface, 36, 44
  volume, 35
Cheng, D. K., 112
Child-Langmuir law, 119
Chipman, R. A., 307
Circuit:
  distributed, 162, 236
  electric, 163
  magnetic, 96–100
Circuit theory from field theory, 162–165
Circulation, 9–11, 51, 142
Coaxial cable:
  capacitance, 103, 147
  cavity, 302
  E field of, 103, 115, 168
  H field of, 82–83
  inductance, 109
  R, G, L, and C of, 300

Coaxial diode, 120
Coefficient of magnetic coupling, 168
Coefficient of reflection, 190, 238, 239
Coefficient of transmission, 190
Coercive mmf, 97
Conductance, 107, 299
Conductivity:
 defined, 71
 of lossy dielectric, 106
 relation to reluctance, 97
 table of values, 373
Conductor, 69
 filamentary, 69
 good, 185, 194, 225
 moving, 138
 perfect, 110, 161
Conformal transformation, 67
Conservation of charge, 71, 126–128, 229
Conservative field, 25, 335
Constant unit vectors, 6
Constitutive relations, 141
Continuous function, 51
Convolution, 39, 49
Coordinate component transformation, 6
Coordinates:
 field, 37
 source, 37
Coordinate system:
 right-handed, 4–7
 transformations between, 6
Corona, 60
Coulomb force, 37, 101
Coulomb gauge, 88
Coulomb's law, 34, 37, 114
Coupling into a waveguide, 329, 357
Critical angle, 213, 345, 352
Curl, 16, 20–24
Current:
 bound, 94
 conduction, 69
 defined, 70, 127
 displacement, 138
 filamentary, 69
Current density, 69–72, 126
 conduction, 69, 128
 convention, 69, 128
 defined, 69
 displacement, 138
 general, 69
 surface, 70
Current element, differential, 70
Cyclotron frequency, 125
Cylindrical coordinate system, 5
 for expressing curl, 22
 for expressing divergence, 16
 for expressing gradient, 18
 transformation to other coordinate systems, 5, 6

Dekker, A. J., 112
Del operator, 15, 62, 88
Diamagnetic material, 92
Dielectiric constant, see Permittivity, relative
Dielectric loading, 365
Dielectric loss angle, 281
Dielectric loss factor, 281
Dielectrics, 56–61, 185, 225
Dielectric strength, 60
Dielectric waveguides, 345
Differential current element, 69
Differential vector length, 8
Differential volume, 8
Diode:
 junction, 124
 tunnel, 258
 vacuum, 115, 119, 120
Dipole:
 antenna, 169
 electrostatic, 52–53, 57, 115
 magnetostatic, 89–91
Dipole moment, 57, 91
Dirac delta function, see Unit impulse function
Directional coupler, 257, 272, 314
Directional derivative, 18
Dispersion, 186, 220, 327
Dispersive medium, 185, 220
 anomalously, 220
 normally, 220
Distance, source point to field point, 37
Distortion, 186, 279
Distortionless line, 279, 285, 287
Divergence, 13–16
Divergence theorem, 26–28, 54, 127
Double stub matching, 275
Duality, 88

Earth, 48
Eddy current, 200
Edminister, J. A., 112
Eigenvalue, 337
Electric circuit analog, for magnetic circuit, 100
Electric field intensity, 37–47
 defined, 38
 of line charge, 40, 42
 of N point charges, 38
 of surface charge, 44
Electric flux:
 defined, 53
 density, 53, 128
Electrode, 102, 106
Electromotive force, see Emf

INDEX

Electron, 57, 91
  charge of, 71
  conduction, 71
  mass of, 372
  spin, 91
  valence, 71
Electrostatic deflection system, 113
Emf, 130
  motional, 134, 136
  transformer, 132, 134
Energy, *see also* Work
  electric, 144, 146, 290
  kinetic, 114, 119
  magnetic, 144, 146, 290
  to move point charge, 47
  potential, 114, 119
  stored in capacitor, 147
  stored in inductor, 147
Equipotential surface, 17, 19, 63
Euler formula, 66
Evanescent field, 213, 322, 359
Even function, 42
Everitt, E. E., 292, 307

Faraday disk generator, 167
Faraday's law, 130–137
Ferrimagnetic material, 92
Ferrite, 92, 220
Ferroelectric material, 57
Ferromagnetic domain, 92
Ferromagnetic material, 57, 92
Fink, D. G., 112
Fluid flow, 16, 23
Flux, 11–13, 128
Flux lines, 77
Flux linkage, 108
Force:
  in air gap, 148
  on charge, 37
  on closed circuit, 101
  on conductor, 101
  on differential current element, 100
Fourier:
  heat conduction law, 20
  inverse transform, 281
  series, 66
  transform, 143, 281

Gauge function, 150
Gaussian surface, 54
Gauss's law, 15, 34, 53–56, 62, 129
  integral form, 53
  point form, 54
Generator, 136, 167
Gradient, 16–20, 51
Gravity, 114

Ground, 48
Ground plane, 67

Half-wave dipole, 169, 170, 395
Hayt, W. H., 112, 302, 307
Helmholtz integral, 49, 88, 153, 155, 158
Helmholtz wave equation, 153, 157, 375
Hertzian dipole, 393
Homogeneous material, 57
Horseshoe electromagnet, 148
Hughes, W. L., 307
Hysteresis:
  dielectric, 59
  loop, 97
  magnetic, 92

Image charge, 68
Image current, 104
Image theory, 67–69, 104, 117, 118
Impedance:
  characteristic, 283
  input, 235, 244
  intrinsic, 181
  load, 246
  surface, 196
  wave, 180, 181, 323
Incidence:
  normal, 187
  oblique, 205
Incident wave, 188
Inductance:
  defined, 108, 147, 234
  external, 108, 299
  internal, 108, 196, 283, 299
  mutual, 108, 110
Inductivity, 281
Inhomogeneous dielectric, 123
Interface, plane, 187
  conductor-perfect conductor, 204
  dielectric-conductor, 188
  dielectric-dielectric, 200, 223
  dielectric-perfect conductor, 201
Ionized region, 226
Irrotational field, 23, 24, 51
Isothermal surface, 17, 19
Isotropic material, 57

Johnk, C. T., 165
Jordan, E. C., 221, 368
Joule's law, 144
Junction, *p-n*, 107, 124

Kirchhoff's current law, 71, 165, 237
Kirchhoff's voltage law, 165, 237, 239
Kraus, J. D., 112

Lamination, 197–200
Laplace's equation, 62–63, 231
Laplace transform, 143
Laplacian operator, 62, 88
Leakage flux, 98
Lenz's law, 130, 133
Liao, S. Y., 166, 301, 308
Lightning rod, 116
Linearity, 38
Linear system, 38
Line charge density, 35
Line integral, 9
Lorentz condition, 152, 335, 336
Lorentz force, 129
Lorrain, P., 112
Loss tangent, 185, 282

McQuistan, R. B., 31
Macroscopic behavior, 57
Magnetic field intensity:
  of coaxial cable, 82
  defined, 74, 129
  of filamentary current, 26, 77
  of solenoid, 83
  of surface current, 79
  of toroid, 119
Magnetic flux, 79, 283
Magnetic flux density, 128, 129
  defined, 73, 129
  remnant, 97
Magnetic loss angle, 281
Magnetic loss factor, 281
Magnetic materials, 91–96
Magnetic vector potential, 87
Magnetization, 92–95, 149
Magnetomotive force, see Mmf
Magnetostatic deflection system, 120
Marcuvitz, N., 368
Material parameters, 373
Maxwell, J. C., 166
Maxwell's equations:
  dynamic, 132, 137, 140–143
  static, 34, 51, 54, 85–86, 111
Measurement:
  of ϵ_R, 305
  of input impedance, transmission line, 244, 256
  of o_d, 305
  of VSWR, 244, 315
Microscopic behavior, 57
Mixed boundary conditions, 63
Mmf, 97
Mobility, 71
Modulation, 187, 312
Moon, P., 112
Moore, R. K., 308

Motor:
  axial gap, 120
  radial gap, 101–102
Mott, H., 221

Neff, H. P., 112, 166
Newton's gravitational law, 114
Newton's second law, 114
Normal component:
  at conductor boundary, 63
  at dielectric-dielectric boundary, 60, 61
  at magnetic-magnetic boundary, 96
  at perfect conductor, 161
Norton equivalent circuit, 239
Nuclear spin, 91
Numerical method:
  for electrostatics, 42, 113
  for magnetostatics, 76

Oblique incidence, 205
Odd function, 42
Ohm's law, 71, 128
Orthogonal surface, 5
Orthogonal unit vectors, 5
Owen, G. E., 166

Paddlewheel, 23, 24
Page, L., 112
Paramagnetic material, 92
Pattern:
  E-plane, 394
  H-plane, 394
Paul, C. R., 113
Permanent magnetization, 92, 97
Permeability, 95, 129
  complex, 281
  of free space, 72
  relative, 95
Permittivity:
  complex, 183, 281
  of free space, 37
  relative, 59, 373
Perturbation, 282, 303, 358
Phase constant, 172, 184
Phasors, 142
Physical constants, 372
Pierce, J. R., 221
Planar transmission line, 122, 229
Plasma frequency, 226
Plonsey, R., 113
Plonus, M. A., 113
Point charge, 35
  E field of, 38
  potential field of, 49
Poisson's equation, 61–62, 88
Polarization:
  antenna, 220

dielectric, 57
horizontal, 205
vertical, 205
wave, 176, 181, 217
 circular, 219
 elliptical, 218
 left-handed, 218
 linear, 176, 218
 right-handed, 219
Polarization vector, 57, 59
Popovic, B. D., 221
Potential:
 absolute, 48
 electrostatic, 47–50
 retarded, 153
 scalar electric, 48, 151
 scalar magnetic, 87, 150–156, 336
 vector electric, 334
 vector magnetic, 87, 150–156, 334
Potential difference, 48, 50, 159
 defined, 48
 relation to voltage, 48, 159
Potential field:
 of coaxial cable, 104
 of dipole, 52, 68
 of inclined planes, 168
 of line charge, 52
 of parallel line charges, 104
 of point charge, 49
 of polarization, 58
Potential integral, 49
Potential reference, 48
Power, 102, 120, 145, 241, 282, 393, 395
Power density, 179
Poynting's theorem, 143–148
Poynting vector, 145, 179
Prism effect, 224
Product solution, 64, 375, 376
Propagation constant, 184, 284
Proton, 57, 372

Quality factor, 290, 303

Radiation, 162, 347, 388
Radiation pattern:
 half-wave dipole, 394
 Hertzian dipole, 394
Ramo, S., 113, 221, 368
Rao, N. N., 166
Reflected wave, 187, 188, 308
Relaxation time, 130
Relay, 148
Reluctance, 97
Resistance:
 characteristic, 234
 defined, 107

high frequency, 196
 radiation, 393
Resonant cavity:
 transmission line, 249, 315
 waveguide, 355, 358, 382
Retarded time, 153
Right-handed coordinate system, 4
Right-hand rule, 23
Ryder, J. D., 308

Schey, H. M., 31
Schelkunoff, S. A., 387
Separation equation, 337, 376
Separation of variables, 64, 375, 376
Sharpe, R. A., 307
Shedd, P. C., 113
Shen, L. C., 166
Single-stub matching, 270
Skilling, H. H., 23, 31, 308, 368
Skin depth, 195, 299
Skin effect, 193–200, 226
Skitek, G. G., 166
Slotted line, 244, 257
Slotted waveguide, 329
Smith, P. H., 257
Smith chart, 257, 267
Snell's law, 209
Solenoid, 83–85, 96
Solenoidal field, 16, 25, 129, 334
Special functions, 375
Spherical Bessel functions, 384
Spherical cavity, 384
Spherical coordinate system, 5
 for expressing curl, 23
 for expressing divergence, 16
 for expressing gradient, 18
 transformation to other coordinate
  systems, 5, 6
Spiegel, M. R., 31
Standing wave ratio, 190, 191, 243
Stokes' theorem, 28–30, 51, 131
Stripline, 122, 301
Superconductor, 72
Superparamagnetic material, 92
Superposition, 38
Surface charge density, 36
Surface guided wave, 353
Susceptibility:
 electric, 59
 magnetic, 95
Symbols and units, 396

Tangential component:
 at conductor boundary, 63
 at dielectric-dielectric boundary, 60, 61
 at magnetic-magnetic boundary, 96

Tangential component *(Continued)*
  at perfect conductor, 161
Tapered transmission line, 311
Taylor's series, 14, 21
TEM mode, 229, 282
TE mode, 217, 229
Temperature, 17, 19, 72
Tensor, 60, 95
Thévenin circuit, transmission line, 239, 285
Thomassen, K. I., 368
Time delay, 153, 187, 278, 327, 328, 368, 369
TM mode, 229
Toroid, 98, 119
Torque, 57
  of electrostatic dipole, 57
  on magnetic dipole, 91
Total internal reflection, 212
Total transmission, 210
Transfer function, 288
Transformations between coordinate systems, 6
Transformer, 122, 168, 250, 308
Transit-time, 119
Transmission line:
  current, 238
  distributed parameters, 233, 299
  equivalent lumped parameters, 249
  load impedances, 246, 254
  matching, 250, 268, 308
  power, 241
  pulses on, 278, 308
  tapered, 311
  voltage, 238
  VSWR, 243
Transmission parameters, 252, 316
Transmitted wave, 189
Triode, 169
Two-wire transmission line, 104, 116, 230, 287

Uniform damped plane wave, 183
Uniform plane wave, 172, 205
Uniqueness, 63, 67
Unit impulse function, 56
Unit vector, 5, 18

Van Duzer, T., 113
Varactor diode, 107, 124
Vector:
  addition, 1
  field, 1
  multiplication, 2–4
  subtraction, 2

Vector identities, *inside front and back covers*, 24–30
Vector operations, *inside front and back covers*
Vector relations, *inside front and back covers*
Velocity:
  of charge, 69, 71
  drift, 71
  of energy flow, 179, 216, 323
  escape, 115
  of group, 186, 187, 216, 278, 327
  of light, 153
  phase, 178, 185, 216, 322
  of propagation, 176
Volume charge density, 35
VSWR, 244, 315

Wave:
  cylindrical, 172
  damped, 183
  evanescent, 213
  plane, 187, 205
  spherical, 172
  standing, 190, 202, 355
  traveling, 176
Wave equation, 153, 157, 238
Waveguide, cylindrical, 344, 377
Waveguide, cylindrical cavity, 358, 382
Waveguide, dielectric, 345
Waveguide, parallel plane, 217, 229
Waveguide, rectangular, 317, 337
  attenuation, 359
  current, 329, 368
  cutoff frequency, 317, 321, 338
  field equations, 320, 337, 339
  field plots, 342
  impedance, 331
  losses, 358
  TE_{10} mode, 317
Waveguide, rectangular cavity, 355
  bandwidth, 366
  losses, 359
  modes, 355, 356
  Q, 363
  resonant frequency, 355, 356
Wavelength, 132, 162, 178, 185, 323
Wavenumber, *see* Phase constant
Whinnery, J. R., 113
Width of minimum method for VSWR, 315
Work, 25

Zahn, M., 113

VECTOR OPERATIONS—SPHERICAL COORDINATES

$$\nabla \alpha = \mathbf{a}_r \frac{\partial \alpha}{\partial r} + \mathbf{a}_\theta \frac{1}{r} \frac{\partial \alpha}{\partial \theta} + \mathbf{a}_\phi \frac{1}{r \sin \theta} \frac{\partial \alpha}{\partial \phi}$$

$$\nabla \cdot \mathbf{A} = \frac{1}{r^2} \frac{\partial}{\partial r}(r^2 A_r) + \frac{1}{r \sin \theta} \frac{\partial}{\partial \theta}(A_\theta \sin \theta) + \frac{1}{r \sin \theta} \frac{\partial A_\phi}{\partial \phi}$$

$$\nabla \times \mathbf{A} = \mathbf{a}_r \frac{1}{r \sin \theta} \left(\frac{\partial}{\partial \theta}(A_\phi \sin \theta) - \frac{\partial A_\theta}{\partial \phi} \right)$$

$$+ \mathbf{a}_\theta \frac{1}{r} \left(\frac{1}{\sin \theta} \frac{\partial A_r}{\partial \phi} - \frac{\partial}{\partial r}(r A_\phi) \right) + \mathbf{a}_\phi \frac{1}{r} \left(\frac{\partial}{\partial r}(r A_\theta) - \frac{\partial A_r}{\partial \theta} \right)$$

$$\nabla^2 \alpha = \frac{1}{r^2} \frac{\partial}{\partial r}\left(r^2 \frac{\partial \alpha}{\partial r}\right) + \frac{1}{r^2 \sin \theta} \frac{\partial}{\partial \theta}\left(\sin \theta \frac{\partial \alpha}{\partial \theta}\right) + \frac{1}{r^2 \sin^2 \theta} \frac{\partial^2 \alpha}{\partial \phi^2}$$

$$\nabla^2 \mathbf{A} = \mathbf{a}_r \left[\nabla^2 A_r - \frac{2}{r^2}\left(A_r + \cot \theta A_\theta + \operatorname{cosec} \theta \frac{\partial A_\phi}{\partial \phi} + \frac{\partial A_\theta}{\partial \theta}\right) \right]$$

$$+ \mathbf{a}_\theta \left[\nabla^2 A_\theta - \frac{1}{r^2}\left(\operatorname{cosec}^2 \theta A_\theta - 2\frac{\partial A_r}{\partial \theta} + 2 \cot \theta \operatorname{cosec} \theta \frac{\partial A_\phi}{\partial \phi}\right) \right]$$

$$+ \mathbf{a}_\phi \left[\nabla^2 A_\phi - \frac{1}{r^2}\left(\operatorname{cosec}^2 \theta A_\phi - 2 \operatorname{cosec} \theta \frac{\partial A_r}{\partial \phi} - 2 \cot \theta \operatorname{cosec} \theta \frac{\partial A_\theta}{\partial \phi}\right) \right]$$

ADDITION AND MULTIPLICATION

$$A^2 = \mathbf{A} \cdot \mathbf{A}$$
$$|A|^2 = \mathbf{A} \cdot \mathbf{A}^*$$
$$\mathbf{A} + \mathbf{B} = \mathbf{B} + \mathbf{A}$$
$$\mathbf{A} \cdot \mathbf{B} = \mathbf{B} \cdot \mathbf{A}$$
$$\mathbf{A} \times \mathbf{B} = -\mathbf{B} \times \mathbf{A}$$
$$(\mathbf{A} + \mathbf{B}) \cdot \mathbf{C} = \mathbf{A} \cdot \mathbf{C} + \mathbf{B} \cdot \mathbf{C}$$
$$(\mathbf{A} + \mathbf{B}) \times \mathbf{C} = \mathbf{A} \times \mathbf{C} + \mathbf{B} \times \mathbf{C}$$
$$\mathbf{A} \times (\mathbf{B} \times \mathbf{C}) = (\mathbf{A} \cdot \mathbf{C})\mathbf{B} - (\mathbf{A} \cdot \mathbf{B})\mathbf{C}$$
$$\mathbf{A} \cdot (\mathbf{B} \times \mathbf{C}) = \mathbf{B} \cdot (\mathbf{C} \times \mathbf{A}) = \mathbf{C} \cdot (\mathbf{A} \times \mathbf{B})$$